Advances in

QUANTUM ELECTRONICS

Volume 3

Advances in

Quantum Electronics

Edited by

D. W. GOODWIN

Department of Physics, University of York, England

VOLUME 3

1975

ACADEMIC PRESS
LONDON AND NEW YORK
A Subsidiary of Harcourt Brace Jovanovich, Publishers

ACADEMIC PRESS INC. (LONDON) LTD.
24/28 Oval Road,
London NW1

United States Edition published by
ACADEMIC PRESS INC.
111 Fifth Avenue
New York, New York 10003

Library of Congress Catalog Card Number: 72–12267
ISBN: 0–12–035003–3

Printed in Great Britain by
William Clowes & Sons, Limited
London, Colchester and Beccles

LIST OF CONTRIBUTORS

L. C. BALLING, *Department of Physics, University of New Hampshire, Durham, New Hampshire, U.S.A.*

C. C. DAVIS, *Physics Department, Schuster Laboratory, University of Manchester, Manchester, England.*

T. A. KING, *Physics Department, Schuster Laboratory, University of Manchester, Manchester, England.*

PREFACE

This, the third volume of Advances in Quantum Electronics contains two major review articles considered to be of complementary interest to laser physicists.

First, an article by Balling reviews theoretical and experimental work undertaken in the field of optical pumping. Paramagnetic atoms can be orientated in their ground state by illumination with circularly polarised resonance radiation and their hyperfine splittings and magnetic moments studied. This has led to both an understanding in depth and the development of frequency standards, magnetometers and masers.

Secondly, Davis and King review the field of gaseous ion lasers. Because of their technological significance attention is concentrated on the noble gas and metal vapour ion lasers and their practical applications. These two review articles should prove of value to those engaged in theoretical and experimental studies in the field of quantum electronics.

February, 1975 D. W. GOODWIN

CONTENTS

Optical Pumping

L. C. BALLING

Gaseous Ion Lasers

C. C. DAVIS and T. A. KING

OPTICAL PUMPING

L. C. BALLING

Department of Physics,
University of New Hampshire, Durham, New Hampshire, USA

I. Introduction

This chapter is primarily intended to provide the reader with an introduction to optical pumping sufficient to enable him to undertake research in the field. I have attempted to strike a balance between a superficial review of all aspects of optical pumping and a detailed discussion of a limited number of topics. I have placed an emphasis on the optical pumping of alkali atoms because of the great variety of experiments which have been and can be performed with alkali optical pumping techniques.

Section II presents an overview of the optical-pumping field on an elementary level; Section III contains a brief review of the use and properties of the density matrix as applied to the statistical behavior of assemblages of atoms or ions. In Sections IV–VI, the density matrix approach is systematically applied to the theory of optical-pumping r.f. spectroscopy and spin-exchange optical pumping. The theoretical discussion is at a level which should be readily understandable to a student who has taken two or three semesters of graduate nonrelativistic quantum mechanics. Because the sections on the theory of optical-pumping experiments contain a straightforward application of nonrelativistic quantum mechanics to the analysis of the behavior of atoms interacting with each other and electromagnetic fields, they might well be of interest to graduate students who are not interested in optical pumping *per se*.

This chapter has been written on the assumption that the sections will be read consecutively. This is particularly true of the first five sections in which the theoretical discussion builds steadily on the development of preceding sections and chapters.

Sections VI and VII deal with the experimental side of optical pumping. Section VII is a review of optical-pumping experiments and contains numerous tables of physical data such as atomic g-factors, hyperfine splittings, hyperfine pressure shifts, spin-exchange cross sections, relaxation times, etc. These tables include data obtained by optical-pumping methods and by other experimental techniques as well, and in a number of cases the data is compared with theoretical calculations. Section VIII is intended to aid a newcomer to the field in the construction and operation of an alkali optical-pumping apparatus.

Although I have attempted to present a reasonably broad view of optical pumping, my choice of topics and the emphasis I have placed on them tends to reflect my own research background and interests. For different views of the subject, the reader is invited to consult the review articles and books which are listed at the end of this chapter on page 162. The bibliography contains only those articles and books which are referred to in the text.

II. Optical Pumping: An Overview

A. INTRODUCTION

In 1950, Kastler proposed a method for orienting paramagnetic atoms in their ground state by illuminating them with circularly-polarized optical resonance radiation. He called this process "optical pumping". His proposal introduced a new and powerful technique for studying the properties of atoms and ions by means of r.f. spectroscopy.

In the succeeding twenty years, optical pumping has been used to measure the hyperfine splittings and magnetic moments of an impressive variety of atoms and ions. In terms of precision and reliability, optical pumping competed favorably as a technique with the far more expensive atomic beam method. In addition, many kinds of interatomic interactions have been studied in optical-pumping experiments. The optical-pumping process itself has been the subject of considerable study, providing as it does the opportunity to investigate in detail the interaction of atoms with resonant and off-resonant light. Optical pumping has also been applied to the construction of frequency standards, magnetometers and masers.

In short, optical pumping is an important and well established experimental technique in atomic physics. A surprising number of different types of experiments can be performed at relatively low cost, because the apparatus is basically quite simple. The theoretical analysis of optical-pumping experiments can be a challenging and satisfying application of non-relativistic quantum mechanics. Despite this, comparatively few physicists have worked in this area.

It is the purpose of this article to introduce the reader to the field of optical pumping. This section is designed to give an overview of the subject on an elementary level. The theory of optical-pumping r.f. spectroscopy and descriptions of optical-pumping experiments and techniques will be treated in detail in succeeding chapters. We will primarily, though not exclusively, be concerned with optical pumping as a means of producing population differences in the ground state sublevels of paramagnetic atoms in order to detect radio frequency transitions between these levels. That is, we will be dealing with magnetic resonance in atoms, and we shall begin our discussion of optical-pumping with a simplified treatment of magnetic resonance familiar to students of NMR and EPR.

B. MAGNETIC RESONANCE

If we place an atom with total angular momentum $\hbar\mathbf{F}$ in a weak magnetic field $H_0\hat{k}$ in the z-direction, it will interact with the field through its magnetic dipole moment $\boldsymbol{\mu}$. The Hamiltonian $\mathscr{H}$ for the interaction is

$$\mathscr{H} = -\boldsymbol{\mu} . H_0 \hat{k}. \qquad \text{...(B.1)}$$

The magnetic moment is related to the total angular momentum by

$$\boldsymbol{\mu} = g_F \mu_0 \mathbf{F}, \qquad \text{...(B.2)}$$

where μ_0 is the Bohr magneton and g_F is the g-factor. The g-factor can be positive or negative, depending on the atom's ground state configuration. The eigenstates of $\mathscr{H}$ are just the eigenstates of F_z. The energy eigenvalues are

$$E_M = -g_F \mu_0 H_0 M, \qquad \text{...(B.3)}$$

where M is the eigenvalue of F_z. The energy difference ΔE between two adjacent levels is

$$\Delta E = g_F \mu_0 H_0. \qquad \text{...(B.4)}$$

If an oscillating magnetic field $2H_1 \hat{\imath} \cos \omega t$ is applied in the x-direction, the Hamiltonian $\mathscr{H}$ becomes

$$\mathscr{H} = -g_F \mu_0 H_0 F_z - g_F \mu_0 2H_1 \cos \omega t F_x. \qquad \text{...(B.5)}$$

First-order perturbation theory for a harmonic perturbation tells us that the atoms will undergo transitions between adjacent magnetic sublevels and that the transition probabilities $\Gamma_{M\to M+1}$ and $\Gamma_{M+1\to M}$ are given by

$$\Gamma_{M\to M+1} = \frac{2\pi}{\hbar}(g_F \mu_0 H_0)^2 \, |\langle M|F_x|\, M+1\rangle|^2 \, \delta(\Delta E - \hbar\omega), \qquad \text{...(B.6)}$$

and

$$\Gamma_{M+1\to M} = \frac{2\pi}{\hbar}(g_F \mu_0 H_0)^2 |\langle M+1| F_x |M\rangle|^2 \delta(-\Delta E + \hbar\omega). \quad \text{...(B.7)}$$

The delta functions in these equations show that transitions will occur only when the resonance condition

$$\hbar\omega = |\Delta E|, \quad \text{...(B.8)}$$

is satisfied. Because the energy levels are equally spaced, transitions between all adjacent sublevels will occur simultaneously. Implicit in these equations is the assumption that the state $|M+1\rangle$ is higher in energy than the state $|M\rangle$ which is only true if g_F is negative. Besides the resonance condition, the important point to notice is that $\Gamma_{M+1\to M} = \Gamma_{M\to M+1}$. This means that if there are equal numbers of atoms in the two energy levels, the number of atoms undergoing transitions from $|M\rangle$ to $|M+1\rangle$ will equal the number of atoms going from $|M+1\rangle$ to $|M\rangle$. If we wish to observe a macroscopic change in the magnetization of an ensemble of atoms, there must be an initial difference in the populations of the two levels.

In NMR and EPR experiments, the Boltzman distribution of the populations of the energy levels of spins in a bulk sample in thermal equilibrium is relied upon to produce the desired population differences. When working with orders of magnitude fewer free atoms, however, one must develop artificial means for producing large population differences and sensitive detection schemes in order to observe magnetic resonance transitions. The optical-pumping process provides the means for achieving the necessary population differences and also the means for detecting the transitions.

Before going on to a discussion of the optical-pumping process, we will look at magnetic resonance from a classical point of view. The classical approach is often more useful for a qualitative understanding of the signals one observes. A classical analysis is possible because, as we shall see in later chapters, $\hbar$ does not appear in the quantum mechanical equations of motion for the operator **F**.

The classical equation of motion for the atomic angular momentum is

$$\hbar \frac{d\mathbf{F}}{dt} = \boldsymbol{\mu} \times H_0 \hat{k}, \quad \text{...(B.9)}$$

or

$$\frac{d\mathbf{F}}{dt} = g_F \frac{\mu_0 H_0}{\hbar} \mathbf{F} \times \hat{k}. \quad \text{...(B.10)}$$

Since $d\mathbf{F}/dt$ is orthogonal to **F**, the torque produced by the field $H_0 \hat{k}$ causes a precession of **F** about the z-axis with angular velocity ω_0. That is,

$$\frac{d\mathbf{F}}{dt} = \omega_0 \hat{k} \times \mathbf{F}, \quad \text{...(B.11)}$$

with

$$\omega_0 = -\frac{g_F \mu_0 H_0}{\hbar}. \qquad \text{...(B.12)}$$

If we view the precessing magnetic moment from a reference frame rotating in the same sense about the z-axis with angular velocity ω, the time derivative $\partial\mathbf{F}/\partial t$ in the rotating frame is related to $\mathrm{d}\mathbf{F}/\mathrm{d}t$ in the laboratory frame by

$$\frac{\mathrm{d}\mathbf{F}}{\mathrm{d}t} = \omega\hat{k} \times \mathbf{F} + \frac{\partial\mathbf{F}}{\partial t}. \qquad \text{...(B.13)}$$

Using equation (B.11), we see that the time dependence of **F** in the rotating frame is

$$\frac{\partial\mathbf{F}}{\partial t} = (\omega_0 - \omega)\,\hat{k} \times \mathbf{F}. \qquad \text{...(B.14)}$$

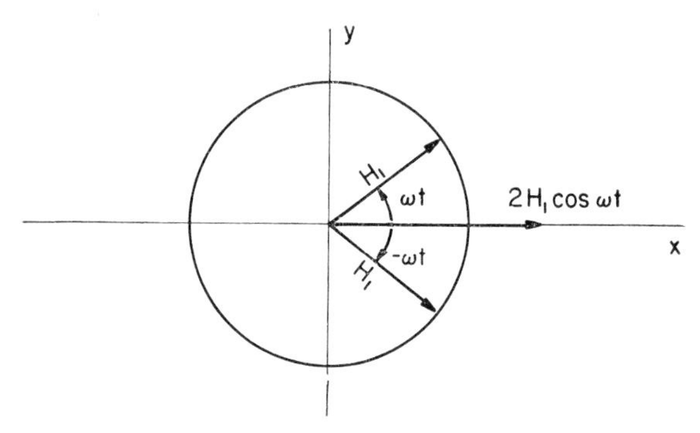

FIG. 1. The superposition of two counter-rotating magnetic fields of equal amplitude.

If we apply a rotating magnetic field H_1 fixed along the x-axis of the rotating coordinate system, the equation of motion in the rotating frame is

$$\frac{\partial\mathbf{F}}{\partial t} = (\omega_0 - \omega)\,\hat{k} \times \mathbf{F} + \omega_1\,\hat{\imath} \times \mathbf{F}, \qquad \text{...(B.15)}$$

where

$$\omega_1 = -\frac{g_F \mu_0 H_1}{\hbar}. \qquad \text{...(B.16)}$$

The rotating field H_1 is equivalent to the linearly oscillating field $2H_1\,\hat{\imath}\cos\omega t$ in the laboratory frame which was considered above. To see this we note that

$$2H_1\,\hat{\imath}\cos\omega t = H_1(\hat{\imath}\cos\omega t + \hat{\jmath}\sin\omega t) + H_1(\hat{\imath}\cos\omega t - \hat{\jmath}\sin\omega t).$$

That is, the oscillating field is the superposition of two fields rotating in opposite directions as shown in Fig. 1.

The component which rotates in the same sense and with nearly the same frequency as the precessing angular momentum will exert a torque on **F**. The counter rotating component will have a negligible effect, however, because of its rapidly changing orientation with respect to the direction of **F**. We, therefore, assume that the linearly oscillating field can be replaced in the analysis by the single rotating field considered in equation (B.15).

At resonance, $\omega = \omega_0$ and equation (B.15) becomes

$$\frac{\partial \mathbf{F}}{\partial t} = \omega_1 \hat{\imath} \times \mathbf{F}. \qquad \text{...(B.17)}$$

In the rotating frame at resonance, the angular momentum precesses with angular velocity ω_1, about the field H_1. The resulting motion of **F** in the laboratory frame is illustrated in Fig. 2.

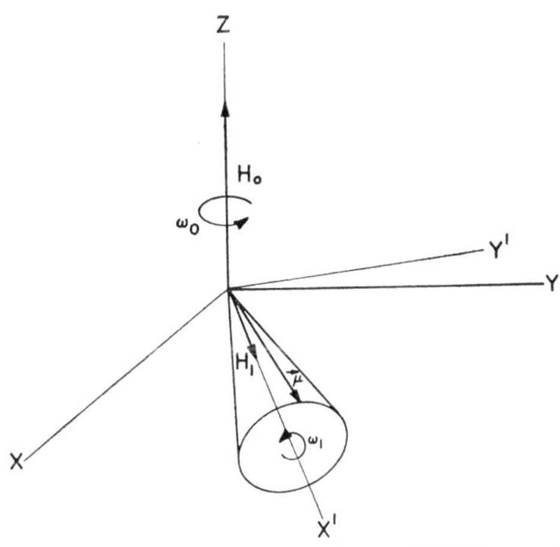

FIG. 2. The precession of a magnetic moment $\boldsymbol{\mu}$ about a field H_1 rotating in the X–Y plane with angular velocity ω_0.

It is apparent from Fig. 2 that the application of the resonant rotating field causes the z-component of **F** to oscillate with angular velocity ω_1 and the x- and y-components of **F** to oscillate with angular velocity ω_0 in the laboratory frame. In the absence of the rotating field, the angular momenta of individual atoms precess about the steady field, but the oscillation of F_x and F_y is not observable because the precessions are not in phase with each other. The effect of the rotating field is to lock the individual angular momenta into coherent precession about the steady field. The resonance condition that $\omega = \omega_0$ is equivalent to the resonance condition obtained quantum mechanically and given by equation (B.8).

Although the resonance condition is the same in the quantum mechanical and classical treatments, the oscillatory behavior of F_z obtained classically is in sharp contrast to the constant transition rate between eigenstates of F_z obtained by 1st-order perturbation theory. The problem is that both discussions were over simplified. We have neglected in the classical analysis any interactions between the magnetic moment **F** and the random time varying fields of neighboring atoms (relaxation effects) which interrupt the coherent precession of **F** about the field H_1. On the other hand, the constant transition rate we obtained from 1st-order perturbation theory would break down as soon as the probability of finding the atom in its initial state was much less than unity. In Section IV we will give a more rigorous quantum mechanical treatment including relaxation effects which will exhibit both features of the simplified discussions given above.

C. OPTICAL PUMPING

We have seen in the previous section that it is possible to induce r.f. transitions between adjacent energy levels arising from the interaction of an atom's total magnetic moment with a weak magnetic field. The existence of these energy levels is just the Zeeman effect. In a strong magnetic field, the atom can no longer be treated adequately by ascribing to it a single magnetic moment, because the nuclear spin becomes decoupled from the electronic spin. Also, the simplified treatment of magnetic resonance did not cover hyperfine transitions between energy levels arising from the mutual interaction between the nuclear and electronic magnetic moments.

R.F. transitions between hyperfine levels or between energy levels of the atom in the presence of a magnetic field which are not described by the simple Zeeman effect alone, can still be understood in basically the same way. The task of the present section is to introduce the reader to the role which optical-pumping plays in making it possible to observe these r.f. transitions in a macroscopic collection of atoms.

As we have already noted, in order for magnetic resonance to produce a macroscopic change in the magnetization of an ensemble of atoms, there must be an initial difference in the populations of the levels involved in the r.f. transition. Optical pumping produces the desired population differences and simultaneously provides the means for detecting the r.f. transitions.

To illustrate the optical-pumping process we will consider a fictitious alkali atom with zero nuclear spin. By neglecting the nuclear spin we will keep things simple while retaining the essential physics. The ground state of this atom is $S_{1/2}$ and the first excited state is $P_{1/2}$. Thus the total electronic angular momentum $J = \frac{1}{2}$ is the same in both states. The two states are connected by the D_1 optical transition. The basic apparatus is shown schematically in Fig. 3.

The alkali atoms are contained in a glass cell filled with an inert buffer gas which prevents disorienting collisions between the atoms and the cell walls. The optical-pumping cell is situated in a steady magnetic field $H_0\hat{k}$. The oscillating field $2H_1\hat{\imath}\cos\omega t$ is produced by a Helmholtz pair surrounding the cell. Circularly polarized D_1 resonance radiation is incident on the cell in the direction of the magnetic field H_0, passes through the cell, and is focused on a photodetector.

When an atom absorbs a photon, it is excited to the $P_{1/2}$ state from which it radiates back down to the ground state. Both the $S_{1/2}$ and $P_{1/2}$ states are split by the magnetic field H_0 into two sublevels which have magnetic quantum numbers $J_z = \pm\frac{1}{2}$. If the radiation is left-circularly polarized, the photons have spin 1 in the direction of the field. Because the photons have spin 1, an atom must make a $J_z = 1$ transition in going from the ground state to the excited state in order to conserve angular momentum. Therefore, only an atom in the $J_z = -\frac{1}{2}$ sublevel can absorb a photon. The atoms in the excited state will fluoresce back

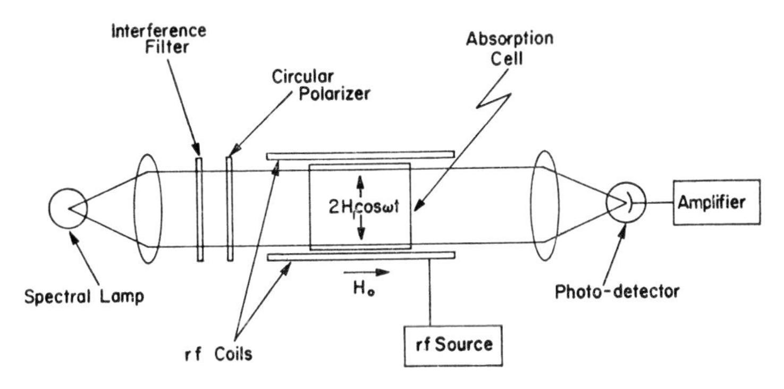

FIG. 3. A basic optical-pumping apparatus.

down to both ground state sublevels, but if an atom lands in the $J_z = -\frac{1}{2}$ state it will be kicked back up to the excited state by absorption of a photon. The absorption and emission cycle is shown in Fig. 4 for the case where collisions with buffer gas atoms completely mix the excited state level populations and for the case of no mixing.

If there is no excited state mixing, the probability of an atom returning from the excited state to a particular ground state level is determined by the square of the matrix element for the electric dipole transition. In this case, the probability of making a transition $(\frac{1}{2} \to -\frac{1}{2})$ is twice that of the transition $(\frac{1}{2} \to \frac{1}{2})$. If there is complete mixing of the excited state sublevels by collisions with the buffer gas in a time short compared to the lifetime of the excited state, then an atom returns to either ground state sublevel with equal probability.

Because atoms in the $+\frac{1}{2}$ sublevel of the ground state cannot absorb the D_1 radiation, the atoms will be pumped out of the $-\frac{1}{2}$ sublevel into the $+\frac{1}{2}$ level via the excited state. Relaxation effects such as spin disorienting collisions between the optically pumped atoms and the buffer gas atoms will tend to destroy the polarization of the atoms, but if the pumping light is sufficiently intense, a substantial difference in the populations of sublevels will be achieved. As the atoms approach an equilibrium polarization under the competing influences of the pumping light and relaxation effects, the atoms become more transparent to the pumping light than they were at thermal equilibrium, since there are fewer atoms in the light-absorbing $J_z = -\frac{1}{2}$ level.

If we now apply a resonant r.f. field to induce transitions between the ground state sublevels, the r.f. will tend to equalize the populations of the two levels,

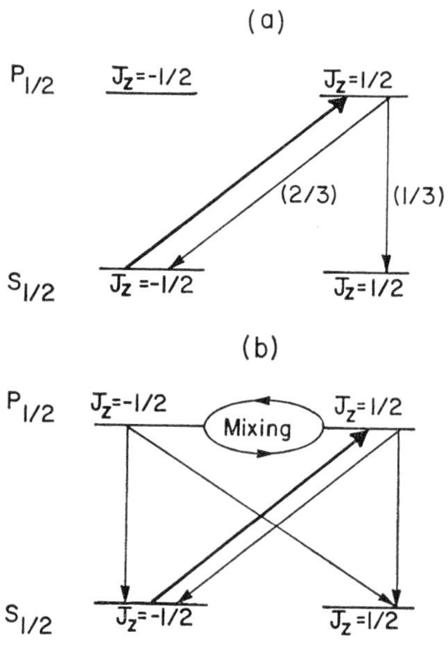

FIG. 4. The optical-pumping cycle for a spin-$\frac{1}{2}$ system (a) for the case of no mixing in the excited state and (b) for the case of excited state mixing.

and the number of atoms which can absorb the pumping light will increase. The intensity of the pumping light transmitted by the optical-pumping cell will drop when the resonant r.f. field is applied. By monitoring the transmitted light intensity with a photodetector, one can detect the r.f. transitions.

A more detailed understanding of the optical-pumping experiment described above can be obtained by means of a simplified mathematical analysis of the changes in the ground state sublevel populations which occur under the influence of the pumping light, the r.f. field, and relaxation effects. Let the total number of alkali atoms in the cell be N, and let $N_\pm$ denote the number of atoms in the $J_z = \pm\frac{1}{2}$ sublevels of the ground state. The effect of the pumping light upon the sublevel populations, assuming complete mixing in the excited state, is described by the equations

$$\frac{dN_+}{dt} = \tfrac{1}{2}\Gamma_p N_-, \qquad \text{...(C.1)}$$

and

$$\frac{dN_-}{dt} = \tfrac{1}{2}\Gamma_p N_- - \Gamma_p N_-, \qquad \text{...(C.2)}$$

where the probability of an atom in the $J_z = -\frac{1}{2}$ level absorbing a photon is Γ_p, and we have assumed complete reorientation in the excited state. The factor $\frac{1}{2}$ reflects the fact that half the atoms raised to the excited state return to the $J_z = -\frac{1}{2}$ level.

When a resonant r.f. field is applied, transitions between the two sublevels occur. If we denote the transition probabilities $\Gamma_{1/2\to-1/2} = \Gamma_{-1/2\to1/2} = \Gamma_{\text{r.f.}}$, the rate equations become

$$\frac{dN_+}{dt} = \tfrac{1}{2}\Gamma_P N_- + \Gamma_{\text{r.f.}}(N_- - N_+), \qquad \text{...(C.3)}$$

and

$$\frac{dN_-}{dt} = -\tfrac{1}{2}\Gamma_P N_- + \Gamma_{\text{r.f.}}(N_+ - N_-) \qquad \text{...(C.4)}$$

Random collisions between the optically pumped atoms and the buffer gas atoms or cell walls tend to restore thermal equilibrium and equalize the populations. In the absence of r.f. or pumping light the return to equilibrium would be exponential with a characteristic time constant τ_1. If we include relaxation in the rate equations, they become

$$\frac{dN_+}{dt} = \tfrac{1}{2}\Gamma_P N_- - \Gamma_{\text{r.f.}}(N_+ - N_-) - \frac{(N_+ - \frac{1}{2}N)}{\tau_1}, \qquad \text{...(C.5)}$$

$$\frac{dN_-}{dt} = -\tfrac{1}{2}\Gamma_P N_- + \Gamma_{\text{r.f.}}(N_+ - N_-) - \frac{(N_- - \frac{1}{2}N)}{\tau_1}. \qquad \text{...(C.6)}$$

We can define the polarization of an ensemble of spin-$\frac{1}{2}$ atoms as the probability of finding an atom with $J_z = \frac{1}{2}$ minus the probability of finding an atom with $J_z = -\frac{1}{2}$. This is just

$$\frac{N_+ - N_-}{N}.$$

The average value of the operator J_z over the collection of atoms is just the sum of its eigenvalues $\pm\frac{1}{2}$ multiplied by the probability of finding that eigenvalue. That is,

$$\overline{\langle J_z \rangle} = \frac{1}{2}\frac{(N_+ - N_-)}{N}. \qquad \text{...(C.7)}$$

Because $N_+ + N_- = N$,

$$2N_- = N(1-2\overline{\langle J_z \rangle}). \qquad \text{...(C.8)}$$

We can subtract equation (C.6) from equation (C.5) to obtain the equation for the time rate of change of $\overline{\langle J_z \rangle}$,

$$\frac{d\overline{\langle J_z \rangle}}{dt} = \tfrac{1}{4}\Gamma_p - \left(\tfrac{1}{2}\Gamma_p + 2\Gamma_{\text{r.f.}} + \frac{1}{\tau_1}\right)\overline{\langle J_z \rangle}. \qquad \text{...(C.9)}$$

In the absence of an r.f. field, $\Gamma_{\text{r.f.}} = 0$ and the equilibrium value of $\overline{\langle J_z \rangle}$ is obtained by setting

$$\frac{d\overline{\langle J_z \rangle}}{dt} = 0.$$

$$\overline{\langle J_z \rangle}_0 = \frac{\tfrac{1}{4}\Gamma_p}{\tfrac{1}{2}\Gamma_p + 1/\tau_1} \qquad \text{...(C.10)}$$

Thus, if there were no relaxation, the atoms would become completely polarized with $\overline{\langle J_z \rangle}_0 = \tfrac{1}{2}$. The greater the pumping light intensity, the larger Γ_p will be. For a given relaxation time τ_1, the greater the intensity of the pumping light, the closer $\overline{\langle J_z \rangle}_0$ approaches $\tfrac{1}{2}$.

When a resonant r.f. field is applied, the new equilibrium value of $\overline{\langle J_z \rangle}$ is reduced to

$$\overline{\langle J_z \rangle} = \frac{\tfrac{1}{4}\Gamma_p}{\tfrac{1}{2}\Gamma_p + 1/\tau_1 + 2\Gamma_{\text{r.f.}}}. \qquad \text{...(C.11)}$$

As we noted earlier, this treatment ignores the coherent precession of the vector $\overline{\langle J \rangle}$ about the rotating field H_1 which was described by our classical analysis. This coherence results in an oscillating $\overline{\langle J_x \rangle}$ and $\overline{\langle J_y \rangle}$ which can be observed in crossed beam experiments (see Subsection I), and which decay due to spin relaxation with a characteristic relaxation time τ_2.

If the number of alkali atoms in the path of the optical-pumping beam is not too great so that the cell is optically thin, the pumping light intensity will be nearly constant over the length of the cell and the amount of light absorbed will be proportional to N_- or equivalently to $(1 - 2\overline{\langle J_z \rangle})$. The simplest way of detecting the r.f. transitions, then, is to monitor the intensity of pumping light transmitted by the cell. The observed signal will be proportional to the difference between the values of $\overline{\langle J_z \rangle}$ with and without the r.f. given by equations (C.10) and (C.11). Even in the more complicated case of a real alkali atom with nuclear spin, it turns out that under certain conditions the absorption of D_1 radiation is proportional to $(1-2\overline{\langle J_z \rangle})$ where $\overline{\langle J_z \rangle}$ is the average value of the z-component of the electronic angular momentum in the ground state.

As can be seen from equation (C.11), $\overline{\langle J_z \rangle}$ approaches zero as the strength of the resonant r.f. field, and hence $\Gamma_{r.f.}$, increases. The r.f. tends to equalize the sublevel populations. For strong r.f. fields, the observed signal will be proportional to $\overline{\langle J_z \rangle_0}$ given by equation (C.10). To maximize the optical-pumping signal, one must maximize the pumping light intensity and minimize relaxation effects.

Another detection scheme is to observe the intensity of fluorescent light emitted by the optically-pumped atoms as they radiate back to the ground state. A decrease in the ground state polarization of the atoms increases the probability that an atom will be excited by the pumping light. Thus the application of the r.f. field produces an increase in the intensity of fluorescent light from the sample.

D. OPTICAL PUMPING OF ALKALI ATOMS

In contrast to our simple spin-$\frac{1}{2}$ model for an alkali atom, all of the alkali atoms have a non-zero nuclear spin. The ground state of the alkali atom is $S_{\frac{1}{2}}$. The electronic angular momentum **J** is just the spin of the valence electron. The dipole interaction between the electronic and nuclear magnetic moments couples the angular momentum **J** to the nuclear spin **I** to form a total angular momentum $\mathbf{F} = \mathbf{I} + \mathbf{J}$. The eigenvalues of F^2 are $F(F+1)$ with $F = I \pm \frac{1}{2}$. The energy of the ground state depends upon the relative orientation of the nuclear and electronic spins. For $F = I + \frac{1}{2}$, the spins are parallel, for $F = I - \frac{1}{2}$ they are antiparallel so that the alkali atom ground state is split into two hyperfine levels characterized by quantum number F. The state of lowest energy is $F = I - \frac{1}{2}$. The separation $h\Delta\nu$ between the two levels is the hyperfine splitting and is given by

$$h\Delta\nu = A\mathbf{I}.\mathbf{J}, \qquad \text{...(D.1)}$$

where A is the dipole interaction constant.

In a weak magnetic field, the two hyperfine levels are split by the Zeeman effect into sublevels characterized by the magnetic quantum number M which is the eigenvalue of F_z. The Hamiltonian for the atom in a field $H_0 \hat{k}$ is

$$\mathscr{H} = A\mathbf{I}.\mathbf{J} - g_I \mu_0 H_0 \hat{k}.\mathbf{I} - g_J \mu_0 H_0 \hat{k}.\mathbf{J}. \qquad \text{...(D.2)}$$

If the magnetic field is sufficiently weak so that **I** and **J** remain tightly coupled, they will each precess about the total angular momentum **F**, as **F** precesses about the field $H_0 \hat{k}$. Since the magnetic moment of the nucleus is much smaller than that of the electron, we can neglect, to first approximation, the second term in equation (D.2). Because the angular momentum **J** is precessing rapidly about **F** compared to the rate at which **F** precesses about the field (the hyperfine interaction is many orders of magnitude larger than the Zeeman interaction), only the component of **J** which lies along **F** effectively interacts with the field

$H_0\hat{k}$. This is illustrated in Fig. 5. Therefore, we can replace equation (D.2) with the approximate Hamiltonian

$$\mathscr{H} \cong A\mathbf{I}.\mathbf{J} - \frac{g_J\,\mu_0\,H_0(\mathbf{J}.\mathbf{F})\,F_z}{F^2}. \qquad \text{...(D.3)}$$

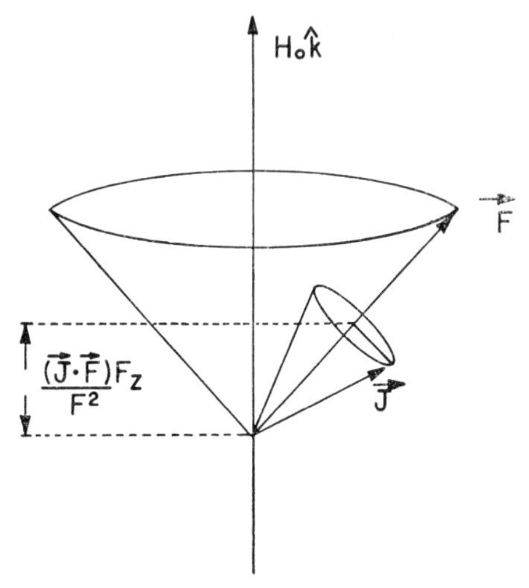

FIG. 5. The precession of the electronic angular momentum **J** about the total angular momentum **F** as **F** precesses about the magnetic field H_0.

If we note that

$$(\mathbf{F}-\mathbf{J})^2 = I^2, \qquad \text{...(D.4)}$$

and

$$(\mathbf{F}-\mathbf{J})^2 = F^2 + J^2 - 2\mathbf{J}.\mathbf{F} \qquad \text{...(D.5)}$$

and that

$$F^2 = (\mathbf{I}+\mathbf{J})^2 = I^2 + J^2 + 2\mathbf{I}.\mathbf{J} \qquad \text{...(D.6)}$$

we can write for the eigenvalues of $\mathscr{H}$

$$E_{F,M} = \frac{A}{2}\left[F(F+1) - I(I+1) - \tfrac{3}{4}\right] - g_F\,\mu_0\,H_0\,M, \qquad \text{...(D.7)}$$

where

$$g_F = \pm\frac{g_J}{2I+1} \qquad \text{for} \qquad F = I \pm \tfrac{1}{2}. \qquad \text{...(D.8)}$$

The alkali atom interacts with a weak magnetic field through a magnetic moment characterized by the g-factor g_F, the sign of which is determined by the hyperfine level of the atom. The change in sign occurs because the direction of the electron spin and hence the direction of the net magnetic moment of the atom is parallel to **F** for $F = I + \frac{1}{2}$ and antiparallel to **F** for $F = I - \frac{1}{2}$.

In this approximation, F and M are good quantum numbers and each hyperfine level splits into $2F+1$ equally spaced sublevels characterized by magnetic quantum number M. The use of the semi-classical vector model of the atom with which we obtained our effective Hamiltonian in equation (D.3) is justified by first-order perturbation theory in Section V.

As the strength of the magnetic field H_0 increases, $\mathbf{I}$ and $\mathbf{J}$ become decoupled from each other, until in the Paschen–Back limit they precess separately about $H_0 \hat{k}$. In this limit the energy levels are determined by the quantum numbers j_z and I_z and the hyperfine interaction $A\mathbf{I}.\mathbf{J}$ is treated as a perturbation.

In order to solve the problem for arbitrary H_0, one must apply degenerate perturbation theory to the $S_{1/2}$ ground state and diagonalize the Hamiltonian in Equation (D.2) in the $|FM\rangle$ representation. The solutions of the determinant yield the sublevel energies $E_{F,M}$ as a function of the field H_0 and the quantum numbers F and M. These solutions are given by the Breit–Rabi formula

$$E_{F,M} = \frac{-\Delta W}{2(2I+1)} - g_I \mu_0 H_0 M \pm \frac{\Delta W}{2} \sqrt{1 + \frac{4M}{2I+1} x + x^2}, \qquad \text{...(D.9)}$$

where

$$\Delta W = \frac{hA}{2}(2I+1), \qquad \text{...(D.10)}$$

and

$$x = \frac{(-g_J + g_I)\mu_0 H_0}{\Delta W} \qquad \text{...(D.11)}$$

The ($\pm$) signs apply for $F = I \pm \frac{1}{2}$. For weak fields where $x \ll 1$, we can expand the square root to first order in x, neglect the second term in equation (D.9), and immediately obtain the Zeeman splittings given by the approximate Hamiltonian in equation (D.7).

The energy level structure of an alkali atom such as Rb^{87} with nuclear spin $I = 3/2$ calculated with the Breit–Rabi formula is shown in Fig. 6. By means of optical pumping, one can polarize alkali atoms in a weak magnetic field,so that there are substantial differences in the populations of the sublevels. This enables one to observe r.f. transitions between Zeeman or hyperfine levels.

Although the energy level structure is more complicated for a real alkali atom than for the hypothetical alkali atom with zero nuclear spin which we treated in the preceding section, the way in which the optical-pumping process is used to polarize the atoms and detect the transitions is essentially the same.

A typical experimental arrangement is shown in Fig. 7. To optically pump Rb^{87}, for instance, the resonance radiation from atoms excited in an electrodeless discharge lamp is passed through a filter which selects the D_1 emission line corresponding to the $5S_{1/2} \leftrightarrow 5P_{1/2}$ transition. The light is circularly polarized, passed through the optical-pumping cell containing Rb metal and a buffer gas, and focused on a photodetector.

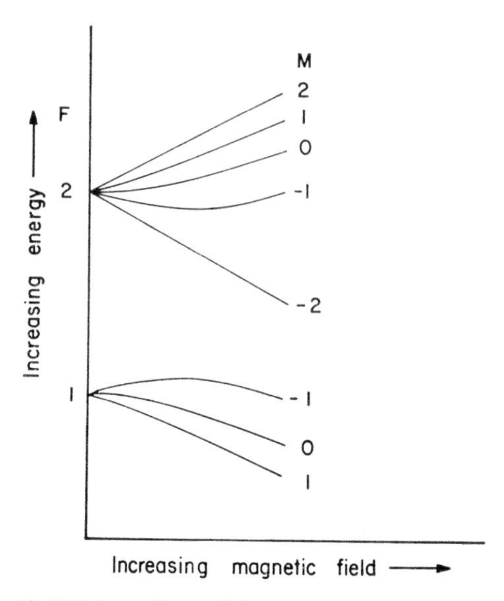

FIG. 6. The magnetic field dependence of the ground state sublevels of an alkali atom with nuclear spin $I = \frac{3}{2}$ as calculated with the Breit–Rabi formula.

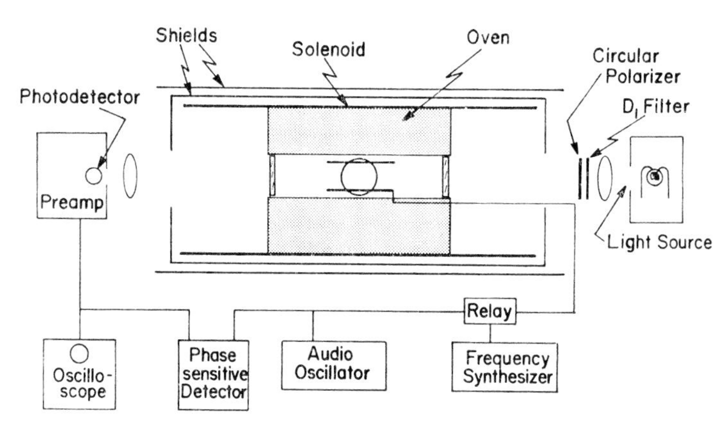

FIG. 7. An alkali optical-pumping apparatus using magnetic shields, transmission monitoring and amplitude modulation of the r.f.

The cell can be filled with an inert buffer gas to prevent disorienting collisions between Rb atoms and the walls making the spin-relaxation time as long as possible. Another technique is to use an evacuated cell with the walls carefully coated with a transparent hydrocarbon compound with which the Rb atoms can collide many times without spin disorientation (Bouchiat and Brossel, 1966). This technique has the advantage that the Rb atoms move rapidly back and forth across the bulb averaging out magnetic field inhomogeneities.

The energy level structure of the $5P_{1/2}$ state has the same arrangement of energy levels as the ground state since $J = \frac{1}{2}$ in both states. If the incident pumping light is left-circularly polarized so that the atoms must make a $\Delta M = +1$ transition when they are excited to the $5P_{1/2}$ state, atoms in any sublevel with the exception of the $F = 2$, $M = 2$ level can absorb a photon. If there were no spin-relaxation in the ground state, all of the atoms would eventually be pumped into the $F = 2$, $M = 2$ level, and the atoms would be completely polarized. Since there is spin relaxation, there is a distribution of atoms among all of the energy levels. The actual population distribution depends upon the degree to which there is collisional mixing of the sublevels in the excited state (Franz and Franz, 1966) and further discussion of this will be deferred until Section V. The important point is, however, that the optical-pumping process leads to population differences in the ground state. If the sign of the circular polarization or the direction of the magnetic field is reversed, the population distribution will reverse.

The use of a D_1 filter was suggested by Franzen and Emslie (1957) and it is important for efficient pumping. If the D_2 line ($S_{1/2} \leftrightarrow P_{3/2}$) is present in the pumping light it tends to counteract the effect of the D_1 line. Since there is an $F = 3$, $M = 3$ level in the $P_{3/2}$ state, ground state atoms in the $F = 2$, $M = 2$ level can also absorb a photon. Under these circumstances, the establishment of population differences depends upon differences in the probability for an atom in a particular ground state sublevel to absorb a photon, none of which is zero. To make matters worse, the D_2 light tends to pump atoms out of the very sublevels which the D_1 pumping cycle is populating. Indeed if there is substantial excited state mixing and the D_1 and D_2 lines are of equal intensity, no polarization of the atoms will be achieved.

A linearly oscillating r.f. field is applied to the optically-pumped atoms by means of a Helmholtz pair surrounding the sample with its axis perpendicular to the steady magnetic field $H_0 \hat{k}$. If the frequency of the r.f. field is adjusted to induce Zeeman transitions ($\Delta F = 0$, $\Delta M = \pm 1$), the populations of the levels coupled by the r.f. field will tend to be equalized with a resulting change (usually a decrease) in the transparency of the sample. If the r.f. is switched on and off by a coaxial relay, the intensity of the pumping light transmitted by the optical-pumping cell will be amplitude modulated at the switching rate. This produces an ac signal from the photodetector which is amplified and displayed on an oscilloscope or phase-sensitive detector.

Hyperfine transitions ($\Delta F = \pm 1$, $\Delta M = \pm 1$) can be observed in the same manner. In addition, if the hyperfine components of the D_1 pumping light are not of equal intensity, differences in the populations of the $M = 0$ states in the two hyperfine levels of the ground state can be established. This allows one to observe the $(F, 0 \leftrightarrow F \pm 1, 0)$ transition (Arditi and Carver, 1961). The advantage of this transition for precision hyperfine measurements is that it is, to first order, magnetic field independent. In order to induce the $0 - 0$ transition, the r.f. coil axis must be parallel to the steady magnetic field.

One way to produce the steady magnetic field H_0 is to construct a large Helmholtz pair with additional coils to cancel the vertical component of the earth's magnetic field. This is a quick and dirty method which has the advantage of low cost and easy access to the optical-pumping cell.

A more sophisticated approach is to utilize a large solenoid equipped with correction coils surrounded by cylindrical magnetic shields (Hanson and Pipkin, 1965). The solenoid-shield system has a number of important advantages. With the aid of correction coils an ultra-homogeneous magnetic field can be produced, and the magnetic shields virtually eliminate the perturbing effects of random magnetic fields which are always present in a laboratory.

In most optical-pumping setups, the density of alkali atoms in the cell is determined by the alkali vapor pressure which has an approximately exponential temperature dependence. The optimum Rb signal is obtained at $\sim 50°C$ and the range of temperatures over which a signal can be detected is 10–90°C. When the cell is too cold there is insufficient density for measurable absorption of the pumping light. When the cell is too hot, the cell becomes optically thick. The optimum temperatures for the other alkali atoms are Cs ($\sim 35°C$), Na ($\sim 150°C$), K ($\sim 80°C$), and Li ($\sim 350°C$). At these temperatures the alkali vapor pressure is $\sim 4 \times 10^{-6}$ Torr. Recently, special techniques for controlling the density of the optically pumped alkali atoms independently of the cell temperature (Balling *et al.*, 1969; Wright *et al.*, 1970) have been developed which extend the operating temperature range by more than an order of magnitude.

Further details of the experimental and theoretical aspects of the optical-pumping of alkali atoms will be discussed in later chapters. With the exception of lithium, for which there are considerable difficulties, all other alkali atoms are easily optically pumped. Because of the relative simplicity of the hydrogen-like atoms, their structure is of considerable theoretical interest, and optical pumping has made possible precision measurements of their ground state hyperfine structure and of their electronic and nuclear magnetic moments. The optical pumping of alkali metal atoms became even more important, however, with the advent of spin-exchange optical pumping.

E. SPIN-EXCHANGE OPTICAL PUMPING

In 1958, Dehmelt revolutionized the field of optical pumping by introducing an extension of optical-pumping technique which greatly increased its usefulness as a spectroscopic tool. He optically pumped sodium atoms in an argon buffer gas and introduced free electrons into the cell by means of a pulsed r.f.

discharge. The free electrons were polarized by spin-exchange collisions with the sodium atoms, and he was able to observe the spin resonance of the free electrons. During an e-Na collision, the incoming electron is indistinguishable from the Na valence electron. If the spins of the two electrons are anti-parallel, there is a large probability that the collision will result in an exchange of spin orientations of the two electrons. Thus if the Na atoms are polarized by optical pumping, the spin-exchange collisions between the electrons and Na atoms will polarize the free electrons.

If an r.f. field is applied at the electron spin resonance frequency, the electrons will be depolarized. The e-Na collisions will then partially depolarize the Na atoms resulting in a decrease in the intensity of the pumping light intensity transmitted by the cell. In this way the electron spin resonance of free electrons can be observed, and precision measurements of the free electron magnetic moment can be made.

The power of the spin-exchange method lies in the fact that not only free electrons, but any atom or ion with an odd number of valence electrons and an S-state ground state can, in principle, be polarized by spin-exchange collisions with optically pumped alkali atoms. At one stroke, the number of atoms and ions which can be studied is tremendously increased. The S-state requirement is a result of the observation that states of higher orbital angular momentum are rapidly disoriented by collisions with cell walls or buffer gas atoms. A list of the atoms and ions which have been studied by spin-exchange optical pumping is given in Table I.

Two limitations on the kinds of atoms which can be spin-exchange optically pumped are the chemical reactivity of the atom with alkali atoms and the degree of difficulty with which a sufficient density of the atoms is obtained. The spin-exchange cross section for an electron–alkali atom collision is $\sim 2 \times 10^{-14}$ cm^2. All of the atom–atom spin-exchange cross sections observed so far are of this same order of magnitude. As it happens, this means that the density of atoms to be spin-exchange polarized must be $\sim 10^{10} - 10^{11}$ atoms/cm^3 for optimum spin-exchange signals.

In many cases the element involved is not a monomer and must be dissociated by a discharge. If the alkali atoms in the cell are eliminated by reaction with the element's molecules before a sufficient density of monomers is present, the experiment will fail. On the other hand, if the atoms to be spin-exchange polarized are from a metal with very low vapor pressure, one has the problem of raising the temperature of the metal to produce sufficient vapor pressure while maintaining the correct density of alkali atoms. A further discussion of this problem and methods of solution will be discussed in later sections.

Not only does spin-exchange optical pumping allow the study of the hyperfine structure and g-factors of an increased number of atoms and ions, but it offers the opportunity to investigate the spin-exchange collision process itself.

TABLE I. *A list of atoms and ions which have been polarized by spin exchange with optically pumped atoms. If an atom or ion can be spin-exchange polarized by one species of alkali atom, it can generally be polarized by another*

Spin-exchange polarized species	Optically pumped species	References*
e^-	Na	Dehmelt, 1958; Balling, 1966
e^-	Rb	Balling *et al.*, 1964; Balling and Pipkin, 1965
e^-	Cs	Balling and Pipkin, 1964
e^-	He($^3S_1$)	Schearer, 1968a
H	Na	Anderson *et al.*, 1958; Ruff and Carver, 1965
H	Rb	Hughes and Robinson, 1969a, b; Anderson *et al.*, 1960
D	Rb	Hughes and Robinson, 1969a, b; Wright *et al.*, 1970
T	Rb	Pipkin and Lambert, 1962; Balling and Pipkin, 1965
Li	Rb	Balling *et al.*, 1969; Wright *et al.*, 1969
K	Rb	Beahn and Bedard, 1972
Li, Na, K, Rb, Cs	Na, K, Rb, Cs	
N	Rb	Lambert and Pipkin, 1963; Weiss *et al.*, 1970
P	Rb	Lambert and Pipkin, 1962
Ag	Rb	Balling *et al.*, 1969
Cu	Cs	Hofmann-Reinecke, 1969
Mn	Rb	Davis *et al.*, 1971
Eu	Cs	Tilgner *et al.*, 1969
Ne (excited states)	He ($^3S_1$)	Schearer, 1968b
He^3 (nuclear polarization)	Rb	Bouchiat *et al.*, 1960
Sr^+	Rb	Gibbs and Churchill, 1971
Cd^+	Rb	Gibbs and Churchill, 1971
He^+	Cs	Schuessler *et al.*, 1969
Hg^+	Rb	Hoverson and Schuessler, 1972
Rb^+ (charge exchange)	Rb	Mitchell and Fortson, 1968
Cs^+ (charge exchange)	Cs	Nienstadt *et al.*, 1972

* The references are representative examples of each case and should not be considered complete.

Values for electron–atom and atom–atom spin-exchange cross sections which can be obtained from optical-pumping experiments are of considerable theoretical interest. A thorough discussion of the theory of spin-exchange optical-pumping will be given in Section VI.

F. OPTICAL PUMPING OF MERCURY AND OTHER $^1S_0$ ATOMS

Much of our understanding of the optical-pumping process, the interaction of optically pumped atoms with the pumping light, and relaxation processes has come from the thorough and imaginative theoretical and experimental work of the optical-pumping group in Paris headed by A. Kastler. In many of their experiments, they have optically pumped the ground state of an odd isotope of mercury.

Although the S_0 configuration of the Hg ground state is diamagnetic, the odd isotopes have a non-zero nuclear spin which can be oriented by optical pumping. The orientation of the nucleus in the ground state permits the precision measurement of the nuclear magnetic moment. In addition, since Hg is easy to work with, it has been used extensively in experimental studies of the optical-pumping cycle (Cohen-Tannoudji, 1962) and to study relaxation processes (Cohen-Tannoudji, 1963). Mercury is convenient to work with for

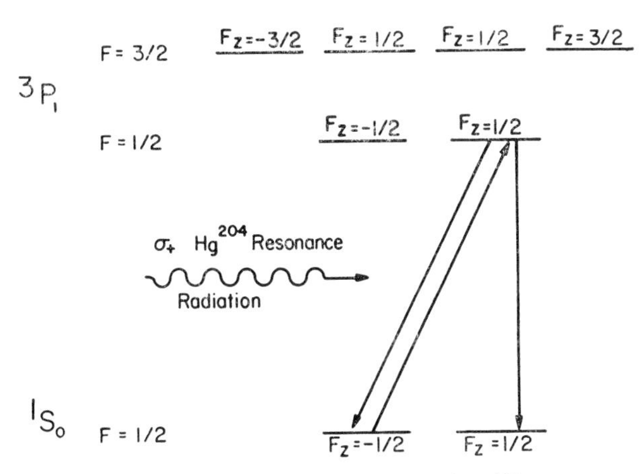

FIG. 8. The optical pumping of Hg^{199}.

several reasons. It has a high vapor pressure, pumping light sources are easy to construct, and no buffer gas or wall coating is required because the atom in its ground state can collide many times with a glass or quartz wall without disorienting the nuclear spin.

The isotopes Hg^{199} and Hg^{201} are easily optically pumped (Cagnac, 1961). Hg^{199} is particularly useful for study since the nuclear spin is $I = \frac{1}{2}$ so that the ground state is a model two-level system.

The Hg atom is excited to the $^3P_1$ state by 2537 Å resonance radiation. The ground state and excited state structures are shown in Fig. 8. To optically pump the ground state, use is made of the fortuitous overlap of emission and absorption lines of different mercury isotopes. For example, left-circularly polarized (σ_+) radiation from a Hg^{204} light source selectively excites the $F = \frac{1}{2}$ level of the $^3P_1$ state in Hg^{199} and produces a $\Delta M = +1$ transition. Thus the atoms are pumped out of the $M = -\frac{1}{2}$ level into $M = +\frac{1}{2}$ level of the ground state. Alternatively, one can selectively excite the $F = 3/2$ level by using a Hg^{199}

light source and filtering out the $\Delta F = 0$ component with a Hg^{204} absorption cell.

By a similar method the odd isotopes of Cd can also be optically pumped (Lehman and Brossel, 1964, 1966; Chaney and McDermott, 1969). Attempts to optically pump Zn^{67} with the singlet resonance line failed, however, and led to a careful analysis of just how the nuclear orientation takes place during the optical-pumping cycle (Lehman, 1964). The nuclear spin is oriented by the $\mathbf{I}.\mathbf{J}$ interaction in the excited state. If the coupling is not sufficiently strong to orient the nucleus before the atom radiates back down to the ground state, the pumping process will fail. Optical pumping of Zn^{67} has been achieved (Spence and McDermott, 1967), however, using an intercombination line.

Analogous pumping schemes have been used to optically pump the odd isotopes of Barium (Olschewski and Otten, 1966) and Ytterbium (Olschewski and Otten, 1967).

G. OPTICAL PUMPING OF HELIUM

A great many experiments have been performed over the past decade which involve the optical pumping of the 2^3S_1 metastable state of He. The pertinent energy level structure of the He^4 atom is shown schematically in Fig. 9.

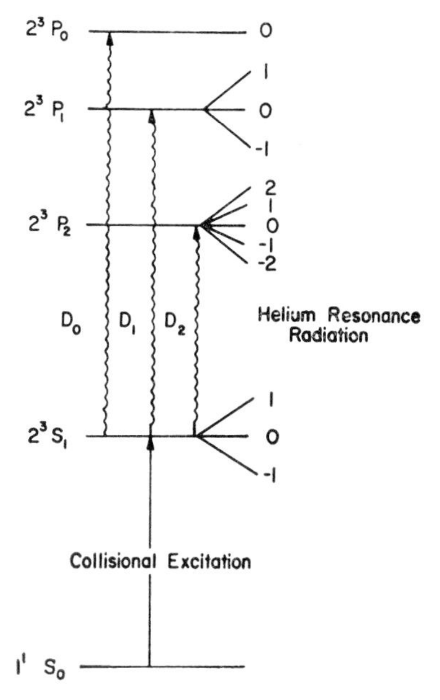

FIG. 9. The energy level structure pertinent to the optical pumping of the $^3S_1$ metastable state of He^4.

In their original experiment, Colegrove and Franken (1960) used unpolarized resonance radiation to optically align the metastable 2^3S_1 state. A few Torr of He gas was contained in an absorption cell in which a weak electrodeless discharge was maintained. The discharge produces metastable He atoms which survive $\sim 10^{-4}$ sec and relax to the ground state through metastability exchange collisions with ground state atoms. The optimum density of metastables is the canonical $\sim 10^{10}$ atoms/c.c.

The optical pumping of the metastable state is completely analogous to the pumping of alkali atoms. The D_0, D_1 and D_2 lines of the pumping light are too close together to be easily filtered. As a result, the population differences which the pumping process is able to produce among the metastable state sublevels is sensitive to the relative intensities of the three components of the pumping light as well as to the degree of mixing in the excited state. In a typical case, the intensities of the D_1 and D_2 lines are equal and two or three times more intense than the D_0 line. As in the case of the alkali atoms, the sublevels which absorb light with the least probability are populated at the expense of the more strongly absorbing levels. Under these conditions it is possible to populate the $m = \pm 1$ levels in He^4 at the expense of the $m = 0$ level with unpolarized light. This situation is called alignment. If circularly polarized light is used (Walters *et al.*, 1962), polarization of the metastable state is achieved. R.f. transitions between magnetic sublevels in the metastable state can be detected by transmission monitoring or resonance fluorescence.

The polarization of the metastable state of He^3 by means of optical pumping can be used to produce large numbers of oriented nuclei in He^3 ground state atoms (Schearer *et al.*, 1963). The nuclei become oriented in the metastable state through the $\mathbf{I}.\mathbf{J}$ interaction in the optically pumped state. When the atoms return to the ground state through a metastability exchange collision, the collision affects only the electronic configuration, leaving the nucleus in its oriented state. Because the relaxation time for nuclear spins in the $^1S_0$ ground state is long, a large nuclear polarization (20–40 %) can be maintained. Indeed, sufficient polarizations for nuclear maser action was obtained in an elegant experiment by Robinson and Myint (1964).

The principal sources of linewidth for r.f. transitions between sublevels in the $^3S_1$ state are the metastability exchange collisions between the metastable and ground state atoms. These collisions limit the lifetime of the metastable state. If precision measurements of the g-factor or hyperfine splitting in the metastable state are to be made, this linewidth must be reduced. At room temperatures, typical linewidths are of the order of 30 kHz. As has been demonstrated experimentally (Colegrove *et al.*, 1964), the metastability exchange cross section drops off sharply with decreasing temperature. Rosner and Pipkin (1970) have taken advantage of this temperature dependence in their precision measurement of the hyperfine splitting in the $^3S_1$ state of He^3.

By optically pumping a cryogenically cooled sample at a temperature of 15°K, they reduced their linewidth to 1–2KHz.

The optical pumping of He^3 at higher pressure, where there is considerable excited state mixing, is more efficient when a He^4 lamp is used. As a result of the isotope shift of the spectral lines, He^4 resonance radiation can only excite the D_0 transition in He^3. This turns out to be more efficient than using a He^3 lamp. At lower pressures the He^3 lamp is more efficient.

Recently, a new technique similar to spin-exchange optical pumping has been developed in which atoms are converted into polarized ions by Penning collisions with optically pumped metastable He atoms. The polarization of the excited ions is monitored by observing the polarized fluorescence from the ions. The technique has been applied to Cd^+ and Zn^+ (Schearer and Holton, 1970).

In a manner completely analogous to the optical pumping of He, the $^3P_2$ states of Ne, Xe, and A have been successfully polarized by optical pumping (Schearer, 1968c, 1969a, b). The interesting point is that the states being polarized are P states which means that there is a large depolarization cross section for collisions with buffer gas atoms and so it is surprising that optical pumping is possible. The success of the experiments immediately implied an anomalously low disorientation cross section for the P states of these atoms.

H. FREQUENCY SHIFTS IN OPTICAL-PUMPING EXPERIMENTS

The interaction of optically pumped atoms with electromagnetic radiation and with other atoms can produce a variety of shifts in the hyperfine and Zeeman transition frequencies of the oriented atoms. Although these shifts are small, they are easily observed in precision optical-pumping experiments, and observation of these shifts has provided a means for studying the interactions which produce them.

1. *Light Shifts*

Barrat and Cohen-Tannoudji (1961) carried through a detailed quantum mechanical analysis of the optical-pumping cycle. Applying 2nd-order perturbation theory to treat the interaction of the optically pumped atoms with the pumping radiation, they showed that the pumping light could be treated as a relaxation effect which broadened r.f. transitions between sublevels of the state being optically pumped. Their theory predicted not only a broadening of these transitions, but a shift in their frequencies as well. The predicted shifts were proportional to the intensity of the pumping light and arose from two distinct physical processes, real and virtual absorption of the pumping radiation.

Photons with a wavelength very close to the wavelength required to optically excite an atom can produce virtual transitions to an excited state which violate conservation of energy within the limits of the uncertainty principle. Barrat and Cohen-Tannoudji showed that these virtual transitions will in general change the self-energy of the ground state sublevels of the absorbing atoms. Since the energy changes vary from level to level the Zeeman and hyperfine transition frequencies are shifted.

Real transitions, on the other hand, do not produce changes in the energies of the ground state sublevels, but they can be the source of shifts in the r.f. transition frequencies nevertheless. When an r.f. field produces magnetic resonance in a collection of optically pumped atoms, the r.f. couples the levels involved in the r.f. transition in a coherent way. As a relaxation process, the absorption of pumping light tends to destroy this coherence. Barrat and Cohen-Tannoudji showed, however, that a fraction of this coherence can be preserved in the optical-pumping cycle if the Zeeman spread of the optical transitions is small compared to the natural width of the excited state. This partial circulation of coherence shifts the observed magnetic resonance frequencies. The magnitude of the shift turns out to depend upon the magnitude of the steady magnetic field H_0.

Light intensity shifts were first observed in alkali hyperfine transitions by Arditi and Carver (1961). Cohen-Tannoudji verified the essential predictions of his theory in an optical-pumping study of Hg^{199} (Cohen-Tannoudji, 1962). Shifts due to both real and virtual transitions have been observed in the optical pumping of He (Schearer, 1962; Schearer and Sinclair, 1968). Extensive theoretical and experimental studies of the interaction between radiation and optically pumped vapors have been carried out by W. Happer and co-workers (Happer and Mathur, 1967; Mathur *et al.*, 1968). Happer employs a semi-classical approach to the problem in which the radiation field is treated classically and the atomic system quantum mechanically. He has published a review of light propagation in optical-pumping experiments (Happer, 1970). In precision measurements of hyperfine splittings or magnetic moments, care must be taken to avoid errors due to the light shift. This is accomplished by making measurements at different light intensities and extrapolating to zero light intensity. The light shift is also a source of error in optically pumped frequency standards which rely upon the hyperfine frequency of an optically pumped atom.

2. *Hyperfine Pressure Shifts*

The hyperfine splitting in the ground state of an atom arises from the multipole interactions between the electronic angular momentum and the nuclear spin. We have already discussed the dipole interaction $A\,\mathbf{I}\,.\,\mathbf{J}$ for the case of the alkalies. In other paramagnetic atoms, higher multipole terms can enter in, but their magnitudes are down by several orders of magnitude.

The dipole interaction constant A depends upon the electronic wavefunction and in particular upon its value at the nucleus. The presence of nearby atoms will perturb the wavefunction and alter the magnitude of A. If only binary atom–atom interactions are considered, the resulting shift in the hyperfine frequency will be a linear function of density.

In optical-pumping experiments in which a buffer gas is used and precision determinations of the hyperfine frequencies of oriented atoms or ions are made, the hyperfine frequency is found to be a linear function of the buffer gas density at pressures below 1 atm. At higher pressures, a deviation from linearity for the pressure shift of the Rb hyperfine splitting in an argon buffer gas has been observed (Ensberg and Putlitz, 1969). Pioneering studies of pressure shifts were made in measurements of the hyperfine structure in the ground state of optically pumped alkali atoms (Bender *et al.*, 1958; Arditi and Carver, 1961).

Unfortunately the shifts are called pressure shifts rather than density shifts because optical-pumping cells are filled at room temperature with a given pressure of buffer gas and the misnomer "pressure shift" has endured. This causes confusion in the literature when the temperature dependence of such shifts is measured. Experimentally, the density of the buffer gas remains constant while the pressure goes up and down with temperature. Theoretical calculations of the temperature dependence of the frequency shifts are often made under the mistaken impression that the experimenter maintains the pressure constant as the temperature is varied.

As in the case of light intensity shifts, pressure shifts must be taken into account in precision measurements and in frequency standards. By making several samples with various buffer gas densities an extrapolation to zero buffer gas density can be made.

The measurement of hyperfine pressure shifts is of intrinsic interest because they provide a handle on low energy interatomic interactions. A number of theoretical calculations of hyperfine pressure shifts have been carried out, but as yet only moderate agreement between experiment and theory has been reached. Recently, the alkali optical-pumping technique has been extended to allow hyperfine pressure shift measurements to be made over a wide range of temperatures (Weiss *et al.*, 1970; Wright *et al.*, 1970; Wright, 1972) providing a more stringent test of pressure shift calculations (Kunik and Kaldor, 1971; Rao *et al.*, 1970; Ray *et al.*, 1970; Ikenberry and Das, 1971). Experimental and theoretical work on hyperfine pressure shifts will be reviewed in more detail in Section VII.

3. *Spin-exchange Frequency Shifts*

In spin-exchange optical-pumping experiments, electrons, atoms, or ions are oriented by spin-exchange collisions with optically pumped atoms. As was discovered in electron-alkali spin-exchange experiments, the exchange collisions not only flip the spin of the electron but also shift the electron spin resonance frequency. A quantum mechanical treatment of the spin-exchange optical-pumping signal (Balling *et al.*, 1964) reveals that the frequency shift is proportional to the density and polarization of the alkali atoms.

A similar shift is predicted for r.f. transitions in atoms undergoing spin-exchange collisions (Bender, 1963, 1964; Balling *et al.*, 1964); but so far spin-exchange frequency shifts due to atom–atom collisions have only been observed for H–H collisions in a hydrogen maser experiment (Crampton *et al.*, 1972).

The spin-exchange frequency shifts are easily eliminated as a source of error in precision optical-pumping measurements because of the dependence upon the polarization of the pumped atoms. Measurements are made with both left- and right-circularly polarized pumping light and the results averaged. The sign of the polarization and hence the sign of the frequency shift changes with the reversal of the polarization of the pumping light.

In spin-exchange optical-pumping experiments involving electrons and alkali atoms, the frequency shift can be used in conjunction with linewidth measurements to obtain information about the electron-alkali atom scattering phase shifts. Details of the quantum theory of spin-exchange optical-pumping signals are given in Section VI and optical-pumping experiments designed to study electron-alkali atom collisions are discussed in Section VII.

I. CROSSED BEAM DETECTION

If alkali atoms are being optically pumped by circularly polarized D_1 radiation incident along a static magnetic field H_0 in the z-direction with hyperfine components of equal intensity, the absorption of the pumping light for an optically thin sample is proportional to $(1-2\overline{\langle J_z \rangle})$, where $2\overline{\langle J_z \rangle}$ is the electronic polarization of the optically pumped atoms. If an r.f. field is applied at the magnetic resonance frequency of the atoms, the components $\overline{\langle J_x \rangle}$ and $\overline{\langle J_y \rangle}$ of the electronic polarization in the laboratory frame oscillate at the precession frequency ω_0.

As was first pointed out by Dehmelt (1957), the oscillation of a transverse component of the polarization can be detected by passing a circularly polarized beam of resonance radiation through the optical-pumping cell in a direction perpendicular to the z-axis. The absorption of a crossed beam in the x-direction will be proportional to $(1-2\langle J_x \rangle)$, so the intensity of the light transmitted by the cell in the x-direction will be modulated at the magnetic resonance frequency. This effect was first observed by Bell and Bloom (1957) and has been used for light modulation experiments and studies of spin relaxation, and is the basis of operation for self oscillating optically pumped magnetometers. The presence of the second beam has two effects upon the absorbing atoms: (i) it tends to polarize them in the x-direction and (ii) to relax the polarization in the z-direction.

J. SPIN RELAXATION

In our elementary treatment of optical pumping in Subsection C, we made use of a time constant τ_1 called the longitudinal relaxation time. If the populations of the two sublevels of a spin-$\frac{1}{2}$ system are displaced from their equilibrium values by optical pumping or some other means, they will tend to relax exponentially to their thermal equilibrium values with a characteristic time constant τ_1. The resulting relaxation of $\overline{\langle J_z \rangle}$ comes about principally through three types of spin disorienting collisions: collisions with buffer gas atoms, collisions with the walls of the optical-pumping cell, and spin-exchange collisions between the optically pumped atoms and another species also present in the cell.

As we have already noted, a resonant r.f. field will produce a coherent precession of the oriented spins about the static magnetic field. In the absence of the r.f. field, spin relaxation will cause this coherence to die away exponentially with a characteristic time constant τ_2. τ_2 is called the transverse relaxation time and is the characteristic time in which the individually precessing spins get out of phase with each other. As we shall see in Section IV, the width of magnetic resonance signal is determined, in part, by the magnitude of τ_2. The maximum polarization of the optically pumped atoms and hence the amplitude of the resonance signal depends upon τ_1. In order to obtain a strong, narrow resonance signal, one must minimize the relaxation times τ_1 and τ_2.

The relaxation times for optically pumped atoms under varying conditions have been widely studied experimentally and theoretically in order to understand the processes underlying spin relaxation. Most of the early experimental investigations of longitudinal relaxation utilized a technique introduced by Franzen (1959) referred to as relaxation in the dark. In this type of experiment, atoms are optically pumped by a circularly polarized beam incident along the static magnetic field in the z-direction. The pumping light is cut off suddenly at $t = 0$ by a shutter, and the atomic polarization decreases under the influence of the spin-disorienting collisions. For a spin-$\frac{1}{2}$ system,

$$\overline{\langle J_z(t) \rangle} = \overline{\langle J_z(0) \rangle}\, e^{-t/\tau_1}, \qquad \text{...(J.1)}$$

where

$$\frac{1}{\tau_1} = \left(\frac{1}{T_1^1} + \frac{1}{T_1^2} + \cdots \frac{1}{T_1^N} \right) \qquad \text{...(J.2)}$$

The relaxation time τ_1 is a composite of the relaxation times $T_1^1, \ldots T_1^N$ due to N different relaxation mechanisms. If the shutter is opened at time t the light absorption is proportional to $(1-2\overline{\langle J_z(t) \rangle})$. If the absorption is measured at the instant the shutter is opened as a function of the time t that the shutter remains closed, the longitudinal relaxation time τ_1 can be determined.

The transverse relaxation time can be measured by using crossed beam detection. A resonant r.f. field is applied to establish coherence and turned off at the same time as the pumping beam in the z-direction. The oscillating $\overline{\langle J_x(t)\rangle}$ will decay with time constant τ_2 and can be monitored by the crossed beam. The monitoring beam must be kept weak, as it itself is an additional relaxation mechanism. Alternatively, the linewidth of the resonance signal can be measured to determine τ_2. τ_2 is, of course, a composite of relaxation times $T_2^1 \ldots T_2^N$. That is

$$\frac{1}{\tau_2} = \frac{1}{T_2^1} + \frac{1}{T_2^2} + \cdots \frac{1}{T_2^N}. \qquad \ldots(\text{J.3})$$

As has been pointed out by Bouchiat and Brossel (1962, 1966), the situation is far more complicated for the relaxation of optically pumped alkali atoms as compared to the simple spin-$\frac{1}{2}$ system. In a weak magnetic field, the nuclear and electronic spins are closely coupled by the hyperfine interaction and both are polarized by the pumping process. Collisions which disorient the electron spin leave the nuclear orientation intact and the $\mathbf{I}.\mathbf{J}$ interaction tends to repolarize the electron spin. This slows down the relaxation of the electronic polarization $\overline{\langle J_z\rangle}$ and it no longer decays as a single exponential.

As Bouchiat and Brossel have also pointed out, care must be taken in applying Franzen's technique to make sure that the absorption of pumping light is actually proportional to $(1-2\overline{\langle J_z\rangle})$. It turns out (see Section V) that this is only true if the two hyperfine components of the D_1 pumping light are of equal intensity. If not, there will be some hyperfine pumping of the alkali atoms, and the intensity of the pumping light absorbed by the sample will depend upon the hyperfine level populations as well as upon $\overline{\langle J_z\rangle}$. Much of the early work on relaxation times is unreliable because of a lack of appreciation of these complications.

Bouchiat and Grossetete (1966) have introduced a technique for measuring longitudinal relaxation times utilizing two separate light beams. One beam is intense and directed along the static magnetic field. The spectral and polarization characteristics of this beam determine the polarization and differences in the hyperfine level populations which are established as initial conditions. A second weak monitoring beam passes through the cell at an angle to the pumping beam, e.g. 45°. The spectral and polarization characteristics of this second beam determine which observable will be monitored. For instance, if the monitoring beam has equal intensity hyperfine components, is circularly polarized, and is the D_1 line it will monitor $\overline{\langle J_z\rangle}$. If the beam is unpolarized and has unequal hyperfine component intensities, it will monitor $\langle \mathbf{I}.\mathbf{J}\rangle$ which is a function of the difference in populations of the hyperfine levels (see

Section V). The pumping light is switched off at time $t = 0$ and the decay of the observable is monitored by the weak beam.

In succeeding chapters, the relationship of relaxation processes to observed optical-pumping signals will be discussed in greater detail. In Section VII we will summarize the physical information which has been obtained from measurements of relaxation times.

III. Density Matrix Methods

A. The Density Operator

If the state of a system can be described at a time t by a Dirac state vector $|\alpha, t\rangle$, then the observable properties of the system are determined by the time evolution of $|\alpha, t\rangle$. The time evolution is given by the Schrödinger equation

$$i\hbar \frac{\partial}{\partial t} |\alpha, t\rangle = \mathscr{H} |\alpha, t\rangle, \qquad \text{...(A.1)}$$

where $\mathscr{H}$ is the Hamiltonian for the system. If an observable represented by the Hermitian operator A is measured in a large number of identical systems at time t, the average of the measurements is just the expectation value of A, where

$$\langle A \rangle = \langle \alpha, t | A | \alpha, t \rangle. \qquad \text{...(A.2)}$$

In an optical-pumping experiment one observes a signal, which is a function of the macroscopic average of some observable, the electronic angular momentum for instance, over a large number of identical atomic systems. In order to calculate the form of the optical-pumping signal it is necessary to calculate the time dependence of this macroscopic average under the influence of the pumping light, r.f. fields, etc. If all the atoms were prepared in some state $|\alpha, t\rangle$, and the complete Hamiltonian were known, we could simply calculate the expectation value of the desired observable as a function of time using the Schrödinger equation.

Obviously, the collection of atoms in the optical-pumping cell cannot be prepared in identical states at some initial time. Indeed, we cannot write down the complete Hamiltonian for an individual atom which is subject to random collisions with other atoms. Nevertheless, it is possible to predict the macroscopic behavior of some observables by applying the techniques of statistical mechanics. To this end, we introduce the density operator ρ.

For simplicity we will first consider a situation where we are interested in one observable A only. The eigenvalues of A are a_i, and we will assume for the moment that the spectrum is discrete. That is,

$$A | a_i, t \rangle = a_i | a_i, t \rangle. \qquad \text{...(A.3)}$$

Although we cannot write down the state vector of a particular atom in our ensemble, we may be able to say what the probability is of finding an atom in the state $|a_i, t\rangle$. If this probability is $P(a_i)$, then the average of a large number of measurements of the observable A is just

$$\overline{\langle A \rangle} = \sum_i a_i P(a_i). \qquad \text{...(A.4)}$$

The bar over the bracket distinguishes this ensemble average from the expectation value $\langle A \rangle$ computed from a precise knowledge of the state of each atom.

We now define the density operator $\rho(A)$ by the equation

$$\rho(t) \equiv \sum_i |a_i, t\rangle P(a_i) \langle a_i, t|. \qquad \text{...(A.5)}$$

To express $\overline{\langle A \rangle}$ in terms of the operators ρ and A, we compute the trace of ρA in the $|a_i, t\rangle$ representation:

$$\begin{aligned} \operatorname{Tr} \rho A &= \sum_j \langle a_j, t|\rho A| a_j, t\rangle = \sum_{j,i} \langle a_j, t|a_i, t\rangle P(a_i) \langle a_i, t|A| a_j, t\rangle \\ &= \sum_{j,i} \delta_{ji} P(a_i) \langle a_i, t|A| a_j, t\rangle = \sum_i a_i P(a_i). \end{aligned} \qquad \text{...(A.6)}$$

Thus, we have the simple relation

$$\overline{\langle A \rangle} = \operatorname{Tr} \rho A = \operatorname{Tr} A \rho. \qquad \text{...(A.7)}$$

Since

$$\sum_i P(a_i) = 1,$$

we have the property

$$\operatorname{Tr} \rho = 1. \qquad \text{...(A.8)}$$

From equation (A.7), we see that in order to compute $\overline{\langle A \rangle}$ as a function of time, we must calculate the time dependence of the density operator ρ. The density operator contains all the physical information which we can obtain from our ensemble of atoms.

If we know the Hamiltonian for a system, we can compute the time rate of change of the density operator. Since

$$i\hbar \frac{\partial}{\partial t} |a_i, t\rangle = \mathscr{H} |a_i, t\rangle, \qquad \text{...(A.9a)}$$

and

$$-i\hbar \frac{\partial}{\partial t} \langle a_i, t| = \langle a_i, t| \mathscr{H}, \qquad \text{...(A.9b)}$$

we have immediately from the definition in equation (A.5)

$$\frac{\partial}{\partial t} \rho(t) = \sum_i \frac{\partial}{\partial t} |a_i, t\rangle P(a_i) \langle a_i, t| + \sum_i |a_i, t\rangle P(a_i) \frac{\partial}{\partial t} \langle a_i, t| = \frac{1}{i\hbar} [\mathscr{H}, \rho(t)]. \qquad \text{...(A.10)}$$

Thus, if the operator A has no explicit time dependence,

$$\frac{\mathrm{d}}{\mathrm{d}t}\overline{\langle A\rangle} = \mathrm{Tr}\frac{\partial\rho}{\partial t}A = \frac{1}{i\hbar}\mathrm{Tr}\,[\mathscr{H},\rho]\,A. \qquad \text{...(A.11)}$$

It is easy to generalize our treatment to handle a system described in terms of more than one observable. Let $|a_i^{(1)}, a_j^{(2)} \dots a_k^{(N)}, t\rangle$ be a complete set of basis vectors where $a_i^{(1)}, a_j^{(2)} \dots a_k^{(N)}$ are the eigenvalues of the commuting set of operators $A^{(1)}, A^{(2)}, \dots A^{(N)}$. That is,

$$A^{(1)}|a_i^{(1)} \dots a_k^{(N)}, t\rangle = a_i^{(1)}|a_i^{(1)} \dots a_i^{(N)}, t\rangle, \qquad \text{...(A.12)}$$

etc. The density operator for the ensemble is

$$\rho(t) = \sum_{i,j,\dots k} |a_i^{(1)} \dots a_k^{(N)}, t\rangle P(a_i^{(1)} \dots a_k^{(N)},) \langle a_i^{(1)} \dots a_k^{(N)}, t|, \qquad \text{...(A.13)}$$

where

$$P(a_i^{(1)}, \dots a_k^{(N)}) = P(a_i^{(1)})P(a_j^{(2)}) \dots P(a_k^{(N)}). \qquad \text{...(A.14)}$$

The ensemble average for a particular observable $\overline{\langle A^{(1)}\rangle}$ is

$$\overline{\langle A^{(1)}\rangle} = \sum_i a_i^{(1)} P(a_i^{(1)}), \qquad \text{...(A.15)}$$

where $P(a_i^{(1)})$ is the probability of finding an atom with eigenvalue $a_i^{(1)}$ irrespective of the other observables. But,

$$\begin{aligned}\mathrm{Tr}\,\rho A^1 &= \sum_{\substack{l,\dots m\\ i,\dots k}} \langle a_l^{(1)} \dots a_m^{(N)}, t|A^{(1)}|\, a_i^{(1)} \dots a_k^{(N)}, t\rangle P(a_i^{(1)} \dots a_k^{(N)}) \\ &\quad \times \langle a_i^{(1)} \dots a_k^{(N)}, t|\, a_l^{(1)} \dots a_m^{(N)}, t)\rangle \\ &= \sum_i a_i^{(1)} \left[\sum_{j,\dots k} P(a_i^{(1)}\, a_j^{(2)} \dots a_k^{(N)})\right].\end{aligned} \qquad \text{...(A.16)}$$

Since

$$\sum_{j,\dots k} P(a_i^{(1)}, a_j^{(2)} \dots a_k^{(N)}) = P(a_i^{(1)})$$

is the probability of finding an atom in a state with eigenvalue $a_i^{(1)}$ irrespective of the other observables, we have

$$\mathrm{Tr}\,A^{(1)}\rho = \sum_i a_i^{(1)} P(a_i^{(1)}) = \overline{\langle A^{(1)}\rangle}. \qquad \text{...(A.17)}$$

We also see from equation (A.16) that the density operator for one observable only is obtained from the general density operator by taking the trace over the unwanted coordinates. That is,

$$\rho^{(1)} = \sum_i |a_i^{(1)}, t\rangle P(a_i^{(1)}) \langle a_i^{(1)}, t| = \mathrm{Tr}_{a_j^{(2)},\dots a_k^{(N)}}\,\rho. \qquad \text{...(A.18)}$$

In essentially the same way we can handle the statistical properties of an ensemble of two species of particles which are statistically independent. If we label the state vectors of particle 1 by $|a_i, t\rangle$ and those of particle 2 by $|b_j, t\rangle$, then the density operator for the combined ensemble is

$$\rho(1,2) = \sum_{i,j} |a_i, t\rangle |b_j, t\rangle P(a_i) P(b_j) \langle b_j, t| \langle a_i, t|. \qquad \text{...(A.19)}$$

We can obtain the density operator for particles 1 and particles 2 separately by tracing over the unwanted particle coordinates. Thus

$$\rho(1) = \mathrm{Tr}_2\, \rho(1,2) = \sum_j \langle b_j, t|\rho(1,2)|b_j, t\rangle = \sum_i |a_i, t\rangle P(a_i) \langle a_i, t|, \qquad \text{...(A.20)}$$

where we have used the fact that

$$\sum_j P(b_j) = 1.$$

Similarly, we have

$$\rho(2) = \mathrm{Tr}_1\, \rho(1,2). \qquad \text{...(A.21)}$$

B. THE DENSITY MATRIX

The matrix elements of the density operator ρ in a particular representation constitute the density matrix, and in many quantum mechanics texts it is the density matrix which is discussed rather than the operator. The density operator, as defined in equation (A.5), obviously forms a diagonal matrix in the $|a_i, t\rangle$ representation. In some other basis, however, the density matrix will not necessarily be diagonal.

When the density matrix, rather than the operator, is the starting point of the statistical mechanical treatment of an ensemble, the discussion usually takes a slightly different approach from the one we have taken up to now. Given an ensemble of identical atoms we could write down a general state vector $|\alpha, t\rangle$ for an atom which is some linear combination of the complete set of eigenstates $|a_i\rangle$ of an observable A. That is,

$$|\alpha, t\rangle = \sum_i C_i(t) |a_i\rangle, \qquad \text{...(B.1)}$$

where

$$A|a_i\rangle = a_i |a_i\rangle. \qquad \text{...(B.2)}$$

The expectation value of A for an atom in a particular state $|\alpha, t\rangle$ is

$$\langle A\rangle = \langle \alpha, t|A|\alpha, t\rangle = \sum_{i,j} C_j^*(t) C_i(t) \langle a_j|A|a_i\rangle. \qquad \text{...(B.3)}$$

Since we have a statistical mixture of states, we can only concern ourselves with the ensemble average of the expectation value. We must therefore take an average over the coefficients $C_j^*(t) C_i(t)$ in equation (B.3).

$$\overline{\langle A\rangle} = \sum_{i,j} \overline{C_j^*(t) C_i(t)} \langle a_j|A|a_i\rangle. \qquad \text{...(B.4)}$$

If we now call the coefficients $\overline{C_j^*(t)\,C_i(t)}$ the elements of the density matrix ρ_{ij}, we have from equation (B.4)

$$\overline{\langle A\rangle} = \sum_{i,j} \rho_{ij} A_{ji} = \operatorname{Tr} \rho A, \qquad \text{...(B.5)}$$

where we have denoted the matrix elements $\langle a_j|A|\,a_i\rangle$ as A_{ji}. We can of course show directly that the coefficients $\overline{C_j^*(t)\,C_i(t)}$ are the elements of the density matrix as defined in Subsection A. We begin by writing down the density operator in the $|\alpha, t\rangle$ representation in which it is diagonal.

$$\rho = \sum_{\alpha} |\alpha, t\rangle P(\alpha) \langle \alpha, t|. \qquad \text{...(B.6)}$$

Knowledge of the probabilities $P(\alpha)$ was implicit in our averaging in equation (B.5). Using equation (B.1), we have

$$\rho(t) = \sum_{\alpha,i,j} C_j^\alpha(t)|a_j\rangle P(\alpha) \langle a_i|\, C_i^{\alpha *}(t), \qquad \text{...(B.7)}$$

where the superscript α on a coefficient C_i^α indicates that it is the expansion coefficient appropriate to the state $|\alpha, t\rangle$. The matrix elements of $\rho(t)$ in the $|a_i\rangle$ representation are

$$\begin{aligned} \rho_{kl} = \langle a_k|\rho|\,a_l\rangle &= \sum_{\alpha,i,j} \langle a_k|C_j^\alpha(t)|\,a_j\rangle P(\alpha) \langle a_i|C_i^{\alpha *}(t)|\,a_l\rangle \\ &= \sum_{\alpha} C_k^\alpha(t)\, C_l^{\alpha *}(t) P(\alpha) = \overline{C_k(t)\, C_l^*(t)}, \end{aligned} \qquad \text{...(B.8)}$$

which is what we set out to show.

If we can calculate the time dependence of the coefficients $C_i(t)$ from Schrödinger's equation we immediately obtain the time dependence of $\rho(t)$. This is completely equivalent to using equation (A.10), but it is often more convenient to calculate the $C_i(t)$ rather than to use equation (A.10).

We also note that if we have the density matrices for two statistically independent ensembles of atoms, we can obtain the density matrix for the combined ensemble by taking the outer product of the two matrices. This is equivalent to forming the matrix elements of the operator $\rho(1, 2)$ as defined in equation (A.19).

Finally, the following relation will prove useful for the calculation of various ensemble averages in later sections. If A and B are two operators, then

$$\operatorname{Tr} AB\rho = \operatorname{Tr} B\rho A = \operatorname{Tr} \rho AB. \qquad \text{...(B.9)}$$

To show this, we write

$$\operatorname{Tr} AB\rho = \sum_{i} \langle a_i|AB\rho|\,a_i\rangle. \qquad \text{...(B.10)}$$

Inserting complete sets of states, we have

$$\begin{aligned}\operatorname{Tr} AB\rho &= \sum_{i,j,k} \langle a_i|A|a_j\rangle\langle a_j|B|a_k\rangle\langle a_k|\rho|a_i\rangle \\ &= \sum_{i,j,k} \langle a_k|\rho|a_i\rangle\langle a_i|A|a_j\rangle\langle a_j|B|a_k\rangle = \operatorname{Tr}\rho AB. \qquad \text{...(B.11)}\end{aligned}$$

One more permutation gives us the second equality.

C. THE DENSITY MATRIX FOR A SPIN-$\frac{1}{2}$ SYSTEM

To illustrate the use of the density matrix, we will consider an ensemble of spin-$\frac{1}{2}$ particles. If we restrict our attention to the spin degree of freedom, the density matrix will be a 2×2 matrix. We will work in the representation where the eigenstates of S_z are the basis vectors. That is,

$$\mathbf{S} = \tfrac{1}{2}\boldsymbol{\sigma}, \qquad \text{...(C.1)}$$

where the matrices of $\boldsymbol{\sigma}$ are the Pauli spin matrices

$$\sigma_x = \begin{pmatrix} 0 & 1 \\ 1 & 0 \end{pmatrix};\ \sigma_y = \begin{pmatrix} 0 & -i \\ i & 0 \end{pmatrix};\ \sigma_z = \begin{pmatrix} 1 & 0 \\ 0 & -1 \end{pmatrix}. \qquad \text{...(C.2)}$$

An arbitrary state $|\chi\rangle$ can be expressed as a superposition of the eigenstates of S_z. In terms of spinors

$$\chi = C_1\begin{pmatrix} 1 \\ 0 \end{pmatrix} + C_2\begin{pmatrix} 0 \\ 1 \end{pmatrix} = \begin{pmatrix} C_1 \\ C_2 \end{pmatrix}. \qquad \text{...(C.3)}$$

In this representation, the density matrix has the form

$$\rho = \begin{pmatrix} \overline{|C_1|^2} & \overline{C_1 C_2^*} \\ \overline{C_2 C_1^*} & \overline{|C_2|^2} \end{pmatrix} = \begin{pmatrix} \rho_{11} & \rho_{12} \\ \rho_{21} & \rho_{22} \end{pmatrix}. \qquad \text{...(C.4)}$$

The physical significance of the density matrix elements is made clear by computing the ensemble average of the spin vector $\mathbf{S}$. Note that $\rho_{12} = \rho_{21}^*$.

$$\langle S_z\rangle = \operatorname{Tr}\rho S_z = \tfrac{1}{2}\operatorname{Tr}\rho\sigma_z = \tfrac{1}{2}(\rho_{11} - \rho_{22}), \qquad \text{...(C.5)}$$

$$\langle S_x\rangle = \operatorname{Tr}\rho S_x = \tfrac{1}{2}(\rho_{21} + \rho_{12}) = \operatorname{Re}\rho_{21}, \qquad \text{...(C.6)}$$

$$\langle S_y\rangle = \operatorname{Tr}\rho S_y = \frac{i}{2}(\rho_{12} - \rho_{21}) = \operatorname{Im}\rho_{21}. \qquad \text{...(C.7)}$$

It is immediately obvious from the definition of the ensemble average and equation (C.5) that ρ_{11} and ρ_{22} give the probability of finding a particle with spin up or down along the z-axis, respectively. In order that there be a macroscopic component of the ensemble spin along either the x- or y-axis, the off-diagonal density matrix elements must be non-zero. This is the case in a magnetic resonance experiment where the rotating magnetic field causes the spins to precess coherently about the static magnetic field in the z-direction.

D. SPIN-EXCHANGE COLLISIONS

In a typical spin-exchange optical-pumping experiment, alkali atoms are polarized by optical pumping, and free electrons or another species of atom or ion are polarized by spin-exchange collisions with the optically-pumped alkali atoms. In order to understand the optical-pumping signals, we must determine how the spin systems of the two sets of particles are coupled by spin-exchange collisions. To do this, we must calculate the effect of these collisions on the density matrices of the two spin systems.

To set up the basic problem we will assume that we have two sets of particles 1 and 2, which have spin, and collide with each other. We will assume that between collisions, the particles are statistically independent. The density operator for the spin for particle 1 can be written as

$$\rho(1) = \sum_{s_1} |s_1\rangle P(s_1) \langle s_1|, \qquad \text{...(D.1)}$$

where $P(s_1)$ is the probability of finding a particle in the spin state $|s_1\rangle$. Similarly, we can write for the second species of particles

$$\rho(2) = \sum_{s_2} |s_2\rangle P(s_2) \langle s_2|. \qquad \text{...(D.2)}$$

After some average time t, all of the particles 1 will have collided with a particle 2. As a result of the collision,

$$\rho(1) = \sum_{s_1} |s_1\rangle P(s_1) \langle s_1| \rightarrow \sum_{s_1} |s_1'\rangle P(s_1) \langle s_1'| = \rho'(1), \qquad \text{...(D.3)}$$

where $|s_1'\rangle$ is the final spin state after a collision for a particle initially in the state $|s_1\rangle$ before the collision. The change in the spin state of particle 1 due to the collision will depend upon the spin of the target particle and the relative momentum of the particles before and after the collision. In order to compute the final density operator $\rho'(1)$, we must express the density operator in a form which includes the spin coordinate of particle 2 and the relative momentum of the colliding particles. If $P(\mathbf{k})$ is the probability of finding the relative momentum $\hbar\mathbf{k}$ of particles 1 and 2 before a collision, we can write

$$\rho(1) = \sum_{\substack{\mathbf{k}, s_2, \\ \overline{\mathbf{k}, s_2}, s_1}} \langle \overline{\mathbf{k}s_2} | \mathbf{k}s_2\rangle | s_1\rangle P(s_1) P(s_2) P(\mathbf{k}) \langle \mathbf{k}s_2 | \langle s_1 | \overline{\mathbf{k}s_2}\rangle, \qquad \text{...(D.4)}$$

because

$$\langle \overline{\mathbf{k}s_2} | \mathbf{k}s_2\rangle |s_1\rangle = \delta_{\bar{s}_2, s_2} \delta_{\bar{\mathbf{k}}, \mathbf{k}} |s_1\rangle, \qquad \text{...(D.5)}$$

and

$$\sum_{s_2, \mathbf{k}} P(s_2) P(\mathbf{k}) = 1. \qquad \text{...(D.6)}$$

But this is just

$$\rho(1) = \mathrm{Tr}_{2,\mathbf{k}}\, \rho(s_1 s_2 \mathbf{k}), \qquad \text{...(D.7)}$$

where $\rho(s_1 s_2 \mathbf{k})$ is the density operator for the combined spin system of particles 1 and 2 including momentum states, and the trace extends over the momentum coordinate $\mathbf{k}$ and spin coordinate of particle 2.

Thus, if we write down the density matrix for the combined spin system including momentum states, we can recover the spin-space density matrix for particles 1 by tracing over the momentum coordinate and spin coordinates of particle 2. Since between collisions the two spin systems are assumed to be statistically independent, the total density matrix is just the outer product of the two spin-space density matrices $\rho(1)$ and $\rho(2)$ and the momentum density matrix.

After a collision,

$$\rho'(1) = \mathrm{Tr}_{2,\mathbf{k}'}\, \rho(s_1' s_1' \mathbf{k}'), \qquad \ldots(\mathrm{D.8})$$

where the primes indicate the coordinates after a collision and the trace is as in equation (D.7).

This means that we can calculate the change in the density matrix $\rho(1)$ due to the collision by calculating the change in the density matrix for the combined system including momentum. To do this we will apply conventional scattering theory utilizing the well known S-matrix. We focus our attention on the single two-body collision between a particle 1 and a particle 2. Viewed in the center of mass frame, the initial two-particle state is $|\mathbf{k}s_1 s_2\rangle$. As a result of the collision, this state evolves into a final state $|\mathbf{k}' s_1' s_2'\rangle$. The two states are connected by a unitary operator $\mathscr{S}$ so that

$$|\mathbf{k}' s_1' s_2'\rangle = \mathscr{S}|\mathbf{k}s_1 s_2\rangle. \qquad \ldots(\mathrm{D.9})$$

To abbreviate the notation in what follows, we will denote the initial state $|\mathbf{k}s_1 s_2\rangle$ as $|\mathbf{k}s\rangle$ where s represents the two-particle spin state. Using this notation,

$$|\mathbf{k}' s'\rangle = \mathscr{S}|\mathbf{k}s\rangle. \qquad \ldots(\mathrm{D.10})$$

The matrix elements of the unitary operator $\mathbf{S}$ defined in equation (D.10) form the S-matrix for the scattering of particle 1 from particle 2.

$$\mathbf{S}_{\substack{\mathbf{k},\mathbf{k}' \\ s,s'}} = \langle \mathbf{k}s|\mathbf{S}|\,\mathbf{k}' s'\rangle \qquad \ldots(\mathrm{D.11})$$

Before the collision,

$$\rho(s,\mathbf{k}) = \sum_{\mathbf{k}s} |\mathbf{k}s\rangle P(s)\, P(\mathbf{k})\, \langle \mathbf{k}s|, \qquad \ldots(\mathrm{D.12})$$

and after the collision

$$\rho'(s',\mathbf{k}') = \sum_{\mathbf{k}s} |\mathbf{k}' s'\rangle P(s)\, P(\mathbf{k})\, \langle \mathbf{k}' \mathbf{s}'|. \qquad \ldots(\mathrm{D.13})$$

Using Equation (D.10), we have

$$\rho'(s',\mathbf{k}') = \mathbf{S}\rho(s,\mathbf{k})\,\mathbf{S}^{\dagger}. \qquad \ldots(\mathrm{D.14})$$

This is the fundamental equation which relates the density matrix for the two-particle system, including spin and momentum coordinates after a collision, to the density matrix before the collision. To calculate the density matrix $\rho'(1)$ for the spin coordinates of particle 1 after the collision, we merely take the trace over the unwanted coordinates. As a result of a single collision between the particles 1 and a particle 2, the spin space density matrix $\rho'(1)$ is

$$\rho'(1) = \mathrm{Tr}_{2,\mathbf{k}'}\, \mathbf{S}\rho(s\mathbf{k})\,\mathbf{S}^{\dagger}, \qquad \text{...(D.15)}$$

where the trace is over the spin coordinate of particle 2 and over the momentum coordinate $\mathbf{k}$. By a completely symmetrical argument we have

$$\rho'(2) = \mathrm{Tr}_{1,\mathbf{k}'}\, \mathbf{S}\rho(s,\mathbf{k})\,\mathbf{S}^{\dagger}, \qquad \text{...(D.16)}$$

for the change in the spin–space density matrix for particles 2 undergoing a single collision with a particle 1.

In Section VI, where the theory of spin-exchange optical pumping will be treated in detail, we will employ equations (D.15) and (D.16) to calculate the time rate of change of the density matrices for the spins of alkali atoms colliding with electrons and with other alkali atoms in terms of the scattering phase shifts for the collisions.

E. SPIN-RELAXATION TIMES

The random collisions between polarized atoms or ions with buffer gas atoms or with the walls of the optical pumping cell disorient the spins of the polarized particles. As is done in the analysis of NMR experiments, this spin relaxation can be handled phenomenologically by assuming an exponential decay of the polarization due to spin disorienting collisions. As an illustration, consider the spin relaxation of a spin-$\frac{1}{2}$ system. The rate of change of the spin-space density matrix due to spin-relaxation is assumed to be

$$\frac{d\rho}{dt} = \begin{pmatrix} \dfrac{\frac{1}{2}-\rho_{11}}{T_1} & \dfrac{-\rho_{12}}{T_2} \\ \dfrac{-\rho_{21}}{T_2} & \dfrac{\frac{1}{2}-\rho_{22}}{T_1} \end{pmatrix}. \qquad \text{...(E.1)}$$

Thus the diagonal density matrix elements both relax exponentially to $\frac{1}{2}$ with time constant T_1, the longitudinal relaxation time. We are ignoring the small population difference between the eigenstates of S_z due to the Boltzman distribution at thermal equilibrium. The off-diagonal density matrix elements relax to zero with time constant T_2, the transverse relaxation time. In thermal equilibrium there is no coherent precession of the spins about the z-axis.

We can calculate the time dependence of the observables $\overline{\langle S_z \rangle}, \overline{\langle S_x \rangle}$, and $\overline{\langle S_y \rangle}$ due to spin relaxation using equation (E.1).

$$\frac{\overline{\mathrm{d}\langle S_z \rangle}}{\mathrm{d}t} = \mathrm{Tr}\left(S_z \frac{\mathrm{d}\rho}{\mathrm{d}t}\right) = \frac{1}{2}\frac{\rho_{22} - \rho_{11}}{T_1} = -\frac{\overline{\langle S_z \rangle}}{T_1}, \qquad \text{...(E.2)}$$

$$\frac{\overline{\mathrm{d}\langle S_x \rangle}}{\mathrm{d}t} = \mathrm{Tr}\left(S_x \frac{\mathrm{d}\rho}{\mathrm{d}t}\right) = -\frac{\overline{\langle S_x \rangle}}{T_2}, \qquad \text{...(E.3)}$$

$$\frac{\overline{\mathrm{d}\langle S_y \rangle}}{\mathrm{d}t} = \mathrm{Tr}\left(S_y \frac{\mathrm{d}\rho}{\mathrm{d}t}\right) = -\frac{\overline{\langle S_y \rangle}}{T_2}. \qquad \text{...(E.4)}$$

IV. Optical Pumping of a Spin-$\frac{1}{2}$ System

Using the density matrix formalism developed in the preceding section, we will now calculate the form of the magnetic resonance signal we would expect to obtain by optically pumping fictitious alkali atoms with zero nuclear spin. The advantage of treating the spin-$\frac{1}{2}$ system first is that it is easy to handle mathematically and the basic physics is not obscured by the complications introduced by consideration of the nuclear spin. The spin-$\frac{1}{2}$ case is physically realized in the optical pumping of Hg^{199} and of the singly-ionized Group II atoms, but it also approximates the behavior of actual alkali atoms reasonably well.

The physical situation which we will analyse is the following. Atoms with total angular momentum $j=\frac{1}{2}$ are situated in a static magnetic field in the z-direction. They are illuminated by circularly polarized pumping radiation incident along the z-direction. A linearly oscillating field in the x-direction is applied to produce magnetic resonance, and the signal is detected by monitoring the intensity of the pumping light transmitted by the absorption cell. Using the density matrix, we will calculate the behavior of the macroscopic angular momentum of the optically pumped atoms under the influence of the pumping light, the static and oscillating magnetic fields, and spin relaxation.

A. The Optical-Pumping Process

We begin by considering the optical pumping of a paramagnetic atom in a steady magnetic field H_0 in the z-direction. In their classic paper, Barrat and Cohen-Tannoudji (1961) used 2nd-order perturbation theory to describe the optical-pumping cycle. We will simplify their discussion by only using 1st-order perturbation theory. We will have to appeal to their 2nd-order treatment to interpret some of our results and to include features lost in the 1st-order calculation. Our task is to compute the time dependence of the density matrix for the ground state angular momentum of the atom due to optical pumping.

We will accomplish this by calculating the time dependence of a state vector for the ground state of the atom in the radiation field of the pumping light. The results of this section apply to an arbitrary paramagnetic atom and not just to the spin-$\frac{1}{2}$ case.

We will write the state vector for the ground state of the optically pumped atom in the radiation field as $|\mu,\mathbf{k}_1,\ldots\mathbf{k}_i,\ldots\mathbf{k}_N,\sigma\rangle$. The radiation field consists of N photons with propagation vectors $\mathbf{k}_i$ and circular polarization σ, with $\sigma=\pm1$ for left- and right-circularly polarized light respectively. The eigenvalues of the atom's angular momentum in the z-direction are denoted by μ. The unperturbed Hamiltonian $\mathscr{H}_0$ for the system includes the energy of the radiation field and of the atom interacting with the static magnetic field $H_0\hat{k}$. The interaction of the atom with the radiation field is treated as a perturbation. If we define the ground state of the atom in the absence of the magnetic field and in the presence of N non-interacting photons as the state of zero energy, then

$$\begin{aligned}\mathscr{H}_0|\mu,\mathbf{k}_1,\ldots\mathbf{k}_i\ldots\mathbf{k}_N,\sigma\rangle &= -g_J\,\mu_0\,H_0\,\mu|\mu,\mathbf{k}_1\ldots\mathbf{k}_i\ldots\mathbf{k}_N,\sigma\rangle\\ &= \hbar\omega_g\,\mu|\mu,\mathbf{k}_1\ldots\mathbf{k}_i\ldots\mathbf{k}_N,\sigma\rangle, \qquad \text{(A.1)}\end{aligned}$$

where g_J is the g-factor for the ground state.

After an atom absorbs a photon of vector $\mathbf{k}_i$ and arrives in an excited state characterized by energy hck_0 and magnetic quantum number m, we have

$$\begin{aligned}\mathscr{H}_0|m,\mathbf{k}_1,\ldots\mathbf{k}_{i-1},\mathbf{k}_{i+1},\ldots\mathbf{k}_N,\sigma\rangle &= (-g_{J'},\mu_0\,H_0\,m+\hbar ck_0-\hbar ck_i)\\ &\quad\times|m,\mathbf{k}_1,\ldots\mathbf{k}_{i-1},\mathbf{k}_{i+1},\ldots\mathbf{k}_N,\sigma\rangle\\ &= (\hbar\omega_e\,m+\hbar ck_0-\hbar ck_i)(m,\mathbf{k}_1,\ldots\mathbf{k}_{i-1},\mathbf{k}_{i+1},\ldots\mathbf{k}_N,\sigma\rangle. \qquad \ldots\text{(A.2)}\end{aligned}$$

In equation (A.2), g_J' is the g-factor for the excited state and ω_e the Larmor frequency for the excited state.

In what follows, we will abbreviate our notation and write the state vector for the atom and radiation field before absorbing a photon as $|\mu\rangle$ and the state vector for the system after absorbing a photon as $|m,-\mathbf{k}_i\rangle$. If we include the interaction of the light with the atom in the full Hamiltonian $\mathscr{H}$ for the system, we have

$$\mathscr{H}=\mathscr{H}_0+\mathscr{H}_I, \qquad \ldots\text{(A.3)}$$

where

$$\mathscr{H}_I=\frac{e}{mc}\sum_i \mathbf{A}\,.\,\mathbf{p}_i, \qquad \ldots\text{(A.4)}$$

in the gauge where $\nabla\,.\,\mathbf{A}=0$.

In these equations, e and m are the charge and mass of an electron, $\mathbf{p}_i$ is the linear momentum of the ith electron in the atom, and $\mathbf{A}$ is the vector potential for the quantized radiation field.

For an arbitrary state vector $|\alpha(t)\rangle$ describing the system, the time dependent Schrödinger equation is

$$i\hbar\frac{\partial}{\partial t}|\alpha(t)\rangle = \mathscr{H}|\alpha(t)\rangle. \qquad \text{...(A.5)}$$

We can extract the time dependence due to the unperturbed Hamiltonian $\mathscr{H}_0$ by going to the interaction representation in which the state vector is $|\phi(t)\rangle$ defined by

$$|\phi(t)\rangle = e^{i\mathscr{H}_0 t/\hbar}|\alpha(t)\rangle. \qquad \text{...(A.6)}$$

We also define

$$\mathscr{H}'_I = e^{i\mathscr{H}_0 t/\hbar}\,\mathscr{H}_I\,e^{-i\mathscr{H}_0 t/\hbar}. \qquad \text{...(A.7)}$$

It follows that

$$\frac{i\hbar\partial|\phi(t)\rangle}{\partial t} = \mathscr{H}'_I|\phi(t)\rangle. \qquad \text{...(A.8)}$$

The next step is to expand $|\phi(t)\rangle$ in the eigenstates of the unperturbed Hamiltonian $\mathscr{H}_0$. The time dependence of $|\phi(t)\rangle$ will be contained in the expansion coefficients. We will not include states which correspond to absorption of a photon followed by emission. These states would be included in a 2nd-order perturbation treatment. That is, we write

$$|\phi(t)\rangle = \sum_{\mu} a_\mu(t)|\mu\rangle + \sum_{m,i} b_{m,i}(t)|\,m, -\mathbf{k}_i\rangle. \qquad \text{...(A.9)}$$

Once we find the time dependence of the coefficients a_μ, we can construct the ensemble average of $a_\mu a^*_{\mu'}$ which will give us the matrix elements of the density matrix for the atom's ground state.

Substituting our expansion into equation (A.8), we obtain

$$i\hbar\frac{\partial}{\partial t}a_\mu(t) = \sum_{m,i}\langle\mu|\mathscr{H}'_I|\,m - \mathbf{k}_i\rangle\, b_{m,i}(t), \qquad \text{...(A.10)}$$

and

$$i\hbar\frac{\partial}{\partial t}b_{m,i}(t) = \sum_{\mu'}\langle m, -\mathbf{k}_i|\mathscr{H}'_I|\,\mu'\rangle\, a_{\mu'}(t). \qquad \text{...(A.11)}$$

Recalling equation (A.7), these equations become

$$i\hbar\frac{\partial}{\partial t}a_\mu(t) = \sum_{m,i}\langle\mu|\mathscr{H}_I|\,m - \mathbf{k}_i\rangle \exp\left[(i\omega_g\mu - \omega_e m + ck_i - ck_0)t\right] b_{m,i}(t), \qquad \text{...(A.12)}$$

and

$$i\hbar\frac{\partial}{\partial t}b_{m,i}(t) = \sum_{\mu'}\langle m, -\mathbf{k}_i|\mathscr{H}_I|\,\mu'\rangle \exp\left[(-i\omega_g\mu' + \omega_e m - ck_i - ck_0)t\right] a_{\mu'}(t). \qquad \text{...(A.13}$$

The matrix elements in these equations are of the form

$$\langle m, -k_i|\,\mathscr{H}_I|\mu\rangle = A(k_i)\,e^{i\mathbf{k}_i.\mathbf{R}}\langle m|\hat{\varepsilon}_\sigma.\mathbf{D}|\,\mu\rangle\,\delta_{m,\mu\pm1}, \qquad \text{...(A.14)}$$

where **R** is the position of the atom's center of mass, and

$$e\mathbf{D} = \sum_i e\mathbf{r}_i$$

is the dipole moment operator for the atom with $\mathbf{r}_i$ being the position of the ith electron in the center of mass frame. $\hat{\varepsilon}_\sigma$ is the polarization vector for the radiation and the $(\pm)$ sign in equation (A.14) goes with $\sigma = \pm 1$. We will assume that $\sigma = +1$. The amplitude $A(k_i)$ is given by Davydov, (1965)

$$A(k_i) = -i\,\mathrm{e}\sqrt{\frac{2\pi\hbar n_{\mathbf{k}_i\sigma}}{\omega_i L^3}}\,\omega_0 \qquad \text{...(A.15)}$$

where $n_{\mathbf{k}_i\sigma}$ is the average number of photons with wave vector $\mathbf{k}_i$ and polarization σ, $\omega_0 = ck_0$, $\omega_i = ck_i$. L is the length of the side of the box used in the periodic boundary condition required to give discrete wave vectors $\mathbf{k}_i$. The factor $\mathrm{e}^{i\mathbf{k}_i.\mathbf{R}}$ is often neglected in elementary treatments of light absorption which are not concerned with the center of mass motion of the atom. If $\mathbf{v}$ is the velocity of the atom, then $\mathbf{R} = \mathbf{v}t$, and this factor will give us the Doppler shift of the absorption line.

If we introduce the abbreviations

$$\Delta_i = [ck_0 - ck_i + \mathbf{k}_i.\mathbf{v} + \omega_e(\mu+1) - \omega_g\,\mu], \qquad \text{...(A.16)}$$

and

$$\hat{\varepsilon}_\sigma.\mathbf{D} = D_+, \qquad \text{...(A.17)}$$

for $\sigma = 1$, equations (A.12) and (A.13) become

$$i\hbar\frac{\partial}{\partial t}a_\mu(t) = \sum_i \langle\mu|D_+|\,\mu+1\rangle\,A^*(k_i)\,\mathrm{e}^{-i\Delta_i t}\,b_{\mu+1,i}(t), \qquad \text{...(A.18)}$$

and

$$i\hbar\frac{\partial}{\partial t}b_{\mu+1,i}(t) = \langle\mu+1|D_+|\,\mu\rangle\,A(k_i)\,\mathrm{e}^{i\Delta_i t}\,a_\mu(t). \qquad \text{...(A.19)}$$

In order to integrate these equations we require the initial conditions at $t = 0$. We will assume that the atom is initially in its ground state so that

$$b_{m,i}(0) = 0. \qquad \text{...(A.20)}$$

If the interaction with the radiation field is sufficiently weak, we can assume that there is a range of t for which $a_\mu(t)$ is essentially constant and

$$a_\mu(t) \cong a_\mu(0), \qquad \text{...(A.21)}$$

the standard assumption in 1st-order perturbation theory. We can now integrate with respect to t and obtain from equation (A.19)

$$b_{\mu+1,i}(t) \cong \frac{\langle\mu+1|D_+|\,\mu\rangle\,A(k)\,(\mathrm{e}^{i\Delta_i t}-1)}{-\Delta_i\hbar}\,a_\mu(t). \qquad \text{...(A.22)}$$

Substituting this into equation (A.18), we obtain

$$i\hbar\frac{\partial}{\partial t}a_\mu(t)=\sum_i\frac{|A(k_i)|^2\,|\langle\mu|\,D_+|\,\mu+1\rangle|^2(1-\mathrm{e}^{-i\Delta_i t})}{-\hbar\Delta_i}\,a_\mu(t). \quad \text{...(A.23)}$$

Multiplying both sides by $a^*_{\mu'}(t)$, we get

$$a^*_{\mu'}(t)\frac{\partial}{\partial t}a_\mu(t)=\sum_i\frac{|A(k_i)|^2\,|\langle\mu|\,D_+|\mu+1\rangle|^2(1-\mathrm{e}^{-i\Delta_i t})}{-i\hbar^2\,\Delta_i}\,a_\mu(t)\,a^*_{\mu'}(t). \quad \text{...(A.24)}$$

The complex conjugate equation is

$$a_\mu(t)\frac{\partial}{\partial t}a^*_{\mu'}(t)=\sum_i\frac{|A(k_i)|^2\,|\langle\mu'|\,D_+|\mu'+1\rangle|^2(1-e^{i\Delta'_i t})}{i\hbar^2\,\Delta'_i}\,a_\mu(t)\,a^*_{\mu'}(t), \quad \text{...(A.25)}$$

where

$$\Delta'_i=(ck_0-ck_i+\mathbf{k}_i\,.\,\mathbf{v}+\omega_e(\mu'+1)-\omega_g\,\mu'). \quad \text{...(A.26)}$$

If the Zeeman splittings in the ground state and excited state of the atom are small compared with the energy separation of the excited and ground states $c\hbar k_0$, we can neglect the Zeeman energy in (A.16) and (A.26), and write

$$\Delta_i\simeq\Delta'_i\simeq c[k_0-k_i+\mathbf{k}_i\,.\,\mathbf{v}/c]=c[k_0-\tilde{k}_i]. \quad \text{...(A.27)}$$

In order to proceed further we introduce the special function $\zeta(x)$, (Heitler, 1954) defined by

$$\zeta(x)\equiv\lim_{t\to\infty}\frac{1-\mathrm{e}^{itx}}{x}=\lim_{t\to\infty}\left(\frac{1-\cos tx}{x}-i\frac{\sin tx}{x}\right). \quad \text{...(A.28)}$$

It is well known that

$$\lim_{t\to\infty}\frac{1}{\pi}\frac{\sin(tx)}{x}=\delta(x). \quad \text{...(A.29)}$$

Furthermore,

$$\lim_{t\to\infty}\frac{1-\cos(tx)}{x}=\frac{\mathscr{P}}{x}, \quad \text{...(A.30)}$$

where $\mathscr{P}/x$ is called the Cauchy principal value of $1/x$. Both this function and the Dirac delta function have meaning only in terms of an integral. Since $\cos(tx)$ rapidly oscillates and does not contribute to an integral, $\mathscr{P}/x$ behaves as $1/x$ for $x\neq0$ and vanishes at $x=0$. Thus,

$$\int_{-a}^{b}f(x)\frac{\mathscr{P}}{x}\,\mathrm{d}x=\int_{-a}^{-\varepsilon}\frac{f(x)}{x}\,\mathrm{d}x+\int_{+\varepsilon}^{b}\frac{f(x)}{x}\,\mathrm{d}x(\varepsilon\to0). \quad \text{...(A.31)}$$

Therefore

$$\zeta(x)=\frac{\mathscr{P}}{x}-i\pi\delta(x). \quad \text{...(A.32)}$$

The complex conjugate of $\zeta(x)$ is

$$\zeta^*(x) = \frac{\mathscr{P}}{x} + i\pi\delta(x). \qquad \text{...(A.33)}$$

Another useful representation for the Dirac delta function is

$$\delta(x) = \frac{1}{\pi} \lim_{\Gamma \to 0} \frac{\Gamma}{x^2 + \Gamma^2}, \qquad \text{...(A.34)}$$

and for the Cauchy principal value of $1/x$

$$\frac{\mathscr{P}}{x} = \lim_{\Gamma \to 0} \frac{x}{x^2 + \Gamma^2}. \qquad \text{...(A.35)}$$

If we allow t to become large compared with Δ_i in equations (A.24) and (A.25) but not so large that our assumption that $a_\mu(t) \simeq a_\mu(0)$ breaks down, these equations become

$$a_{\mu'}^*(t) \frac{\partial}{\partial t} a_\mu(t) = \sum_i \frac{|A(k_i)|^2 \, |\langle\mu| \, D_+ |\mu + 1\rangle|^2}{-i\hbar^2} \zeta^*(\Delta_i) \, a_{\mu'}^*(t) \, a_\mu(t), \qquad \text{...(A.36)}$$

and

$$a_\mu(t) \frac{\partial}{\partial t} a_{\mu'}^*(t) = \sum_i \frac{|A(k_i)|^2 \langle\mu'| \, D_+ |\mu' + 1\rangle|^2}{i\hbar^2} \zeta(\Delta_i) \, a_\mu(t) \, a_{\mu'}^*(t). \qquad \text{...(A.37)}$$

Adding these equations, we obtain

$$\begin{aligned} \frac{\partial}{\partial t} [a_{\mu'}^*(t) \, a_\mu(t)] = &- \sum_i |A(k_i)|^2 \, [|\langle\mu| \, D_+ |\mu + 1\rangle|^2 + |\langle\mu'| \, D_+ |\mu' + 1\rangle|^2] \\ &\times \frac{\pi}{\hbar^2} \delta(\Delta_i) \, a_{\mu'}^*(t) \, a_\mu(t) + \frac{i}{\hbar^2} \sum_i |A(k_i)|^2 \, [|\langle\mu| \, D_+ |\mu + 1\rangle|^2 \\ &- |\langle\mu'| \, D_+ |\mu' + 1\rangle|^2] \frac{\mathscr{P}}{\Delta_i} a_{\mu'}^*(t) \, a_\mu(t), \end{aligned} \qquad \text{...(A.38)}$$

or

$$\begin{aligned} \frac{\partial}{\partial t} [a_{\mu'}^*(t) \, a_\mu(t)] = &- \frac{\pi}{c\hbar^2} \sum_i |A(k_i)|^2 \, [|\langle\mu| \, D_+ |\mu + 1\rangle|^2 + |\langle\mu'| \, D_+ |\mu' + 1\rangle|^2] \\ &\times \delta(k_0 - \tilde{k}_i) \, a_{\mu'}^*(t) \, a_\mu(t) + \frac{i}{c\hbar^2} \sum_i |A(k_i)|^2 \, [|\langle\mu| \, D_+ |\mu + 1\rangle|^2 \\ &- |\langle\mu'| \, D_+ |\mu' + 1\rangle|^2] \frac{\mathscr{P}}{k_0 - \tilde{k}_i} a_{\mu'}^*(t) \, a_\mu(t). \end{aligned} \qquad \text{...(A.39)}$$

Both the delta function and the principal value function are defined in terms of an integration over their argument. Therefore we must pass from a summation over the discrete propagation vectors $\mathbf{k}_i$ to an integration over a continuum of vectors **k**. The discrete plane wave photon states arise from periodic boundary conditions which require that

$$\mathbf{k}_i = k_{ix}\hat{\imath} + k_{iy}\hat{\jmath} + k_{iz}\hat{k} = \frac{n_x 2\pi}{L}\hat{\imath} + \frac{n_y 2\pi}{L}\hat{\jmath} + \frac{n_z 2\pi}{L}\hat{k}, \qquad \text{...(A.40)}$$

where n_x, n_y, n_z are integers and L is the length of one side of the box. As the length $L \to \infty$, the discrete spectrum approaches a continuum and

$$\sum_i \mathbf{k}_i \to \frac{L^3}{(2\pi)^3}\int \mathrm{d}^3 k = \frac{L^3}{(2\pi)^3}\int k^2 \,\mathrm{d}k\,\mathrm{d}\Omega. \qquad \text{...(A.41)}$$

The first term on the right hand side of equation (A.39) becomes

$$\begin{aligned} -\frac{\pi}{c\hbar^2}\frac{L^3}{(2\pi)^3}\int k^2\,\mathrm{d}k\,\mathrm{d}\Omega\,|A(k)|^2\,[|\langle\mu|\,D_+|\mu+1\rangle|^2 \\ +|\langle\mu'|\,D_+|\mu'+1\rangle|^2]\,\delta(k_0-\tilde{k})\,a^*_{\mu'}(t)\,a_\mu(t) \\ = -\frac{\pi}{c\hbar^2}\frac{L^3}{(2\pi)^3}\int k^2\,\mathrm{d}k\,\mathrm{d}\Omega\frac{e^2\,2\pi\hbar n_{k\sigma}}{\omega L^3}\,\omega_0^2[|\langle\mu|\,D_+|\mu+1\rangle|^2 \\ +|\langle\mu'|\,D_+|\mu'+1\rangle|^2]\,\delta(\tilde{k}_0-k)\,a^*_{\mu'}(t)\,a_\mu(t). \end{aligned} \qquad \text{...(A.42)}$$

We have used equation (A.15) and noted that since $v/c \ll k_0$ we can replace $\delta\,(k_0-\tilde{k})$ with $\delta\,(\tilde{k}_0-k)$ where

$$\tilde{k}_0 = k_0 + \frac{\mathbf{k}_0\,.\,\mathbf{v}}{c}.$$

If we switch to ω as the variable of integration and observe that the number of photons $I(\omega)$ in a frequency interval $\mathrm{d}\omega$ propagating in the direction subtended by $\mathrm{d}\Omega$ is given by

$$I(\omega) = \frac{n_{\omega\sigma}\,\omega^2}{(2\pi)^3\,c^2}\,\mathrm{d}\Omega, \qquad \text{...(A.43)}$$

equation (A.42) becomes

$$\begin{aligned} -2\pi^2\,\tilde{\omega}_0\,\alpha\int_0^\infty I(\omega)\,[1|\langle\mu|\,D_+|\mu+1\rangle|^2 \\ +|\langle\mu'|\,D_+|\mu'+1\rangle|^2]\,\delta(\tilde{\omega}_0-\omega)\,\mathrm{d}\omega\,a^*_{\mu'}(t)\,a_\mu(t), \end{aligned}$$

where $\alpha = e^2/hc$ is the fine structure constant, and $\tilde{\omega}_0 = c\tilde{k}_0$. As is shown in standard quantum mechanics texts, the cross section $\sigma_\mu(\omega)$ for absorption of circularly polarized light by an atom in the ground state sublevel μ is just

$$\sigma_\mu(\omega) = 4\pi^2\,\tilde{\omega}_0\,\alpha|\langle\mu|\,D_+|\mu+1\rangle|^2\,\delta(\tilde{\omega}_0-\omega). \qquad \text{...(A.44)}$$

The absorption line is infinitely sharp which is a defect resulting from the use of 1st-order perturbation theory, and it is centered at the Doppler-shifted frequency $\tilde{\omega}_0$.

If we note that

$$\frac{\mathscr{P}}{k_0 - \tilde{k}} = \frac{-\mathscr{P}}{k - \tilde{k}_0},$$

we can write equation (A.39) in the form

$$\frac{\partial}{\partial t}[a_{\mu'}^*(t), a_\mu(t)] = -\tfrac{1}{2}\int_0^\infty I(\omega)[\sigma_\mu(\omega) + \sigma_{\mu'}(\omega)]\,\mathrm{d}\omega\, a_{\mu'}^*(t)\, a_\mu(t) - i\frac{\Delta E_{\mu\mu'}}{\hbar} a_{\mu'}^*(t)\, a_\mu(t), \quad \text{...(A.45)}$$

where

$$\Delta E_{\mu\mu'} = \frac{L^3}{\hbar(2\pi)\,3c}\int k^2|A(k)|^2\,[|\langle\mu|\, D_+|\mu + 1\rangle - |\langle\mu'|\, D_+|\mu' + 1\rangle|^2]\frac{\mathscr{P}}{(k - \tilde{k}_0)}\,\mathrm{d}k\,\mathrm{d}\Omega. \quad \text{...(A.46)}$$

The term $\Delta E_{\mu\mu'}$ represents a change in the Zeeman splitting between two sublevels μ and μ' of the ground state due to virtual transitions from the ground state to the excited state. This interpretation of the expression in equation (A.46) is based upon the 2nd-order calculation of Barrat and Cohen-Tannoudji in which the non-physical delta function and principal value function do not appear and strict conservation of energy is not required within the limits of the uncertainty principal. They obtain an equation essentially the same as equation (A.45), but instead of a delta function they have a function of the form

$$\frac{\Gamma/2}{[k - \tilde{k}_0]^2 + \dfrac{\Gamma^2}{4}},$$

where Γ is the lifetime of the excited state and $\tilde{k}_0$ is shifted slightly from our value due to a change in the self-energy of the excited state resulting from the emission process. In the limit $\Gamma \to 0$, their result reduces to ours, since

$$\lim_{\Gamma\to 0}\frac{\Gamma/2}{[k - \tilde{k}_0]^2 + \dfrac{\Gamma^2}{4}} = \delta(k - \tilde{k}_0)$$

Similarly, in place of the principal value function they have a term

$$\frac{k - \tilde{k}_0}{[k - \tilde{k}_0]^2 + \dfrac{\Gamma^2}{4}}$$

which approaches $\mathscr{P}/(k - \tilde{k}_0)$ as $\Gamma \to 0$. This term represents a change in the self-energy of the ground state sublevels due to transitions in which $k \neq \tilde{k}_0$. In general the shift varies from one sublevel to another resulting in a shift in the Zeeman splitting between levels.

The magnitude and sign of $\Delta E_{\mu\mu'}$ due to virtual transitions depends upon the details of the line profile of the pumping light i.e. the convolution of $|A(k)|^2$ and

$$\frac{k-\tilde{k}_0}{[k-k_0]^2+\dfrac{\Gamma^2}{4}}$$

in the integral in equation (A.46). In particular, if $|A(k)|^2$ is centered at $\tilde{k}_0$ and is an even function of $(k-\tilde{k}_0)$, $\Delta E_{\mu\mu'}$ will vanish since

$$\frac{k-\tilde{k}_0}{[k-\tilde{k}_0]^2+\dfrac{\Gamma^2}{4}}$$

is odd. Although quantitative calculations of the shifts due to virtual transitions in the optical pumping of alkali atoms are not usually practicable because of the complicated level structure and emission line profiles, the shifts can be eliminated as a source of systematic error by extrapolating to zero pumping light intensity. $\Delta E_{\mu\mu'}$ is proportional to $|A(k)|^2$ and hence linearly dependent upon the intensity of the pumping light.

In addition, Barrat and Cohen-Tannoudji obtain a shift in the energy levels due to real transitions in which coherence between two sublevels in the ground state is partially conserved during the optical-pumping cycle. This effect is completely missed in a 1st-order perturbation calculation. This shift is significantly large only if the Larmor frequencies of the excited state and ground state are sufficiently close and if there is no collisional mixing of the excited state sublevels. This is true for the optical pumping of mercury, but we will assume complete excited state disorientation for alkali atoms in a buffer gas and neglect the shift due to real transitions.

Equation (A.45) allows us to calculate the time dependence of the density matrix describing the angular momentum in the ground state of the optically pumped atoms due to the pumping process. If we denote a particular matrix element of the density matrix in the interaction representation as $\rho^I_{\mu\mu'}$, then

$$\rho^I_{\mu\mu'}(t)=\overline{a_\mu(t)\,a^*_{\mu'}(t)}, \qquad \text{...(A.47)}$$

where the bar indicates an ensemble average. We can calculate $d\rho^I_{\mu\mu'}/dt$ by performing an ensemble average of equation (A.45). Since the center frequency $\tilde{\omega}_0$ of the absorption line of the atoms depends upon

$$\mathbf{k}_0\cdot\frac{\mathbf{v}}{c}=\frac{k_0 v_z}{c},$$

we must average over the z component of the atom velocities as well as over the spin coordinates. If we assume a Maxwell–Boltzmann velocity distribution, the probability of finding an atom with velocity v_z in an interval $\mathrm{d}v_z$ is proportional to

$$e^{-Mv_z^2/2kT},$$

where M is the mass of the atom, T is the temperature of the optical-pumping cell, and k is the Boltzmann constant. Since

$$\tilde{\omega}_0 = \omega_0\left(1+\frac{v_z}{c}\right), \qquad \text{...(A.48)}$$

$$v_z = \frac{\tilde{\omega}_0 - \omega_0}{\omega_0}c, \qquad \text{...(A.49)}$$

and

$$\exp\left(-\frac{Mv_z^2}{2kT}\right) = \exp\left(-\frac{Mc^2}{2kT}\right)\frac{(\tilde{\omega}_0-\omega_0)^2}{\omega_0^2} \qquad \text{...(A.50)}$$

Thus our average over v_z becomes one over $\tilde{\omega}_0$ and we obtain from equation (A.45)

$$\frac{\mathrm{d}\rho_{\mu\mu'}(t)}{\mathrm{d}t} = \frac{-\frac{1}{2}\int_0^\infty\int_0^\infty I(\omega)[\sigma_\mu(\omega)+\sigma_{\mu'}(\omega)]\exp\left(-\frac{Mc^2}{2kT}\frac{(\tilde{\omega}_0-\omega_0)^2}{\omega_0^2}\right)\mathrm{d}\tilde{\omega}_0\,\mathrm{d}\omega}{\int_0^\infty \exp\left(\frac{-Mc^2}{2kT}\frac{(\tilde{\omega}_0-\omega_0)^2}{\omega_0^2}\right)\mathrm{d}\omega_0}\rho_I^{\mu\mu'}(t)$$

$$+\frac{i\overline{\Delta E}_{\mu\mu'}}{\hbar}\rho_{\mu\mu'}^I(t) \qquad \text{...(A.51)}$$

where $\overline{\Delta E}_{\mu\mu'}$ is the velocity average of $\Delta E_{\mu\mu'}$. Making use of the delta function in $\sigma_\mu(\omega)$, we obtain

$$\frac{\mathrm{d}\rho_{\mu\mu'}^I(t)}{\mathrm{d}t} = -\left[\tfrac{1}{2}\int_0^\infty I(\omega)[\sigma_{\mu 0}+\sigma_{\mu' 0}]\exp\left(-\frac{Mc^2}{2kT}\frac{(\omega-\omega_0)^2}{\omega_0^2}\right)\mathrm{d}\omega + \frac{i\Delta E_{\mu\mu'}}{\hbar}\right]\rho_{\mu\mu'}^I(t), \qquad \text{...(A.52)}$$

where

$$\sigma_{\mu 0} = \frac{4\pi^2\alpha\omega_0|\langle\mu+1|\,D_+|\mu\rangle|^2}{\int_0^\infty \exp(-Mc^2(\omega-\omega_0)^2/2\omega_0^2kT)\,\mathrm{d}\omega}. \qquad \text{...(A.53)}$$

If we set $\mu=\mu'$ in equation (A.52) and note that $\Delta E_{\mu\mu}=0$, we can see that a great deal of arithmetic has led us to a physically obvious result.

$$\frac{\mathrm{d}\rho_{\mu\mu}^I(t)}{\mathrm{d}t} = \left[-\int_0^\infty I(\omega)\,\sigma_{\mu 0}\exp\left(-\frac{Mc^2(\omega-\omega_0)^2}{2kT\omega_0^2}\right)\mathrm{d}\omega\right]\rho_{\mu\mu}^I(t) \qquad \text{...(A.54)}$$

The diagonal matrix element $\rho_{\mu\mu}^I(t)$ gives us the probability of finding an atom in the sublevel μ. In other words $\rho_{\mu\mu}^I$ is the fractional number of atoms in our optical-pumping cell in this sublevel. The rate of depletion is proportional to the number of atoms in the level and the integral over ω of the product of the Doppler-broadened cross section and the light intensity.

Our 1st-order calculation has not included the effect of spontaneous emission back to the ground state, however, so that we must add in, after the fact, a term to describe the repopulation of the ground state sublevels due to the return of the atoms from the excited state. This is automatically included in the 2nd-order calculation of Barrat and Cohen-Tannoudji. If we assume complete reorientation in the excited state, then an atom which is raised to the excited state from a sublevel μ can return to μ or any other sublevel μ' with equal probability. That is,

$$\frac{\mathrm{d}\rho^I_{\mu\mu}(t)}{\mathrm{d}t} = -\left[\int_0^\infty I(\omega)\,\sigma_{\mu 0}\exp\left(-\frac{Mc^2(\omega-\omega_0)^2}{2kT\omega_0^2}\right)\mathrm{d}\omega\right]\rho^I_{\mu\mu}(t) + \left[\frac{\sum_{\mu'}\int_0^\infty I(\omega)\,\sigma_{\mu' 0}\exp\left(-\frac{Mc^2(\omega-\omega_0)^2}{2kT\omega_0^2\,\mathrm{d}\omega}\right)}{N}\right]\rho^I_{\mu'\mu'}(t), \quad \ldots(\text{A.55})$$

where N is the number of sublevels in the ground state.

If we write

$$\frac{1}{T_{\mu\mu'}} = \tfrac{1}{2}\int_0^\infty I(\omega)[\sigma_{\mu 0}+\sigma_{\mu' 0}]\exp\left(-\frac{Mc^2(\omega-\omega_0)^2}{2kT\omega_0^2}\right)\mathrm{d}\omega, \quad \ldots(\text{A.56})$$

and assume that there is no preservation of coherence in the optical-pumping cycle, then we have for an arbitrary matrix element $\rho^I_{\mu\mu'}(t)$

$$\frac{\mathrm{d}\rho^I_{\mu\mu'}(t)}{\mathrm{d}t} = -\left[\frac{1}{T_{\mu\mu'}} + \frac{i\Delta E_{\mu\mu'}}{\hbar}\right]\rho^I_{\mu\mu'}(t) + \left[\frac{\sum_{\mu''}\frac{1}{T_{\mu''\mu''}}\rho^I_{\mu''\mu''}(t)}{N}\right]\delta_{\mu\mu'}. \quad \ldots(\text{A.57})$$

Thus far we have calculated everything in the interaction picture, but it is a simple matter to translate our results back into the Schrödinger picture. Recall that the state of an atom $|\phi(t)\rangle$ in the interaction picture is $\mathrm{e}^{iH_0t/\hbar}|\alpha(t)\rangle$, where $|\alpha(t)\rangle$ is the state of the atom in the Schrödinger picture. We expanded $|\phi(t)\rangle$ in eigenstates of the Hamiltonian $\mathscr{H}_0$ and formed the density matrix from an ensemble average of products of the expansion coefficients. We can do the same thing for the state $|\alpha(t)\rangle$.

If

$$|\phi(t)\rangle = \sum_\mu a_\mu(t)|\,\mu\rangle + \sum_{m,k} b_{m,-k}(t)|m,-k\rangle,$$

and

$$|\alpha(t)\rangle = \sum_\mu c_\mu(t)|\,\mu\rangle + \sum_{m,k} d_{m,-k}(t)|m,-k\rangle,$$

then

$$\sum_\mu a_\mu(t)|\,\mu\rangle + \sum_{m,k} b_{m,-k}(t)|\,m,-k\rangle = \exp\left(\frac{i\mathscr{H}_0 t}{\hbar}\right) \times\left[\sum_\mu c_\mu(t)|\,\mu\rangle + \sum_{m,k} d_{m,-k}(t)|\,m,-k\rangle\right]. \quad \ldots(\text{A.58})$$

Multiplying through by $\langle \mu' |$, we have

$$a_{\mu'}(t) = \left\langle \mu' \left| \exp\left(\frac{i\mathscr{H}_0 t}{\hbar}\right) \right| \mu' \right\rangle c_{\mu'}(t), \qquad \ldots(\text{A}.59)$$

or

$$c_{\mu'}(t) = \mathrm{e}^{-i\omega_g \mu' t}\, a_{\mu'}(t). \qquad \ldots(\text{A}.60)$$

The density matrix element $\rho_{\mu\mu'}(t)$ in the Schrödinger picture is therefore

$$\begin{aligned} \rho_{\mu\mu'}(t) &= \overline{c_\mu(t)\, c^*_{\mu'}(t)} \\ &= \mathrm{e}^{i\omega_g(\mu'-\mu)t}\, \overline{a_\mu(t)\, a^*_{\mu'}(t)} \\ &= \mathrm{e}^{i\omega_g(\mu'-\mu)t}\, \rho^I_{\mu\mu'}(t). \end{aligned} \qquad \ldots(\text{A}.61)$$

Thus

$$\frac{\mathrm{d}\rho_{\mu\mu'}(t)}{\mathrm{d}t} = \mathrm{e}^{i\omega_g(\mu'-\mu)t}\frac{\mathrm{d}\rho^I_{\mu\mu'}}{\mathrm{d}t} + i\omega_g(\mu'-\mu)\,\mathrm{e}^{i\omega_g(\mu-\mu)t}\,\rho^I_{\mu\mu'}(t). \qquad \ldots(\text{A}.62)$$

This can also be written as

$$\frac{\mathrm{d}\rho_{\mu\mu'}(t)}{\mathrm{d}t} = \mathrm{e}^{i\omega_g(\mu'-\mu)t}\frac{\mathrm{d}\rho^I_{\mu\mu'}(t)}{\mathrm{d}t} + \frac{1}{i\hbar}[\mathscr{H}_0, \rho]_{\mu\mu'}. \qquad \ldots(\text{A}.63)$$

Combining equations (A.57) and (A.63), we obtain for the time dependence of the density matrix elements $\rho_{\mu\mu'}(t)$ in the Schrödinger picture due to the pumping light and the static magnetic field the equation

$$\frac{\mathrm{d}\rho_{\mu\mu'}(t)}{\mathrm{d}t} = -\left(\frac{1}{T_{\mu\mu'}} + \frac{\mathrm{i}\overline{\Delta E_{\mu\mu'}}}{\hbar}\right)\rho_{\mu\mu'}(t) + \left(\frac{\sum\limits_{\mu''}\dfrac{1}{T_{\mu''\mu''}}\rho_{\mu''\mu''}}{N}\right)\delta_{\mu\mu'} + \frac{1}{i\hbar}[\mathscr{H}_0, \rho]_{\mu\mu'}. \qquad \ldots(\text{A}.64)$$

In other words, when we go back to the Schrödinger picture we simply add in the time derivative of the density matrix due to the static magnetic field.

B. OPTICAL PUMPING OF A SPIN-$\frac{1}{2}$ ATOM

Equation (A.64) applies to the optical pumping of an arbitrary paramagnetic atom which undergoes complete reorientation in the excited state. We will now specialize our results to the simplest case of atom with an $S_{1/2}$ ground state and zero nuclear spin. This is an approximation to a real alkali atom if we neglect its nuclear spin. Since we are assuming complete reorientation in the excited state, equation (A.64) does not apply to the optical pumping of Hg^{199} for which there is no reorientation in the excited state, and the term describing the repopulation of the ground state sublevels is not a constant independent of μ as it is in equation (A.64). Also, there can be preservation of coherence between two sub-levels during the optical-pumping cycle. Nevertheless, the differences are relatively minor and we will obtain equations for the optical-pumping signal similar to those obtained by Cagnac (1961) for Hg^{199}.

For our $S_{1/2}$ atom with zero nuclear spin, there are only two orientations $j_z = \pm\frac{1}{2}$ of the atom's angular momentum with respect to the z-axis. If we use left-circularly polarized pumping light, exciting the atoms to the $P_{1/2}$ state, only atoms in the $j_z = -\frac{1}{2}$ level can absorb a photon. If there is complete reorientation in the excited state, the atoms return with equal probability to either the $j_z = +\frac{1}{2}$ or $-\frac{1}{2}$ level.

In equation (A.64) the index μ can be 1 or 2 with ρ_{11} being the probability of the atom having $j_z = \frac{1}{2}$ and ρ_{22} being the probability of the atom having $j_z = -\frac{1}{2}$. From equation (A.56) we have

$$\frac{1}{T_{11}} = 0, \quad \text{...(B.1)}$$

$$\frac{1}{T_{22}} = \int_0^\infty I(\omega)\,\sigma_0 \exp\left(-\frac{Mc^2(\omega-\omega_0)^2}{2kT\omega_0^2}\right)\mathrm{d}\omega, \quad \text{...(B.2)}$$

with

$$\sigma_0 = \frac{4\pi^2\,\alpha\omega_0|\langle P,\frac{1}{2}|\,D_+|S,-\frac{1}{2}\rangle|^2}{\int_0^\infty \exp(-Mc^2(\omega-\omega_0)^2/2\omega_0^2kT)\,\mathrm{d}\omega} \quad \text{...(B.3)}$$

where $\langle P,\frac{1}{2}|$ is the excited $P_{1/2}$ state with $j_z = \frac{1}{2}$ and $|S,-\frac{1}{2}\rangle$ is the ground state with $j_z = -\frac{1}{2}$. Also,

$$\frac{1}{T_{12}} = \frac{1}{T_{21}} = \frac{1}{2}\int_0^\infty I(\omega)\,\sigma_0 \exp\left(-\frac{Mc^2(\omega-\omega_0)^2}{2kT\omega_0^2}\right)\mathrm{d}\omega. \quad \text{...(B.4)}$$

Neglecting for the moment the effect of the static magnetic field, the equations of motion for the density matrix elements of our spin-$\frac{1}{2}$ atom are, from equation (A.64),

$$\frac{\mathrm{d}\rho_{11}}{\mathrm{d}t} = \frac{1}{2}\frac{1}{T_{22}}\rho_{22}, \quad \text{...(B.5)}$$

$$\frac{\mathrm{d}\rho_{22}}{\mathrm{d}t} = -\frac{1}{2}\frac{1}{T_{22}}\rho_{22}, \quad \text{...(B.6)}$$

$$\frac{\mathrm{d}\rho_{12}}{\mathrm{d}t} = -\left(\frac{1}{T_{12}} + \frac{i\Delta E}{\hbar}\right)\rho_{12}, \quad \text{...(B.7)}$$

$$\frac{\mathrm{d}\rho_{21}}{\mathrm{d}t} = -\left(\frac{1}{T_{21}} - \frac{i\Delta E}{\hbar}\right)\rho_{21}, \quad \text{...(B.8)}$$

with

$$\Delta E = \overline{\Delta E_{12}} = -\overline{\Delta E_{21}}. \quad \text{...(B.9)}$$

Until now, we have tacitly assumed that the pumping light intensity $I(\omega)$ which appears in equations (B.2) and (B.4) is constant over the volume of the optical-pumping cell. This is obviously not true, for as the pumping light traverses the cell in the z-direction, its intensity is attenuated by absorption. This implies that the density matrix elements will also be a function of z, the distance from the front of the cell. Since only atoms with $j_z = -\frac{1}{2}$ can absorb a photon, the attenuation of the light intensity $I(\omega, z)$, as a function of z, is given by

$$\frac{\partial I(\omega, z)}{\partial z} = -N\sigma(\omega)\, I(\omega, z)\, \rho_{22}(z), \qquad \text{...(B.10)}$$

where

$$\sigma(\omega) = \sigma_0 \exp\left(-\frac{Mc^2(\omega - \omega_0)^2}{2kT\omega_0^2}\right), \qquad \text{...(B.11)}$$

and N is the density of spin-$\frac{1}{2}$ atoms. Integrating equation (B.10) we obtain

$$I(\omega, z) = I(\omega, 0) \exp\left(-\int_0^z N\sigma(\omega)\, \rho_{22}(\xi)\, \mathrm{d}\xi\right), \qquad \text{...(B.12)}$$

where $I(\omega,0)$ is the light intensity at the front of the cell. Substituting th s expression in equation (B.5) gives

$$\frac{\mathrm{d}}{\mathrm{d}t}\rho_{11}(z) = \tfrac{1}{2}\int_0^\infty I(\omega, 0)\, \sigma(\omega) \exp\left(-\int_0^z N\sigma(\omega)\, \rho_{22}(\xi)\, \mathrm{d}\xi\right) \rho_{22}(z)\, \mathrm{d}\omega. \qquad \text{...(B.13)}$$

If we define the average value of $\rho_{11}(z)$ as

$$\langle \rho_{11} \rangle = \frac{1}{z_0}\int_0^{z_0} \rho_{11}\, \mathrm{d}z, \qquad \text{...(B.14)}$$

where z_0 is the length of the cell, then

$$\frac{\mathrm{d}}{\mathrm{d}t}\langle \rho_{11} \rangle = \frac{1}{2z_0}\int_0^{z_0}\int_0^\infty I(\omega, 0)\, \sigma(\omega) \exp\left(-\int_0^z N\sigma(\omega)\, \rho_{22}(\xi)\, \mathrm{d}\xi\right) \rho_{22}(z)\, \mathrm{d}\omega\, \mathrm{d}z. \qquad \text{...(B.15)}$$

If we note that

$$\frac{\mathrm{d}}{\mathrm{d}z}\left[\int_0^z N\sigma(\omega)\, \rho_{22}(\xi)\, \mathrm{d}\xi\right] = N\sigma(\omega)\, \rho_{22}(z), \qquad \text{...(B.16)}$$

we can integrate the right hand side of equation (B.15) with respect to z to obtain

$$\frac{\mathrm{d}}{\mathrm{d}t}\langle \rho_{11} \rangle = \frac{1}{2Nz_0}\int_0^\infty I(\omega, 0)\, [1 - \exp(-Nz_0\, \sigma(\omega)\, \langle \rho_{22} \rangle)]\, \mathrm{d}\omega. \qquad \text{...(B.17)}$$

If the product $Nz_0\sigma_0$ is sufficiently small, we can expand the exponential and equation (B.17) becomes

$$\frac{\mathrm{d}}{\mathrm{d}t}\langle \rho_{11} \rangle = \tfrac{1}{2}\int_0^\infty I(\omega, 0)\, \sigma(\omega)\, \langle \rho_{22} \rangle\, \mathrm{d}\omega. \qquad \text{...(B.18)}$$

Similarly,

$$\frac{\mathrm{d}}{\mathrm{d}t}\langle\rho_{22}\rangle = -\tfrac{1}{2}\int_0^\infty I(\omega,0)\,\sigma(\omega)\,\mathrm{d}\omega\langle\rho_{22}\rangle. \qquad \text{...(B.19)}$$

For the off-diagonal elements we have for the real part of their time dependence

$$\frac{\mathrm{d}\langle\rho_{12}\rangle}{\mathrm{d}t} = \frac{1}{2z_0}\int_0^{z_0}\int_0^\infty I(\omega,0)\,\sigma(\omega)\exp\left(-\int_0^z N\sigma(\omega)\,\rho_{22}(\xi)\,\mathrm{d}\xi\right)\rho_{12}(z)\,\mathrm{d}\omega\,\mathrm{d}z. \qquad \text{...(B.20)}$$

and

$$\frac{\mathrm{d}\langle\rho_{21}\rangle}{\mathrm{d}t} = \frac{1}{2z_0}\int_0^{z_0}\int_0^\infty I(\omega,0)\,\sigma(\omega)\exp\left(-\int_0^z N\sigma(\omega)\,\rho_{22}(\xi)\,\mathrm{d}\xi\right)\rho_{21}(z)\,\mathrm{d}\omega\,\mathrm{d}z. \qquad \text{...(B.21)}$$

Unfortunately equations (B.20) and (B.21) cannot be integrated with respect to z. In order to obtain a simple expression for the effect of the pumping light on the off-diagonal elements we must assume that $Nz_0\sigma$ and hence the light absorption is sufficiently small so that we can replace the exponential by unity. That is, for low absorption

$$\frac{\mathrm{d}\langle\rho_{12}\rangle}{\mathrm{d}t} \simeq -\tfrac{1}{2}\int_0^\infty I(\omega,0)\,\sigma(\omega)\,\mathrm{d}\omega\langle\rho_{12}\rangle, \qquad \text{...(B.22)}$$

and

$$\frac{\mathrm{d}\langle\rho_{21}\rangle}{\mathrm{d}t} \simeq -\tfrac{1}{2}\int_0^\infty I(\omega,0)\,\sigma(\omega)\,\mathrm{d}\omega\langle\rho_{21}\rangle. \qquad \text{...(B.23)}$$

In other words, for the off-diagonal elements we are completely neglecting the variation of the light intensity over the cell. We must make the same approximation for the imaginary term in their time dependence.

If we now define the pumping time T_p by

$$\frac{1}{T_p} = \tfrac{1}{2}\int_0^\infty I(\omega,0)\,\sigma(\omega)\,\mathrm{d}\omega, \qquad \text{...(B.24)}$$

we have for the time dependence of the density matrix ρ averaged over z due to the pumping light

$$\left[\frac{\mathrm{d}\rho}{\mathrm{d}t}\right]_{\text{light}} = \begin{bmatrix} \dfrac{\rho_{22}}{T_p} & -\left(\dfrac{1}{T_p}+\dfrac{i\Delta E}{\hbar}\right)\rho_{12} \\ -\left(\dfrac{1}{T_p}-\dfrac{i\Delta E}{\hbar}\right)\rho_{21} & -\dfrac{\rho_{22}}{T_p} \end{bmatrix}, \qquad \text{...(B.25)}$$

where we have dropped the bracket indicating an average over z.

The equation describing the time dependence of ρ under the influence of the pumping light relaxation effects, and the static and r.f. fields has the form

$$\frac{d\rho}{dt} = \left[\frac{d\rho}{dt}\right]_{\text{light}} + \left[\frac{d\rho}{dt}\right]_{\text{relaxation}} + \left[\frac{d\rho}{dt}\right]_{\text{fields}} \qquad \text{...(B.26)}$$

The effect of spin-relaxation collisions with the walls and the buffer gas is treated phenomenologically by the equations

$$\frac{d\rho}{dt_{\text{relaxation}}} = \begin{bmatrix} \dfrac{\frac{1}{2} - \rho_{11}}{T_1} & \dfrac{-\rho_{12}}{T_2} \\ \dfrac{-\rho_{21}}{T_2} & \dfrac{\frac{1}{2} - \rho_{22}}{T_1} \end{bmatrix}. \qquad \text{...(B.27)}$$

The effect of the static and r.f. magnetic fields is given by

$$\left[\frac{d\rho}{dt}\right]_{\text{fields}} = \frac{1}{i\hbar}[\mathscr{H}, \rho], \qquad \text{...(B.28)}$$

where $\mathscr{H}$ is the Hamiltonian describing the interaction of the spin-$\frac{1}{2}$ atoms with the fields $H_0 \hat{k}$ and $2H_1 \hat{\imath} \cos\omega t$.

$$\mathscr{H} = -g_J \mu_0 H_0 J_z - 2g_J \mu_0 H_1 J_x \cos\omega t, \qquad \text{...(B.29)}$$

or

$$\mathscr{H} = \hbar\omega_0 J_z + 2\hbar\omega_1 J_x \cos\omega t, \qquad \text{...(B.30)}$$

where

$$\omega_0 = -\frac{g_J \mu_0 H_0}{\hbar}, \qquad \text{...(B.31)}$$

and

$$\omega_1 = -\frac{g_J \mu_0 H_1}{\hbar}. \qquad \text{...(B.32)}$$

The contribution $1/(i\hbar)[\mathscr{H}, \rho]$ to the time derivative of ρ is most easily calculated in a coordinate system rotating about the z-axis with angular velocity ω. In this coordinate system, the Hamiltonian is time independent.

Given a state vector $|a\rangle$ describing an atom in the laboratory frame, then the same state viewed from a frame rotated through an angle ωt about the z-axis is described by the ket $|a'\rangle$

$$|a'\rangle = e^{i\omega t J_z}|a\rangle. \qquad \text{...(B.33)}$$

The Hamiltonian $\mathscr{H}'$ viewed from this rotating frame is obtained by noting that

$$\langle a'|\mathscr{H}'|a'\rangle = \langle a|H|a\rangle. \qquad \text{...(B.34)}$$

But

$$\langle a|\mathscr{H}|a\rangle = \langle a|e^{-i\omega t J_z} e^{+i\omega t J_z} \mathscr{H} e^{-i\omega t J_z} e^{i\omega t J_z}|a\rangle = \langle a'|e^{i\omega t J_z} \mathscr{H} e^{-i\omega t J_z}|a'\rangle. \qquad \text{...(B.35)}$$

Thus

$$\mathscr{H}' = e^{i\omega t J_z} \mathscr{H} e^{-i\omega t J_z}. \quad \ldots(B.36)$$

If we refer to the definition of the density operator given in Section III, equation (A.5), it is immediately clear that the density matrix ρ' in rotating frame is related to ρ in the laboratory frame by

$$\rho' = e^{i\omega t J_z} \rho e^{-i\omega t J_z}. \quad \ldots(B.37)$$

If we ignore the counter rotating component of the linearly oscillating field, we can rewrite equation (B.30) as

$$\mathscr{H} = \hbar\omega_0 J_z + \hbar\omega_1 [J_x \cos\omega t + J_y \sin\omega t], \quad \ldots(B.38)$$

or

$$\mathscr{H} = \hbar\omega_0 J_z + \frac{\hbar\omega_1}{2} [J_+ e^{-i\omega t} + J_- e^{i\omega t}], \quad \ldots(B.39)$$

where

$$J_\pm = J_x \pm iJ_y. \quad \ldots(B.40)$$

From the commutation relations for angular momentum, we have

$$e^{i\omega t J_z} J_+ e^{-i\omega t J_z} = e^{i\omega t} J_+, \quad \ldots(B.41)$$

$$e^{i\omega t J_z} J_- e^{-i\omega t J_z} = e^{-i\omega t} J_-, \quad \ldots(B.42)$$

and

$$e^{i\omega t J_z} J_z e^{-i\omega t J_z} = J_z. \quad \ldots(B.43)$$

Therefore the Hamiltonian $\mathscr{H}'$ in the rotating coordinate system is

$$\begin{aligned} \mathscr{H}' &= \hbar\omega_0 J_z + \frac{\hbar\omega_1}{2}(J_+ + J_-) \\ &= \hbar\omega_0 J_z + \hbar\omega_1 J_x. \end{aligned} \quad \ldots(B.44)$$

Differentiating equation (B.37) with respect to time, we obtain

$$\frac{d\rho'}{dt} = i\omega[J_z, \rho'] + e^{i\omega t J_z} \frac{d\rho}{dt} e^{-i\omega t J_z}. \quad \ldots(B.45)$$

$$\frac{d\rho'}{dt} = i\omega[J_z, \rho'] + \frac{e^{i\omega t J_z}}{i\hbar} [\mathscr{H}, \rho] e^{-i\omega t J_z} + e^{i\omega t J_z} \left(\left[\frac{d\rho}{dt}\right]_{\text{light}} + \left[\frac{d\rho}{dt}\right]_{\text{relaxation}} \right) e^{-i\omega t J_z}. \quad \ldots(B.46)$$

Considering the second term on the right hand side we find that

$$\begin{aligned} \frac{1}{i\hbar} e^{i\omega t J_z} [\mathscr{H}, \rho] e^{-i\omega t J_z} &= \frac{1}{i\hbar} [e^{i\omega t J_z} \mathscr{H} e^{-i\omega t J_z}, e^{i\omega t J_z} \rho e^{-i\omega t J_z}], \\ &= \frac{1}{i\hbar} [\mathscr{H}', \rho']. \end{aligned} \quad \ldots(B.47)$$

Since $\mathbf{J} = \frac{1}{2}\boldsymbol{\sigma}$, we obtain from equation (B.44) the relation

$$\frac{1}{i\hbar}[\mathscr{H}', \rho'] = -\frac{i\omega_0}{2}[\sigma_z, \rho'] - \frac{i\omega_1}{2}[\sigma_x, \rho']. \qquad \text{...(B.48)}$$

Therefore

$$\frac{\mathrm{d}\rho'}{\mathrm{d}t} = \frac{i(\omega - \omega_0)}{2}[\sigma_z, \rho'] - \frac{i\omega_1}{2}[\sigma_x, \rho'] + \exp\left(\frac{i\omega t\sigma_z}{2}\right)\left(\left[\frac{\mathrm{d}\rho}{\mathrm{d}t}\right]_{\text{light}} + \left[\frac{\mathrm{d}\rho}{\mathrm{d}t}\right]_{\text{relaxation}}\right)\exp\left(\frac{-i\omega t\sigma_z}{2}\right). \qquad \text{...(B.49)}$$

We also have from equations (B.25) and (B.27)

$$\exp\left(i\frac{\omega t}{2}\sigma_z\right)\left[\frac{\mathrm{d}\rho}{\mathrm{d}t}\right]_{\text{light}}\exp\left(-i\frac{\omega t}{2}\sigma_z\right) = \begin{bmatrix} \dfrac{\rho'_{22}}{T_p} & -\left(\dfrac{1}{T_p} + \dfrac{i\Delta E}{\hbar}\right)\rho'_{12} \\ -\left(\dfrac{1}{T_p} - \dfrac{i\Delta E}{\hbar}\right)\rho'_{21} & -\dfrac{\rho'_{22}}{T_p} \end{bmatrix}, \qquad \text{...(B.50)}$$

and

$$\exp\left(i\frac{\omega t}{2}\sigma_z\right)\left[\frac{\mathrm{d}\rho}{\mathrm{d}t}\right]_{\text{relaxation}}\exp\left(-i\frac{\omega t}{2}\sigma_z\right) = \begin{bmatrix} \dfrac{\frac{1}{2} - \rho'_{11}}{T_1} & -\dfrac{\rho'_{12}}{T_2} \\ \dfrac{\rho'_{21}}{T_2} & \dfrac{\frac{1}{2} - \rho'_{22}}{T_1} \end{bmatrix}. \qquad \text{...(B.51)}$$

In order to calculate the time dependence of the ensemble average of σ_x, σ_y, and σ_z in the rotating coordinate system due to the pumping light, the applied fields, and spin relaxation we use the relations

$$\frac{\mathrm{d}}{\mathrm{d}t}\overline{\langle\sigma_z\rangle}' = \mathrm{Tr}\,\sigma_z\frac{\mathrm{d}\rho'}{\mathrm{d}t}, \qquad \text{...(B.52)}$$

$$\frac{\mathrm{d}}{\mathrm{d}t}\overline{\langle\sigma_x\rangle}' = \mathrm{Tr}\,\sigma_x\frac{\mathrm{d}\rho'}{\mathrm{d}t}, \qquad \text{...(B.53)}$$

$$\frac{\mathrm{d}}{\mathrm{d}t}\overline{\langle\sigma_y\rangle}' = \mathrm{Tr}\,\sigma_y\frac{\mathrm{d}\rho'}{\mathrm{d}t}. \qquad \text{...(B.54)}$$

Combining equations (B.49), (B.50) and (B.51), we obtain

$$\frac{\mathrm{d}}{\mathrm{d}t}\overline{\langle\sigma_z\rangle}' = \frac{i(\omega - \omega_0)}{2}\mathrm{Tr}\,\sigma_z[\sigma_z, \rho'] - \frac{i\omega_1}{2}\mathrm{Tr}\,\sigma_z[\sigma_x, \rho'] + \frac{1}{T_p} - \overline{\langle\sigma_z\rangle}'\left(\frac{1}{T_p} + \frac{1}{T_1}\right). \qquad \text{...(B.55)}$$

To obtain the last two terms on the right hand side of equation (B.55) we evaluated the traces of σ_z times the expressions in equations (B.50) and (B.51) by matrix multiplications and used the fact that

$$2\rho'_{22} = \rho'_{11} + \rho'_{22} - (\rho'_{11} - \rho'_{22}) = 1 - \overline{\langle\sigma_z\rangle}'. \qquad \text{...(B.56)}$$

In order to evaluate the remaining traces in equation (B.55), we remind the reader that

$$\operatorname{Tr} A\rho = \operatorname{Tr} \rho A, \qquad \text{...(B.57)}$$

$$\operatorname{Tr} AB\rho = \operatorname{Tr} \rho AB = \operatorname{Tr} B\rho A, \qquad \text{...(B.58)}$$

$$[\sigma_x, \sigma_y] = 2i\sigma_z \qquad \text{...(B.59)}$$

with cyclic permutations, and

$$\sigma_z^2 = 1. \qquad \text{...(B.60)}$$

Hence,

$$\operatorname{Tr} \sigma_z[\sigma_z, \rho'] = \operatorname{Tr} \sigma_z^2 \rho' - \operatorname{Tr} \rho' \sigma_z^2 = 0, \qquad \text{...(B.61)}$$

and

$$\operatorname{Tr} \sigma_z[\sigma_x, \rho'] = \operatorname{Tr} [\sigma_z, \sigma_x] \rho' = 2i \operatorname{Tr} \sigma_y \rho'. \qquad \text{...(B.62)}$$

Thus,

$$\frac{\mathrm{d}}{\mathrm{d}t} \overline{\langle \sigma_z \rangle}' = \omega_1 \overline{\langle \sigma_y \rangle}' + \frac{1}{T_p} - \left(\frac{1}{T_p} + \frac{1}{T_1} \right) \overline{\langle \sigma_z \rangle}'. \qquad \text{...(B.63)}$$

Similarly, one finds that

$$\frac{\mathrm{d}}{\mathrm{d}t} \overline{\langle \sigma_y \rangle}' = [(\omega_0 + \delta\omega) - \omega] \overline{\langle \sigma_x \rangle}' - \omega_1 \overline{\langle \sigma_z \rangle}' - \left(\frac{1}{T_2} + \frac{1}{T_p} \right) \overline{\langle \sigma_y \rangle}', \qquad \text{...(B.64)}$$

and

$$\frac{\mathrm{d}}{\mathrm{d}t} \overline{\langle \sigma_x \rangle}' = -[(\omega_0 + \delta\omega) - \omega] \overline{\langle \sigma_y \rangle}' - \left(\frac{1}{T_2} + \frac{1}{T_p} \right) \langle \sigma_x \rangle'. \qquad \text{...(B.65)}$$

where

$$\delta\omega = \frac{\Delta E}{\hbar}, \qquad \text{...(B.66)}$$

is the shift in the resonance frequency due to the pumping light.

C. THE EQUILIBRIUM TRANSMISSION SIGNAL

In a typical experiment, where the change in the polarization $\overline{\langle \sigma_z \rangle}$ of the optically pumped atoms in response to a resonant r.f. field is detected by monitoring the intensity of the pumping light transmitted by the cell, the r.f. is switched on and off at a fixed rate and this results in an observable modulation of the transmitted light intensity. In this section, we will assume that the r.f. remains on long enough for the optically pumped system to come to equilibrium polarization $\overline{\langle \sigma_z \rangle}_e$, and we will calculate the resulting change in the light intensity transmitted by the cell after the equilibrium polarization has been reached. In the next section, we will calculate the way in which the polarization approaches equilibrium and calculate the form of the transient signal.

The solution of the coupled differential equations (B.63), (B.64), and (B.65) will give the time dependence of the optically pumped atom's angular momentum at magnetic resonance. We will first solve equations (B.63–B.65) for this equilibrium situation. At equilibrium in the rotating frame

$$\frac{\mathrm{d}\overline{\langle\sigma_z\rangle}'}{\mathrm{d}t} = \frac{\mathrm{d}\overline{\langle\sigma_x\rangle}'}{\mathrm{d}t} = \frac{\mathrm{d}\overline{\langle\sigma_y\rangle}'}{\mathrm{d}t} = 0. \qquad \text{...(C.1)}$$

Hence

$$0 = \omega_1 \overline{\langle\sigma_y\rangle}'_e - \frac{\overline{\langle\sigma_z\rangle}'_e}{\tau_1} + \frac{1}{T_p}, \qquad \text{...(C.2)}$$

$$0 = \Delta\omega\langle\sigma_x\rangle'_e - \omega_1\overline{\langle\sigma_z\rangle}'_e - \frac{\overline{\langle\sigma_y\rangle}'_e}{\tau_2}, \qquad \text{...(C.3)}$$

$$0 = -\Delta\omega\overline{\langle\sigma_y\rangle}'_e - \frac{\overline{\langle\sigma_x\rangle}'_e}{\tau_2}, \qquad \text{...(C.4)}$$

where

$$\Delta\omega = [(\omega_0 + \delta\omega) - \omega], \qquad \text{...(C.5)}$$

$$\frac{1}{\tau_1} = \frac{1}{T_1} + \frac{1}{T_p}, \qquad \text{...(C.6)}$$

and

$$\frac{1}{\tau_2} = \frac{1}{T_2} + \frac{1}{T_p}. \qquad \text{...(C.7)}$$

The subscript e denotes the state of equilibrium.

If we denote the values of $\overline{\langle\sigma_z\rangle}'$ in the absence of an applied r.f. field ($\omega_1 = 0$) as $\overline{\langle\sigma_z\rangle}'_0$, equation (C.2) tells us that

$$\overline{\langle\sigma_z\rangle}'_0 = \frac{\tau_1}{T_p}, \qquad \text{...(C.8)}$$

Therefore

$$\overline{\langle\sigma_z\rangle}'_e = \omega_1\,\tau_1\overline{\langle\sigma_y\rangle}'_e + \overline{\langle\sigma_z\rangle}'_0, \qquad \text{...(C.9)}$$

$$\overline{\langle\sigma_y\rangle}'_e = -\frac{\overline{\langle\sigma_x\rangle}'_e}{\Delta\omega\tau_2} \qquad \text{...(C.10)}$$

and

$$\overline{\langle\sigma_x\rangle}'_e = \frac{\omega_1\overline{\langle\sigma_z\rangle}'_e}{\Delta\omega} + \frac{\overline{\langle\sigma_y\rangle}'_e}{\Delta\omega\tau_2}. \qquad \text{...(C.11)}$$

Combining these equations gives

$$\overline{\langle\sigma_z\rangle}'_e = \overline{\langle\sigma_z\rangle}'_0\left[1 - \frac{\omega_1^2\,\tau_2\,\tau_1}{1 + \omega_1^2\,\tau_2\,\tau_1 + \Delta\omega^2\,\tau_2^2}\right]. \qquad \text{...(C.12)}$$

Equation (C.12) holds in the laboratory frame of reference as well as in the rotating coordinate system because the rotation is about the z-axis, and $\rho_{11} = \rho'_{11}$; $\rho_{22} = \rho'_{22}$. Therefore $\overline{\langle\sigma_z\rangle}' = \overline{\langle\sigma_z\rangle}$.

To derive an expression for the transmission signal, i.e. the change in light intensity δI due to the application of the r.f., we must relate the equilibrium polarization $\overline{\langle\sigma_z\rangle}_e$ to the transmitted light intensity. Recalling equation (B.12), the transmitted light intensity $I(\omega, z_0)$ is given by

$$\begin{aligned} I(\omega, z_0) &= I(\omega, 0)\exp\left(-\int_0^{z_0} N\sigma(\omega)\,\rho_{22}(\xi)\,\mathrm{d}\xi\right) \\ &= I(\omega, 0)\exp\left(-z_0\,N\sigma(\omega)\,\langle\rho_{22}\rangle\right) \qquad \text{...(C.13)} \end{aligned}$$

Since

$$\overline{\langle\sigma_z\rangle} = \rho_{11} - \rho_{22}, \qquad \text{...(C.14)}$$

and

$$\rho_{11} + \rho_{22} = 1, \qquad \text{...(C.15)}$$

$$\rho_{22} = \tfrac{1}{2}(1 - \overline{\langle\sigma_z\rangle}), \qquad \text{...(C.16)}$$

where we have left out the brackets around ρ_{11} and ρ_{22} which indicate the average over z. Assuming once again that $Nz_0\sigma$ is small, we expand the exponential in equation (C.13). The total light intensity I transmitted by the cell is therefore

$$I = \int_0^\infty I(\omega, 0)\left[1 - \frac{z_0\,N\sigma(\omega)}{2}(1 - \overline{\langle\sigma_z\rangle})\right]\mathrm{d}\omega. \qquad \text{...(C.17)}$$

Thus the amount of light absorbed by the sample is proportional to $(1 - 2\overline{\langle J_z\rangle})$.

If we denote the transmitted intensity in the absence of r.f. as I_0 and the intensity at equilibrium with r.f. as I_e, then the observed signal

$$\delta I = I_e - I_0$$

is given by

$$\delta I = \int_0^\infty I(\omega, 0)\,\sigma(\omega)\,\mathrm{d}\omega\,\frac{Nz_0}{2}\,[\overline{\langle\sigma_z\rangle}_e - \overline{\langle\sigma_z\rangle}_0], \qquad \text{...(C.18)}$$

and using equation (C.12),

$$\delta I = \frac{-Nz_0}{2}\int_0^\infty I(\omega, 0)\,\sigma(\omega)\,\mathrm{d}\omega\,\overline{\langle\sigma_z\rangle}_0\,\frac{\omega_1^2\,\tau_1\,\tau_2}{1 + \omega_1^2\,\tau_1\,\tau_2 + \Delta\omega^2\,\tau_2^2}. \qquad \text{...(C.19)}$$

With the aid of equation (B.24) and (C.8), we have for the transmission signal

$$\delta I = -Nz_0\frac{\tau_1}{(T_p)^2}\,\frac{\omega_1^2\,\tau_1\,\tau_2}{1 + \omega_1^2\,\tau_1\,\tau_2 + \Delta\omega^2\,\tau_2^2}. \qquad \text{...(C.20)}$$

The equilibrium resonance signal described by equation (C.20) has a Lorentzian line shape with a full width at half maximum given by

$$\Delta\omega^2 = \frac{1}{\tau_2^2} + \frac{\tau_1}{\tau_2}\,\omega_1^2. \qquad \text{...(C.21)}$$

The center of the line is at $\omega_0 + \delta\omega$. The amplitude of the signal is seen to be proportional to the square of the incident light intensity due to the factor $(1/T_p)^2$ and it is also proportional to the product Nz_0.

When the absorption of the pumping light is appreciable, we cannot expand the exponential in equation (C.17) or in equation (B.17) which gave a constant pumping time independent of the polarization $\overline{\langle\sigma_z\rangle}$. If one continues to ignore the problem with the relaxation of the off-diagonal matrix elements due to the pumping light [equation (B.21)], it is possible (Balling *et al.*, 1964) to obtain a Lorentzian line shape by considering only small *changes* in the polarization $\overline{\langle\sigma_z\rangle}$ due to the application of an r.f. field.

The signal described by equation (C.20) was derived under the assumption of negligible inhomogeneity in the steady magnetic field. If the magnitude of the field varies over the optical-pumping cell, the observed signal will be some superposition of Lorentzians centered at various values of ω_0.

When the r.f. is switched off, the polarization, which initially has the value $\overline{\langle\sigma_z\rangle_e}$ is pumped back up to $\overline{\langle\sigma_z\rangle_0}$. This can be seen by setting $\omega_1 = 0$ in equation (B.63). That is,

$$\frac{\mathrm{d}}{\mathrm{d}t}\overline{\langle\sigma_z\rangle} = \frac{1}{T_p} - \frac{1}{\tau_1}\overline{\langle\sigma_z\rangle}. \qquad \ldots(\text{C.22})$$

If we note that

$$\frac{1}{T_p} = \frac{\overline{\langle\sigma_z\rangle_0}}{\tau_1}, \qquad \ldots(\text{C.23})$$

the appropriate solution to equation (C.22) is

$$\overline{\langle\sigma_z\rangle} = \overline{\langle\sigma_z\rangle_0} + (\overline{\langle\sigma_z\rangle_e} - \overline{\langle\sigma_z\rangle_0})\,\mathrm{e}^{-t/\tau_1}. \qquad \ldots(\text{C.24})$$

The signal $\delta I(t) = I(t) - I_0$ is

$$\delta I(t) = \int_0^\infty I(\omega,0)\,\sigma(\omega)\frac{Nz_0}{2}\,\mathrm{d}\omega[\overline{\langle\sigma_z\rangle} - \langle\sigma_z\rangle_0]. \qquad \ldots(\text{C.25})$$

Using equation (C.24),

$$\delta I(t) = \int_0^\infty I(\omega,0)\,\sigma(\omega)\,\mathrm{d}\omega\frac{Nz_0}{2}[\overline{\langle\sigma_z\rangle_e} - \overline{\langle\sigma_z\rangle_0}]\,\mathrm{e}^{-t/\tau_1}. \qquad \ldots(\text{C.26})$$

But this is just

$$\delta I(t) = \delta I\,\mathrm{e}^{-t/\tau_1}, \qquad \ldots(\text{C.27})$$

where δI is the equilibrium signal given by equations (C.18) and (C.20).

D. THE SPIN-$\frac{1}{2}$ TRANSIENT TRANSMISSION SIGNAL

In the preceding section, we calculated the transmission signal when an r.f. field was applied to the cell and an equilibrium polarization was established. We shall now investigate the transient signal we obtain immediately after the r.f. is turned on as the sample approaches its equilibrium polarization. To keep things simple we will restrict ourselves to the case where the r.f. is tuned exactly to the resonant frequency so that $\Delta\omega = 0$.

For exact resonance, equations (B.63) and (B.64) become

$$\frac{d}{dt}\overline{\langle\sigma_z\rangle}' = \omega_1\overline{\langle\sigma_y\rangle}' + \frac{1}{T_p} - \frac{1}{\tau_1}\overline{\langle\sigma_z\rangle}', \quad \text{...(D.1)}$$

$$\frac{d}{dt}\overline{\langle\sigma_y\rangle}' = -\omega_1\overline{\langle\sigma_z\rangle}' - \frac{1}{\tau_2}\overline{\langle\sigma_y\rangle}', \quad \text{...(D.2)}$$

Differentiating equation (D.1) and using equation (D.2),

$$\frac{d^2\overline{\langle\sigma_z\rangle}}{dt^2} + \left(\frac{1}{\tau_1} + \frac{1}{\tau_2}\right)\frac{d\overline{\langle\sigma_z\rangle}}{dt} + \left(\omega_1^2 + \frac{1}{\tau_1\tau_2}\right)\overline{\langle\sigma_z\rangle} = \frac{1}{T_p\tau_2}, \quad \text{...(D.3)}$$

Where we have dropped the prime because $\overline{\langle\sigma_z\rangle}' = \overline{\langle\sigma_z\rangle}$.

The solution to equation (D.3) is obtained by adding the general solution of the homogeneous reduced equation to a particular solution of the complete equation. A particular solution is obtained by solving

$$\left(\omega_1^2 + \frac{1}{\tau_1\tau_2}\right)\overline{\langle\sigma_z\rangle} = \frac{1}{T_p\tau_2} \quad \text{...(D.4)}$$

Rearranging equation (D.4) gives

$$\overline{\langle\sigma_z\rangle} = \frac{\tau_1}{T_p}\left(\frac{1}{\omega_1^2\tau_1\tau_2 + 1}\right) \quad \text{...(D.5)}$$

as a particular solution for equation (D.3). But this is identical to equation (C.12) with $\Delta\omega = 0$ which shows that the particular solution to equation (D.3) is

$$\overline{\langle\sigma_z\rangle} = \overline{\langle\sigma_z\rangle}_e$$

where $\overline{\langle\sigma_z\rangle}_e$ is the equilibrium polarization when an r.f. field is applied exactly at resonance.

The general solution to the homogeneous reduced equation obtained from equation (D.3) is

$$\overline{\langle\sigma_z\rangle} = \exp\left(-\frac{1}{2}\left(\frac{1}{\tau_1} + \frac{1}{\tau_2}\right)t\right)$$
$$\times\left[C\exp\left(i\sqrt{\omega_1^2 - \frac{1}{4}\left(\frac{1}{\tau_1} - \frac{1}{\tau_2}\right)^2}\, t\right) + D\exp\left(-i\sqrt{\omega_1^2 - \frac{1}{4}\left(\frac{1}{\tau_1} - \frac{1}{\tau_2}\right)^2}\right)t\right], \quad \text{...(D.6)}$$

where C and D are arbitrary constants. The solution to equation (D.3) is therefore

$$\overline{\langle\sigma_z\rangle} = \overline{\langle\sigma_z\rangle}_e + \exp\left(-\frac{1}{2}\left(\frac{1}{\tau_1}+\frac{1}{\tau_2}\right)t\right)$$
$$\times\left[C\exp\left(i\sqrt{\omega_1^2-\frac{1}{4}\left(\frac{1}{\tau_1}-\frac{1}{\tau_2}\right)^2}\,t\right)+D\exp-i\sqrt{\omega_1^2-\frac{1}{4}\left(\frac{1}{\tau_1}-\frac{1}{\tau_2}\right)^2}\,t\right], \quad \text{...(D.7)}$$

which approaches the equilibrium solution $\overline{\langle\sigma_z\rangle}_e$ as $t\to\infty$.

It is instructive to determine the limiting form of the solution to equation (D.7) for the case where the r.f. field is strong and for the case where it is very weak. If the field is strong, so that

$$\omega_1^2 \gg \frac{1}{4}\left(\frac{1}{\tau_1}-\frac{1}{\tau_2}\right)^2,$$

then the solution becomes

$$\overline{\langle\sigma_z\rangle} = \overline{\langle\sigma_z\rangle}_e + \exp\left(-\frac{1}{2}\left(\frac{1}{\tau_1}+\frac{1}{\tau_2}\right)t\right)[A\cos\omega_1 t + B\sin\omega_1 t.] \quad \text{...(D.8)}$$

To evaluate A and B we must apply the initial conditions. If the r.f. is switched on at $t=0$, then at $t=0$.

$$\overline{\langle\sigma_z\rangle} = \overline{\langle\sigma_z\rangle}_0,$$

and

$$A = \overline{\langle\sigma_z\rangle}_0 - \overline{\langle\sigma_z\rangle}_e. \quad \text{...(D.9)}$$

Also at $t=0$

$$\frac{\mathrm{d}\overline{\langle\sigma_z\rangle}}{\mathrm{d}t} = 0.$$

Hence,

$$B = \frac{\left(\frac{1}{\tau_1}+\frac{1}{\tau_2}\right)}{2\omega_1}(\overline{\langle\sigma_z\rangle}_0 - \overline{\langle\sigma_z\rangle}_e). \quad \text{...(D.10)}$$

Therefore,

$$\overline{\langle\sigma_z\rangle} = \overline{\langle\sigma_z\rangle}_e + [\overline{\langle\sigma_z\rangle}_0 - \overline{\langle\sigma_z\rangle}_e]\exp\left(-\frac{1}{2}\left(\frac{1}{\tau_1}+\frac{1}{\tau_2}\right)t\right)\left[\cos\omega_1 t + \frac{\frac{1}{\tau_1}+\frac{1}{\tau_2}}{2\omega_1}\sin\omega_1 t\right]. \quad \text{...(D.11)}$$

From equation (D.11) we see that at resonance, for a strong r.f. field, the polarization $\overline{\langle\sigma_z\rangle}$ and hence the transmitted light intensity undergoes damped oscillation with angular frequency ω_1. The amplitude of the polarization decays exponentially to the equilibrium polarization $\overline{\langle\sigma_z\rangle}_e$ given by equation (C.12) with $\Delta\omega = 0$.

This corresponds to the qualitative description of the behavior of an ensemble of spins at resonance given in Section II. The angular momenta of the atoms precess about the rotating magnetic field with frequency ω_1 until the atomic angular momenta get out of phase with each other due to relaxation effects.

To obtain an explicit expression for the transmission signal as a function of time, we make use of equation (C.17). The total pumping light intensity transmitted by the cell is

$$I(t) = \int_0^\infty I(\omega, 0)\left[1 - \frac{z_0 N\sigma(\omega)}{2}(1 - \langle\sigma_z\rangle)\right] d\omega. \qquad \text{...(D.12)}$$

The signal is the change in the intensity $\delta I(t)$ as a result of applying the r.f. field.

$$\delta I(t) = I(t) - I_0. \qquad \text{...(D.13)}$$

Using equations (D.13) and (D.12), we have

$$\delta I(t) = \delta I\left[1 - \exp\left(-\frac{1}{2}\left(\frac{1}{\tau_1} + \frac{1}{\tau_2}\right)t\right)\left(\cos\omega_1 t + \frac{\frac{1}{\tau_1} + \frac{1}{\tau_2}}{2\omega_1}\sin\omega_1 t\right)\right]. \qquad \text{...(D.14)}$$

where δI is given by equation (C.20) with $\Delta\omega = 0$.

The oscillations of the light intensity are called the "wiggles", but they should not be confused with the "wiggles" seen in NMR experiments. In an NMR experiment, the wiggles are caused by the beating of the e.m.f. induced by precessing nuclear moments with the applied r.f. field as the steady magnetic field H_0 is swept through resonance.

If the r.f. is sufficiently weak, so that

$$\omega_1^2 \ll \frac{1}{4}\left(\frac{1}{\tau_1} - \frac{1}{\tau_2}\right)^2,$$

the wiggles will be damped out in a time short compared to the period of oscillation. In this case our general solution is

$$\overline{\langle\sigma_z\rangle} = \overline{\langle\sigma_z\rangle_e} + A\,e^{-t/\tau_1} + B\,e^{-t/\tau_2} \qquad \text{...(D.15)}$$

Applying the initial conditions gives

$$\overline{\langle\sigma_z\rangle} = \overline{\langle\sigma_z\rangle_e} + \frac{(\overline{\langle\sigma_z\rangle_e} - \overline{\langle\sigma_z\rangle_0})}{\tau_2 - \tau_1}(\tau_1\,e^{-t/\tau_1} - \tau_2\,e^{-t/\tau_2}). \qquad \text{...(D.16)}$$

Therefore,

$$\delta I(t) = \delta I\left[1 + \frac{1}{\tau_2 - \tau_1}(\tau_1\,e^{-t/\tau_1} - \tau_2\,e^{-t/\tau_2})\right], \qquad \text{...(D.17)}$$

where δI is given by equation (C.20) with $\Delta\omega = 0$.

Although the expressions for the optical-pumping signals which we have just derived are for a fictitious spin-$\frac{1}{2}$ alkali atom, they are essentially correct for the optical-pumping of Hg^{199} with only minor modifications for the pumping rate due to no reorientation in the excited state (Cagnac, 1961). They also describe the qualitative behavior of the signals obtained for real alkali atoms with nuclear spin.

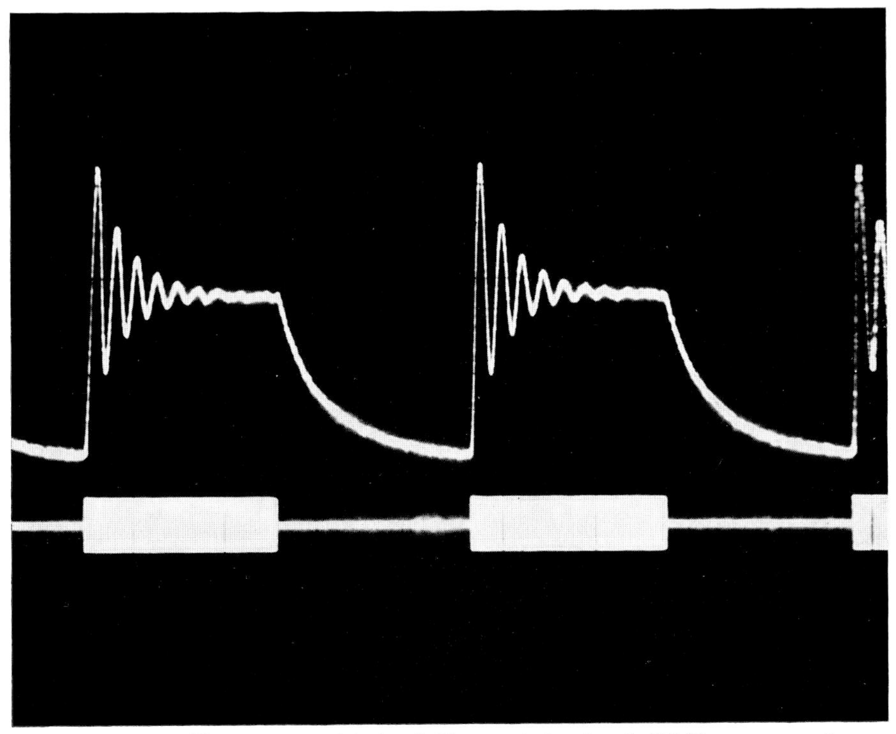

FIG. 10. An oscilloscope trace of the low field transmission signal of Rb^{87} at resonance for a strong r.f. field. The lower trace is synchronized with the upper trace and displays the amplitude modulation (7 Hz) of the applied r.f.

Figure 10 is an oscilloscope trace of the transmission optical-pumping signal for Rb^{87} atoms at exact resonance with a strong r.f. field. The r.f. is switched on and off at a rate of 10 Hz. The upper trace shows the response of the system to the applied r.f. the modulation of which is shown for reference on the lower trace. One sees the decaying wiggles qualitatively described by equation (D.14) and the return of the light intensity to its initial value when the r.f. is turned off as described by equation (C.27). The amplifier has inverted the sign of the signal. Figure 11 shows a similar trace for the weak field case described by equation (D.17).

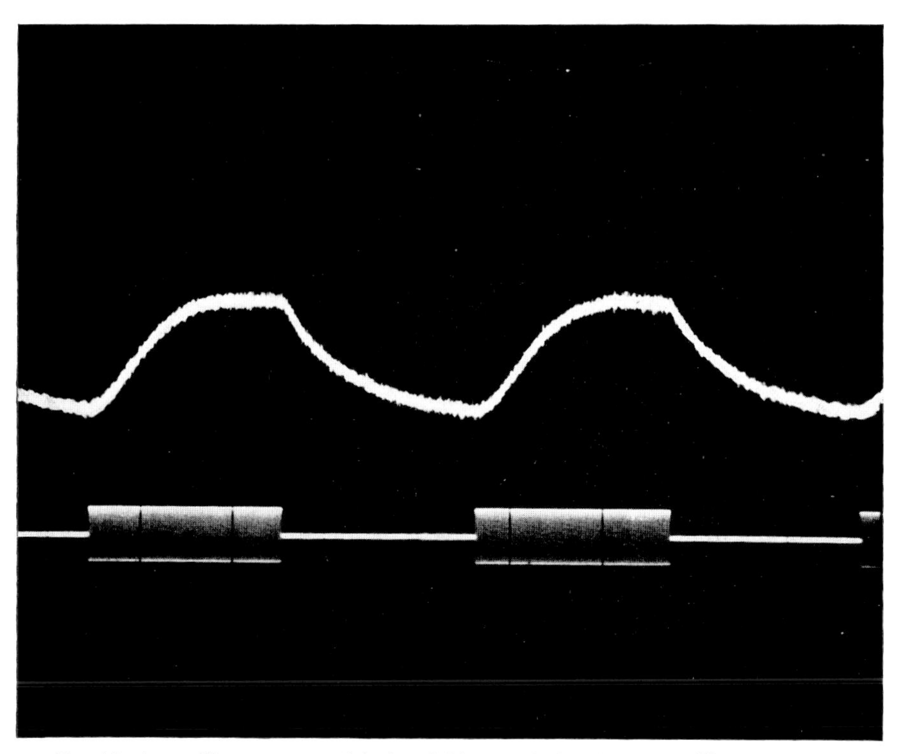

FIG. 11. An oscilloscope trace of the low field transmission signal of Rb^{87} at resonance for a weak r.f. field. The lower trace is synchronized with the upper trace and displays the amplitude modulation (7 Hz) of the applied r.f.

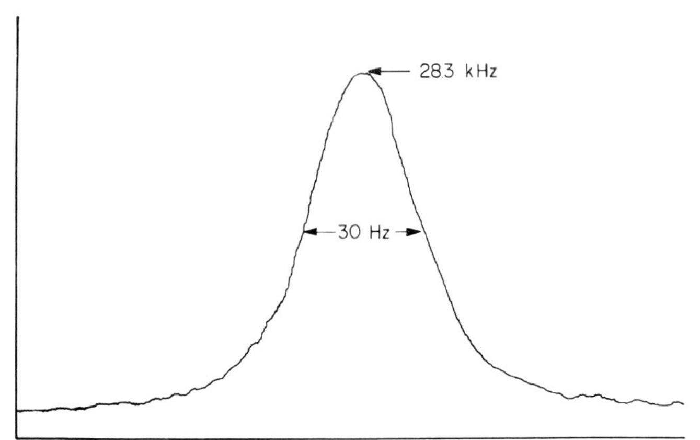

FIG. 12. A chart recording of the low field Rb^{87} transmission signal as a function of r.f. frequency for a weak r.f. field. The signal is obtained from the voltage output of a phase-sensitive detector. The r.f. is chopped at a rate of 10 Hz which is the reference frequency for the phase-sensitive detector.

Figure 12 is a plot of the equilibrium transmission signal obtained from the output of a phase-sensitive detector as a function of r.f. frequency as described by equation (C.20).

E. THE EQUILIBRIUM CROSSED-BEAM SIGNAL

If instead of monitoring the intensity of the pumping light transmitted by the cell, one passes weak circularly-polarized D_1 radiation through the cell in a direction perpendicular to the steady magnetic field, then upon application of a resonant r.f. field, one observes the coherent precession of the atomic angular momenta about the steady field.

If the beam is left-circularly polarized and passes through the cell in the y-direction, the intensity of the transmitted light is given by equation (C.17) if we replace $\overline{\langle\sigma_z\rangle}$ by $\overline{\langle\sigma_y\rangle}$. The intensity $I(\omega,0)$ is now the intensity of the crossed beam. In order to compute the time dependence of $\overline{\langle\sigma_y\rangle}$ we will first compute it in the rotating frame and then transform back to the laboratory frame.

After the r.f. field has been applied and equilibrium established, we have in the rotating frame from equations (C.10) and (C.11)

$$\overline{\langle\sigma_x\rangle}'_e = \frac{\omega_1 \overline{\langle\sigma_z\rangle}'_e}{\Delta\omega} - \frac{\overline{\langle\sigma_x\rangle}'_e}{\Delta\omega^2 \tau_2^2}, \qquad \text{...(E.1)}$$

or

$$\overline{\langle\sigma_x\rangle}'_e = \frac{\omega_1 \tau_2^2 \Delta\omega}{1+\Delta\omega^2 \tau_2^2} \overline{\langle\sigma_z\rangle}'_e. \qquad \text{...(E.2)}$$

With the aid of equation (C.12), this becomes

$$\overline{\langle\sigma_x\rangle}'_e = \frac{\overline{\langle\sigma_z\rangle}'_0 \omega_1 \tau_2^2 \Delta\omega}{1+\omega_1^2 \tau_2 \tau_1 + \Delta\omega^2 \tau_2}. \qquad \text{...(E.3)}$$

Exactly at resonance $\Delta\omega = 0$ and $\overline{\langle\sigma_x\rangle}'_e = 0$ in the rotating frame.

From equation (C.10) we also have

$$\overline{\langle\sigma_y\rangle}'\mathrm{e} = -\frac{\overline{\langle\sigma_z\rangle}'_0 \omega_1 \tau_2}{1+\omega_1^2 \tau_2 \tau_1 + \Delta\omega^2 \tau_2^{2\prime}}, \qquad \text{...(E.4)}$$

which takes on a maximum value at resonance.

Recalling that

$$\rho = \exp\left(-i\frac{\omega t}{2}\sigma_z\right)\rho' \exp\left(i\frac{\omega t}{2}\sigma_z\right), \qquad \text{...(E.5)}$$

and that

$$\exp\left(i\frac{\omega t}{2}\sigma_z\right) = \begin{bmatrix} \exp\left(i\frac{\omega t}{2}\right) & 0 \\ 0 & \exp\left(-i\frac{\omega t}{2}\right) \end{bmatrix}$$

we have

$$\rho_{12} = \rho'_{12}\,\mathrm{e}^{-i\omega t}, \qquad \text{...(E.7)}$$

and

$$\rho_{21} = \rho'_{21}\,\mathrm{e}^{i\omega t}. \qquad \text{...(E.8)}$$

Therefore at resonance

$$\overline{\langle\sigma_y\rangle}_e = \overline{\langle\sigma_y\rangle}'_e \cos\omega t + \overline{\langle\sigma_x\rangle}'_e \sin\omega t = \overline{\langle\sigma_y\rangle}'_e \cos\omega t, \qquad \text{...(E.9)}$$

and

$$\overline{\langle\sigma_x\rangle}_e = \overline{\langle\sigma_x\rangle}'_e \cos\omega t - \overline{\langle\sigma_y\rangle}'_e \sin\omega t = -\overline{\langle\sigma_y\rangle}'_e \sin\omega t. \qquad \text{...(E.10)}$$

In the laboratory frame at resonance, $\overline{\langle\sigma_y\rangle}_e$ oscillates sinusoidally at the resonance frequency with an amplitude given by equation (E.4) with $\Delta\omega = 0$. Therefore the intensity of the crossed beam is also modulated at this frequency. The modulation is 90° out of phase with the applied r.f. field $2H_1\,\hat{\imath}\cos\omega t$. If the light modulation is converted to an A.C. voltage, shifted in phase by 90°, and fed back to the r.f. coils, the system can be made to self oscillate. This is the principle on which self-oscillating magnetometers operate.

V. Optical Pumping of Alkali Atoms

In this section we shall consider the optical pumping of an alkali atom, including the effect of nuclear spin. The hyperfine coupling of the nuclear and electronic angular momenta considerably complicates the analysis of the optical-pumping signal, and this section is not intended to be a definitive treatment of the subject. Nevertheless, it should give the reader an understanding of the theoretical problems introduced by the inclusion of nuclear spin and of the degree to which the simpler spin-$\frac{1}{2}$ model of the preceding section approximates the behavior of the alkali-atom optical-pumping signal.

A. Effective Hamiltonian for an Alkali Atom in a Weak Magnetic Field

In Section II we used a semi-classical argument to arrive at an expression for the approximate Hamiltonian for an alkali atom in a weak magnetic field. We can put the discussion on a firmer footing by using 1st-order perturbation theory.

The ground state of the alkali atom is $S_{1/2}$. The electronic angular momentum $\mathbf{J}$ is just the spin of the valence electron. The nuclear spin of the atom is $\mathbf{I}$. If we place the atom in a weak magnetic field $H_0\hat{k}$ in the z-direction, the Hamiltonian for the atom is

$$\mathscr{H} = A\mathbf{I}.\mathbf{J} - g_J\,\mu_0\,H_0\,\mathbf{J}.\hat{k} - g_I\,\mu_0\,H_0\,\mathbf{I}.\hat{k}, \qquad \text{...(A.1)}$$

where μ_0 is the Bohr magneton, A is the hyperfine interaction constant, g_J is the electronic g-factor, and g_I is the nuclear g-factor.

Since $g_I \ll g_J$ we can approximate $\mathscr{H}$ by

$$\mathscr{H} \cong A\mathbf{I}.\mathbf{J} - g_J \mu_0 H_0 \mathbf{J}.\hat{k}. \qquad \text{...(A.2)}$$

In the absence of an external magnetic field, the eigenstates of the Hamiltonian are the states $|F, M, I, J\rangle$ where

$$\mathbf{F} = \mathbf{I} + \mathbf{J}, \qquad \text{...(A.3)}$$

$$F_z = I_z + J_z, \qquad \text{...(A.4)}$$

and

$$F^2|F, M, I, J\rangle = F(F+1)\,|\,F, M, I, J\rangle, \qquad \text{...(A.5)}$$

$$F_z|F, M, I, J\rangle = M\,|\,F, M, I, J\rangle. \qquad \text{...(A.6)}$$

The ground state is thus split into two hyperfine levels characterized by quantum number $F = I \pm \frac{1}{2}$. In what follows, we shall abbreviate our notation and write $|F, M, I, J\rangle$ as $|F, M\rangle$. If $g_J \mu_0 H_0 \ll A$, we can treat the interaction with the external field as a perturbation. Using 1st-order perturbation theory, we have

$$\mathscr{H}|F, M\rangle = E_{F,M}|\,F, M\rangle, \qquad \text{...(A.7)}$$

with

$$E_{F,M} \cong E_{F,0} - \langle F, M|\,g_J \mu_0 H_0 J_z|\,F, M\rangle. \qquad \text{...(A.8)}$$

Using the properties of coupled angular momentum states and the phase convention of Condon and Shortley (1935), it is easy to show that

$$\langle F, M|\mathbf{J}|\,F, M\rangle = \left[\frac{J(J+1) - I(I+1) + F(F+1)}{2F(F+1)}\right] M\hat{k}. \qquad \text{...(A.9)}$$

Thus,

$$E_{F,M} = E_{F,0} - g_J \mu_0 H_0 M \left[\frac{F(F+1) + J(J+1) - I(I+1)}{2F(F+1)}\right]. \qquad \text{...(A.10)}$$

If we substitute $F = I \pm \frac{1}{2}$ in equation (A.10) we obtain

$$E_{F,M} = E_{F,0} \pm \frac{\hbar\omega_0}{(2I+1)} M, \qquad \text{...(A.11)}$$

where the $\pm$ sign corresponds to $F = I \pm \frac{1}{2}$ and

$$\hbar\omega_0 = -g_J \mu_0 H_0. \qquad \text{...(A.12)}$$

To make the connection with the vector model of the alkali atom given in Section II, we note that the energy levels given by equations (A.10) and (A.11) are the eigenvalues of the effective Hamiltonian

$$\mathscr{H} = A\mathbf{I}.\mathbf{J} - g_J \mu_0 \frac{\mathbf{J}.\mathbf{F}}{|F|^2} F_z H_0, \qquad \text{...(A.13)}$$

since

$$(\mathbf{J}.\mathbf{F}) = \tfrac{1}{2}[F^2 + J^2 - (F-J)^2] = \tfrac{1}{2}[F^2 + J^2 - I^2], \qquad \text{...(A.14)}$$

and

$$\frac{\frac{1}{2}[F^2+J^2-I^2]}{|F|^2}|F,M\rangle = \frac{[F(F+1)+J(J+1)-I(I+1)]}{2F(F+1)}|F,M\rangle$$

$$= \pm\frac{1}{2I+1}|F,M\rangle, \qquad \text{...(A.15)}$$

where the $\pm$ sign refers to $F = I \pm \frac{1}{2}$.

The angular momenta **I** and **J** are coupled tightly together to form the resultant angular momentum **F** which precesses about the static field $H_0\hat{k}$. The operator **J**.**F** changes sign for $F = I \pm \frac{1}{2}$ so that atoms in different hyperfine levels precess in opposite directions about the field.

B. DENSITY MATRIX FOR THE ALKALI ATOM GROUND STATE

There are $(2)(2I+1)$ linearly independent states $|F, M\rangle$. The density matrix for the alkali atom ground state is a $2(2I+1) \times 2(2I+1)$ matrix the diagonal elements of which specify the probability of finding an atom in a particular state $|F', M'\rangle$. The off-diagonal elements represents any existing coherence between two states $|F, M\rangle$ and $|F', M'\rangle$.

We will be restricting ourselves to consideration of Zeeman transitions $(F, M \leftrightarrow F, M \pm 1)$ in what follows, so we will assume that no coherence exists between states in different hyperfine levels. Thus any off-diagonal elements which connect different hyperfine levels will be set equal to zero, and the density matrix will have a form similar to that shown in Fig. 13 for the case

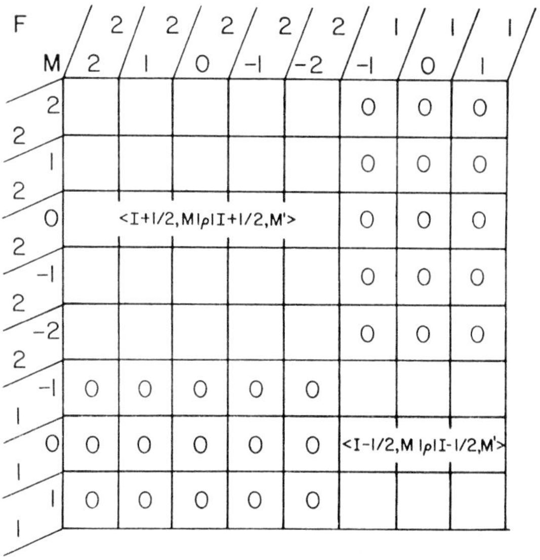

F, M	2, 2	2, 1	2, 0	2, −1	2, −2	1, −1	1, 0	1, 1
2, 2						0	0	0
2, 1						0	0	0
2, 0	$\langle I+1/2, M \vert \rho \vert I+1/2, M'\rangle$					0	0	0
2, −1						0	0	0
2, −2						0	0	0
1, −1	0	0	0	0	0			
1, 0	0	0	0	0	0	$\langle I-1/2, M \vert \rho \vert I-1/2, M'\rangle$		
1, 1	0	0	0	0	0			

FIG. 13. The structure of the density matrix for Rb^{87} in the $|F, M\rangle$ representation. Off-diagonal elements connecting states with different F values vanish.

of Rb^{87}, where the full density matrix is divided into two sub-matrices (+) and (−) describing the state populations and coherence between the sublevels in the $F = I \pm \frac{1}{2}$ hyperfine multiplets respectively. The matrix elements of the null matrices are the vanishing off-diagonal elements connecting states in different hyperfine multiplets. Basically, we are dividing the ensemble of alkali atoms into two groups characterized by $F = I \pm \frac{1}{2}$ and assuming that the two groups are statistically independent.

Thus, we can express the full density matrix ρ as

$$\rho = \rho_+ + \rho_-,$$

where ρ_+ and ρ_- are $2(2I+1) \times 2(2I+1)$ dimensional matrices with all matrix elements zero except for the elements corresponding to $F = I + \frac{1}{2}$ for ρ_+ and $F = I - \frac{1}{2}$ for ρ_-. This is shown schematically in Fig. 14. We use the

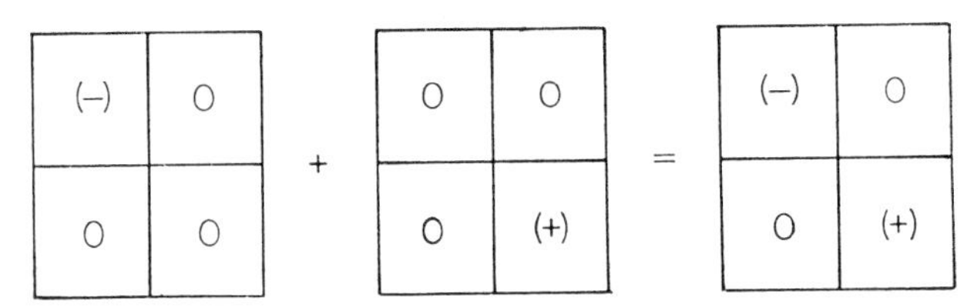

FIG. 14. The Rb^{87} density matrix can be expressed as the sum of two matrices ρ_+ and ρ_- describing the $I + \frac{1}{2}$ and $I - \frac{1}{2}$ sublevels separately.

density matrix to compute the ensemble averages of the relevant observables. The ensemble average $\langle \mathbf{F} \rangle$ is

$$\overline{\langle \mathbf{F} \rangle} = \mathrm{Tr}\, \mathbf{F}\rho = \sum_{\substack{F,M \\ F',M'}} \langle F, M | \mathbf{F} | F', M' \rangle \langle F', M' | \rho | F, M \rangle. \qquad \text{...(B.1)}$$

However, we are requiring that

$$\langle F', M' | \rho | F, M \rangle = 0, \qquad \text{if} \qquad F' \neq F,$$

so

$$\overline{\langle \mathbf{F} \rangle} = \sum_{F,M,M'} \langle F, M | \mathbf{F} | F, M' \rangle \langle F, M' | \rho | F, M \rangle, \qquad \text{...(B.2)}$$

If we note that

$$F_x = \frac{F_+ + F_-}{2}, \qquad \text{...(B.3)}$$

and

$$F_y = \frac{F_+ - F_-}{2i}, \qquad \text{...(B.4)}$$

where $F\pm$ are the ladder operators defined by

$$F_\pm | F, M \rangle = \sqrt{(F \mp M)(F \pm M + 1)}\, | F, M \pm 1 \rangle, \qquad \text{...(B.5)}$$

we obtain from equation (B.2) the relations

$$\overline{\langle F_z \rangle} = \sum_{F,M} M \langle F, M | \rho | F, M \rangle, \qquad \text{...(B.6)}$$

and

$$\overline{\langle F_x \rangle} = \sum_{F,M} \tfrac{1}{2}\sqrt{(F-M)(F+M+1)} \times [\langle F, M | \rho | F, M+1 \rangle + \langle F, M+1 | \rho | F, M \rangle], \qquad \text{...(B.7)}$$

$$\overline{\langle F_y \rangle} = \sum_{F,M} \frac{1}{2i}\sqrt{(F-M)(F+M+1)} \times [\langle F, M | \rho | F, M+1 \rangle - \langle F, M+1 | \rho | F, M \rangle]. \qquad \text{...(B.8)}$$

The ensemble average for the electronic angular momentum $\overline{\langle \mathbf{J} \rangle}$ is

$$\overline{\langle \mathbf{J} \rangle} = \mathrm{Tr}\, \mathbf{J}\rho = \sum_{\substack{F,M \\ F',M'}} \langle F, M | \mathbf{J} | F', M' \rangle \langle F' M' | \rho | F, M \rangle. \qquad \text{...(B.9)}$$

Since we are requiring that $\langle F' M' | \rho | FM \rangle = 0$ for $F \neq F'$,

$$\overline{\langle J_z \rangle} = \sum_{\substack{F,M \\ M'}} \langle F, M | J_z | FM' \rangle \langle FM' | \rho | FM \rangle. \qquad \text{...(B.10)}$$

From the properties of angular momentum (Condon and Shortley, 1935), we have

$$\overline{\langle J_z \rangle} = \sum_M \langle I + \tfrac{1}{2} M | \rho | I + \tfrac{1}{2}, M \rangle \frac{M}{2I+1} - \sum_M \langle I - \tfrac{1}{2}, M | \rho | I - \tfrac{1}{2}, M \rangle \frac{M}{2I+1}, \qquad \text{...(B.11)}$$

C. MAGNETIC RESONANCE IN A WEAK FIELD

As can be seen from equation (A.11), the energy levels of an alkali atom in a weak magnetic field $H_0 \hat{k}$ are equally spaced. This means that a linearly oscillating field $2H_1 \hat{\imath} \cos \omega t$ can induce Zeeman transitions between all of the adjacent sublevels simultaneously. The effective Hamiltonian for the atom in the presence of the static and oscillating fields is

$$\mathscr{H} \cong A\mathbf{I}.\mathbf{J} + \hbar\omega_0 \frac{\mathbf{J}.\mathbf{F}}{|F|^2} F_z + 2\hbar\omega_1 \frac{\mathbf{J}.\mathbf{F}}{|F|^2} F_x \cos \omega t, \qquad \text{...(C.1)}$$

where

$$\hbar\omega_1 = -g_J \mu_0 H_1. \qquad \text{...(C.2)}$$

The effect of the Hamiltonian upon the density matrix is given by

$$\frac{\mathrm{d}\rho}{\mathrm{d}t} = \frac{1}{i\hbar}[\mathscr{H}, \rho]. \qquad \text{...(C.3)}$$

As we found for the spin-$\frac{1}{2}$ model in the preceding chapter, it is convenient to transform our equations to a coordinate system rotating with angular velocity ω about the z-axis. Because atoms with $F=I+\frac{1}{2}$ precess in the opposite direction from atoms with $F=I-\frac{1}{2}$, the sign of ω must be different in the two cases in order to achieve resonance. Since the Hamiltonian in equation (C.1) does not connect states with different F quantum numbers, we can calculate the time dependence of the matrices ρ_+ and ρ_- separately.

We will transform to a coordinate system rotating with angular velocity ω about the z-axis and compute $d\bar{\rho}_+/dt$ in this coordinate system

$$\bar{\rho}_+ = e^{iF_z\omega t}\rho_+ e^{-iF_z\omega t}, \qquad \text{...(C.4)}$$

and transform to a frame rotating in the opposite direction to compute $d\bar{\rho}_-/dt$.

$$\bar{\rho}_- = e^{-iF_z\omega t}\rho_- e^{iF_z\omega t}. \qquad \text{...(C.5)}$$

Differentiating equations (C.4) and (C.5) with respect to time and using equation (C.3), we obtain

$$\frac{d\bar{\rho}_\pm}{dt} = \pm i\omega[F_z, \bar{\rho}_\pm] + \frac{1}{i\hbar}[e^{\pm iF_z\omega t}\mathscr{H}e^{\mp iF_z\omega t}, \bar{\rho}_\pm]. \qquad \text{...(C.6)}$$

The next step is to recognize that the effect of the linearly oscillating field

$$2H_1\hat{\imath}\cos\omega t$$

is equivalent to a rotating field

$$H_1[\hat{\imath}\cos\omega t + \hat{\jmath}\sin\omega t]$$

for the atoms with $F=I+\frac{1}{2}$ and equivalent to a field

$$H_1[\hat{\imath}\cos\omega t - \hat{\jmath}\sin\omega t]$$

rotating in the opposite direction for the atoms with $F=I-\frac{1}{2}$. Therefore we replace the linearly oscillating term in the Hamiltonian with

$$\frac{\hbar\omega_1}{2}[F_+e^{\mp i\omega t} + F_-e^{\pm i\omega t}],$$

where the (+) sign is used in the computation of $d\bar{\rho}_+/dt$ and the (–) sign is used to compute $d\bar{\rho}_-/dt$.

Noting that

$$[F_z, \mathbf{J}.\mathbf{F}] = [F_z, \mathbf{I}.\mathbf{J}] = 0, \qquad \text{...(C.7)}$$

and replacing the linearly oscillating field by a rotating field, we obtain from equation (C.6) the result

$$\frac{d\bar{\rho}_\pm}{dt} = \pm i\omega[F_z, \bar{\rho}_\pm] - i\omega_0\left[\frac{\mathbf{J}.\mathbf{F}}{|F|^2}F_z, \bar{\rho}_\pm\right] + \frac{A}{i\hbar}[\mathbf{I}.\mathbf{J}, \bar{\rho}_\pm] - i\omega_1\left[\frac{\mathbf{J}.\mathbf{F}}{|F|^2}F_x, \bar{\rho}_\pm\right]. \qquad \text{...(C.8)}$$

The time evolution of $\langle\mathbf{F}\rangle$ due to the static and r.f. magnetic fields is given by

$$\frac{\mathrm{d}\overline{\langle\mathbf{F}\rangle}}{\mathrm{d}t} = \mathrm{Tr}\,\mathbf{F}\frac{\mathrm{d}\rho_+}{\mathrm{d}t} + \mathrm{Tr}\,\mathbf{F}\frac{\mathrm{d}\rho_-}{\mathrm{d}t} \quad \text{...(C.9)}$$

With the aid of equation (C.8) and the commutation relations

$$[\mathbf{F}, \mathbf{I}.\mathbf{J}] = 0, \quad \text{...(C.10)}$$

$$[F_i, F_j] = iF_k, \quad \text{...(C.11)}$$

$$[\mathbf{F}, \mathbf{J}.\mathbf{F}] = 0, \quad \text{...(C.12)}$$

we have in the rotating frame

$$\frac{\mathrm{d}\overline{\langle F_z\rangle}_+}{\mathrm{d}t} = \mathrm{Tr}\,F_z\frac{\mathrm{d}\bar{\rho}_+}{\mathrm{d}t} = \omega_1\,\mathrm{Tr}\frac{\mathbf{J}.\mathbf{F}}{|F|^2}F_y\bar{\rho}_+, \quad \text{...(C.13)}$$

and

$$\frac{\mathrm{d}\overline{\langle F_z\rangle}_-}{\mathrm{d}t} = \omega_1\,\mathrm{Tr}\frac{\mathbf{J}.\mathbf{F}}{|F|^2}F_y\bar{\rho}_-. \quad \text{...(C.14)}$$

Using equation (A.15),

$$\frac{\mathrm{d}\overline{\langle F_z\rangle}_\pm}{\mathrm{d}t} = \pm\frac{\omega_1}{2I+1}\overline{\langle F_y\rangle}_\pm. \quad \text{...(C.15)}$$

Similarly,

$$\frac{\mathrm{d}\overline{\langle F_x\rangle}_\pm}{\mathrm{d}t} = \pm\left(\omega - \frac{\omega_0}{2I+1}\right)\overline{\langle F_y\rangle}_\pm \quad \text{...(C.16)}$$

and

$$\frac{\mathrm{d}\overline{\langle F_y\rangle}_\pm}{\mathrm{d}t} = \mp\left(\omega - \frac{\omega_0}{2I+1}\right)\overline{\langle F_x\rangle}_\pm \mp \frac{\omega_1}{2I+1}\overline{\langle F_z\rangle}_\pm. \quad \text{...(C.17)}$$

Equations (C.15)–(C.17) describe the time evolution of $\overline{\langle\mathbf{F}\rangle}_+$ and $\overline{\langle\mathbf{F}\rangle}_-$ in the rotating frame due to the application of an r.f. field.

D. THE OPTICAL-PUMPING CYCLE

If the alkali atoms are illuminated with left-circularly polarized D_1 radiation incident along the z-direction, the atoms can undergo a ($\Delta F = 0, \pm 1, \Delta M = 1$) optical transition to the $P_{1/2}$ excited state. The multiplet structure of the excited state is the same as that of the ground state. The possible absorption paths are shown schematically in Fig. 15 with the magnetic quantum number suppressed.

If we assume complete reorientation in the excited state, the atoms will radiate down to any one of the ground state sublevels with equal probability.

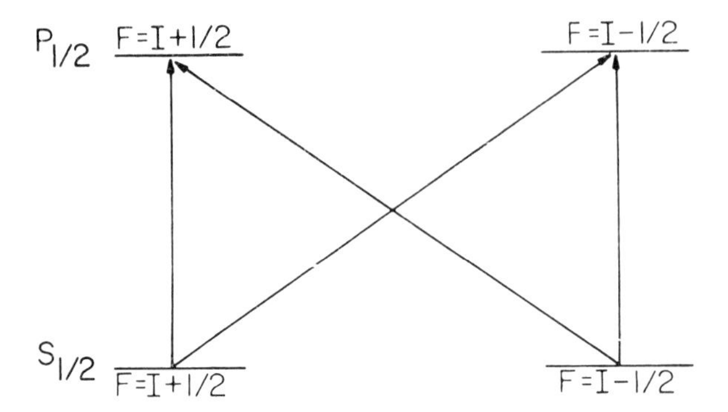

FIG. 15. The optical absorption paths for the $S_{1/2} \rightarrow P_{1/2}$ transition in Rb^{87} with the Zeeman structure suppressed.

Applying equation (A.54) in Section IV to our present case, we have, for the time rate of change of a diagonal density matrix element $\langle F, M|\rho|F, M\rangle$ due to absorption of the pumping light, the equation

$$\frac{\mathrm{d}}{\mathrm{d}t}\langle F, M|\rho|F, M\rangle = -\left[\int_0^\infty I(\omega)\,\sigma_{0F,M}\exp\left(\frac{-Mc^2(\omega-\omega_0)^2}{2\omega_0^2 kT}\right)\mathrm{d}\omega + \int_0^\infty I(\omega)\,\sigma'_{0F,M}\exp\left(\frac{-Mc^2(\omega-\omega_0')^2}{2\omega_0^2 kT}\right)\mathrm{d}\omega\right]\langle F, M|\rho|F, M\rangle, \quad \text{...(D.1)}$$

where

$$\sigma_{0F,M} = \frac{4\pi^2\alpha\omega_0|\langle F, M+1, P_{1/2}|\, D_+|F, M, S_{1/2}\rangle|^2}{\int_0^\infty \exp(-Mc^2(\omega-\omega_0)^2/2\omega_0^2 kT)\,\mathrm{d}\omega} \quad \text{...(D.2)}$$

and

$$\sigma'_{0F.M} = \frac{4\pi^2\alpha\omega_0|\langle F', M+1, P_{1/2}|\, D_+|F, M, S_{1/2}\rangle|^2}{\int_0^\infty \exp(-Mc^2(\omega-\omega_0')^2/2\omega_0'^2 kT)} \quad \text{...(D.3)}$$

and we assume negligible variation of $I(\omega)$ over the volume of the optical-pumping cell.

In these equations, a diagonal element $\langle F, M|\rho|F, M\rangle$ corresponds to a particular state $|F, M\rangle$, ω_0 is the center frequency for a $(F, M \leftrightarrow F, M+1)$ transition and ω_0' is the center frequency for a $(F, M \rightarrow F', M+1)$ transition.

The dipole moment operator D_+ is a tensor operator which transforms under rotations like the spherical harmonic $Y_1^1(\Omega)$ and it operates upon the spatial part of the wavefunction. As a result of the Wigner–Eckart theorem, the dependence upon M of the matrix elements in equations (D.2) and (D.3) factors out. The reader is referred to the classic text on angular momentum by Edmonds (1957). Using Edmond's equation (5.41), we find that

$$|\langle F', M+1, P_{1/2}|\, D_+|F, M, S_{1/2}\rangle|^2 = \begin{pmatrix} F' & 1 & F \\ -(M+1) & 1 & M \end{pmatrix}^2 \times |\langle F', I, J|\,|D_+|\,|F, I, J\rangle|^2, \quad \text{...(D.4)}$$

since I and J are the same in the ground and excited states. The right-hand side of equation (D.4) is the product of a 3-j symbol and a reduced matrix element. We can further reduce the matrix element, because D_+ operates only on the electronic angular momentum. Using Edmonds's equation (7.1.7), we have

$$|\langle F', I, J|\,|D_+|\,|F, I, J\rangle|^2 = (2F+1)(2F'+1)\begin{pmatrix} j & F' & I \\ F & J & 1 \end{pmatrix}^2 |\langle J|\,|D_+|\,|J\rangle|^2, \quad \text{...(D.5)}$$

where the bracket on the right hand side is a 6-j symbol. Combining equations (D.4) and (D.5) and observing that $j = \frac{1}{2}$, we can use Edmonds's tables for 3-j and 6-j symbols to obtain

$$\frac{\mathrm{d}}{\mathrm{d}t}\langle F, M|\rho|\, F, M\rangle = -\frac{1}{T_p}\left[1 \mp \frac{2M}{2I+1}\right]\langle F, M|\rho|\, F, M\rangle. \quad \text{...(D.6)}$$

The (−) sign is used for a state with $F = I + \frac{1}{2}$ and the (+) sign is used for a state with $F = I - \frac{1}{2}$, and

$$\frac{1}{T_p} = \frac{\int_0^\infty I(\omega)\exp(-Mc^2(\omega-\omega_0)^2/2kT\omega_0^2)\, 4\pi^2\alpha\omega_0|\langle J|\,|D_+|\,|J\rangle|^2\,\mathrm{d}\omega}{\int_0^\infty \exp(-Mc^2(\omega-\omega_0)^2/2kT\omega_0^2)\,\mathrm{d}\omega} \quad \text{...(D.7)}$$

We have tacitly assumed that the pumping radiation has hyperfine components of equal intensity so that

$$\int_0^\infty I(\omega)\exp\left(\frac{-Mc^2(\omega-\omega_0)^2}{2\omega_0^2 kT}\right)\mathrm{d}\omega = \int_0^\infty I(\omega)\exp\left(\frac{-Mc^2(\omega-\omega_0')^2}{2\omega_0'^2 kT}\right)\mathrm{d}\omega. \quad \text{...(D.8)}$$

Adding in the effect of atoms radiating back down to the ground state under conditions of complete excited state reorientation, we have

$$\frac{\mathrm{d}}{\mathrm{d}t}\langle F, M|\rho|\, F, M\rangle = -\frac{1}{T_p}\left[1 \mp \frac{2M}{2I+1}\right]\langle F, M|\rho|\, F, M\rangle + \frac{1}{2(2I+1)}\sum_{F'M'}\frac{1}{T_p}\left(1 \mp \frac{2M'}{2I+1}\right)\langle F', M'|\rho|\, F', M'\rangle. \quad \text{...(D.9)}$$

The 2nd term on the right hand side is just the rate at which atoms are excited, divided by the number of ground state sub-levels.

In a similar fashion, we can calculate the rate of destruction of an off-diagonal matrix element $\langle F, M|\rho|F, M'\rangle$ due to absorption starting with equation (A.57) in Section IV. The result is

$$\frac{\mathrm{d}}{\mathrm{d}t}\langle F, M|\rho|F, M'\rangle = -\frac{1}{T_p}\left[1 \mp \frac{M+M'}{2I+1}\right]\langle F, M|\rho|F, M'\rangle + \frac{i\Delta E_{MM'}}{\hbar}\langle F, M|\rho|F, M'\rangle, \qquad \text{...(D.10)}$$

where $\Delta E_{MM'}$ is the energy shift due to virtual transitions. The (−) sign in equation (D.10) refers again to a state with $F = I + \frac{1}{2}$ and the (+) sign to a state with $F = I - \frac{1}{2}$.

The energy shift $\Delta E_{MM'}$ depends on the details of the spectral profile of the pumping lamp as was discussed in Section IV. In general, $\Delta E_{MM'}$ will have different values from one level to another.

In the discussion of the optical-pumping signal in a weak magnetic field we will ignore the level shifts due to the pumping light. The signal one observes is essentially a superposition of simultaneous transitions between all the Zeeman sublevels and little if any physical information can be extracted from the net light shift of the low field resonance signal.

The effect of the pumping light upon the observables $\overline{\langle F_z\rangle}_\pm$ is obtained from equations (D.9) and (B.6).

$$\frac{\mathrm{d}\overline{\langle F_z\rangle}_\pm}{\mathrm{d}t} = \sum_M M\frac{\mathrm{d}}{\mathrm{d}t}\langle I \pm \tfrac{1}{2}, M|\rho|I \pm \tfrac{1}{2}, M\rangle. \qquad \text{...(D.11)}$$

$$\frac{\mathrm{d}\overline{\langle F_z\rangle}_\pm}{\mathrm{d}t} = \sum_M -\frac{M}{T_p}\left(1 \mp \frac{2M}{2I+1}\right)\langle I \pm \tfrac{1}{2}, M|\rho|I \pm \tfrac{1}{2}, M\rangle + \frac{1}{2(2I+1)}\sum_{F'M'}\frac{1}{T_p} \times\left(1 \mp \frac{2M'}{2I+1}\right)\langle F' M'|\rho|F' M'\rangle \sum_{M=-(I\pm 1/2)}^{I\pm 1/2} M. \qquad \text{...(D.12)}$$

Since

$$\sum_{M=-(I\pm 1/2)}^{I\pm 1/2} M = 0,$$

$$\frac{\mathrm{d}\overline{\langle F_z\rangle}_\pm}{\mathrm{d}t} = -\frac{\overline{\langle F_z\rangle}_\pm}{T_p} \pm \frac{2}{2I+1}\frac{\overline{\langle F_z^2\rangle}_\pm}{T_p}, \qquad \text{...(D.13)}$$

where

$$\langle F_z^2\rangle_\pm = \sum_M \langle I \pm \tfrac{1}{2}, M|\rho|I \pm \tfrac{1}{2}, M\rangle M^2. \qquad \text{...(D.14)}$$

Comparison of equation (D.13) with the analogous equation (C.22) in Section IV describing the effect of the pumping light on the polarization of a spin-$\frac{1}{2}$ system reveals one of the difficulties in an analysis which includes nuclear spin. The term on the right hand side of equation (D.13) which described the pumping of $\overline{\langle F_z \rangle}_+$ and $\overline{\langle F_z \rangle}_-$, is not a constant as was the case for the spin-$\frac{1}{2}$ system.

Since $\overline{\langle F_z^2 \rangle}$ is positive definite, one sees from equation (D.13) that the pumping process tends to make $\overline{\langle F_z \rangle}_+$ positive and $\overline{\langle F_z \rangle}_-$ negative. That is, the populations of the positive M levels are increased in the $I+\frac{1}{2}$ multiplet, and the negative M levels are populated in the $I-\frac{1}{2}$ multiplet. This is partially a result of our assumption of complete reorientation in the excited state. If we had assumed that there was no reorientation in the excited state, the populations of the positive M levels would be increased in both levels (Franzen and Emslie, 1957).

Since

$$\overline{\langle J_z \rangle} = \frac{1}{2I+1}[\overline{\langle F_z \rangle}_+ - \overline{\langle F_z \rangle}_-], \qquad \text{...(D.15)}$$

we see that the assumption of complete excited state reorientation leads to an increase in $\overline{\langle J_z \rangle}$ due to the pumping process.

Turning to the transverse components of the angular momentum $\langle \mathbf{F} \rangle$, we can use equations (B.7), (B.8) and (D.10) to calculate

$$\frac{\mathrm{d}\overline{\langle F_x \rangle}_\pm}{\mathrm{d}t} \quad \text{and} \quad \frac{\mathrm{d}\overline{\langle F_y \rangle}_\pm}{\mathrm{d}t}.$$

$$\frac{\mathrm{d}\overline{\langle F_x \rangle}_\pm}{\mathrm{d}t} = \sum_M \tfrac{1}{2}\sqrt{(I\pm\tfrac{1}{2}-M)(I\pm\tfrac{1}{2}+M+1)} \times \left(\frac{\mathrm{d}}{\mathrm{d}t}\langle I\pm\tfrac{1}{2},M|\rho|I\pm\tfrac{1}{2},M+1\rangle + \frac{\mathrm{d}}{\mathrm{d}t}\langle I\pm\tfrac{1}{2},M+1|\rho|I\pm\tfrac{1}{2},M\rangle\right). \qquad \text{...(D.16)}$$

Ignoring the energy shift terms, we have

$$\frac{\mathrm{d}\overline{\langle F_x \rangle}_\pm}{\mathrm{d}t} = -\frac{1}{2T_p}\sum_M \sqrt{(I\pm\tfrac{1}{2}-M)(I\pm\tfrac{1}{2}+M+1)}\left(1 \mp \frac{2M+1}{2I+1}\right) \times (\langle I\pm\tfrac{1}{2},M|\rho|I\pm\tfrac{1}{2},M+1\rangle + \langle I\pm\tfrac{1}{2},M+1|\rho|I\pm\tfrac{1}{2},M\rangle). \qquad \text{...(D.17)}$$

Similarly,

$$\frac{\mathrm{d}\overline{\langle F_y \rangle}_\pm}{\mathrm{d}t} = -\frac{1}{2iT_p}\sum_M \sqrt{(I\pm\tfrac{1}{2}-M)(I\pm\tfrac{1}{2}+M+1)}\left(1 \mp \frac{2M+1}{2I+1}\right) \times (\langle I\pm\tfrac{1}{2},M|\rho|I\pm\tfrac{1}{2},M+1\rangle - \langle I\pm\tfrac{1}{2},M+1|\rho|I\pm\tfrac{1}{2},M\rangle). \qquad \text{...(D.18)}$$

The difficulty with these equations is immediately apparent. The right hand side of equation (D.17), for instance, is not simply a constant multiplying $\overline{\langle F_x \rangle}$. The effect of the pumping light on the transverse angular momentum cannot be described in terms of a single relaxation time as was the case for the spin-$\frac{1}{2}$ model of Section IV.

E. Absorption of the Pumping Light

We must now calculate the way in which changes in the ensemble average of the angular momentum affect the amount of pumping light absorbed by the optical-pumping cell. In the spin-$\frac{1}{2}$ case, we found that the amount of light absorbed was proportional to $(1 - 2\overline{\langle J_z\rangle}$ for an optically thin sample. It turns out that this is also true for a real alkali atom with nuclear spin.

The intensity of the pumping light $I(\omega, z)$ as a function of frequency ω and distance z from the front of the cell is governed by the equation

$$\frac{\partial I(\omega, z)}{\partial z} = -I(\omega, z)\, N \sum_{F,M} \sigma_{F,M}(\omega) \langle F, M|\rho(z)|\,|F, M\rangle, \qquad \text{...(E.1)}$$

where N is the density of alkali atoms and $\sigma_{F,M}(\omega)$ is defined by

$$\sigma_{F,M}(\omega) = \sigma_{0F,M} \exp\left(\frac{-Mc^2(\omega-\omega_0)^2}{2kT\omega_0^2}\right) + \sigma'_{0F,M} \exp\left(\frac{-Mc^2(\omega-\omega'_0)^2}{2kT\omega_0^2}\right). \qquad \text{...(E.2)}$$

Integrating equation (E.1), we obtain for the light transmitted by the cell

$$I(\omega, z_0) = I(\omega, 0) \exp\left(-\int_0^{z_0} N \sum_{F,M} \sigma_{F,M}(\omega) \langle F, M|\rho(\xi)|\, F, M\rangle \, \mathrm{d}\xi\right)$$
$$= I(\omega, 0) \exp\left(-Nz_0 \sum_{F,M} \sigma_{F,M}(\omega) \overline{\langle F, M|\rho|\, F, M\rangle}\right), \qquad \text{...(E.3)}$$

where $\overline{\langle F, M|\rho|\, F, M\rangle}$ is an average of the diagonal matrix element over z, and z_0 is the length of the cell. If the cell is optically thin, we can expand the exponential in equation (E.3) and obtain the intensity $I(\omega, z_0)$ transmitted by the cell

$$I(\omega, z_0) \simeq I(\omega, 0)\left[1 - z_0 N \sum_{F,M} \sigma_{F,M}(\omega) \overline{\langle F, M|\rho|\, F, M\rangle}\right]. \qquad \text{...(E.4)}$$

The total light intensity I transmitted by the cell is

$$I = \int_0^\infty I(\omega, z_0)\, \mathrm{d}\omega = \int_0^\infty I(\omega, 0)\, \mathrm{d}\omega - \frac{z_0 N}{T_p} \sum_{F,M} \left[1 \mp \frac{2M}{(2I+1)}\right] \overline{\langle F, M|\rho|\, F, M\rangle}, \qquad \text{...(E.5)}$$

where the $(\mp)$ signs refer to the $(I \pm \frac{1}{2})$ multiplets and $1/T_p$ is given by equation (D.7) with $I(\omega) = I(\omega, 0)$. But this is just

$$I = \int_0^\infty I(\omega, 0)\, \mathrm{d}\omega - \frac{z_0 N}{T_p}[1 - 2\overline{\langle J_z\rangle}], \qquad \text{...(E.6)}$$

where $\overline{\langle J_z \rangle}$ is averaged over the length of the cell. This is exactly the same result as was obtained for our spin-$\frac{1}{2}$ model.

F. SPIN RELAXATION

The principal relaxation mechanisms for $\overline{\langle \mathbf{F} \rangle}$ in the ground state of an alkali atom are the pumping light, which we have already considered, collisions with the buffer gas atoms, and collisions with the walls of the optical-pumping cell.

The coupling of the electronic and nuclear spins through the hyperfine interaction complicates the analysis of spin relaxation. It has been demonstrated experimentally (Bouchiat and Brossel, 1966; Franz, 1966; Beverini *et al.*, 1971) that collisions with an inert buffer gas or with a paraffin-coated wall randomize the electron spin without directly affecting the nuclear spin. The hyperfine interaction brings about a disorientation of the nuclear spin between collisions, and the nuclear spin in turn acts as a flywheel of polarization and slows down the relaxation of the electron spin. If we consider for the moment the randomization of the electron spin due to collisions with buffer gas atoms, it can be shown (Masnou-Seeuws and Bouchiat, 1967) that

$$\frac{\mathrm{d}\overline{\langle J_z \rangle}}{\mathrm{d}t} = -\frac{\overline{\langle J_z \rangle}}{T_g} + \frac{2\overline{\langle I_z \rangle}}{(2I+1)^2 T_g}, \qquad \text{...(F.1)}$$

and

$$\frac{\mathrm{d}\overline{\langle I_z \rangle}}{\mathrm{d}t} = -\frac{2\overline{\langle I_z \rangle}}{(2I+1)^2 T_g}, \qquad \text{...(F.2)}$$

where

$$\frac{1}{T_1} = vN_g \sigma_g, \qquad \text{...(F.3)}$$

In equation (F.3), v is the velocity of the alkali atom relative to the buffer gas atoms, N_g is the density of buffer gas atoms, and σ_g is the cross section for the randomization of the alkali atom's electron spin in a collision with a buffer gas atom. If the nuclear spin were zero, equation (F.1) would reduce to the simple relaxation equation for a spin-$\frac{1}{2}$ particle used in Section IV.

Equations (F.1) and (F.2) can be added to obtain

$$\frac{\mathrm{d}\overline{\langle F_z \rangle}}{\mathrm{d}t} = -\frac{\overline{\langle J_z \rangle}}{T_g}, \qquad \text{...(F.4)}$$

This equation would apply for the case of a very high buffer gas pressure where collisions with the cell wall are unimportant.

Relaxation by electron-randomizing collisions is effectively the same as relaxation by spin-exchange collisions with a second species of alkali atoms whose electronic polarization is maintained equal to zero. In our treatment of spin-exchange relaxation in Section VI, we will discuss this mode of relaxation and derive equations of the same form as equations (F.1) and (F.2).

In an evacuated cell with paraffin-coated walls, relaxation also occurs via electron randomization (Masnou-Seeuws and Bouchiat, 1967), and the relaxation of $\overline{\langle F_z \rangle}$ is described by

$$\frac{\mathrm{d}\overline{\langle F_z \rangle}}{\mathrm{dt}} = -\frac{\overline{\langle J_z \rangle}}{T_w}, \qquad \text{...(F.5)}$$

where T_w is the relaxation time for electron-randomizing collisions with the wall.

A third interesting situation is a cell with a low buffer gas pressure and uncoated walls. The collision of an alkali atom with an uncoated wall produces "uniform" relaxation in which any Zeeman level can be reached with equal probability in a single collision. The equations of motion of the polarizations $\overline{\langle I_z \rangle}$ and $\overline{\langle J_z \rangle}$ are (Masnou-Seeuws and Bouchiat, 1967)

$$\frac{\mathrm{d}\overline{\langle J_z \rangle}}{\mathrm{d}t} = -D\nabla^2\overline{\langle J_z \rangle} - \frac{\overline{\langle J_z \rangle}}{T_g} + \frac{2}{(2I+1)^2}\frac{\overline{\langle I_z \rangle}}{T_g}. \qquad \text{...(F.6)}$$

and

$$\frac{\mathrm{d}\overline{\langle I_z \rangle}}{\mathrm{d}t} = -D\nabla^2\overline{\langle I_z \rangle} - \frac{2}{(2I+1)^2}\frac{\overline{\langle I_z \rangle}}{T_g}, \qquad \text{...(F.7)}$$

with the boundary condition that

$$\overline{\langle F_z \rangle} = \overline{\langle I_z \rangle} = \overline{\langle J_z \rangle} = 0$$

at the walls of the cell. D is the diffusion coefficient for a given buffer gas and buffer gas pressure. The solution of the diffusion equation including all diffusion modes (Beverini *et al.*, 1971; Minguzzi *et al.*, 1966b) can be usually well approximated by employing Franzen's approximation (Franzen, 1959; Franz, 1972) and solving for the 1st diffusion mode only. When this is done the effect of diffusion to the walls can be described by a single effective relaxation time τ_w so that equations (F.6) and (F.7) become

$$\frac{\mathrm{d}\overline{\langle J_z \rangle}}{\mathrm{d}t} = -\frac{\overline{\langle J_z \rangle}}{\tau_w} + \frac{-\overline{\langle J_z \rangle}}{T_g} + \frac{2}{(2I+1)^2}\frac{\overline{\langle I_z \rangle}}{T_g}, \qquad \text{...(F.8)}$$

and

$$\frac{\mathrm{d}\overline{\langle I_z \rangle}}{\mathrm{d}t} = -\frac{\overline{\langle I_z \rangle}}{\tau_w} - \frac{2}{(2I+1)^2}\frac{\overline{\langle I_z \rangle}}{T_g}. \qquad \text{...(F.9)}$$

Adding these equations gives

$$\frac{\mathrm{d}\overline{\langle F_z \rangle}}{\mathrm{d}t} = -\frac{\overline{\langle J_z \rangle}}{T_g} - \frac{\overline{\langle F_z \rangle}}{\tau_w}. \qquad \text{...(F.10)}$$

Appealing in advance to the results for spin-exchange relaxation, we can write for the transverse relaxation of $\overline{\langle \mathbf{F} \rangle}$ due to electron-randomizing collisions with an inert buffer gas and an uncoated wall

$$\frac{\mathrm{d}\overline{\langle F_x \rangle}}{\mathrm{d}t} = -\frac{1}{T_g}\overline{\langle J_x \rangle} - \frac{\overline{\langle F_x \rangle}}{T_w}, \qquad \text{...(F.11)}$$

and

$$\frac{\mathrm{d}\overline{\langle F_y \rangle}}{\mathrm{d}t} = -\frac{1}{T_g}\overline{\langle J_y \rangle} - \frac{\overline{\langle F_y \rangle}}{T_w}. \qquad \text{...(F.12)}$$

G. THE LOW-FIELD OPTICAL-PUMPING SIGNAL

As we have seen, the response of real alkali atoms to the pumping light, the r.f. field, and spin-disorienting collisions is more complicated than was the case for the spin-$\frac{1}{2}$ model used in Section IV. A rigorous solution to the problem would require us to solve the equations of motion for the individual populations of the ground state sublevels with the aid of a computer. The form of the optical-pumping signal derived in Section IV on the basis of the spin-$\frac{1}{2}$ model, however, provides a good qualitative description of the actual alkali signals observed in the laboratory.

If we make appropriate approximations in the equations describing the optical pumping and spin relaxation of the alkali atoms, we can obtain equations of motion for the polarizations $\overline{\langle F_z \rangle}_{\pm}$, $\overline{\langle F_x \rangle}_{\pm}$, and $\overline{\langle F_y \rangle}_{\pm}$ of the same form as equations (B.63)–(B.65) in Section IV. The validity of the expression for the signal which we obtain will depend on the extent to which the approximations which we make are justified by experimental conditions.

The first of the complications which we must circumvent is the fact that the effect of the pumping light on $\overline{\langle F_x \rangle}_{\pm}$ and $\overline{\langle F_y \rangle}_{\pm}$ in equations (D.18) and (D.19) cannot be described by a single relaxation time because of the factor

$$\left(1 \mp \frac{2M+1}{2I+1}\right)$$

multiplying each density matrix element on the right hand side of these equations. There is little one can do at this point but ignore the M-dependence of this factor and describe the relaxation of $\overline{\langle F_x \rangle}_{\pm}$ and $\overline{\langle F_y \rangle}_{\pm}$ due to the pumping process by an "average" relaxation time.

Equations (F.10)–(F.12) describe the relaxation of $\overline{\langle F_z \rangle}$, $\overline{\langle F_x \rangle}$, and $\overline{\langle F_y \rangle}$ due to collisions with the walls and buffer gas atoms. The presence of the terms $\overline{\langle J_z \rangle}$, $\overline{\langle J_x \rangle}$, and $\overline{\langle J_y \rangle}$ in these equations complicates the analysis. If the buffer gas pressure is low, however, wall collisions will dominate and $\overline{\langle F_z \rangle}_{\pm}$, $\overline{\langle F_x \rangle}_{\pm}$ and $\overline{\langle F_y \rangle}_{\pm}$ will all relax independently with a single relaxation time T_w.

With these simplifications, we can combine equations (C.15)–(C.17), (D.13), (D.16), (D.17) and (F.10)–(F.12) to obtain the *approximate* equations of motion

$$\frac{d\overline{\langle F_z\rangle}_\pm}{dt} \pm \cong \pm\frac{\omega_1}{2I+1}\overline{\langle F_y\rangle}_\pm - \overline{\langle F_z\rangle}_\pm\left[\frac{1}{T_p}+\frac{1}{T_w}\right] \pm \frac{2}{2I+1}\frac{\overline{\langle F_z^2\rangle}_\pm}{T_p}, \quad \text{...(G.1)}$$

$$\frac{d\overline{\langle F_x\rangle}_\pm}{dt} \cong \pm\left(\omega-\frac{\omega_0}{2I+1}\right)\overline{\langle F_y\rangle}_\pm - \overline{\langle F_x\rangle}_\pm\left[\frac{1}{T'_{p_\pm}}+\frac{1}{T_w}\right], \quad \text{...(G.2)}$$

and

$$\frac{d\overline{\langle F_y\rangle}_\pm}{dt} \cong \mp\left(\omega-\frac{\omega_0}{2I+1}\right)\overline{\langle F_x\rangle}_\pm \mp \frac{\omega_1}{2I+1}\overline{\langle F_z\rangle}_\pm - \overline{\langle F_y\rangle}_\pm\left[\frac{1}{T'_{p_\pm}}+\frac{1}{T_w}\right], \quad \text{...(G.3)}$$

where we have assumed that wall relaxation is predominant and approximated the effect of the pumping light on $\overline{\langle F_x\rangle}_\pm$ and $\overline{\langle F_y\rangle}_\pm$ by a single "average" relaxation time $T'_{p_\pm}$.

With the exception of the pumping term

$$\frac{2}{2I+1}\frac{\overline{\langle F_z^2\rangle}_\pm}{T_p}$$

in equation (G.1), these equations now have the same form as equations (B.63)–(B.65) in Section IV. The pumping term is not simply a constant as was the case in the equation of motion for $\overline{\langle\sigma_z\rangle}$ for the spin-$\frac{1}{2}$ model in Section IV. Therefore the response of the system to the application of the r.f. field will in general be nonlinear. To obtain the equilibrium transmission signal, we set the derivatives in equations (G.1)–(G.3) equal to zero. The equilibrium polarizations $\overline{\langle F_z\rangle}_{e\pm}$ in the presence of the r.f. field are

$$\overline{\langle F_z\rangle}_{e\pm} = \pm\frac{\tau_1}{T_p}\frac{2}{2I+1}\overline{\langle F_z^2\rangle}_{e\pm}\left(1-\frac{\omega_1'^2\tau_1\tau_{2\pm}}{1+\Delta\omega^2\tau_{2\pm}^2+\omega_1'^2\tau_1\tau_{2_\pm}}\right), \quad \text{...(G.4)}$$

where

$$\omega_1' = \frac{\omega_1}{2I+1}, \quad \text{...(G.5)}$$

$$\tau_1 = \frac{1}{T_p}+\frac{1}{T_w}, \quad \text{...(G.6)}$$

$$\tau_{2_\pm} = \frac{1}{T'_{p_\pm}}+\frac{1}{T_w}, \quad \text{...(G.7)}$$

$$\Delta\omega = \omega - \frac{\omega_0}{2I+1}, \quad \text{...(G.8)}$$

and $\overline{\langle F_z^2\rangle}_{e\pm}$ is the equilibrium value of $\overline{\langle F_z^2\rangle}$.

In the absence of the r.f. field, the equilibrium values of $\overline{\langle F_z\rangle}_\pm$ and $\overline{\langle F_z^2\rangle}_\pm$ are $\overline{\langle F_z\rangle}_{0\pm}$ and $\overline{\langle F_z^2\rangle}_{0\pm}$ and we have

$$\overline{\langle F_z\rangle}_{0\pm} = \pm\frac{\tau_1}{T_p}\frac{2}{2I+1}\overline{\langle F_z^2\rangle}_{0\pm}. \qquad \text{...(G.9)}$$

The change in polarization $\delta\overline{\langle F_z\rangle}_\pm$ due to the application of the r.f. field is

$$\delta\overline{\langle F_z\rangle}_\pm = \overline{\langle F_z\rangle}_{e\pm} - \overline{\langle F_z\rangle}_{0\pm}. \qquad \text{...(G.10)}$$

If the r.f. field is weak or if the polarization is small, the change in the value of $\overline{\langle F_z^2\rangle}$ in response to the r.f. will be small, and to first approximation we can replace $\overline{\langle F_z^2\rangle}_e$ by $\overline{\langle F_z^2\rangle}_0$ in equation (G.4). If we do this, we can insert equation (G.10) into equation (G.4) to obtain the approximate expression

$$\delta\overline{\langle F_z\rangle}_\pm \cong \mp\frac{\tau_1}{T_p}\frac{2\overline{\langle F_z^2\rangle}_{0\pm}}{(2I+1)}\frac{\omega_1'^2\tau_1\tau_{2\pm}}{1+\omega_1'^2\tau_1\tau_{2\pm}+\Delta\omega^2\tau_{2\pm}^2} \qquad \text{...(G.11)}$$

From equation (E.6), we see that the change in light intensity δI transmitted by the cell due to a change in polarization is

$$\delta I = \frac{2NZ_0}{T_p}\delta\overline{\langle J_z\rangle} = \frac{2NZ_0}{T_p(2I+1)}(\delta\overline{\langle F_z\rangle}_+ - \delta\overline{\langle F_z\rangle}_-). \qquad \text{...(G.12)}$$

Equations (G.11) and (G.12) indicate that for small changes in polarization, the signal we observe is approximated by the superposition of two Lorentzian lineshapes.

The signal predicted on the basis of equations (G.12) and (G.13) is centered at the Zeeman frequency $\omega_0/(2I+1)$ of the alkali atoms and has an amplitude for optically thin samples proportional to the square of the light intensity and to the product NZ_0.

If we ignore the nonlinearities resulting from the variation of $\overline{\langle F_z^2\rangle}_\pm$ in the pumping term of equation (G.1) we could also obtain expressions for the transient optical-pumping signals similar to the expressions derived in Section IV. In particular, for strong r.f. fields we expect to observe "wiggles" occurring at a frequency $\omega_1/(2I+1)$ which is in fact what one observes.

Detailed observation of the transient signals and the lineshape of the steady state signal immediately reveals nonlinearities and the breakdown of the approximations made in the description of the effect of the pumping light and spin disorientation. However, the expressions we have arrived at are completely adequate to give qualitative understanding of the low field signals. This can be quite useful in optimizing the performance of an optical-pumping apparatus.

H. ALKALI ATOMS IN A MAGNETIC FIELD OF INTERMEDIATE STRENGTH: RESOLVED ZEEMAN TRANSITIONS

In the previous sections, we studied the behavior of optically pumped alkali atoms in a weak magnetic field. In many alkali optical-pumping experiments, however, the magnetic field is of intermediate strength. In measurements of magnetic moment ratios, for example, it is desirable to minimize the ratio of the linewidth to the resonance frequency and this is accomplished by increasing the field strength. In a field of intermediate strength, individual Zeeman transitions ($\Delta F = 0$, $\Delta M = \pm 1$) are resolved and the amplitudes of the signals can be related to the population differences between Zeeman levels.

The energy levels of an alkali atom in its ground state in a magnetic field $H_0 \hat{k}$ of intermediate strength are given by the Breit–Rabi formula.

$$E_{F,M} = -\frac{\Delta E}{2(2I+1)} - g_I \mu_0 H_0 M \pm \frac{\Delta E}{2}\sqrt{1 + \frac{4MX}{2I+1} + X^2}, \quad \text{...(H.1)}$$

where

$$X = \frac{-g_J \mu_0 + g_I \mu_0}{\Delta E} H_0, \quad \text{...(H.2)}$$

and ΔE is the difference in energy between the two hyperfine levels. The ($\pm$) signs in equation (H.1) apply for $F = I \pm \frac{1}{2}$. Because the energy levels are not a linear function of the magnetic field strength, the resonant r.f. frequency which induces Zeeman transitions between adjacent Zeeman sublevels is different for each transition. When the splitting of these transitions becomes larger than the width of the Zeeman resonances, one can resolve the individual transitions.

To get a rough idea of how the Zeeman transition frequencies behave with increasing field strength, we can expand the square root in equation (H.1) retaining terms of order X^2.

$$E_{F,M} \cong -\frac{\Delta E}{2(2I+1)} - g_I \mu_0 H_0 M \pm \frac{\Delta E}{2}\left[1 + \frac{2MX}{2I+1} + \frac{X^2}{2}\left(1 - \frac{4M^2}{(2I+1)^2}\right)\right]. \quad \text{...(H.3)}$$

The resonant frequency ω of a Zeeman transition ($F, M \leftrightarrow F, M-1$) is

$$|\omega| = \frac{|E_{F,M} - E_{F,M-1}|}{\hbar}, \quad \text{...(H.4)}$$

or approximately

$$|\omega| \cong \frac{\omega_0}{2I+1} + \frac{\omega_0^2}{(2I+1)^2}\left[\frac{1-2M}{\omega_{HFS}}\right] \pm \frac{g_I}{g_J}\omega_0 \quad \text{...(H.5)}$$

where

$$\omega_{HFS} = \frac{\Delta E}{\hbar} \qquad \text{...(H.6)}$$

$$\omega_0 = \frac{|g_J|\,\mu_0\,H_0}{\hbar}, \qquad \text{...(H.7)}$$

and the $(\mp)$ sign applies to $F = I \pm \frac{1}{2}$. We can use equation (H.5) to make a quick estimate of the splitting between adjacent Zeeman transitions.

$$|\omega|_{F,M\to F,M-1} - |\omega|_{F,M-1\to F,M-2} = -\frac{2\omega_0^2}{\omega_{HFS}(2I+1)^2} \cong -\frac{2\omega^2}{\omega_{HFS}} \qquad \text{...(H.8)}$$

Thus, the splitting of the Zeeman transitions increases approximately as twice the ratio of the square of the transition frequency to the hyperfine splitting.

The relative amplitudes of the resolved Zeeman transitions depend upon the populations of the magnetic sublevels and upon the matrix elements for the transitions. We can use time-dependent perturbation theory to calculate these amplitudes. The transition rate $\Gamma_{F,M\leftrightarrow F,M-1}$ is proportional to

$$|\langle F, M|J_x|F, M-1\rangle^2 \Delta n,$$

where Δn is the difference in the populations of the two sublevels. We can calculate these matrix elements for any F or M. As an example, the matrix elements for an alkali atom with $I = 3/2$ are listed in Table II.

TABLE II. *Values of* $|\langle F, M|J_x|F, M-1\rangle|^2$ *for* $I = 3/2$ *for different values of F and M*

F	$M \leftrightarrow$	$F,$	$M-1$	$\lvert\langle F, M\lvert J_x\rvert F, M-1\rangle\rvert^2$
2	2	2	1	2/32
2	1	2	0	3/32
2	0	2	−1	3/32
2	−1	2	−2	3/32
1	−1	1	−0	2/32
1	0	1	1	1/32
				1/32

The application of the resonant r.f. will tend to reduce the difference Δn in the population of the two levels, resulting in a change in the intensity of the pumping light transmitted by the optical-pumping cell. From equation (E.5) we see that a change in the populations of the two levels results in a change δI in the transmitted light given by

$$\delta I = \pm\frac{2Z_0 N}{T_p(2I+1)}[M\delta\langle F, M|\rho|F, M\rangle + (M-1)\,\delta\langle F, M-1|\rho|F, M-1\rangle], \qquad \text{...(H.9)}$$

where the ($\pm$) sign applies for $F = I \pm \frac{1}{2}$. Since

$$\delta\langle F, M|\rho|F, M\rangle = -\delta\langle F, M-1|\rho|F, M-1\rangle,$$

we have

$$\delta I = \pm \frac{2Z_0 N}{T_p(2I+1)} \delta\langle F, M|\rho|F, M\rangle. \quad \text{...(H.10)}$$

If the applied r.f. field is weak so that the resonance is not saturated, the change in populations $\delta\langle F, M|\rho|F, M\rangle$ will be proportional to the transition rate $\Gamma_{F,M\leftrightarrow F,M-1}$ and the observed signal will be proportional to the $|\langle F, M|J_x|F, M-1\rangle|^2$ and to the difference Δn in the populations of the two levels. The sign of the change δI in the light intensity indicates which level has the greater population. By measuring the relative amplitudes of the resolved Zeeman transitions, the population differences produced by the optical-pumping process can be studied.

If the strength of the r.f. field is increased to the point where 2nd-order time dependent perturbation theory is appropriate, one can observe double-quantum (Winter, 1959) transitions ($F, M \leftrightarrow F, M-2$). These transitions occur at frequencies intermediate between the frequencies of the ($F, M \leftrightarrow F, M-1$) and ($F, M-1 \leftrightarrow F, M-2$) transitions and are distinguished from the single-quantum transitions by their amplitude and width. The single-quantum transitions are broader and have an amplitude proportional to the r.f. power, whereas the double-quantum transition amplitude is proportional to the square of the r.f. power. Recorder tracings of the resolved Zeeman transitions for Rb^{87} ($I = 3/2$) are shown in Figs. 16 and 17.

As we have seen, the amplitudes of the resolved Zeeman transitions depend upon the relative populations of the various sublevels. The particular distribution of atoms among the various sublevels which occurs in an experiment depends of course on the characteristics of the pumping light, spin relaxation, and the degree of reorientation in the excited state (Franz, 1966, 1969, 1970; Franz and Franz, 1966; Fricke *et al.*, 1967; Violino, 1968; Papp and Franz, 1972). Franzen and Emslie (1957) were the first to compute equilibrium sublevel populations for alkali atoms optically pumped by circularly polarized D_1 radiation. They made their computation for $I = 3/2$ assuming long relaxation times and high light intensity, and they considered two ideal cases: no reorientation in the excited state, and complete reorientation in the excited state. Although their assumptions do not usually correspond closely with experimental conditions, it is instructive to note that the assumption of no reorientation in the excited state leads to an inversion in the ordering of the populations of the lower hyperfine multiplet relative to the ordering obtained assuming complete reorientation. For complete reorientation, the $|F = 1, M = -1\rangle$ has the largest population in the lower multiplet and the $|F = 1, M = 1\rangle$ state the least, as we noted in Subsection D. With no reorientation, the return of the atoms from the excited state to the ground state depends upon their initial ground state sublevel and the result is an equilibrium population ordering in the lower multiplet with the $|F = 1, M = 1\rangle$ state having the largest population.

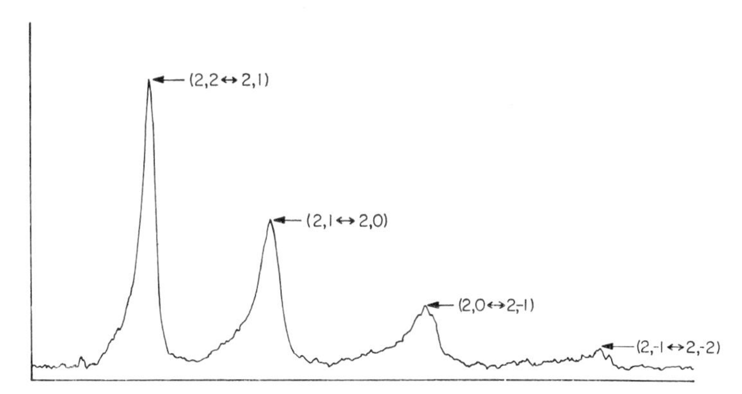

FIG. 16. A chart recording of the resolved Zeeman transitions in the $F = 2$ multiplet of Rb^{87}. The Zeeman frequencies are ~ 1·67 MHz.

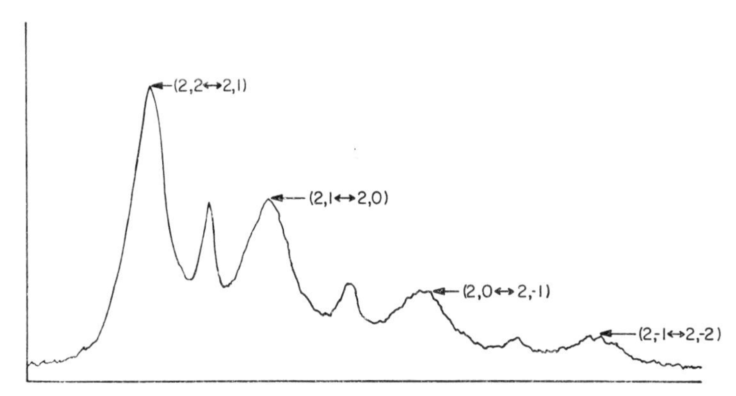

FIG. 17. A chart recording of the same Zeeman transitions as in Fig. 16 with an increase in r.f. power of 18 dB. The narrow double-quantum transitions which occur between the single quantum transitions are easily seen.

The practical situation does not usually conform to either of these two extremes. The assumption of complete reorientation in the excited state is probably better for Na than for the heavier alkalis such as Rb and Cs.

In the foregoing discussion, we have tacitly assumed that the changes in the populations of two sublevels coupled by a resonant r.f. field do not affect the populations of other levels. If the alkali atom density is increased to the point where spin-exchange collisions between the alkali atoms become an important factor, this assumption breaks down (along with our assumption that the sample is optically thin). Although a detailed discussion of spin-exchange collisions will be given in Section VI, it is appropriate at this point to note one of the effects of strong spin-exchange coupling between the optically-pumped alkali atoms.

If the time between spin-exchange collisions is short compared to the pumping time (Anderson and Ramsey, 1963), the spin-exchange collisions bring about an ordering of the populations of the lower hyperfine mutliplet so that the state with the highest value of M has the largest population even if the pumping process tends to populate the states in the opposite order. In this situation, the least absorbing level has the highest population. Anderson and Ramsey (1963) have performed an interesting experiment under these conditions. They continuously applied an r.f. field and swept the magnetic field through the individual Zeeman resonances. Upon sweeping through the transitions in the lower hyperfine multiplet, they observed an initial increase in the transmitted light intensity due to the inverse ordering of the populations followed by a decrease in light intensity. The decrease in light intensity is brought about by the spin-exchange coupling of the populations of the levels involved in the r.f. transition to the populations of the additional sublevels. The applied r.f. alters the populations of levels not directly involved in the transition, via spin-exchange collisions, and this results in a decrease in the light intensity transmitted by the cell. The time delay between the transitory increase in transparency and the subsequent decrease is of the order of the spin-exchange collision time.

I. OPTICAL-PUMPING SIGNALS UNDER VARYING PUMPING LIGHT CONDITIONS

The calculations of the optical-pumping signals given in this section thus far have been based on the assumptions that the pumping radiation contained hyperfine components of equal intensity, was incident along the z-axis, was left circularly polarized (σ_+), and was entirely D_1 radiation. With these restrictions, the amount of light absorbed in an optically thin sample of alkali atoms was found to be proprotional to $(1 - 2\overline{\langle J_z \rangle})$.

It is important to know what observables are represented by the absorption signal under other pumping light conditions. Bouchiat (1965) has computed the absorption of pumping light by an optically thin alkali vapor in a weak magnetic field for a number of different spectral profiles of the pumping radiation and her results are summarized in Table III.

Of particular interest is the situation where the intensities of the hyperfine components of the incident radiation are not equal. In this case, even if there is no Zeeman pumping (unpolarized light) the absorption of radiation pumps the atoms out of one hyperfine level and into the other. The absorption signal depends upon the difference in the populations of the two hyperfine levels. This

TABLE III. *The absorption of the pumping light under various light source conditions*

Spectral profile of pumping radiation	Polarization and direction of pumping radiation	Absorption proportional to
D-1 with equal intensity hyperfine components	$\sigma\pm$ in z-direction	$(1 \mp 2\overline{\langle J_z \rangle})$
	$\sigma\pm$ in x-direction	$(1 \mp 2\overline{\langle J_x \rangle})$
D-2 with equal intensity hyperfine components	$\sigma\pm$ in z-direction	$(1 \pm \overline{\langle J_z \rangle})$
	$\sigma\pm$ in x-direction	$(1 \pm \overline{\langle J_x \rangle})$
D-1 or D-2 with unequal hyperfine components	Unpolarized	$L_0 + K\overline{\langle \mathbf{J}.\mathbf{I} \rangle}$*
	$\sigma\pm$ in z-direction	Linear combination of $\overline{\langle \mathbf{J}.\mathbf{I} \rangle}$ $\mathrm{Tr}\,\rho_+ J_z$, $\mathrm{Tr}\,\rho_- J_z$
	$\sigma\pm$ in x-direction	Linear combination of $\overline{\langle \mathbf{J}.\mathbf{I} \rangle}$ $\mathrm{Tr}\,\rho_+ J_x$, $\mathrm{Tr}\,\rho_- J_x$

* The constant L_0 is the amount of light absorbed when the *alkali* atoms are in thermal equilibrium ($\overline{\langle \mathbf{J}\cdot\mathbf{I} \rangle} = 0$) and K is proportional to the difference in intensities of the two hyperfine components.

population difference is simply related to the ensemble average $\overline{\langle \mathbf{J}.\mathbf{I} \rangle}$ as can be seen in the following way.

$$\begin{aligned}\overline{\langle \mathbf{J}.\mathbf{I} \rangle} &= \overline{\langle \tfrac{1}{2}[F^2 - I^2 - J^2] \rangle} \\ &= \sum_{F,M} \langle F, M | \tfrac{1}{2}[F(F+1) - I(I+1) - J(J+1)]\,\rho | F, M \rangle \\ &= \sum_M \tfrac{1}{2} I \langle I+\tfrac{1}{2}, M | \rho | I+\tfrac{1}{2}, M \rangle - \sum_M \tfrac{1}{2}(I+1) \langle I-\tfrac{1}{2}, M | \rho | I-\tfrac{1}{2}, M \rangle.\end{aligned} \quad \text{...(I.1)}$$

Hence,

$$\overline{\langle \mathbf{J}.\mathbf{I} \rangle} = \tfrac{1}{2} I P_+ - \tfrac{1}{2}(I+1) P_-, \quad \text{...(I.2)}$$

where P_+ and P_- are the fractional populations of the upper and lower hyperfine levels, respectively. Since

$$P_+ + P_- = 1, \quad \text{...(I.3)}$$

we have

$$\overline{\langle \mathbf{J}.\mathbf{I} \rangle} = \tfrac{1}{4}[(2I+1)(P_+ - P_-) - 1]. \quad \text{...(I.4)}$$

In thermal equilibrium, all of the magnetic sublevels are equally populated and

$$P_+ - P_- = \frac{1}{2I+1}, \quad \text{...(I.5)}$$

and

$$\overline{\langle \mathbf{J}.\mathbf{I} \rangle} = 0. \qquad \text{...(I.6)}$$

If the pumping radiation (D_1 or D_2) is unpolarized, or if Zeeman pumping is nullified by application of an r.f. field at the Zeeman frequency, the signal will depend only on $\overline{\langle \mathbf{J}.\mathbf{I} \rangle}$. If the radiation is circularly polarized and there is also hyperfine pumping, the signal will be some linear combination of $\overline{\langle \mathbf{J}.\mathbf{I} \rangle}$ and $\overline{\langle J_z \rangle}$ evaluated in each hyperfine level. Particularly in experiments designed to measure relaxation times, it is important to know just which observables play a role in the absorption signal in order to interpret the data correctly.

J. HYPERFINE TRANSITIONS

The signals obtained from hyperfine transitions can be understood in the same way as resolved Zeeman transitions. Even if the hyperfine components of the pumping radiation are equal so that there is no hyperfine pumping

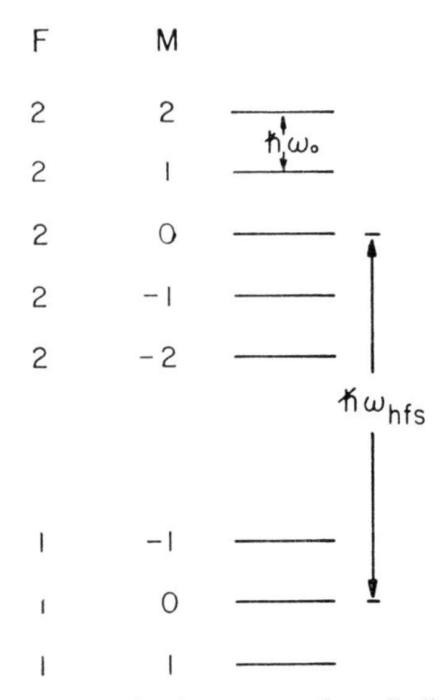

FIG. 18. The ground state energy level structure of an alkali atom with $I = \frac{3}{2}$ in a weak magnetic field is shown schematically.

($F, M \leftrightarrow F \pm 1, M \pm 1$) transitions can be observed because of the polarization of the sample. In a weak magnetic field, some of these transitions will be degenerate. The case of $I = 3/2$ is shown schematically in Fig. 18. The transition frequencies for $\Delta F = \pm 1$, $\Delta M = \pm 1$ with $I = 3/2$ are given in Table IV.

As the field is increased to the point where individual Zeeman transitions can be resolved, the degeneracy in the hyperfine transition frequencies is lifted.

TABLE IV. *Hyperfine transition frequencies* ($\Delta M = \pm 1$) *for* $I = 3/2$

Transition				Frequency
F	$M \leftrightarrow$	F'	M'	
2	2	1	1	$\omega_{\text{HFS}} + 3\omega_0$
2	1	1	0	$\omega_{\text{HFS}} + \omega_0$
2	0	1	−1	$\omega_{\text{HFS}} - \omega_0$
2	−1	1	0	$\omega_{\text{HFS}} - \omega_0$
2	−2	1	−1	$\omega_{\text{HFS}} - 3\omega_0$

If the hyperfine components of the pumping radiation are unequal, there will be hyperfine pumping and it is possible to observe $\Delta F = \pm 1$, $\Delta M = 0$ transitions. To induce these transitions, one uses an r.f. field which oscillates in the z-direction. Of particular interest is the 0–0 transition ($I + \frac{1}{2}, 0 \leftrightarrow I - \frac{1}{2}, 0$). As can be seen from equation (H.1), the 1st-order field dependence of this transition frequency vanishes. Therefore, any broadening of the resonance by inhomogeneities in the steady magnetic field is only a 2nd-order effect.

In summary, the hyperfine structure considerably complicates the analysis of optical pumping signals for alkali atoms. The low-field alkali atom signal, which is the superposition of several simultaneous Zeeman transitions, is qualitatively the same as that calculated for a spin-$\frac{1}{2}$ system but this description is approximate at best. For precision magnetic moment measurements at higher fields and for hyperfine frequency measurements, one generally deals with individual transitions between two sublevels. To fully understand the form and amplitudes of these signals, one must understand the details of the optical pumping cycle including what goes on in the excited state as well as the effects of relaxation and spin-exchange on the populations of the individual sublevels. This is a nontrivial task.

VI. SPIN-EXCHANGE OPTICAL PUMPING

In the preceding sections we used the density matrix formalism to arrive at theoretical expressions for the signals one observes in a standard optical-pumping experiment. In this chapter we will make use of the same basic technique to calculate the signals obtained in spin-exchange optical-pumping experiments. Although the principles of spin-exchange optical pumping were sketched in Section II it is important to understand in detail the effect of spin-exchange collisions upon the observed resonance signals.

Not only are spin-exchange collisions an important relaxation mechanism, but they also can shift the frequency of r.f. transitions. A theoretical understanding of these effects minimizes a potential source of error in precision experiments. In addition, the spin-exchange collision process itself has considerable intrinsic interest to the atomic physicist, and with a proper understanding of the spin-exchange optical-pumping signal, experimental studies of spin-exchange collisions can be undertaken.

A great number of papers have been written on the subject of spin-exchange collisions, and a single section will not suffice to cover everything contained in the literature. It is hoped, however, that this section will equip the reader to pursue the subject further.

A. SPIN EXCHANGE BETWEEN TWO SPECIES OF SPIN-$\frac{1}{2}$ PARTICLES

Once again we will begin our discussion by considering the particularly simple case of a spin-$\frac{1}{2}$ system. We imagine that fictitious alkali atoms with zero nuclear spin are optically pumped in a magnetic field $H_0\hat{k}$ and that free electrons are polarized by spin-exchange collisions with the alkali atoms.

The spin-exchange collisions between the free electrons and the optically pumped alkali atoms polarize the electrons. An r.f. field $2H_1\hat{\imath}\cos\omega t$ is applied to disorient the electrons, and the spin-exchange collisions partially depolarize the alkali atoms producing a change in the intensity of the pumping light transmitted by the cell. Our goal is to calculate the form of this resonance signal. To do this, we must calculate the time dependence of the spin-space density matrix for the electron-alkali atom system under the influence of the spin-exchange collisions, the static and r.f. magnetic fields, additional spin-relaxation mechanisms, and the pumping light. Actually, the only process with which we have not yet dealt is spin exchange.

The density matrices describing the electron and fictitious alkali atom spin systems are both 2×2 matrices. Because the spin of the atoms and electrons are assumed to be statistically independent between collisions, the density matrix $\rho(e, A)$ for the combined electron–atom spin system is just the outer product of the individual electron and alkali-atom density matrices $\rho(e)$ and $\rho(A)$. That is, $\rho(e, A)$ will be a 4×4 matrix given by the outer product

$$\rho(e, A) = \rho(e) \times \rho(A), \qquad \text{...(A.1)}$$

and

$$\rho(e) = \mathrm{Tr}_A\, \rho(e, A), \qquad \text{...(A.2)}$$

and

$$\rho(A) = \mathrm{Tr}_e\, \rho(e, A), \qquad \text{...(A.3)}$$

where Tr_A and Tr_e are traces taken over the atom and electron spin coordinates, respectively.

The pumping light is assumed to pass through the cell parallel to the static magnetic field $H_0\hat{k}$. Therefore the z-axis is the axis of quantization, and the density matrix for the atom spin system has the form

$$\rho(A) = \begin{bmatrix} \rho_{11}(A) & 0 \\ 0 & \rho_{22}(A) \end{bmatrix}, \qquad \text{...(A.4)}$$

with

$$\overline{\langle \sigma_z(A) \rangle} = \rho_{11}(A) - \rho_{22}(A). \qquad \text{...(A.5)}$$

$\overline{\langle \sigma_z(A) \rangle}$ is the polarization of the alkali atoms produced by the optical-pumping process. The off-diagonal elements are zero because there is no resonant r.f. field to produce a coherent precession of the atoms about the static field. That is,

$$\overline{\langle \sigma_x(A) \rangle} = \overline{\langle \sigma_y(A) \rangle} = 0.$$

The density matrix for the electron system, on the other hand, will have non-vanishing off-diagonal elements as a result of the r.f. field applied at the electron spin resonance frequency.

$$\rho(e) = \begin{bmatrix} \rho_{11}(e) & \rho_{12}(e) \\ \rho_{21}(e) & \rho_{22}(e) \end{bmatrix}, \qquad \text{...(A.6)}$$

with

$$\overline{\langle \sigma_z(e) \rangle} = \rho_{11}(e) - \rho_{22}(e). \qquad \text{...(A.7)}$$

The spin-space density matrix $\rho(e, A)$ for the electron-alkali atom spin system is given by

$$\rho(e, A) = \begin{bmatrix} \rho_{11}(e)\,\rho_{11}(A) & 0 & \rho_{12}(e)\,\rho_{11}(A) & 0 \\ 0 & \rho_{11}(e)\,\rho_{22}(A) & 0 & \rho_{12}(e)\,\rho_{22}(A) \\ \rho_{21}(e)\,\rho_{11}(A) & 0 & \rho_{22}(e)\,\rho_{11}(A) & 0 \\ 0 & \rho_{21}(e)\,\rho_{22}(A) & 0 & \rho_{22}(e)\,\rho_{22}(A) \end{bmatrix}. \qquad \text{...(A.8)}$$

We will use equations (D.15) and (D.16) of Section III to calculate the time rate of change of $\rho(e)$ and $\rho(A)$ due to spin-exchange collisions. The first step is to expand $\rho(e, A)$ to include momentum states for the relative momentum of the colliding particles. If we denote the density matrix including momentum states before a collision as $\rho(e, A, \mathbf{k})$ and after a collision as $\rho'(e, A, \mathbf{k})$, then the change in $\rho(e)$ due to a collision between each electron and a single alkali atom is given by equation (D.15) in Section III.

$$\rho'(e) = \mathrm{Tr}_{A,\mathbf{k}'}\, \mathbf{S}\rho(e, A, \mathbf{k})\mathbf{S}^{\dagger} = \mathrm{Tr}_{A,\mathbf{k}'}\, \rho'(e, A, \mathbf{k}') = \mathrm{Tr}_A\, \rho'(e, A), \qquad \text{...(A.9)}$$

where Tr_A is a trace over the spin coordinates of the alkali atoms and $\rho'(e)$ is the spin-space density matrix for the electrons after the collision. After a single collision between each alkali atom and an electron, the density matrix $\rho'(A)$ for the alkali atom spin is given by

$$\rho'(A) = \mathrm{Tr}_e\, \rho'(e, A), \qquad \text{...(A.10)}$$

where the trace is over the electron spin coordinate.

We will assume that the collision is elastic and make the usual reduction of the two-body problem to an effective one-body problem by introducing the reduced mass μ and working with the relative coordinate $\mathbf{r}$ and relative momentum $\hbar\mathbf{k}$. The problem is now one of a single particle scattering off a fixed target located at the origin. Since the alkali atom is orders of magnitude more massive than the electron, the reduced mass and relative momentum are essentially the mass and momentum of the free electron in the laboratory reference frame.

Since an arbitrary solution to the Schrödinger equation can be constructed by a superposition of stationary states, is it sufficient to consider asymptotic solutions to the time-independent Schrödinger equation of the form

$$\psi(\mathbf{r},\mathbf{k})\,|s\rangle = \frac{1}{L^{3/2}}\left\{ e^{ik_0 r}|s_0\rangle + \frac{e^{ikr}}{r} M_{s,s_0}(\mathbf{k},\mathbf{k}_0)|\,s_0\rangle \right\}, \qquad \ldots(\text{A.11})$$

where $\hbar\mathbf{k}_0$ is the momentum of the incoming electron, $\hbar\mathbf{k}$ is the momentum of the outgoing electron, $|s_0\rangle$ is the initial electron–alkali atom spin state and $|s\rangle$ is the final electron–alkali atom spin state. The normalization is one electron in a box of length L. The matrix $M_{s,s_0}(\mathbf{k},\mathbf{k}_0)$ is a generalization of the scattering amplitude used in elementary treatments of scattering which neglect the effect of spin.

Because the wavefunction for the electron–alkali atom system must be antisymmetric in the electron coordinates, the scattering amplitude will depend upon whether the incoming electron and valence electron spins form a singlet or a triplet state. The matrix $M_{s,s_0}(\mathbf{k},\mathbf{k}_0)$ is a matrix in spin space. If we assume there is no spin-orbit coupling in the collision, then

$$M = f_3(\theta)\,P_3 + f_1(\theta)\,P_1, \qquad \ldots(\text{A.12})$$

where $f_3(\theta)$ and $f_1(\theta)$ are the triplet and singlet scattering amplitudes, respectively, θ is the angle between $\mathbf{k}$ and $\mathbf{k}_0$, and P_3 and P_1 are the projection operators for the triplet and singlet spin states. In terms of the Pauli spin matrices,

$$P_3 = \tfrac{1}{4}[3 + \boldsymbol{\sigma}(e).\boldsymbol{\sigma}(A)], \qquad \ldots(\text{A.13})$$

and

$$P_1 = \tfrac{1}{4}[1 - \boldsymbol{\sigma}(e).\boldsymbol{\sigma}(A)], \qquad \ldots(\text{A.14})$$

The operators $\sigma(e)$ and $\sigma(A)$ operate on the free electron and alkali atom spin coordinates, respectively.

By expanding $\psi(\mathbf{r},\mathbf{k})\,|s\rangle$ in partial waves, we can write the scattering amplitudes in the form

$$f_3(\theta) = \frac{1}{2ik}\sum_{l=0}^{\infty}(2l+1)\,(e^{2i\delta_l^3} - 1)\,P_l(\cos\theta), \qquad \ldots(\text{A.15})$$

and

$$f_1(\theta) = \frac{1}{2ik}\sum_{l=0}^{\infty}(2l+1)\,(e^{2i\delta_l^1} - 1)\,P_l(\cos\theta), \qquad \ldots(\text{A.16})$$

where

$$k = |\mathbf{k}| = |\mathbf{k}_0|,$$

and δ_l^3 and δ_l^1 are the triplet and singlet phase shifts for the l-th partial wave.

The matrix elements of the S-matrix are related to the matrix elements of M by (Merzbacher, 1961)

$$\langle \mathbf{k}' s' |\mathbf{S}| \mathbf{k} s\rangle = \delta_{\mathbf{k}',\mathbf{k}}\,\delta_{s',s} + 2\pi i\left(\frac{2\pi\hbar^2}{\mu L^3}\right)\delta(E-E')\,M_{s',s}(\mathbf{k}',\mathbf{k}), \quad \ldots(\text{A.17})$$

where $\delta(E-E')$ is a Dirac delta function of energy with

$$E = \frac{(\hbar k)^2}{2u}\cdot \quad \ldots(\text{A.18})$$

We can now proceed to calculate the time derivatives of $\rho(e)$ and $\rho(A)$ due to spin-exchange collisions. Using equation (A.9), we have for a particular matrix element $\rho'_{s\bar{s}}(e,A)$ of the electron–alkali atom spin space density matrix after a collision, the expression

$$\rho'_{s\bar{s}}(e,A) = \sum_{s',s''}\sum_{\mathbf{k},\mathbf{k}',\mathbf{k}''} \langle \mathbf{k}s|\mathbf{S}|\,\mathbf{k}'s'\rangle\langle \mathbf{k}'s'|\rho|\,\mathbf{k}''s''\rangle\langle \mathbf{k}''s''|\mathbf{S}^\dagger|\,\mathbf{k}s\rangle. \quad \ldots(\text{A.19})$$

To simplify the arithmetic, we will assume that all electrons move with the same velocity so that

$$\langle \mathbf{k}'s'|\rho|\,\mathbf{k}''s''\rangle = \delta_{\mathbf{k}',\mathbf{k}_0}\,\delta_{\mathbf{k}_0,\mathbf{k}''}\,\rho_{s's''}(e,A). \quad \ldots(\text{A.20})$$

If we insert equations (A.20) and (A.17) into equation (A.19), we obtain

$$\begin{aligned}\rho'_{s\bar{s}}(e,A) = \sum_{s's''k}\Big[&\delta_{\mathbf{k},\mathbf{k}_0}\,\delta_{\mathbf{k},\mathbf{k}_0}\,\delta_{ss'}\,\rho_{s's''}(e,A)\,\delta_{s''\bar{s}}\\ &+2\pi i\left(\frac{2\pi\hbar^2}{uL^3}\right)\delta(E-E_0)\,M_{ss'}(\mathbf{k}_0,\mathbf{k}_0)\,\rho_{s's''}(e,A)\,\delta_{s''\bar{s}}\,\delta_{\mathbf{k},\mathbf{k}_0}\\ &-2\pi i\left(\frac{2\pi\hbar^2}{uL^3}\right)\delta(E-E_0)\,\delta_{ss'}\,\rho_{s's''}(e,A)\,M^\dagger_{s''\bar{s}}(\mathbf{k}_0;\mathbf{k}_0)\,\delta_{\mathbf{k}_0,\mathbf{k}}\\ &+(2\pi)^2\left(\frac{2\pi\hbar^2}{uL^3}\right)^2\delta(E-E_0)\,\delta(E-E_0)\,M_{ss'}(\mathbf{k},\mathbf{k}_0)\,\rho_{s's''}(e,A)\,M^\dagger_{s''\bar{s}(\mathbf{k}_0,\mathbf{k})}\Big]\cdot\end{aligned} \quad \ldots(\text{A.21})$$

When we sum over $\mathbf{k}$, we obtain terms involving $\delta(E_0-E_0)=\delta(0)$. In order to interpret this expression, we note that in the derivation of equation (A.17) the delta function arises from the time integral

$$\delta(E-E_0) = \lim_{t\to\infty}\frac{1}{2\pi}\int_{-\infty}^{t} e^{-i(E-E_0)t/\hbar}\,\frac{\mathrm{d}t}{\hbar}\cdot \quad \ldots(\text{A.22})$$

In accordance with this equation, we can interpret $\delta(0)$ as being $T/2\pi\hbar$, where T is the time elapsed in the scattering process. When evaluating

$$\sum_{\mathbf{k}} \delta(E-E_0)\,\delta(E-E_0),$$

we let $L \to \infty$, so that

$$\sum_{\mathbf{k}} \to L^3 \int \frac{d^3 k}{(2\pi)^3} = \frac{L^3}{(2\pi)^3} \int \left(\frac{2uE}{\hbar^2}\right)^{1/2} \frac{u}{\hbar^2}\, dE\, d\Omega. \qquad \text{...(A.23)}$$

Having summed equation (A.21) over $\mathbf{k}$, we obtain

$$\begin{aligned} \rho'_{s\bar{s}}(e,A) = \rho_{s\bar{s}}(e,A) + T\left(\frac{\hbar k}{uL^3}\right) \\ \times \left\{\frac{2\pi i}{k}\left[\sum_{s'} M_{ss'}(\mathbf{k}_0,\mathbf{k}_0)\,\rho_{s'\bar{s}}(e,A) - \sum_{s''} \rho_{ss''}(e,A)\,M^{\dagger}_{s''\bar{s}}(\mathbf{k}_0,\mathbf{k}_0)\right]\right. \\ \left. + \int d\Omega \sum_{s''s'} M_{ss'}(\mathbf{k},\mathbf{k}_0)\,\rho_{s's''}(e,A)\,M^{\dagger}_{s'\bar{s}''}(\mathbf{k}_0,\mathbf{k})\right\}, \qquad \text{...(A.24)} \end{aligned}$$

where the angular integral is over the angle between $\mathbf{k}$ and $\mathbf{k}_0$ and $|\mathbf{k}| = |\mathbf{k}_0|$.

We can rewrite this as

$$\begin{aligned} \rho'(e,A) = \rho(e,A) + \frac{Tv}{L^3} \\ \times \left\{\frac{2\pi i}{k}[M(0)\,\rho(e,A) - \rho(e,A)\,M^{\dagger}(0)] + \int d\Omega\, M(\theta)\,\rho(e,A)\,M^{\dagger}(\theta)\right\}, \end{aligned} \qquad \text{...(A.25)}$$

where v is the velocity of the free electrons. The change in $\rho(e)$ due to a collision between each electron and a single alkali atom is

$$\rho'(e) - \rho(e) = \mathrm{Tr}_A[\rho'(e,A) - \rho(e,A)]. \qquad \text{...(A.26)}$$

If we divide by the collision time T and multiply by the number of alkali atoms in the volume L^3, we obtain for the time rate of change of $\rho(e)$ due to spin-exchange collision

$$\begin{aligned} \frac{d\rho(e)}{dt} = N_A v\, \mathrm{Tr}_A \\ \times \left\{\frac{2\pi i}{k}[M(0)\,\rho(e,A) - \rho(e,A)\,M^{\dagger}(0)] + \int d\Omega\, M(\theta)\,\rho(e,A)\,M^{\dagger}(\theta)\right\}, \end{aligned} \qquad \text{...(A.27)}$$

where N_A is the density of alkali atoms, Similarly,

$$\begin{aligned} \frac{d\rho(A)}{dt} = N_e v\, \mathrm{Tr}_e \\ \times \left\{\frac{2\pi i}{k}[M(0)\,\rho(e,A) - \rho(e,A)\,M^{\dagger}(0)] + \int d\Omega\, M(\theta)\,\rho(e,A)\,M^{\dagger}(\theta)\right\}, \end{aligned} \qquad \text{...(A.28)}$$

where N_e is the density of free electrons.

If we use equation (A.12) and introduce the spin-exchange cross section σ_{SE} and define a shift parameter κ with the equations

$$\sigma_{\text{SE}} = \tfrac{1}{4} \int \mathrm{d}\Omega |f_3 - f_1|^2 = \frac{\pi}{k^2} \sum_{l=0}^{\infty} (2l+1) \sin^2(\delta_l^3 - \delta_l^1), \qquad \text{...(A.29)}$$

and

$$\kappa = \frac{1}{\sigma_{\text{SE}}} \frac{\pi}{2k^2} \sum_{l=0}^{\infty} (2l+1) \sin 2(\delta_l^3 - \delta_l^1), \qquad \text{...(A.30)}$$

we can rewrite equation (A.27) in the form

$$\frac{\mathrm{d}\rho(e)}{\mathrm{d}t} = \tfrac{1}{4} v N_A \sigma_{\text{SE}} \operatorname{Tr}_A [-3\rho(e, A) + (1 + 2i\kappa)\boldsymbol{\sigma}(e) . \boldsymbol{\sigma}(A) \rho(e, A)$$
$$+ (1 - 2i\kappa)\rho(e, A)\boldsymbol{\sigma}(e) . \boldsymbol{\sigma}(A) + \boldsymbol{\sigma}(e) . \boldsymbol{\sigma}(A) \rho(e, A) \boldsymbol{\sigma}(e) . \boldsymbol{\sigma}(A)]. \qquad \text{...(A.31)}$$

This form is convenient for calculational purposes. For our spin-$\frac{1}{2}$ system,

$$\boldsymbol{\sigma}(e) . \boldsymbol{\sigma}(A) = \begin{bmatrix} 1 & 0 & 0 & 0 \\ 0 & -1 & 2 & 0 \\ 0 & 2 & -1 & 0 \\ 0 & 0 & 0 & 0 \end{bmatrix}. \qquad \text{...(A.32)}$$

After some matrix multiplication using the matrix for $\rho(e, A)$ given in equation (A.8), we obtain from equation (A.31)

$$\frac{\mathrm{d}\rho(e)}{\mathrm{d}t} = \begin{bmatrix} \dfrac{\overline{\langle\sigma_z(A)\rangle} - \overline{\langle\sigma_z(e)\rangle}}{2T_e} & -\dfrac{1 - i\kappa\overline{\langle\sigma_z(A)\rangle}}{T_e} \rho_{12}(e) \\ -\dfrac{1 + i\kappa\overline{\langle\sigma_z(A)\rangle}}{T_e} \rho_{21}(e) & \dfrac{\overline{\langle\sigma_z(e)\rangle} - \overline{\langle\sigma_z(A)\rangle}}{2T_e} \end{bmatrix}, \qquad \text{...(A.33)}$$

where

$$\frac{1}{T_e} = N_A v \sigma_{\text{SE}}. \qquad \text{...(A.34)}$$

Similarly, for $\mathrm{d}\rho(A)/\mathrm{d}t$ we have

$$\frac{\mathrm{d}\rho(A)}{\mathrm{d}t} = \begin{bmatrix} \dfrac{\overline{\langle\sigma_z(e)\rangle} - \overline{\langle\sigma_z(A)\rangle}}{2T_{eA}} & 0 \\ 0 & \dfrac{\overline{\langle\sigma_z(A)\rangle} - \overline{\langle\sigma_z(e)\rangle}}{2T_{eA}} \end{bmatrix}, \qquad \text{...(A.35)}$$

where

$$\frac{1}{T_{eA}} = N_e v \sigma_{\text{SE}}. \qquad \text{...(A.36)}$$

Equations (A.33) and (A.35) clearly show the coupling of the free electron and alkali atom polarizations by the spin-exchange collisions. A surprising feature of equation (A.33) is the presence of the imaginary terms in the off-diagonal elements of $\mathrm{d}\rho(e)/\mathrm{d}t$. As was the case for the light intensity shifts, these terms will lead to a shift in the frequency of the observed spin resonance.

We can now use equations (A.33) and (A.35) to compute the time derivatives of the longitudinal and transverse polarizations of the electrons and alkali atoms due to spin-exchange collisions.

$$\frac{\mathrm{d}\overline{\langle\sigma_z(e)\rangle}}{\mathrm{d}t} = \mathrm{Tr}\,\sigma_z \frac{\mathrm{d}\rho(e)}{\mathrm{d}t} = \frac{\overline{\langle\sigma_z(A)\rangle} - \overline{\langle\sigma_z(e)\rangle}}{T_e}. \qquad \text{...(A.37)}$$

Similarly,

$$\frac{\overline{\langle\sigma_x(e)\rangle}}{\mathrm{d}t} = -\frac{\overline{\langle\sigma_x(e)\rangle}}{T_e} + \kappa \frac{\overline{\langle\sigma_y(e)\rangle}\,\overline{\langle\sigma_z(A)\rangle}}{T_e}, \qquad \text{...(A.38)}$$

and

$$\frac{\mathrm{d}\overline{\langle\sigma_y(e)\rangle}}{\mathrm{d}t} = -\frac{\overline{\langle\sigma_y(e)\rangle}}{T_e} - \kappa \frac{\overline{\langle\sigma_x(e)\rangle}\,\overline{\langle\sigma_z(A)\rangle}}{T_e}. \qquad \text{...(A.39)}$$

For the alkali atom polarization, we have

$$\frac{\mathrm{d}\overline{\langle\sigma_z(A)\rangle}}{\mathrm{d}t} = \frac{\overline{\langle\sigma_z(e)\rangle} - \overline{\langle\sigma_z(A)\rangle}}{T_{eA}}. \qquad \text{...(A.40)}$$

B. THE SPIN-EXCHANGE OPTICAL-PUMPING SIGNAL FOR THE SPIN-$\frac{1}{2}$ SYSTEM

Having calculated the effect of spin-exchange collisions on the electron and alkali density matrices, we have established the way in which the polarizations of the two spin systems are coupled by spin exchange. This coupling is described by equations (A.37)–(A.40). If we now include the effects of the pumping light, the r.f. field, and spin relaxation on the electron and alkali polarizations, we will arrive at an expression for the electron resonance signal. Bearing in mind that the pumping light directly affects only the alkali atom density matrix and the r.f. field affects only the electron density matrix, we can, in analogy to equations (B.63)–(B.65) of Section IV, immediately write down equations for the time rate of change of $\overline{\langle\boldsymbol{\sigma}(e)\rangle}$ and $\overline{\langle\boldsymbol{\sigma}(A)\rangle}$ in a coordinate system rotating about the z-axis with angular velocity ω. Using primes to denote the values of the polarizations in a rotating frame, we have

$$\frac{\mathrm{d}\overline{\langle\sigma_z(e)\rangle}'}{\mathrm{d}t} = \frac{\overline{\langle\sigma_z(A)\rangle}' - \overline{\langle\sigma_z(e)\rangle}'}{T_e} - \frac{\overline{\langle\sigma_z(e)\rangle}'}{T_{1e}} + \omega_1\overline{\langle\sigma_y(e)\rangle}', \qquad \text{...(B.1)}$$

$$\frac{\mathrm{d}\overline{\langle\sigma_y(e)\rangle}'}{\mathrm{d}t} = (\omega_0 - \delta\omega_0 - \omega)\overline{\langle\sigma_x(e)\rangle}' - \omega_1\overline{\langle\sigma_z(e)\rangle}' - \overline{\langle\sigma_y(e)\rangle}'\left(\frac{1}{T_e} + \frac{1}{T_{2e}}\right), \qquad \text{...(B.2)}$$

$$\frac{\mathrm{d}\overline{\langle\sigma_x(e)\rangle}'}{\mathrm{d}t} = -(\omega_0 - \delta\omega_0 - \omega)\overline{\langle\sigma_y(e)\rangle}' - \overline{\langle\sigma_x(e)\rangle}'\left(\frac{1}{T_e} + \frac{1}{T_{2e}}\right), \qquad \text{...(B.3)}$$

where the spin-exchange frequency shift $\delta\omega_0$ is given by

$$\delta\omega_0 = \frac{\kappa\overline{\langle\sigma_z(A)\rangle}'}{T_e}, \qquad \text{...(B.4)}$$

and

$$\omega_1 = -g_J\,\mu_0\,H_1/\hbar, \qquad \text{...(B.5)}$$

$$\omega_0 = -g_J\,\mu_0\,H_0/\hbar. \qquad \text{...(B.6)}$$

T_{1e} and T_{2e} are the longitudinal and transverse relaxation times for the disorientation of the free electron spin due to buffer gas relaxation and wall collisions.

For the alkali atom polarization, we have

$$\frac{\mathrm{d}\overline{\langle\sigma_z(A)\rangle}'}{\mathrm{d}t} = \frac{1}{T_p} + \frac{\overline{\langle\sigma_z(e)\rangle}'}{T_{eA}} - \left(\frac{1}{T_p} + \frac{1}{T_{eA}} + \frac{1}{T_{1A}}\right)\overline{\langle\sigma_z(A)\rangle}', \qquad \text{...(B.7)}$$

where T_p is the pumping time and T_{1A} is the longitudinal relaxation time for the alkali atom spin due to buffer gas and wall relaxation.

At equilibrium in the presence of the r.f. field, all time derivatives vanish and equations (B.1)–(B.3) can be combined to give

$$\frac{\overline{\langle\sigma_z(A)\rangle}'}{T_e} = \overline{\langle\sigma_z(e)\rangle}'\left[\frac{1}{T_{1e}} + \frac{1}{T_e} + \frac{\omega_1^2\,\tau_2}{1 + \Delta\omega^2\,\tau_2^2}\right], \qquad \text{...(B.8)}$$

where

$$\frac{1}{\tau_2} = \frac{1}{T_{2e}} + \frac{1}{T_e}, \qquad \text{...(B.9)}$$

and

$$\Delta\omega = \omega_0 - \delta\omega_0 - \omega. \qquad \text{...(B.10)}$$

At equilibrium, equation (B.7) becomes

$$\frac{\overline{\langle\sigma_z(e)\rangle}'}{T_{eA}} = \overline{\langle\sigma_z(A)\rangle}'\left(\frac{1}{T_p} + \frac{1}{T_{eA}} + \frac{1}{T_{1A}}\right) - \frac{1}{T_p}, \qquad \text{...(B.11)}$$

Combining equations (B.8) and (B.11), we obtain

$$\overline{\langle\sigma_z(A)\rangle}'\left[\frac{1}{T_e T_{eA}} - \left(\frac{1}{T_{1e}} + \frac{1}{T_e}\right)\left(\frac{1}{T_p} + \frac{1}{T_{eA}} + \frac{1}{T_{1A}}\right) - \frac{\omega_1^2\,\tau_2\left(\frac{1}{T_p} + \frac{1}{T_{eA}} + \frac{1}{T_{1A}}\right)}{1 + \Delta\omega^2\,\tau_2^2}\right]$$
$$= -\frac{1}{T_p}\left[\frac{1}{T_{1e}} + \frac{1}{T_e} + \frac{\omega_1^2\,\tau_2}{1 + \Delta\omega^2\,\tau_2^2}\right]. \qquad \text{...(B.12)}$$

The polarization $\overline{\langle\sigma_z(A)\rangle}_0'$ of the alkali atoms in the absence of an applied r.f. field is obtained by setting $\omega_1 = 0$ in equation (B.12).

$$\overline{\sigma_z\langle(A)\rangle}_0'\left[\frac{1}{T_e T_{eA}} - \left(\frac{1}{T_{1e}} + \frac{1}{T_e}\right)\left(\frac{1}{T_p} + \frac{1}{T_{eA}} + \frac{1}{T_{1A}}\right)\right] = -\frac{1}{T_p}\left(\frac{1}{T_{1e}} + \frac{1}{T_e}\right). \qquad \text{...(B.13)}$$

If we define

$$\tau_1 = \frac{\dfrac{1}{T_p}+\dfrac{1}{T_{eA}}+\dfrac{1}{T_{1A}}}{\left(\dfrac{1}{T_{1e}}+\dfrac{1}{T_e}\right)\left(\dfrac{1}{T_p}+\dfrac{1}{T_{1A}}\right)+\dfrac{1}{T_{1e}T_{eA}}}, \quad \ldots\text{(B.14)}$$

we have

$$\overline{\langle\sigma_z(A)\rangle}_0' = \frac{\left(\dfrac{1}{T_{1e}}+\dfrac{1}{T_e}\right)}{T_p}\frac{\tau_1}{\left(\dfrac{1}{T_p}+\dfrac{1}{T_{eA}}+\dfrac{1}{T_{1A}}\right)}. \quad \ldots\text{(B.15)}$$

We can write $\overline{\langle\sigma_z(A)\rangle}$ in the laboratory frame in the presence of the r.f. field as

$$\overline{\langle\sigma_z(A)\rangle} = \overline{\langle\sigma_z(A)\rangle}' = \overline{\langle\sigma_z(A)\rangle}_0' + \delta\overline{\langle\sigma_z(A)\rangle}' = \overline{\langle\sigma_z(A)\rangle}_0 + \delta\overline{\langle\sigma_z(A)\rangle} \quad \ldots\text{(B.16)}$$

where $\delta\overline{\langle\sigma_z(A)\rangle}$ is the change in the alkali atom polarization due to the application of the r.f. field. Substituting this into equation (B.12) gives

$$\delta\overline{\langle\sigma_z(A)\rangle}\left[\frac{1}{T_eT_{eA}}-\left(\frac{1}{T_{1e}}+\frac{1}{T_e}\right)\left(\frac{1}{T_p}+\frac{1}{T_{eA}}+\frac{1}{T_{1A}}\right)-\frac{\omega_1^2\tau_2\left(\dfrac{1}{T_p}+\dfrac{1}{T_{eA}}+\dfrac{1}{T_{1A}}\right)}{1+\Delta\omega^2\tau_2^2}\right]$$

$$-\overline{\langle\sigma_z(A)\rangle}_0\frac{\omega_1^2\tau_2\left(\dfrac{1}{T_p}+\dfrac{1}{T_{eA}}+\dfrac{1}{T_{1A}}\right)}{1+\Delta\omega^2\tau_2^2} = -\frac{1}{T_p}\frac{\omega_1^2\tau_2}{1+\Delta\omega^2\tau_2^2}. \quad \ldots\text{(B.17)}$$

Rearranging,

$$\delta\overline{\langle\sigma_z(A)\rangle}\,[1+\Delta\omega^2\tau_2^2+\omega_1^2\tau_1\tau_2] = \omega_1^2\tau_1\tau_2\left[\frac{1}{T_p}\frac{1}{\left(\dfrac{1}{T_p}+\dfrac{1}{T_{eA}}+\dfrac{1}{T_{1A}}\right)}-\overline{\langle\sigma_z(A)\rangle}_0\right]. \quad \ldots\text{(B.18)}$$

Using equation (B.15), we have

$$\delta\overline{\langle\sigma_z(A)\rangle} = \frac{1}{T_p}\frac{1}{\left(\dfrac{1}{T_p}+\dfrac{1}{T_{eA}}+\dfrac{1}{T_{1A}}\right)}\left[1-\tau_1\left(\frac{1}{T_{1e}}+\frac{1}{T_e}\right)\right]\frac{\omega_1^2\tau_1\tau_2}{1+\omega_1^2\tau_1\tau_2+\Delta\omega^2\tau_2^2}$$

$$= -\frac{1}{T_p}\frac{\tau_1\left(\dfrac{1}{T_eT_{eA}}\right)}{\left(\dfrac{1}{T_p}+\dfrac{1}{T_{eA}}+\dfrac{1}{T_{1A}}\right)^2}\frac{\omega_1^2\tau_1\tau_2}{1+\omega_1^2\tau_1\tau_2+\Delta\omega^2\tau_2^2}. \quad \ldots\text{(B.19)}$$

In an optically thin sample, the amount of pumping light transmitted by the alkali atoms is proportional to $(1-\overline{\langle\sigma_z(A)\rangle}) = (1-2\overline{\langle J_z(A)\rangle})$. The transmission signal δI is given by

$$\delta I = -N_A Z_0\frac{1}{T_p^2}\frac{\tau_1\left(\dfrac{1}{T_eT_{eA}}\right)}{\left(\dfrac{1}{T_p}+\dfrac{1}{T_{eA}}+\dfrac{1}{T_{1A}}\right)^2}\frac{\omega_1^2\tau_1\tau_2}{1+\Delta\omega^2\tau_2^2+\omega_1^2\tau_1\tau_2}, \quad \ldots\text{(B.20)}$$

where we have used equations (C.18) and (B.24) of Section IV.

The electron spin resonance signal given by equation (B.20) is a Lorentzian line characterized by time constants τ_1 and τ_2 with a center frequency $\omega_0 - \delta\omega_0$. Since the frequency shift $\delta\omega_0$ is proportional to the alkali atom polarization $\overline{\langle\sigma_z(A)\rangle}$, it can be eliminated as a source of error in precision measurements of the electron spin resonance frequency by making measurements with left- and right-circularly polarized pumping light. This reverses the sign of $\overline{\langle\sigma_z(A)\rangle}$ and cancels out the effect of the frequency shift. In our treatment of the effect of spin-exchange collisions on $\rho(e, A)$, we have assumed that there is no transfer of coherence from one collision to the next. Ensberg and Morgan (1971) have explored the consequence of dropping this restriction. One result is the possibility of a frequency shift which does not reverse with a change in the sign of $\overline{\langle\sigma_z(A)\rangle}$. As yet there is no definite evidence that this is an important effect.

Measurements of the electron–alkali atom spin-exchange cross section can be made by measuring the linewidth of the electron resonance signal. If the density of alkali atoms in the cell is sufficiently high, the spin-exchange collisions are the dominant relaxation mechanism for the electron spin and

$$\frac{1}{\tau_2} \cong \frac{1}{T_e}.$$

From equation (B.20), one sees that the width $\Delta\omega$ at half maximum of the electron resonance is given by

$$\Delta\omega^2 = \frac{1}{\tau_2^2} + \omega_1^2 \frac{\tau_1}{\tau_2}. \qquad \text{...(B.21)}$$

By plotting $\Delta\omega^2$ vs. ω_1^2 at fixed alkali density N_A, one obtains a straight line with a zero-r.f. field intercept

$$\Delta\omega^2_{0-\text{r.f.}} = \frac{1}{\tau_2^2} \cong \frac{1}{T_e^2}, \qquad \text{...(B.22)}$$

The spin-exchange time T_e defined by equation (A.34) was for the unphysical situation where all electrons have identical speeds. If one assumes that the electrons are in thermal equilibrium with the buffer gas, then

$$\frac{1}{T_e} = N_A \langle v\sigma_{\text{SE}} \rangle, \qquad \text{...(B.23)}$$

where the bracket indicates an average over the velocity distribution of the electrons. With a knowledge of the alkali atom density N_A and a measurement of $\Delta\omega_{0-\text{r.f.}}$, one immediately obtains the thermally averaged spin-exchange cross section.

Additional information about the scattering phase shifts is obtained from measurements of the alkali atom polarization $\overline{\langle\sigma_z(A)\rangle}$, the frequency shift $\delta\omega_0$, and the zero-r.f. linewidth $\Delta\omega_{0-\text{r.f.}}$.

$$\frac{\delta\omega_0}{\Delta\omega_{0-\text{r.f.}}} = \frac{\left\langle\dfrac{\kappa}{T_e}\right\rangle\overline{\langle\sigma_z(A)\rangle}}{\dfrac{1}{T_e}} = \frac{\left\langle v\dfrac{\pi}{2k^2}\displaystyle\sum_{l=0}^{\infty}(2l+1)\sin 2(\delta_l^3-\delta_l^1)\right\rangle\langle\sigma_z(A)\rangle}{\langle v\sigma_{\text{SE}}\rangle}, \qquad \text{...(B.24)}$$

where again the brackets indicate an average over the electron velocity distributions.

C. THE EFFECT OF NUCLEAR SPIN ON ELECTRON–ALKALI ATOM SPIN-EXCHANGE COLLISIONS

The entire discussion of the preceding section was based on the fiction that the alkali atoms had zero nuclear spin. The time development of the alkali atom's spin–space density matrix was not complicated by the ground state hyperfine interaction between the alkali atom's valence electron and the nucleus.

The duration of an individual spin-exchange collision between an electron and an alkali atom is sufficiently short so that the nuclear spin plays no part in the collision process, and the **S**-matrix for the collision has the same form as for the fictitious spin-$\frac{1}{2}$ alkali atom. This is also true for spin-exchange collisions between atoms. During the time between collisions, however, the nuclear spin is coupled to the valence electron and this alters the time averaged effect of the spin-exchange collisions (Grossetete, 1964, 1968; Gibbs, 1965; Crampton, 1971) upon the atom's spin–space density matrix.

Between collisions, the spin–space density matrix $\rho(A)$ for an alkali atom has a time dependence governed by the Liouville equation

$$i\hbar\frac{\partial\rho(A)}{\partial t} = [\mathscr{H}, \rho(A)], \qquad \text{...(C.1)}$$

where $\mathscr{H}$ is the Hamiltonian for the ground state of the atom. An individual matrix element $\rho(A)_{ij}$ coupling states of energy E_i and E_j has the time dependence

$$\rho_{ij}(A,t) = \exp\left(-\frac{i}{\hbar}(E_i - E_j)\,t\right)\rho_{ij}(A,0). \qquad \text{...(C.2)}$$

If we take as our Hamiltonian in a weak magnetic field

$$\mathscr{H} = A\mathbf{I}.\mathbf{J} + \hbar\omega_0\frac{\mathbf{J}.\mathbf{F}}{|F|^2}F_z \qquad \text{...(C.3)}$$

we see immediately that eigenstates of F_z are stationary. If we work in the $|F,M\rangle$ representation, the diagonal elements of the density matrix

$\langle F, M|\rho(A)|F, M\rangle$ are stationary, and off-diagonal elements oscillate in time according to equation (C.2).

Because the nuclear spin does not come into play during a spin-exchange collision, we can use equation (A.27) to compute the time rate of change of the electron spin–space density matrix. When we apply the analogous equation (A.28) to obtain the time rate of change of the alkali spin–space density matrix, however, we must discard any off-diagonal terms which are created as a result of spin-exchange. Since the spin-exchange collisions occur at random times, off-diagonal terms which are created by spin exchange and oscillate according to equation (C.2) will average to zero. Therefore, the alkali atom density matrix is initially diagonal and remains diagonal.

Our object in this section is to obtain the equations of motion for the free electron polarization and the electronic polarization of the alkali atoms due to spin-exchange collisions when the effect of the nuclear spin is taken into account. To do this, we could write down the combined density matrix $\rho(e, A)$, apply equation (A.27), and obtain $d\rho(e)/dt$ by matrix multiplication. To compute $d\rho(A)/dt$ we could apply equation (A.28) using explicit matrix multiplication and discard any off-diagonal elements in $d\rho(A)/dt$ created by spin exchange. This would be a cumbersome procedure because of the size of the matrices. The matrix elements $\langle F, M|\rho(A)|F, M\rangle$ form a $2(2I+1) \times (2I+1)2$ dimensional matrix and $\rho(e, A) = \rho(e) \times \rho(A)$ is twice as large.

As it turns out we can obtain the equations of motion for the electron and alkali polarizations without going through explicit matrix multiplication by following a procedure introduced by Lambert (1970). To begin with, we calculate the time derivative of the free electron polarization due to spin exchange.

We can compute the time rate of change of the free electron polarization using equation (A.31) without modification, because the nuclear spin does not enter into the collision process. The electron–alkali atom spin system is described by the combined density matrix $\rho(e, A)$. The component $\overline{\langle\sigma_i(e)\rangle}$ of the electron polarization along a particular axis of a Cartesian coordinate system is

$$\overline{\langle\sigma_i(e)\rangle} = \mathrm{Tr}_e\, \sigma_i(e)\, \rho(e) = \mathrm{Tr}_e\, \sigma_i(e)\, \mathrm{Tr}_A\, \rho(e, A) = \mathrm{Tr}\, \sigma_i(e)\, \rho(e, A), \quad \ldots(\mathrm{C.4})$$

where Tr denotes a trace over the spin coordinates e and A. Similarly

$$\overline{\langle\sigma_i(A)\rangle} = \mathrm{Tr}\, \sigma_i(A)\, \rho(e, A). \quad \ldots(\mathrm{C.5})$$

The operators $\sigma_i(e)$ and $\sigma_i(A)$ in equations (C.4) and (C.5) operate only on the electronic spin coordinates of the free electrons and of the alkali atoms, respectively. We will also need the relation

$$\overline{\langle\sigma_i(e)\rangle}\,\overline{\langle\sigma_j(A)\rangle} = [\mathrm{Tr}_e\, \sigma_i(e)\, \rho(e)]\, [\mathrm{Tr}_A\, \sigma_j(A)\rho(A)] = \mathrm{Tr}\, \sigma_i(e)\, \sigma_j(A)\, \rho(e, A) \quad \ldots(\mathrm{C.6})$$

Applying equation (A.31), we have

$$\frac{d\overline{\langle\sigma_m(e)\rangle}}{dt} = \frac{1}{4T_e}\operatorname{Tr}[-3\sigma_m(e)\,\rho(e,A) + (1+2i\kappa)\,\sigma_m(e)\,\sigma_n(e)\,\sigma_n(A)\,\rho(e,A)$$
$$+ (1-2i\kappa)\,\sigma_m(e)\,\rho(e,A)\,\sigma_n(e)\,\sigma_n(A)$$
$$+ \sigma_m(e)\,\sigma_n(e)\,\sigma_n(A)\,\rho(e,A)\,\sigma_s(e)\,\sigma_s(A)], \qquad \text{...(C.7)}$$

where

$$1/T_e = vN_A\,\sigma_{\mathrm{SE}}, \qquad \text{...(C.8)}$$

and we sum over repeated indices.

To reduce equation (C.7) to something understandable, we will need the following properties of the Pauli spin matrices.

$$\sigma_m\,\sigma_n = \delta_{mn} + i\varepsilon_{mns}\,\sigma_s, \qquad \text{...(C.9)}$$

where ε_{mns} is the totally antisymmetric unit tensor, and

$$\varepsilon_{ijk}\,\varepsilon_{inm} = \delta_{jn}\,\delta_{km} - \delta_{jm}\,\delta_{kn}. \qquad \text{...(C.10)}$$

From equations (C.9) and (C.10) we also have

$$\sigma_s\,\sigma_m\,\sigma_n = \delta_{sm}\,\sigma_n + i\varepsilon_{smn} + \delta_{mn}\,\sigma_s - \sigma_m\,\delta_{sn}. \qquad \text{...(C.11)}$$

The first term on the right hand side of equation (C.7) is just $-3\overline{\langle\sigma_m(e)\rangle}$. To evaluate the second and third terms, we use equation (C.9).

$$\operatorname{Tr}\sigma_m(e)\,\sigma_n(e)\,\sigma_n(A)\,\rho(e,A) = \overline{\langle\sigma_m(A)\rangle} + i\varepsilon_{mns}\overline{\langle\sigma_s(e)\rangle}\,\overline{\langle\sigma_n(A)\rangle}, \qquad \text{...(C.12)}$$

and

$$\operatorname{Tr}\sigma_m(e)\,\rho(e,A)\,\sigma_n(e)\,\sigma_n(A) = \operatorname{Tr}\sigma_n(e)\,\sigma_m(e)\,\sigma_n(A)\,\rho(e,A)$$
$$= \overline{\langle\sigma_m(A)\rangle} + i\varepsilon_{nms}\overline{\langle\sigma_s(e)\rangle}\,\overline{\langle\sigma_n(A)\rangle}, \qquad \text{...(C.13)}$$

For the last term on the right hand side of equation (C.7) we use equations (C.10) and (C.11) to write it in the form

$$\operatorname{Tr}\sigma_s(e)\,\sigma_m(e)\,\sigma_n(e)\,\sigma_s(A)\,\sigma_n(A)\,\rho(e,A) = -\overline{\langle\sigma_m(e)\rangle} + 2\overline{\langle\sigma_m(A)\rangle}. \qquad \text{...(C.14)}$$

We have finally

$$\frac{d\overline{\langle\sigma_m(e)\rangle}}{dt} = \frac{1}{T_e}[\overline{\langle\sigma_m(A)\rangle} - \overline{\langle\sigma_m(e)\rangle} + \kappa\varepsilon_{mns}\overline{\langle\sigma_n(e)\rangle}\,\overline{\langle\sigma_s(A)\rangle}], \qquad \text{...(C.15)}$$

or equivalently

$$\frac{d\overline{\langle\boldsymbol{\sigma}(e)\rangle}}{dt} = \frac{\overline{\langle\boldsymbol{\sigma}(A)\rangle} - \overline{\langle\boldsymbol{\sigma}(e)\rangle}}{T_e} + \overline{\langle\boldsymbol{\sigma}(e)\rangle} \times \frac{\kappa}{T_e}\overline{\langle\boldsymbol{\sigma}(A)\rangle}. \qquad \text{...(C.16)}$$

Since there is no resonant r.f. field applied to the alkali atoms, the atoms have no transverse polarization, and

$$\overline{\langle \sigma(A) \rangle} = \overline{\langle \sigma_z(A) \rangle}\, \hat{k}. \qquad \ldots(\mathrm{C.17})$$

Equation (C.15) is therefore seen to be identical to equations (A.37)–(A.39) describing the time derivative of the electron polarization due to spin-exchange collisions with fictitious alkali atoms with zero nuclear spin. This is not surprising since the spin-exchange collision does not involve nuclear spin effects.

We turn now to the effect of spin exchange on the alkali spin system. In order to avoid matrix multiplication and the need to identify the off-diagonal elements in $\mathrm{d}\rho(A)/\mathrm{d}t$ which must be discarded after equation (A.28) has been applied, we make the following observation. If Q is any observable which commutes with the Hamiltonian in equation (C.3), then the equation

$$\frac{\mathrm{d}\overline{\langle Q \rangle}}{\mathrm{d}t} = \mathrm{Tr}\, \frac{Q\, \mathrm{d}\rho(A)}{\mathrm{d}t}, \qquad \ldots(\mathrm{C.18})$$

involves only the diagonal elements of $\mathrm{d}\rho(A)/\mathrm{d}t$. Therefore, if we apply equation (A.28) to the calculation of $\mathrm{d}\overline{\langle Q \rangle}/\mathrm{d}t$ we will obtain the correct result without further modification. The trick is to express observables in which we are interested in terms of observables which commute with the Hamiltonian. In what follows we will denote the free electron polarization $\overline{\langle \sigma_z(e) \rangle}$ as $2\overline{\langle S_z \rangle}$ and the electronic polarization of the alkali atoms $\overline{\langle \sigma_z(A) \rangle}$ as $2\overline{\langle J_z \rangle}$. We wish to calculate $\mathrm{d}\overline{\langle J_z \rangle}/\mathrm{d}t$ due to spin-exchange with the free electrons, so we must find the appropriate linear combination of observables Q which equal $\overline{\langle J_z \rangle}$ for a diagonal $\rho(A)$.

There is an analogy between the calculation of spin-exchange relaxation and relaxation due to spin-randomizing collisions with a buffer gas or a paraffin-coated wall. Indeed, spin-exchange collision with electrons whose polarization is maintained equal to zero are essentially spin-randomizing collisions. In theoretical studies of relaxation due to spin-randomizing collisions, it was found (Masnou-Seeuws and Bouchiat, 1967) that the relaxation of the observable $\overline{\langle Q \rangle}$ given by

$$\overline{\langle Q \rangle} = \overline{\langle J_z \rangle} - \frac{2\overline{\langle I_z \rangle}}{(2I+1)^2 - 2}, \qquad \ldots(\mathrm{C.19})$$

where

$$\langle I_z \rangle = \sum_M \frac{2IM}{2I+1} \langle I+\tfrac{1}{2}, M | \rho(A) | I+\tfrac{1}{2}, M \rangle$$
$$+ \sum_M \frac{2I+2}{2I+1} M \langle I-\tfrac{1}{2}, M | \rho(A) | I-\tfrac{1}{2}, M \rangle, \qquad \ldots(\mathrm{C.20})$$

was a single exponential decay with a time constant independent of nuclear spin. This suggests that we might be able to write the observable $\overline{\langle Q\rangle}$ in equation (C.19) in a way in which it is obvious that it commutes with the Hamiltonian in equation (C.3) and yet is equal to the right hand side of equation (C.19) for a diagonal density matrix $\rho(A)$.

To this end, let us define the observable Q by

$$Q = \frac{4(\mathbf{I}.\mathbf{J})F_z - F_z}{(2I+1)^2 - 2}. \qquad \text{...(C.21)}$$

In terms of the density matrix elements of $\rho(A)$, the ensemble average of Q is

$$\overline{\langle Q\rangle} = \sum_M \frac{(2I-1)M\langle I+\tfrac{1}{2}, M|\rho(A)|I+\tfrac{1}{2}, M\rangle}{(2I+1)^2-2} - \sum_M \frac{(2I+3)M\langle I-\tfrac{1}{2}, M|\rho(A)|I-\tfrac{1}{2}, M\rangle}{(2I+1)^2-2} \qquad \text{...(C.22)}$$

Equation (C.22) can be rewritten as

$$\overline{\langle Q\rangle} = \sum_M \frac{\left[(2I+1)^2\frac{M}{2I+1} - 2M\right]\langle I+\tfrac{1}{2}, M|\rho(A)|I+\tfrac{1}{2}, M\rangle}{(2I+1)^2} - \frac{\sum_M \left[(2I+1)^2\frac{M}{2I+1} + 2M\right]\langle I-\tfrac{1}{2}, M|\rho(A)|I-\tfrac{1}{2}, M\rangle}{(2I+1)^2-2}. \qquad \text{...(C.23)}$$

But for a diagonal density matrix, equation (C.23) is equivalent to

$$\overline{\langle Q\rangle} = \frac{\overline{\langle J_z\rangle}(2I+1)^2 - 2\overline{\langle F_z\rangle}}{(2I+1)^2-2} = \overline{\langle J_z\rangle} - \frac{2\overline{\langle I_z\rangle}}{(2I+1)^2-2}. \qquad \text{...(C.24)}$$

The observable Q defined by equation (C.21) commutes with the Hamiltonian in equation (C.3) and is a linear combination of $\overline{\langle J_z\rangle}$ and $\overline{\langle I_z\rangle}$ for a diagonal matrix $\rho(A)$.

Since we wish to compute $\mathrm{d}\overline{\langle J_z\rangle}/\mathrm{d}t$ by expressing $\overline{\langle J_z\rangle}$ as a linear combination of observables which commute with the Hamiltonian, we need a second such observable which is a linear combination of $\overline{\langle I_z\rangle}$ and $\overline{\langle J_z\rangle}$. An obvious choice is

$$\overline{\langle F_z\rangle} = \overline{\langle I_z\rangle} + \overline{\langle J_z\rangle}. \qquad \text{...(C.25)}$$

If we compute $\mathrm{d}\overline{\langle F_z\rangle}/\mathrm{d}t$ and $\mathrm{d}\overline{\langle Q\rangle}/\mathrm{d}t$, we can combine the results to obtain $\mathrm{d}\overline{\langle J_z\rangle}/\mathrm{d}t$ and $\mathrm{d}\overline{\langle I_z\rangle}/\mathrm{d}t$ due to spin-exchange with the free electrons. We start with the calculation of $\mathrm{d}\overline{\langle F_z\rangle}/\mathrm{d}t$.

$$\frac{\mathrm{d}\overline{\langle F_z\rangle}}{\mathrm{d}t} = \mathrm{Tr}\, I_z \frac{\mathrm{d}\rho(A)}{\mathrm{d}t} + \mathrm{Tr}\, J_z \frac{\mathrm{d}\rho(A)}{\mathrm{d}t}. \qquad \text{...(C.26)}$$

Since $\overline{\langle F_z \rangle}$ commutes with the Hamiltonian, we can immediately apply equation (A.28) to the calculation of $\mathrm{d}\rho(A)/\mathrm{d}t$ which is equivalent to using equation (A.31) with the indices e and A interchanged and T_e replaced by T_{eA}. Since I_z commutes with $\boldsymbol{\sigma}(e)$ and $\boldsymbol{\sigma}(A)$, the 1st term on the right hand side of equation (C.26) vanishes. The 2nd term is

$$\operatorname{Tr} J_z \frac{\mathrm{d}\rho(A)}{\mathrm{d}t} = \tfrac{1}{2} \operatorname{Tr} \sigma_z(A) \frac{\mathrm{d}\rho(A)}{\mathrm{d}t} = \frac{1}{2} \frac{\overline{\langle \sigma_z(e) \rangle} - \overline{\langle \sigma_z(A) \rangle}}{T_{eA}} \qquad \text{...(C.27)}$$

where we have followed the same calculational procedure which led to equation (C.16) and have used the fact that $\overline{\langle \boldsymbol{\sigma}(A) \rangle} = \overline{\langle \sigma_z(A) \rangle}$. Therefore, we have the result

$$\frac{\mathrm{d}\overline{\langle F_z \rangle}}{\mathrm{d}t} = \frac{\overline{\langle S_z \rangle} - \overline{\langle J_z \rangle}}{T_{eA}}, \qquad \text{...(C.28)}$$

due to spin-exchange collisions.

The next step is to calculate $\mathrm{d}\overline{\langle Q \rangle}/\mathrm{d}t$. We have calculated the $\mathrm{d}\overline{\langle F_z \rangle}/\mathrm{d}t$ part of $\mathrm{d}\overline{\langle Q \rangle}/\mathrm{d}t$. To calculate $\mathrm{d}\overline{\langle \mathbf{I}.\mathbf{J}F_z \rangle}/\mathrm{d}t$, we write

$$\frac{\mathrm{d}\overline{\langle \mathbf{I}.\mathbf{J}F_z \rangle}}{\mathrm{d}t} = \frac{\mathrm{d}\overline{\langle \mathbf{I}.\mathbf{J}J_z \rangle}}{\mathrm{d}t} + \frac{\mathrm{d}\overline{\langle \mathbf{I}.\mathbf{J}I_z \rangle}}{\mathrm{d}t} = \frac{\mathrm{d}}{\mathrm{d}t} \frac{\overline{\langle I_m \sigma_m(A) \sigma_3(A) \rangle}}{4} + \frac{\mathrm{d}}{\mathrm{d}t} \frac{\overline{\langle I_m \sigma_m(A) I_3 \rangle}}{2}, \qquad \text{...(C.29)}$$

and we sum over repeated indices. Bearing in mind that $\mathbf{I}$ commutes with the operators $\sigma(e)$ and $\boldsymbol{\sigma}(A)$, and using equation (A.31) and the same techniques which led to equations (C.15) and (C.16), we have

$$\frac{\mathrm{d}}{\mathrm{d}t} \frac{\overline{\langle I_m \sigma_m(A) I_3 \rangle}}{2} = \frac{1}{2T_{eA}} \times [\overline{\langle I_m I_3 \rangle}\, \overline{\langle \sigma_m(e) \rangle} - \overline{\langle I_m \sigma_m(A) I_3 \rangle} + \kappa \overline{\langle I_m \varepsilon_{mns} \sigma_n(A) \rangle}\, \overline{\langle \sigma_s(e) \rangle}]. \qquad \text{...(C.30)}$$

For a diagonal density matrix,

$$\overline{\langle I_m \varepsilon_{mns} \sigma_n(A) \rangle} = 0, \qquad \text{...(C.31)}$$

because

$$\overline{\langle I_1 \sigma_2(A) \rangle} = \overline{\langle I_2 \sigma_1(A) \rangle}, \qquad \text{...(C.32)}$$

and

$$\overline{\langle I_1 \sigma_3(A) \rangle} = \overline{\langle I_2 \sigma_3(A) \rangle} = 0. \qquad \text{...(C.33)}$$

Also,

$$\overline{\langle I_1 I_3 \rangle} = \overline{\langle I_2 I_3 \rangle} = 0. \qquad \text{...(C.34)}$$

Therefore

$$\frac{\mathrm{d}}{\mathrm{d}t} \overline{\langle \mathbf{I}.\mathbf{J}I_z \rangle} = \frac{\overline{\langle I_z^2 \rangle}\, \overline{\langle S_z \rangle} - \overline{\langle \mathbf{I}.\mathbf{J}I_z \rangle}}{T_{eA}} \qquad \text{...(C.35)}$$

To calculate

$$\frac{d}{dt}\overline{\langle \mathbf{I}.\mathbf{J}J_z\rangle},$$

we use equation (C.9).

$$\frac{d}{dt}\overline{\langle \mathbf{I}.\mathbf{J}J_z\rangle} = \frac{1}{4}\frac{d}{dt}\overline{\langle I_3\rangle} + \frac{i}{4}\frac{d}{dt}\overline{\langle I_m\,\varepsilon_{m3s}\,\sigma_s(A)\rangle}. \qquad \ldots(C.36)$$

The 1st term on the right hand side of equation (C.36) vanishes. The 2nd term is, using equation (A.31) with interchanged indices,

$$\frac{d}{dt}i\frac{\overline{\langle I_2\,\sigma_1(A)\rangle}}{4} - \frac{d}{dt}i\frac{\overline{\langle I_1\,\sigma_2(A)\rangle}}{4}$$

$$= \frac{1}{4T_{eA}}[i\overline{\langle I_2\rangle}\,\overline{\langle\sigma_1(e)\rangle} - i\overline{\langle I_1\rangle}\,\overline{\langle\sigma_2(e)\rangle} - i\overline{\langle I_2\,\sigma_1(A)\rangle} + i\overline{\langle I_1\,\sigma_2(A)\rangle}]$$

$$+ \frac{\kappa}{4T_{eA}}[i\overline{\langle I_2\,\varepsilon_{1ns}\,\sigma_n(A)\rangle}\,\overline{\langle\sigma_s(e)\rangle} - i\overline{\langle I_1\,\varepsilon_{2ns}\,\sigma_n(A)\rangle}\,\overline{\langle\sigma_s(e)\rangle}]. \qquad \ldots(C.37)$$

For a diagonal density matrix,

$$\overline{\langle I_1\rangle} = \overline{\langle I_2\rangle} = 0. \qquad \ldots(C.38)$$

Equations (C.32), (C.33) and (C.38) imply that the right hand side of equation (C.37) vanishes.

Our final result is

$$\frac{d\overline{\langle Q\rangle}}{dt} = \frac{4\overline{\langle I_z^2\rangle}\,\overline{\langle S_z\rangle} - 4\overline{\langle \mathbf{I}.\mathbf{J}I_z\rangle} - \overline{\langle S_z\rangle} + \overline{\langle J_z\rangle}}{T_{eA}[(2I+1)^2-2]} \qquad \ldots(C.39)$$

We note that for a diagonal density matrix,

$$4\overline{\langle I.JJ_z\rangle} = \overline{\langle I_z\rangle}, \qquad \ldots(C.40)$$

which means that we can rewrite equation (C.39) in the form

$$\frac{d\overline{\langle Q\rangle}}{dt} = \frac{4\overline{\langle I_z^2\rangle}\,\overline{\langle S_z\rangle} - 4\overline{\langle \mathbf{I}.\mathbf{J}\mathbf{F}_z\rangle} - \overline{\langle S_z\rangle} + \overline{\langle F_z\rangle}}{T_{eA}[(2I+1)^2-2]}$$

$$= \frac{4\overline{\langle I_z^2\rangle}\,\overline{\langle S_z\rangle} - \overline{\langle S_z\rangle}}{T_{eA}[(2I+1)^2-2]} - \frac{\overline{\langle Q\rangle}}{T_{eA}}. \qquad \ldots(C.41)$$

We have arrived at equation (C.41) by applying equation (C.31) to the observable Q while neglecting nuclear spin effects. Equation (C.41) is valid because Q is an observable which only involves diagonal density matrix elements.

Using equations (C.24) and (C.41), we have

$$\frac{\mathrm{d}\overline{\langle J_z\rangle}}{\mathrm{d}t}-\frac{2\dfrac{\mathrm{d}\overline{\langle I_z\rangle}}{\mathrm{d}t}}{(2I+1)^2-2}=\frac{4\overline{\langle I_z^2\rangle}\,\overline{\langle S_z\rangle}-\overline{\langle S_z\rangle}}{T_{eA}[(2I+1)^2-2]}-\frac{\overline{\langle J_z\rangle}}{T_{eA}}+\frac{2\overline{\langle I_z\rangle}}{[(2I+1)^2-2]\,T_{eA}}. \quad \ldots(\mathrm{C}.42)$$

Combining equations (C.42) and (C.28), we have

$$\frac{\mathrm{d}\overline{\langle J_z\rangle}}{\mathrm{d}t}=\frac{(4\overline{\langle I_z^2\rangle}+1)\,\overline{\langle S_z\rangle}}{(2I+1)^2\,T_{eA}}-\frac{\overline{\langle J_z\rangle}}{T_{eA}}+\frac{2\overline{\langle I_z\rangle}}{(2I+1)^2\,T_{eA}}, \quad \ldots(\mathrm{C}.43)$$

and

$$\frac{\mathrm{d}\overline{\langle I_z\rangle}}{\mathrm{d}t}=\frac{[(2I+1)^2-1-4\overline{\langle I_z^2\rangle}]\,\overline{\langle S_z\rangle}}{(2I+1)^2\,T_{eA}}-\frac{2\overline{\langle I_z\rangle}}{(2I+1)^2\,T_{eA}}. \quad \ldots(\mathrm{C}.44)$$

Equations (C.43) and (C.44) describe the time evolution of the electronic and nuclear spin polarizations due to spin-exchange collisions with free electrons. The inclusion of nuclear spin has considerably complicated the situation.

D. THE SPIN-EXCHANGE ELECTRON RESONANCE SIGNAL WHEN THE EFFECTS OF NUCLEAR SPIN ARE CONSIDERED

When we attempt to arrive at a closed expression for the spin-exchange electron resonance signal, we run into difficulties which are similar to those encountered in Section V when nuclear spin was included in the analysis of the low field alkali optical-pumping signal. At the time of this writing, the problem has not been solved exactly. It appears that a computer calculation of the time dependence of the individual matrix elements $\langle F, M|\rho(A)|FM\rangle$ under the influence of the pumping light, spin relaxation, and spin exchange is necessary.

On the other hand, experiments do not reveal a sensitivity of the electron resonance lineshape to variations in experimental conditions. The observed lineshape is well described by a Lorentzian and the behavior of the signal as a function of r.f. amplitude and alkali density is what one expects from the signal derived in Subsection B neglecting nuclear spin. Large variations in the buffer gas pressure, which should influence the degree of excited state mixing and the mode of spin relaxation in the ground state (spin-randomization versus uniform relaxation at the walls) and therefore alter the equations of motion of the alkali density matrix elements, do not appear to alter the form of the observed electron resonance signal. In addition, no difference in the electron resonance linewidth was observed for collisions with isotopically pure Rb^{87} ($I=3/2$) and Rb^{85} ($I=5/2$) (Davis and Balling, 1972). All of the experiments with spin-exchange polarized electrons, however, have one thing in common:

the *changes* in the alkali polarization $\overline{\langle J_z \rangle}$ due to r.f. disorientation of the free electrons are small compared to the alkali polarization itself, because of the low electron density in the cell. In what follows, we will attempt to exploit this fact.

Using the results of the preceding subsections and the results of Section IV we can write down the equations for the equilibrium polarizations of the electrons and alkali atoms.

$$0 = \frac{d\overline{\langle S_z \rangle}}{dt} = \frac{\overline{\langle J_z \rangle} - \overline{\langle S_z \rangle}}{T_e} - \frac{\overline{\langle S_z \rangle}}{T_{1e}} - \frac{\omega_1^2 \tau_2 \overline{\langle S_z \rangle}}{1 + \Delta\omega^2 \tau_2^2}, \quad \text{...(D.1)}$$

$$0 = \frac{d\overline{\langle F_z \rangle}}{dt} = \frac{\overline{\langle S_z \rangle} - \overline{\langle J_z \rangle}}{T_{eA}} - \frac{\overline{\langle F_z \rangle}}{T_p} + \frac{2\overline{\langle F_z J_z \rangle}}{T_p} - \frac{\overline{\langle F_z \rangle}}{\tau_w} - \frac{\overline{\langle J_z \rangle}}{T_g}, \quad \text{...(D.2)}$$

where the pumping time T_p, the buffer gas relaxation time T_g, and the wall relaxation time τ_w were defined in Section V, and we have assumed complete reorientation in the excited state. These equations contain the observables $\overline{\langle S_z \rangle}$, $\overline{\langle J_z \rangle}$, $\overline{\langle I_z \rangle}$ and $\overline{\langle I_z J_z \rangle}$. Clearly we need more equations than we derived in the previous subsection. Indeed it appears that if we continue along these same lines, we will end up solving the same number of equations as we would if we solved for the individual density matrix elements.

The transmission signal δI which we observe is proportional to the change $\delta\overline{\langle J_z \rangle}$ in the alkali polarization due to a change $\delta\overline{\langle S_z \rangle}$ in the free electron polarization when the r.f. field is turned on. For the case of the fictitious alkali atom with zero nuclear spin, $\delta\overline{\langle J_z \rangle}$ was proportional to $\delta\overline{\langle S_z \rangle}$. At this point we cannot write down the functional relationship between $\delta\overline{\langle J_z \rangle}$ and $\delta\overline{\langle S_z \rangle}$ for real alkali atoms, and from equations (C.43) and (C.44) it is clear it will not be linear.

If we assume, however, that $\delta\overline{\langle J_z \rangle}$ is an analytic function of $\delta\overline{\langle S_z \rangle}$ and expand $\delta\overline{\langle J_z \rangle}$ in a power series in the variable $\delta\overline{\langle S_z \rangle}$, we can argue that for small polarization changes, the higher order terms in the series can be neglected. Thus, if $\overline{\langle S_z \rangle}_0$ and $\overline{\langle J_z \rangle}_0$ are the equilibrium polarizations in the absence of an r.f. field, we can write

$$\overline{\langle S_z \rangle} = \overline{\langle S_z \rangle}_0 + \delta\overline{\langle S_z \rangle}, \quad \text{...(D.3)}$$

and

$$\overline{\langle J_z \rangle} = \overline{\langle J_z \rangle}_0 + \delta\overline{\langle J_z \rangle}, \quad \text{...(D.4)}$$

for the equilibrium polarizations with the r.f. field turned on. We have from equation (D.1)

$$\frac{\overline{\langle J_z \rangle}_0}{T_e} = \frac{\overline{\langle S_z \rangle}_0}{T_e} + \frac{\overline{\langle S_z \rangle}_0}{T_{1e}}. \quad \text{...(D.5)}$$

If we approximate $\delta\overline{\langle J_z\rangle}$ by

$$\delta\overline{\langle J_z\rangle} \cong \left(\frac{\partial\overline{\langle J_z\rangle}}{\partial\delta\overline{\langle S_z\rangle}}\right)_0 \delta\overline{\langle S_z\rangle} = \frac{1}{\tau_{eA}}\,\delta\overline{\langle S_z\rangle}, \qquad \text{...(D.6)}$$

Equation (D.1) gives us

$$\frac{\overline{\langle J_z\rangle_0}}{T_e} + \frac{1}{T_e}\frac{1}{\tau_{eA}}\,\delta\overline{\langle S_z\rangle} = \frac{\overline{\langle S_z\rangle_0}}{T_e} + \frac{\overline{\langle S_z\rangle_0}}{T_{1e}} + \frac{\omega_1^2\tau_2\overline{\langle S_z\rangle_0}}{1+\Delta\omega^2\tau_2^2}$$

$$+\frac{\delta\overline{\langle S_z\rangle}}{T_e} + \frac{\delta\overline{\langle S_z\rangle}}{T_{1e}} + \frac{\omega_1^2\tau_2\,\delta\overline{\langle S_z\rangle}}{1+\Delta\omega^2\tau_2^2}. \qquad \text{...(D.7)}$$

Therefore,

$$\delta\overline{\langle S_z\rangle}\,\omega_1^2\tau_2 + (1+\Delta\omega^2\tau_2^2)\left(\frac{1}{T_e}+\frac{1}{T_{1e}}-\frac{1}{T_e\tau_{eA}}\right) = -\overline{\langle S_z\rangle_0}\,\omega_1^2\tau_2. \qquad \text{...(D.8)}$$

Defining

$$\frac{1}{\tau_1} = \frac{1}{T_e} + \frac{1}{T_{1e}} - \frac{1}{T_e\tau_{eA}}, \qquad \text{...(D.9)}$$

we obtain

$$\delta\overline{\langle S_z\rangle} = -\frac{\overline{\langle S_z\rangle_0}\,\omega_1^2\tau_1\tau_2}{1+\omega_1^2\tau_1\tau_2+\Delta\omega^2\tau_2^2}. \qquad \text{...(D.10)}$$

Combining this with equation (D.6) we have the approximate expression for the transmission signal

$$\delta I = \frac{2N_A z_0}{T_p}\,\delta\overline{\langle J_z\rangle} \cong -\frac{2N_A z_0}{T_p}\frac{1}{\tau_{eA}}\frac{\overline{\langle S_z\rangle_0}\,\omega_1^2\tau_1\tau_2}{1+\omega_1^2\tau_1\tau_2+\Delta\omega^2\tau_2^2}. \qquad \text{...(D.11)}$$

The linewidth of this reasonance is determined by τ_2 which is the same as that defined in Subsection B. As long as this linearization procedure is valid, one can expect to observe a Lorentzian line with a width related to the spin-exchange cross section in the same way as was derived neglecting nuclear spin. As soon as nonlinearities become important, the line will no longer be Lorentzian. The onset of non-linearity can be checked experimentally by varying the amplitude of the applied r.f. to change the amplitude of $\delta\overline{\langle J_z\rangle}$ and monitoring the line shape.

What we have done here is essentially arm waving, and a rigorous analysis of the signals one expects to see under varying experimental conditions is desirable.

E. THE EFFECT OF NUCLEAR SPIN ON SPIN-EXCHANGE COLLISIONS BETWEEN ALKALI ATOMS

The technique of spin-exchange optical pumping is not limited to the polarization of free electrons. In principle, one can polarize any paramagnetic atom or ion by spin-exchange collisions with optically pumped alkali atoms. The study of spin-exchange collisions between alkali atoms is of particular theoretical interest because of the relative simplicity of the alkali atom–alkali atom collision problem. The effect of spin-exchange collisions on the polarizations of hydrogen-like atoms has been widely discussed (Bender, 1963, 1964; Balling *et al.*, 1964; Grossetete, 1964, 1968; Gibbs, 1965, 1971; Crampton, 1971).

The usual experimental technique used to measure the spin-exchange cross section is to observe the relaxation of $\overline{\langle J_z \rangle}$ or $\overline{\langle \mathbf{I}.\mathbf{J} \rangle}$ due to spin-exchange collisions with a second species of alkali atoms whose polarization is maintained equal to zero by continuous application of a resonant r.f. field. $\overline{\langle \mathbf{I}.\mathbf{J} \rangle}$ was defined in Section V and is the hyperfine polarization. With the same experimental technique one can also observe the relaxation of $\overline{\langle J_z \rangle}$ and $\overline{\langle \mathbf{I}.\mathbf{J} \rangle}$ due to self-exchange collisions. It is the purpose of this section to calculate the way in which these observables relax due to alkali spin exchange.

We can apply the technique used in Subsection C to calculate the time dependence of $\overline{\langle \mathbf{I}.\mathbf{J} \rangle}$ and $\overline{\langle J_z \rangle}$ due to spin-exchange between alkali atoms, because the **S**-matrix for a collision between alkali atoms does not involve the nuclear spin coordinates, but simply the spin coordinates of the alkali atoms' valence electrons. The **S**-matrix has the same form as for the collision between two spin-$\frac{1}{2}$ particles. The effect on an alkali atom's polarization of collisions with another alkali atom is essentially the same as the effect of collisions with free electrons. Nuclear spin effects only come into play between collisions. Therefore, we can apply equation (A.31) to calculate the time rate of change of the density matrix for alkali atom species 1 due to spin-exchange collisions with atoms of alkali species 2. For the moment we are not worrying about nuclear spin effects between collisions. If we define the combined density matrix $\rho(1,2)$, we have for the time rate of change of some observable $A(1)$ due to spin-exchange collisions with atoms of species 2

$$\begin{aligned}\frac{\mathrm{d}\overline{\langle A(1)\rangle}}{\mathrm{d}t} &= \mathrm{Tr}\, A(1)\frac{\mathrm{d}\rho(1)}{\mathrm{d}t}\\ &= \frac{1}{4T_{12}}\mathrm{Tr}\,[-3A(1)\,\rho(1,2) + (1+2i\kappa)\,A(1)\,\sigma_n(1)\,\sigma_n(2)\,\rho(1,2)\\ &\quad + (1-2i\kappa)\,A(1)\,\rho(1,2)\,\sigma_n(1)\,\sigma_n(2)\\ &\quad + A(1)\,\sigma_n(1)\,\sigma_n(2)\,\rho(1,2)\,\sigma_s(1)\,\sigma_s(2)],\end{aligned} \qquad \text{...(E.1)}$$

where

$$\frac{1}{T_{12}} = vN_2\,\sigma_{\mathrm{SE}}, \qquad \text{...(E.2)}$$

and v is the relative velocity of the alkali atoms, N_2 is the density of species 2, and σ_{SE} is the spin-exchange cross section, and we have used the notation of equation (C.7).

If we use equation (E.1) to calculate the time rate of change of some observable which does not commute with the Hamiltonian of equation (C.3), we will have to strike out off-diagonal density matrix elements produced by equation (E.1). We avoid this difficulty when dealing with the observable $\overline{\langle \mathbf{I}.\mathbf{J} \rangle}$. Since $\overline{\langle \mathbf{I}(1).\mathbf{J}(1) \rangle}$ commutes with the Hamiltonian, we can use equation (E.1) without modification to compute $\mathrm{d}/\mathrm{d}t\overline{\langle \mathbf{I}.\mathbf{J} \rangle}$.

$$\frac{\mathrm{d}\overline{\langle \mathbf{I}(1).\mathbf{J}(1) \rangle}}{\mathrm{d}t} = \tfrac{1}{2}\,\mathrm{Tr}_1\, I_m(1)\,\sigma_m(1)\frac{\mathrm{d}\rho(1)}{\mathrm{d}t}. \qquad \text{...(E.3)}$$

We will consider a particularly simple experimental situation. The atoms of species 1 are optically pumped to produce a hyperfine polarization $\overline{\langle \mathbf{I}(1).\mathbf{J}(1) \rangle} \neq 0$. They collide with alkali atoms of species 2 whose electronic polarization $\overline{\langle J_z(2) \rangle} = 0$ due to continuous application of a resonant r.f. field. The pumping light is turned off, and the alkali atoms of species 1 relax in the dark due to collisions with the buffer gas and spin-exchange collisions with the unpolarized alkali atoms of species 2.

Applying equation (E.1) and using the operator techniques employed in Subsection C, we immediately obtain the simple result

$$\frac{\mathrm{d}\overline{\langle \mathbf{I}(1).\mathbf{J}(1) \rangle}}{\mathrm{d}t} = -\frac{\overline{\langle \mathbf{I}(1).\mathbf{J}(1) \rangle}}{T_{12}}. \qquad \text{...(E.4)}$$

for the relaxation due to spin-exchange collisions. If we include the effect of buffer gas collisions and wall relaxation, we have

$$\frac{\mathrm{d}\overline{\langle \mathbf{I}(1).\mathbf{J}(1) \rangle}}{\mathrm{d}t} = -\overline{\langle \mathbf{I}(1).\mathbf{J}(1) \rangle}\left[\frac{1}{T_g} + \frac{1}{T_w} + \frac{1}{T_{12}}\right], \qquad \text{...(E.5)}$$

where T_g and T_w are gas- and wall-relaxation times.

The relaxation of $\overline{\langle \mathbf{I}.\mathbf{J} \rangle}$ due to spin-randomizing collisions is a single exponential because these collisions are equivalent to spin-exchange collisions with unpolarized alkali atoms.

If the polarization $\overline{\langle J_z(2) \rangle} \neq 0$, the relaxation of $\overline{\langle \mathbf{I}(1).\mathbf{J}(1) \rangle}$ due to spin exchange is no longer simply exponential.

$$\frac{\mathrm{d}}{\mathrm{d}t}\overline{\langle \mathbf{I}(1).\mathbf{J}(1) \rangle} = \frac{\mathrm{d}}{\mathrm{d}t}\frac{\overline{\langle I_m\,\sigma_m(1) \rangle}}{2},$$

and we immediately obtain from equation (E.1)

$$\frac{\mathrm{d}}{\mathrm{d}t}\overline{\langle \mathbf{I}(1).\mathbf{J}(1) \rangle} = \frac{\overline{\langle I_z(1) \rangle}\,\overline{\langle J_z(2) \rangle} - \overline{\langle \mathbf{I}(1).\mathbf{J}(1) \rangle}}{T_{12}}. \qquad \text{...(E.6)}$$

A second situation of interest is the relaxation in the dark of $\overline{\langle \mathbf{I}.\mathbf{J} \rangle}$ for a single species of alkali atoms due to self-exchange collisions. Identity effects can be neglected (Grossetete, 1968) and we can use equation (E.1) and eliminate the distinction between indices 1 and 2 at the end of the calculation. If $\overline{\langle J_z \rangle} = 0$ (continuous application of a resonant r.f. field at the Zeeman frequency or hyperfine pumping with unpolarized light) we immediately obtain

$$\frac{\mathrm{d}\overline{\langle \mathbf{I}.\mathbf{J} \rangle}}{\mathrm{d}t} = -\frac{\overline{\langle \mathbf{I}.\mathbf{J} \rangle}}{T_{\mathrm{ex}}}, \qquad \text{...(E.7)}$$

where the self-exchange time T_{ex} is given by

$$\frac{1}{T_{\mathrm{ex}}} = vN\sigma_{\mathrm{SE}}, \qquad \text{...(E.8)}$$

and N is the alkali density, v is the relative velocity of the atoms, and σ_{SE} is the cross section for spin-exchange between the identical atoms. If we add in the effect of buffer gas collisions and wall relaxation we have

$$\frac{\mathrm{d}\overline{\langle \mathbf{I}.\mathbf{J} \rangle}}{\mathrm{d}t} = -\overline{\langle \mathbf{I}.\mathbf{J} \rangle}\left[\frac{1}{T_{\mathrm{ex}}} + \frac{1}{T_g} + \frac{1}{T_w}\right]. \qquad \text{...(E.9)}$$

The relaxation in the dark of $\overline{\langle J_z(1) \rangle}$ due to collisions with a 2nd species of unpolarized alkali atoms is more complicated, because $\overline{\langle J_z \rangle}$ does not commute with the ground state Hamiltonian. The equations of motion were originally worked out independently by Grossetete (1964) and Gibbs (1965). We have already worked out the equations of motion for $\overline{\langle J(z) \rangle}$ of an alkali atom due to spin-exchange collisions with free electrons. They are equations (C.43) and (C.44). Since there is no distinction between collisions with unpolarized electrons and collisions with unpolarized alkali atoms, these equations apply to the present case if we set $\overline{\langle S_z \rangle} = 0$ in equations (C.43) and (C.44).

$$\frac{\mathrm{d}\overline{\langle J_z(1) \rangle}}{\mathrm{d}t} = -\frac{\overline{\langle J_z(1) \rangle}}{T_{12}} + \frac{2\overline{\langle I_z(1) \rangle}}{(2I_1+1)^2 T_{12}}, \qquad \text{...(E.10)}$$

$$\frac{\mathrm{d}\overline{\langle I_z(1) \rangle}}{\mathrm{d}t} = -\frac{2\overline{\langle I_z(1) \rangle}}{(2I_1+1)^2 T_{12}}. \qquad \text{...(E.11)}$$

These equations lead to the double exponential decay of $\overline{\langle J_z(1) \rangle}$ due to spin-exchange collisions with the unpolarized atoms of species 2 given by

$$\frac{\overline{\langle J_z(1) \rangle}}{\overline{\langle J_z(1) \rangle}_0} = (1-a)\,e^{-t/T_{12}} + a\,e^{-t/T'_{12}}, \qquad \text{...(E.12)}$$

with

$$a = \frac{2\overline{\langle I_z(1) \rangle}_0}{\overline{\langle J_z(1) \rangle}_0} \frac{1}{[4I_1^2 + 4I_1 - 1]}, \qquad \text{...(E.13)}$$

and

$$T'_{12} = \frac{(2I_1 + 1)^2}{2} T_{12}. \qquad \text{...(E.14)}$$

The validity of the equations of motion for $\overline{\langle \mathbf{I}.\mathbf{J} \rangle}$ and $\overline{\langle J_z \rangle}$ due to alkali spin-exchange collisions has been verified experimentally. It is clear that it is preferable to observe the relaxation of $\overline{\langle \mathbf{I}.\mathbf{J} \rangle}$ rather than $\overline{\langle J_z \rangle}$, because it is simpler to fit the data to a single exponential than to a double exponential.

Another possibility is to perform the measurements in a strong magnetic field in which $\langle \mathbf{I} \rangle$ and $\langle \mathbf{J} \rangle$ are decoupled and precess independently about the field (Paschen–Back limit). Under these conditions (Grossetete, 1964), the nuclear spin does not affect the equation of motion of $\overline{\langle J_z \rangle}$ and it relaxes as a single exponential. In principle, one could use the same approach in measurements of the electron-alkali spin-exchange cross section. It would be difficult, however, to avoid magnetic field broadening of the electron resonance signal which would mask the spin-exchange contribution to the linewidth.

It would also be difficult to measure the electron-alkali atom spin-exchange cross section by directly measuring the relaxation of $\overline{\langle J_z \rangle}$ or $\overline{\langle \mathbf{I}.\mathbf{J} \rangle}$ due to collisions with unpolarized electrons. The relaxation time $1/T_{eA}$ is small because of the low density of electrons in the optical pumping cell. In addition the measurement of the electron density would be difficult.

F. APPLICATION OF SPIN-EXCHANGE RESULTS TO THE RELAXATION OF THE ALKALI SPIN BY SPIN-RANDOMIZING COLLISIONS

As we have already noted, it is plausible that the relaxation of the alkali atom polarization due to spin-randomizing collisions is equivalent to spin-exchange collisions with spin-$\frac{1}{2}$ particles whose polarization is maintained equal to zero. Calculations of relaxation due to spin randomization (Masnou-Seeuws and Bouchiat, 1967) bear this out.

If we set $\overline{\langle S_z \rangle} = 0$ in equations (C.43) and (C.44), we obtain the coupled differential equations

$$\frac{\mathrm{d}\overline{\langle J_z \rangle}}{\mathrm{d}t} = -\frac{\overline{\langle J_z \rangle}}{T_{eA}} + \frac{2\overline{\langle I_z \rangle}}{(2I+1)^2 T_{eA}}, \qquad \text{...(F.1)}$$

and

$$\frac{\mathrm{d}\overline{\langle I_z \rangle}}{\mathrm{d}t} = -\frac{2\overline{\langle I_z \rangle}}{(2I+1)^2 T_{eA}}. \qquad \text{...(F.2)}$$

We can add these equations to obtain

$$\frac{\mathrm{d}\overline{\langle F_z \rangle}}{\mathrm{d}t} = -\frac{\overline{\langle J_z \rangle}}{T_{eA}}, \qquad \text{...(F.3)}$$

which we could also obtain by setting $\overline{\langle S_z \rangle} = 0$ in equation (C.28). By replacing the spin-exchange relaxation time T_{eA} with T_g the spin-randomization time T_g for collisions with a buffer gas, we have the equations we used to describe buffer gas relaxation in Section V.

The relaxation of the transverse components of $\overline{\langle \mathbf{F} \rangle}$ can be calculated in the same way as was done for $\overline{\langle F_z \rangle}$. For spin-exchange with unpolarized electrons we can quickly obtain the equations

$$\frac{\mathrm{d}\overline{\langle F_x \rangle}}{\mathrm{d}t} = -\frac{\overline{\langle J_x \rangle}}{T_{eA}}, \qquad \text{...(F.4)}$$

$$\frac{\mathrm{d}\overline{\langle F_y \rangle}}{\mathrm{d}t} = -\frac{\overline{\langle J_y \rangle}}{T_{eA}}. \qquad \text{...(F.5)}$$

These equations can be taken to describe the relaxation of $\overline{\langle F_x \rangle}$ and $\overline{\langle F_y \rangle}$ due to spin-randomizing collisions with a buffer gas if we replace T_{eA} with T_g.

VII. Optical-Pumping Experiments

Optical pumping has been used to perform a wide variety of experiments in atomic physics. A number of the most common types of optical-pumping experiments will be discussed in this section. The emphasis will be on those experiments which yield numerical results which can be calculated on the basis of theoretical wavefunctions.

In terms of the number of different kinds of experiments which can be performed, the optical pumping of alkali atoms is perhaps the most powerful optical-pumping technique. The usefulness of this technique has been enhanced by the development of methods for optically pumping alkali atoms over a wide range of temperatures.

A. Alkali Optical Pumping at High and Low Temperatures

In the conventional alkali optical-pumping setup, the alkali metal is present on the walls of the optical-pumping cell, and the density of alkali atoms in the cell is determined by the vapor pressure of the alkali metal at the cell operating temperature. Since the vapor pressure is nearly an exponential function of the temperature, the cell operating temperature is restricted to a very narrow range of less than 100°C. The temperature range for optimum signals is of the order of 50°C or less.

By controlling the alkali density independently of the temperature of the optical-pumping cell, one can greatly increase the range of temperatures over which measurements can be made. Not only does this permit the study of the energy dependence of hyperfine pressure shifts, spin-exchange cross sections, and spin-relaxation times, but it makes possible the spin-exchange polarization of paramagnetic metal atoms with very low vapor pressures. By raising the temperature of the cell to the point where there is a sufficient density of the paramagnetic metal atoms to observe spin-exchange signals, one can measure the ground state magnetic moments and hyperfine structure of these atoms. This makes spin-exchange optical pumping much more competitive with atomic beam methods by increasing the variety of atoms which can be studied.

One can solve the problem of spin-exchange polarizing low vapor pressure metal atoms in other ways. Low vapor pressure metals can be introduced into the cell in a flow system or the metal can be heated by a small oven within the cell (Hofmann-Reinecke *et al.*, 1969). Neither of these methods is particularly convenient and the use of an oven inside the cell has some serious drawbacks. Spin-exchange optical pumping could also be tried using optically-pumped metastable helium atoms, but the signal-to-noise ratio should be better with optically pumped alkali atoms.

The technique for controlling the alkali density (Balling *et al.*, 1969) is described in detail in the next section. The basic idea is to keep the alkali metal in a sidearm attached to the optical-pumping cell. As the cell is heated to high temperatures the sidearm is cooled. If the cell is cooled to low temperatures, the sidearm is heated. The number of alkali atoms which diffuse out of the sidearm into the cell is controlled by the temperature of the sidearm. This technique has been successfully used to obtain spin-exchange optical-pumping signals at temperatures as high as 800°C and as low as −135°C.

B. PRECISION MEASUREMENTS

One of the standard methods for studying the structure of atoms and ions is precision r.f. spectroscopy. Precision measurements of the magnetic moments and hyperfine splittings of the ground states and metastable states of paramagnetic atoms and ions can be made using optical-pumping techniques. In most instances, optical-pumping measurements are competitive with atomic beam or maser measurements of these same quantities.

It is not always clear whether a new measurement of a hyperfine splitting or magnetic moment, which has a higher precision than previous measurements, is really useful if the existing experimental number is already several significant figures beyond the best theoretical value for it. There is sometimes a tendency to downplay precision measurements which appear to offer no more than another significant figure to a number that is already "known precisely enough". This can sometimes be a deceptive argument. As the precision of a measurement improves, new effects may be observed which were too small to be seen previously. Hyperfine pressure shifts, light intensity shifts, and spin-exchange frequency shifts are examples. The measurement of hyperfine splittings in the ground state of a complicated atom may not only add unwanted significant figures to a dipole interaction constant but also give the first significant figure of a very small octupole interaction constant which can be calculated.

The apparent usefulness of certain kinds of precision measurements will oscillate as the precision improves. A measurement of g_J for the hydrogen isotopes with a precision of a few parts in 10^8, for instance, exceeds the precision of theoretical values of g_J calculated to 1st order in α and may not appear to be of much use. Yet it is a stepping stone to measurements of even greater precision which can be used to check calculations to 2nd order in α.

1. *Magnetic Moment Ratios*

Early measurements of the g-factors of atoms were undertaken to obtain values for the anomalous magnetic moment of the electron. The free electron g-factor can be obtained much more precisely by measuring directly the difference between the Larmor and cyclotron frequencies of a free electron (Wesley, 1971), and now g_J values are measured primarily to test the theory of relativistic contributions to atomic magnetism and to test the accuracy of atomic wavefunctions.

Spin-exchange optical-pumping was introduced by Dehmelt (1958) in order to measure the ratio of the g-factor for the free electron to the g-factor of the $S_{1/2}$ ground state of Na^{23}. His initial measurements were not very precise because of the inhomogeneity of the static magnetic fields. The introduction by Pipkin of the magnetically shielded precision-wound solenoid (Hanson and Pipkin, 1965) to achieve highly stable and homogeneous magnetic fields made possible the construction of an inexpensive spin-exchange optical-pumping apparatus capable of ultra precise magnetic moment ratio measurements.

This was first demonstrated (Balling and Pipkin, 1965) in measurements of the ratios $g(e)/g_J(Rb^{85})$, and $g_J(H)/g_J(Rb^{85})$ which were precise to 1 and 0·1 ppm, respectively. In these experiments, free electrons and hydrogen atoms were polarized by spin-exchange collisions with optically-pumped Rb atoms in a field of 60 G. The principal limitation on the precision was the inhomogeneity of the magnetic field.

In a precision g-factor experiment, the object is to minimize the linewidth of the observed Zeeman resonances relative to the Zeeman transition frequencies. If the width of the resonance is determined by the broadening due to spin relaxation or the pumping light, it pays to increase the static magnetic field and thus the Zeeman frequency until the magnetic field inhomogeneity is the principal source of linewidth.

For this reason, magnetic moment ratio measurements are performed at intermediate fields in which the individual Zeeman transitions are well separated, and the Breit–Rabi formula must be used to extract values for g_J from measurements of the Zeeman transitions. Auxiliary constants which are required for the calculation are the nuclear g-factor g_I of the atom and the hyperfine splitting.

Robinson and co-workers have refined the spin-exchange optical-pumping technique for the measurement of g-factor ratios to the point where a precision of parts in 10^9 is possible (White *et al.*, 1968; Hughes and Robinson, 1969a, b). He uses a carefully shimmed, shielded solenoid with a field-stabilized power supply which utilizes an optical-pumping signal as the field sensor. Where possible, evacuated cells with coated walls are used. The absence of a buffer gas allows the atoms to bounce around in the cell and average out the residual field inhomogeneity.

In coated cells with volumes of the order of 100 cm^3 his Zeeman resonances have linewidths of the order of 20 Hz in a field of $\sim$50 G. The part of the linewidth due to field inhomogeneity is $<$1 Hz. In cells filled with buffer gas, the magnetic linewidth (25 Hz for Rb^{87}) amounts to $\sim 5 \times 10^{-7}$ of the Zeeman frequency. The short term field drift is only $\sim 1 \times 10^{-9}$.

At this level of precision his results must be extrapolated to zero light intensity to avoid errors due to light intensity shifts of the Zeeman frequencies. The circular polarization of the pumping light must be reversed to avoid errors due to spin-exchange frequency shifts. The lineshape is carefully monitored to check for possible asymmetries due to inhomogeneity in the static and r.f. fields. When this kind of care is taken, optical pumping is a difficult technique to beat.

A list of measured g_J ratios for atoms with $S_{1/2}$ ground states is given in Table V. Although many of the g_J ratios for the alkali atoms have been obtained from atomic beam experiments, more precise values could probably be obtained in optical-pumping experiments. The most interesting g_J ratios, however, are those involving the hydrogen isotopes and the free electron. Recent calculations of g_J for the ground state of an isotope of hydrogen predict (Faustov, 1970; Close and Osborn, 1971; Grotch and Hegstrom, 1971) that

$$\frac{g_J(\mathrm{H})}{g(e)} = 1 - \frac{\alpha^2}{3}\left[1 - \frac{3}{2}\frac{m}{M} + \frac{3m^2}{M^2}\right] + \frac{1}{4\pi}\alpha^3\left[1 - \frac{5m}{3M} + \frac{7}{3}\frac{m^2}{M^2}\right], \quad \ldots(\mathrm{B.1})$$

where m is the mass of the electron, M is the mass of the proton and α is the fine structure constant.

Optical-pumping and hydrogen maser experiments have been performed to measure $g_J(\mathrm{H})/g_J(\mathrm{D})$. The results are given in Table VI. The results of the optical-pumping experiment are in good agreement with the theoretical value obtained from equation B.1 of

$$g_J(\mathrm{H})/g_J(D) = 1 + (7.22 \times 10^{-9}).$$

TABLE V. *A list of measured g_J ratios for atoms with an $S_{1/2}$ ground state*

Ratio	Value	Experimental method*	References
$g_J(H)/g_J(D)$	1·0000000072 (30)	OP	Hughes and Robinson, 1969a, b
	1·0000000094 (14)	HM	Larson *et al.*, 1969
$g_J(H)/g_J(T)$	1·0000001 (3)	OP	Balling and Pipkin, 1965
$g_J(Li)/g_J(K)$	1·0000034 (3)	AB	Böklen *et al.*, 1967
$g_J(Na)/g_J(K)$	1·0000012 (5)	AB	Böklen *et al.*, 1967
	1·0000007 (2)	AB	Vanden Bout, *et al.*, 1968
$g_J(Rb)/g_J(K)$	1·0000184 (4)	AB	Böklen *et al.*, 1967
	1·0000182 (2)	AB	Vanden Bout *et al.*, 1968
	1·00001844 (6)	OP	Beahn and Bedard, 1972
$g_J(Cs)/g_J(K)$	1·0001228 (3)	AB	Böklen *et al.*, 1967
	1·0001231 (3)	AB	Vanden Bout *et al.*, 1968
$g_J(Rb)/g(e)$	1·00000590 (10)	OP	Tiedeman and Robinson, 1972
$g_J(Rb)/g_J(H)$	1·0000235855 (6)	OP	Hughes and Robinson, 1969a, b
$g_J(Ag)/g_J(K)$	1·0000260 (20)	AB	Dahmen and Penselin, 1967
$g_J(Au)/g_J(K)$	1·0005076 (20)	AB	Dahmen and Penselin, 1967
	1·0005040 (20)	AB	Vanden Bout *et al.*, 1967
$g_J(Cu)/g_J(Cs)$	1·0000 (5)	AB	Ting and Lew, 1957

* The experimental method is indicated by the code AB, atomic beam; OP, optical pumping; HM, hydrogen maser.

TABLE VI. *A list of g_J values for hydrogen-like atoms calculated from Table V using the value $g\,(e) = 2{\cdot}0023193154\ (7)$*

	Experimental	Theoretical*	References
$g_J(H)$	2·0022839	2·002283864	Grotch and Hegstrom, 1971; Faustov, 1970; Close and Osborn, 1971
$g_J(Li)$	2·002301	2·002297	Phillips, 1952; Perl, 1953
$g_J(Na)$	2·002296	2·002293	Phillips, 1952; Perl, 1953
$g_J(K)$	2·002294	2·00230	Phillips, 1952; Perl, 1953
$g_J(Rb)$	2·002331	2·00234	Phillips, 1952; Perl, 1953
$g_J(Cs)$	2·002540	2·00244	Phillips, 1952; Perl, 1953

* The theoretical g_J values for the alkali atoms have been taken from a table prepared by V. W. Hughes (1959) and includes a correction of an error in the original calculations of Perl (1953).

Robinson is attempting to provide a further check of equation (B.1) by obtaining $g_J(\mathrm{H})/g(e)$ from measurements of $g_J(\mathrm{Rb})/g(e)$ (Tiedeman and Robinson, 1972) and $g_J(\mathrm{Rb})/g_J(\mathrm{H})$ (Hughes and Robinson, 1969a, b). The measurements of $g_J(\mathrm{Rb})/g(e)$ are in progress. It is difficult to obtain the highest precision in measurements of the electron resonance because a buffer gas must be used and because of spin-exchange broadening.

Using the latest experimental value for $g(e)$ (Wesley, 1971), g_J values for the hydrogen-like atoms are tabulated alongside theoretical calculations of g_J in Table VI. At present, the precision of g_J calculations for the alkali atoms lags far behind the precision of existing measurements.

By measuring appropriate Zeeman transitions within the same atoms, one can measure g_I/g_J for that atom. This ratio is of fundamental interest in the case of hydrogen and it has been measured very precisely in a hydrogen maser experiment (Winkler *et al.*, 1972). The result is $g_J(\mathrm{H})/g_p(\mathrm{H}) = -658{\cdot}210706(6)$. Calculations (Grotch and Hegstrom, 1971) of the bound state correction to the electron-proton g-factor ratio predict that

$$\frac{g_J(\mathrm{H})}{g_P(\mathrm{H})} = \frac{g(e)}{g_P}(1 + 2{\cdot}8 \times 10^{-8}).$$

In more complicated atoms, precision measurements of g_I/g_J are of interest for two reasons. It is an auxiliary constant which is required to extract a g_J ratio from measurements of Zeeman transitions. Secondly, the values of g_I for the free atom can be compared with NMR measurements to obtain accurate values for chemical shifts. In Table VII values of g_I/g_J for the hydrogen-like atoms are presented.

TABLE VII. *g_I/g_J values for the hydrogen-like atoms*

	g_I/g_J	References
H	$-1{\cdot}519270335\ (14) \times 10^{-3}$	Winkler *et al.*, 1972
D	$-2{\cdot}3321733 \times 10^{-4}$	Hughes and Robinson, 1969a, b
T	$-1{\cdot}62051430\ (17) \times 10^{-4}$	Winkler *et al.*, 1972; Duffy, 1959
Li^6	$-2{\cdot}2356978\ (10) \times 10^{-4}$	Beckmann *et al.*, 1974
Li^7	$-5{\cdot}9042719\ (10) \times 10^{-4}$	Beckmann *et al.*, 1974
Na^{23}	$-4{\cdot}0184406\ (40) \times 10^{-4}$	Beckmann *et al.*, 1974
K^{39}	$-0{\cdot}7088613\ (6) \times 10^{-4}$	Beckmann *et al.*, 1974
K^{41}	$-0{\cdot}3890837\ (4) \times 10^{-4}$	Beckmann *et al.*, 1974
Rb^{85}	$-1{\cdot}4664908\ (31) \times 10^{-4}$	White *et al.*, 1968
Rb^{87}	$-4{\cdot}9699147\ (45) \times 10^{-4}$	White *et al.*, 1968
Cs^{133}	$-1{\cdot}9917405\ (30) \times 10^{-4}$	Robinson *et al.*, 1968

As yet no pressure shifts of atomic g-factors have been observed and they are expected to be small. On the other hand, the current precision of optical-pumping experiments should be sufficient to observe the relativistic shift of the magnetic moment of the free electron as a function of temperature due to its motion in the magnetic field.

2. *Hyperfine Structure*

The hyperfine structure of an atomic ground state is a result of the magnetic dipole, electric quadrupole, and higher multipole interactions between the electronic and nuclear charge distributions. The dominant non-central interaction is the magnetic dipole term.

Schwartz (1955) has presented a thorough theoretical treatment of hyperfine structure in which the hyperfine interactions are represented in terms of a multipole expansion of the potentials. Perturbation theory is used to calculate the energies of the hyperfine levels. The 1st-order energy W_F of each hyperfine level can be expressed in the form (Schwartz, 1955)

$$W_F = \sum_k A_k M(I, J, F, k),$$

where A_k is the $2k$-pole interaction constant to be determined by experiment and the coefficients $M(I,J,F,k)$ are given by formulas derived by Schwartz. The series terminates when $k = 2I$ or $k = 2J$, whichever is smaller. Therefore, only a dipole interaction term can be observed for an atom with an $S_{1/2}$ ground state. In conventional notation, the magnetic dipole, electric quadrupole, and magnetic octupole interaction constants are denoted by the letters A, B, and C, respectively.

Because the magnitude of the interaction energy decreases rapidly as the multipole order of the interaction increases, 2nd-order perturbation terms, which mix excited states into the calculation of the dipole and quadrupole interaction energies, may appear as a contribution to the 1st-order magnetic octupole interaction energy.

It is a straightforward matter to measure the hyperfine splittings in the ground state of atoms by means of optical pumping if the atoms can be pumped directly or spin-exchange polarized. Since the measurements can be made at very low magnetic fields, optical-pumping measurements of the hyperfine splittings do not require accurate knowledge of g_I or g_J. The low field also makes it easy to reduce the contribution to the linewidth due to field inhomogeneity. If the atoms are optically pumped directly and the hyperfine components of the pumping light are unequal (Bender *et al.*, 1958; Ernst and Strumia, 1968), hyperfine pumping will occur and 0–0 transitions ($F, M = 0 \leftrightarrow F \pm 1, M = 0$) can be observed. The advantage of the 0–0 transition is that it is field independent to 1st order, and field inhomogeneities will be even less of a problem.

Fundamental sources of systematic error are buffer gas pressure shifts, light intensity shifts, and spin-exchange frequency shifts. Because the buffer gas pressure shifts are often large and because it is difficult to precisely determine the pressure of buffer gas in the cell, these shifts are usually the primary source of error. The temperature dependence of the buffer gas density shift will lead to line broadening and asymmetries in the lineshape when there are significant temperature gradients across the optical-pumping cell.

The hyperfine frequencies of atoms with $S_{1/2}$ ground states which are accessible to optical-pumping technique are given in Table VIII. The precision

TABLE VIII. *Measured hyperfine frequencies for atoms and ions with an $S_{1/2}$ ground state. The hyperfine splitting of Cs^{133} defines the time scale*

Atom	I	Hyperfine frequency (MHz)	Experimental method*	References
H	1/2	1420·405751800 (28)	HM	Crampton *et al.*, 1963
D	1	327·3843523 (25)	HM	Crampton *et al.*, 1966
T	1/2	1516·7014708076 (50)	HM	Mathur *et al.*, 1966
Li^6	1	228·205261 (12)	OP	Wright *et al.*, 1969
Li^7	3/2	803·5040866 (10)	AB	Beckmann *et al.*, 1974
Na^{23}	3/2	1771·6261288 (10)	AB	Beckmann *et al.*, 1974
K^{39}	3/2	461·7197202 (14)	AB	Beckmann *et al.*, 1974
K^{41}	3/2	254·0138720 (20)	OP	Bloom and Carr, 1960
Cu^{63}	3/2	11735·83 (1)	AB	Ting and Lew, 1957
Cu^{65}	3/2	12568·81 (1)	AB	Ting and Lew, 1957
Rb^{85}	5/2	3035·732439 (5)	AB	Penselin *et al.*, 1962
Rb^{87}	3/2	6834·682614 (3)	AB	Penselin *et al.*, 1962
Ag^{107}	1/2	1712·512111 (18)	AB	Dahman and Penselin, 1967
Ag^{109}	1/2	1976·932075 (17)	AB	Dahman and Penselin, 1967
Cs^{133}	7/2	9192·631770	Frequency standard	
Au^{197}	3/2	6099·320184 (13)	AB	Dahman and Penselin, 1967
$^3He+$	1/2	8665·649867 (10)	IT	Schuessler *et al.*, 1969
$^{135}Ba+$	3/2	7183·3412 (5)	OP	Sichart *et al.*, 1970
$^{137}Ba+$	3/2	8037·7422 (8)	OP	Sichart *et al.*, 1970

* Abbreviations: OP, optical pumping; HM, maser; AB, atomic beam; IT, ion trap.

of the measurements is much greater than the accuracy of theoretical calculations. The value of precision optical-pumping measurements of these frequencies lies primarily in the opportunity they afford for studying the small perturbations of the hyperfine structure due to buffer gas pressure shifts and light intensity shifts.

TABLE IX. *The hyperfine structure and g_J values in the ground states of atoms with $J > \frac{1}{2}$ which can be polarized in optical-pumping experiments*

Atom	J	I	g_J	A^* (MHz)	B^*	C^*
N^{14}	3/2	1	2·002134 (5) (Zak and Shugart, 1972)	10·45092906 (19) (Crampton *et al.*, 1970)	1·32 (20) Hz (Crampton *et al.*, 1970) 1·30 (60) Hz (Weiss *et al.*, 1970)†	
N^{15}	3/2	1/2	2·002134 (5) (Zak and Shuggart, 1972)	−14·645457 (5) (Lambert and Pipkin, 1963)†		
P^{31}	3/2	1/2	2·00165 (40) (Pendlebury and Smith, 1964)	55·055691 (8) (Lambert and Pipkin, 1962)†		
As^{75}	3/2	3/2	1·9965 (8) (Pendlebury and Smith, 1964)	−66·204 (1) (Pendlebury and Smith, 1964	−0·535 (3) MHz (Pendlebury and Smith, 1964)	
Sb^{121}	3/2	5/2	1·9705 (2) (Fernando *et al.*, 1960)	−299·034 (4) (Fernando *et al.*, 1960)	−3·68 (2) MHz (Fernando *et al.*, 1960)	
Sb^{123}	3/2	7/2	1·9705 (2) (Fernando *et al.*, 1960)	−162·451 (3) (Fernando *et al.*, 1960)	−4·67 (3) MHz (Fernando *et al.*, 1960)	

Bi^{209}	3/2	9/2	1·6433 (2) (Title and Smith, 1960)	−446·942 (1) (Hull and Brink, 1970)	−304·654 (2) MHz (Hull and Brink, 1970)	0·0165 (1) MHz (Hull and Brink, 1970)
Mn^{55}	5/2	5/2	2·0012 (1) (Childs *et al.*, 1961)	−72·420836 (15) (Davis *et al.*, 1971)†	−0·019031 (17) MHz (Davis *et al.*, 1971)†	−0·7 (11) Hz (Davis *et al.*, 1971)†
Eu^{151}	7/2	5/2	1·99340 (7) (Pichanick and Woodgate, 1961)	−20·0523 (2) (Sanders and Woodgate, 1960)	−0·7012 (35) MHz (Sanders and Woodgate, 1960)	
Eu^{153}	7/2	5/2	1·99340 (7) (Pichanick and Woodgate, 1961)	−8·8532 (2) (Sanders and Woodgate, 1960)	−1·7852 (35) MHz (Sanders and Woodgate, 1960)	

* The constants *A*, *B*, and *C* are the dipole, quadrupole and octupole interaction constants.
† An optical-pumping experiment.

In the case of S-state paramagnetic atoms with more than two hyperfine levels, however, very precise measurements of the hyperfine frequencies allow one to observe the quadrupole and octupole contributions to the hyperfine structure. Table IX gives the hyperfine frequencies and relevant physical constants of a number of such atoms which can be spin-exchange polarized. Again spin-exchange optical pumping is competitive in precision with atomic beam and hydrogen maser techniques.

C. HYPERFINE PRESSURE (DENSITY) SHIFTS

Measurements of hyperfine pressure shifts in optical-pumping experiments are often by-products of experiments designed to measure the hyperfine splitting of an unperturbed paramagnetic atom. The hyperfine splitting of the atom is measured in a number of cells filled with various pressures of buffer gas. The hyperfine frequency is plotted as a function of buffer gas pressure, and the result is extrapolated to zero pressure. The dependence upon pressure of the Li^7 hyperfine splitting in He, Ne, and A buffer gases is shown as an example in Fig. 19. At pressures below 1 atm, the pressure shifts have always been found to be linear.

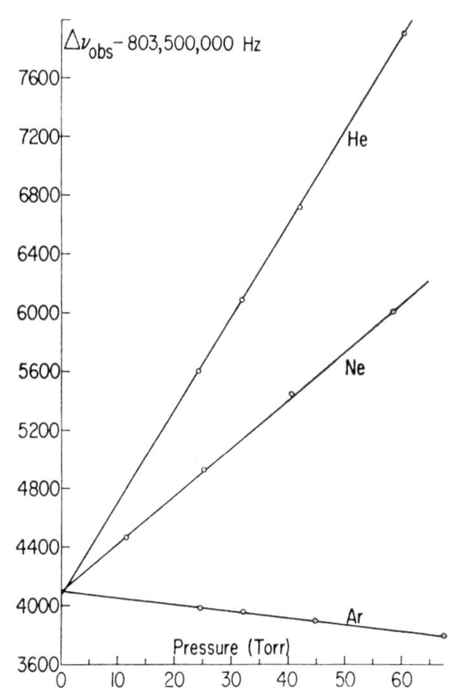

FIG. 19. The hyperfine splitting of Li^7 as a function of buffer gas pressure (Wright *et al.*, 1969).

In recent years, the hyperfine pressure shifts themselves have aroused increasing theoretical interest as they provide a convenient means of testing our understanding of the weak low-energy interactions between atoms. The term "pressure shift" is a misnomer and it should really be called a density shift. Optical-pumping cells are filled with a given pressure of gas at a certain temperature and sealed off. In the course of an experiment, it is the buffer gas density and not the pressure which remains constant. This has been a source of confusion in a number of theoretical papers, particularly in discussions of the temperature dependence of the density shifts.

The density shift results from the perturbation of the paramagnetic atom's wavefunction due to neighboring buffer gas atoms. The hyperfine interaction is dominated by the dipole interaction which is essentially proportional to the electron spin density at the nucleus. At long range, the Van der Waals interaction between the buffer gas atom and the paramagnetic atom pulls the wavefunction away from the nucleus. At close range, the overlap force (Pauli exclusion) between the two electrons has the opposite effect. The net shift is a result of the competition between these two effects. The usual calculational procedure is to calculate the shift $\Delta A(R)$ in the dipole interaction constant due to a buffer gas atom at a distance R and then to weight the shift with the probability $\exp(-V(R)/kT)$ of finding a buffer gas atom a distance R away. $V(R)$ is the interatomic potential. The net shift is written as

$$\langle \Delta A \rangle = \rho \int e^{-V(R)/kT} \Delta A(R) \, dT, \qquad \text{...(C.1)}$$

where ρ is the buffer gas density.

A summary of experimental hyperfine pressure shift measurements for a number of paramagnetic atoms in various buffer gases is given in Table X. The fractional pressure shift $1/A \; \partial A/\partial P$ is listed along with the filling temperature of the cells. The approximate operating temperature at which the measurements were made is also given.

Theoretical calculations of the observed pressure shifts have been moderately successful, and the most recent are listed in Table XI. The striking agreement achieved by some workers (Das and Ray, 1970; Ray *et al.*, 1970) for the pressure shifts of H, Li, and Na in He has been challenged (Ikenberry and Das, 1971) as being fortuitous.

The development of a technique for alkali optical pumping over a wide range of temperatures has increased the usefulness of buffer gas pressure shift measurements. The temperature dependence of the shift at constant buffer gas density provides a more stringent test of theoretical calculations and provides clues as to how they might be improved. The observed temperature dependences of the pressure shift (fixed density) of the deuterium hyperfine splitting (Wright *et al.*, 1970; Wright, 1972) in He, Ne and A buffer gases are

TABLE X. *Measurements of fractional pressure shifts* $1/A\ \partial A/\partial P$ *of the dipole interaction constant in units of* 10^{-9} *Torr*$^{-1}$ *for various atoms*

	He	Ne	A	Kr	Xe	H_2	N_2	Filling temp. (°C)	Operating temp. (°C)	References
H	4·8 ± 0·09	2·88 ± 0·05	−4·78 ± 0·03	−10·4 ± 0·2	−20 ± 2	−0·56 ± 0·10		0	45	Ensberg and Morgan, 1968; Pipkin and Lambert, 1962; Brown and Pipkin, 1968
D	4·1 ± 0·2							27	50	Wright *et al.*, 1970
T		3·24 ± 0·09	−4·78 ± 0·03					0	45	Pipkin and Lambert, 1962; Brown and Pipkin, 1968
Li	77·7 ± 1·0	40·5 ± 1·0	−5·34 ± 0·5					27	390	Wright *et al.*, 1969
Na	73	45	2·8	−42	−85	62	49	22	120	Ramsey and Anderson, 1964
K	93 ± 9	52 ± 4	−1 ± 3	−91 ± 11		71 ± 6		65	65	Bloom and Carr, 1960
Rb	105 ± 2·0	57·4 ± 1·2	−7·5 ± 0·1	−84·9 ± 7·3		96·6 ± 2·0	76·1 ± 1·5	25	25	Bender *et al.*, 1958
Ag		26 ± 3	16 ± 3					25	800	Chase and Lambert, 1973
Cs	174	71	−27	−141	−261	207	101	30	30	Arditi and Carver, 1958
N	20 ± 1	48 ± 1					182 ± 10	27	70	Weiss *et al.*, 1970
P	65·6 ± 4	122 ± 3						0	50	Lambert and Pipkin, 1962
Mn		26 ± 2						24	700	Davis *et al.*, 1971
$(^3S_1)He^3$	−7·4 ± 3·0							0	−258	Rosner and Pipkin, 1970

TABLE XI. *Recent theoretical calculations of fractional pressure shifts* $1/A\ \partial A/\partial P$ *in units of* 10^{-9} *Torr*$^{-1}$

	He	Ne	A	Kr	Xe	Temperature °C*	References
H	5					50	Kunik and Kaldor, 1971
	4·97†					50	Das and Ray, 1970
	1·7	0·11	−3·1	−4·3	−6·6	0	Rao *et al.*, 1970
Li	78·9†					114	Ray *et al.*, 1970
Na	78·1†					20	Ray *et al.*, 1970
N	50					50	Ray *et al.*, 1968a; Rao and Das, 1969
He($^3S_1$)	−15						Dutta *et al.*, 1973

* The specified temperature is the operating and filling temperature.
† The numbers indicated by a dagger have been challenged (Ikenberry and Das, 1971) as being in fortuitous agreement with experiment.

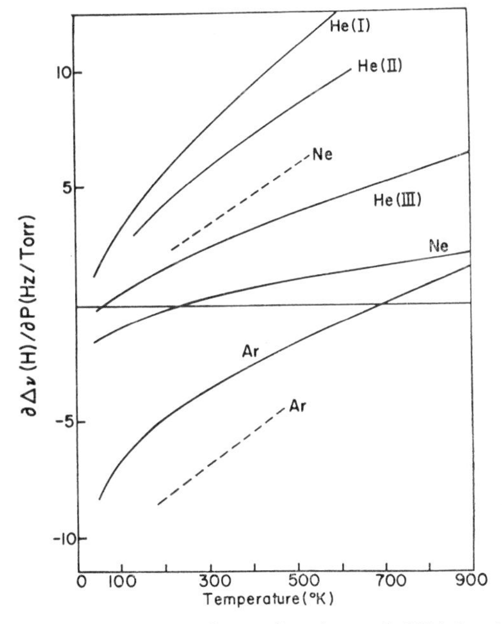

FIG. 20. A comparison of the experimentally observed (Wright, 1972) temperature dependence of the hyperfine pressure shift in He, Ne, and Ar with theory. The solid curves for Ne, Ar, and He (III) are the calculations of Rao *et al.* (1970). The He (I) curve gives the calculations of Kunik and Kaldar (1971). The dashed Ne, and Ar curves and the solid He (II) curve are experimental results.

compared to theoretical calculations of the temperature dependence (Rao *et al.*, 1970; Kunik and Kaldor, 1971) in Fig. 20. It is clear from Table XI and Fig. 20 that the theoretical calculations are on the right track but can stand improvement.

The pressure shifts of the hydrogen isotopes in an argon buffer gas are of particular interest because of their application to the analysis of experiments in which the hyperfine splitting of muonium is measured in the presence of argon. The mass dependence of the fractional hyperfine pressure shift should be sufficiently small (Clarke, 1962) so that it is fair to require that the observed pressure shift of the muonium hyperfine splitting in argon agree with the optical-pumping results for the fractional pressure shift of the hydrogen isotopes. Brown and Pipkin (1968) have found experimentally that the ratio of the fractional pressure shift of tritium in argon to that of hydrogen is $1{\cdot}007 \pm 0{\cdot}012$, and current values for the muonium pressure shift are in agreement with the optical-pumping values.

The muonium experiments are carried out at high pressures which introduces the possibility of a nonlinear pressure shift due to 3-body interactions. Ensberg and zu Putlitz (1969) have observed a nonlinear pressure shift of the Rb hyperfine splitting in argon. Because of the shift in the Rb absorption line in the high density buffer gas they used a high intensity white light source to optically pump the Rb atoms. They made measurements at pressures of 4170 Torr (0°C) and 7390 Torr (°C) at an operating temperature of 80°C and found a fractional density shift $1/A\ \partial A/\partial P = a + bP$, with $a = (-11{\cdot}5 \pm 0{\cdot}3) \times 10^{-9}$/Torr (0°C) and $b = (10 \pm 4) \times 10^{-14}/\text{Torr}^2$ (0°C).

Recently, Ray and Kaufman (1972) have calculated the quadratic term b and obtained qualitative agreement with experiment.

Parenthetically, we note that the presence of a buffer gas can also shift the observed g_J values of paramagnetic atoms. The effect is sufficiently small, however, that it is difficult to study such shifts experimentally. Herman (1968) has calculated the fractional shift in g_J for Rb atoms in a He buffer gas to be

$$\frac{\Delta g_J}{g_J} = 5 \times 11^{-11}/\text{Torr He},$$

which is in agreement with the observations of Robinson and co-workers (Herman, 1968).

D. ELECTRON–ALKALI ATOM SPIN-EXCHANGE COLLISIONS

In recent years, atomic physicists have succeeded in bringing theoretical calculations of electron–alkali atom collision cross sections and experimental measurements of these cross sections into agreement in the low energy range between 0·5 and 3 eV. Unfortunately, the same can not yet be said for the electron–alkali atom spin-exchange cross section at thermal energies (~0·03 eV) as measured in spin-exchange optical-pumping experiments. Thus far optical-pumping measurements have provided the only information on electron–alkali atom scattering phase shifts at these very low energies, and the results are not in agreement with existing phase shift calculations.

The total spin-exchange cross section averaged over the thermal velocity distribution of the electrons is obtained from measurements of the linewidth of the electron spin resonance. As was shown in Section VI, the electron signal δI has the form

$$\delta I = A \frac{\omega_1^2 \tau_1 \tau_2}{1 + \omega_1^2 \tau_1 \tau_2 + (\omega - \omega_0 + \delta\omega_0)^2 \tau_2^2}, \qquad \text{...(D.1)}$$

with

$$\frac{1}{\tau_2} = \frac{1}{T_{2e}} + \frac{1}{T_e}, \qquad \text{...(D.2)}$$

where T_e is the spin-exchange relaxation time and T_{2e} is the transverse relaxation time for the electron spin due to all other relaxation mechanisms. The spin-exchange relaxation time T_e is related to the spin-exchange cross section σ_{SE} by

$$\frac{1}{T_e} = N\langle v\sigma_{SE}\rangle, \qquad \text{...(D.3)}$$

where N is the density of the alkali atoms in the cell, v is the velocity of an electron, and the bracket indicates an average over the electron velocity distribution.

The width $\Delta\omega$ of the electron resonance is given by

$$\Delta\omega^2 = \frac{1}{\tau_2^2} + \omega_1^2 \frac{\tau_1}{\tau_2}. \qquad \text{...(D.4)}$$

By plotting the square of the electron linewidth at fixed density against the r.f. power ω_1^2 one obtains a straight line of slope τ_1/τ_2 with a zero-r.f. intercept $1/\tau_2^2$. From the intercept and a knowledge of the alkali density, one can determine $\langle v\sigma_{SE}\rangle$. The value of ω_1^2 at a given r.f. power can be determined by measuring the frequency of the wiggles on the alkali optical-pumping signal. The alkali atom density can be determined by direct optical absorption measurements (Jarrett, 1964; Rozwadwski and Lipworth, 1965; Gibbs and Hull, 1967; Ioli *et al.*, 1971) or by use of vapor pressure curves for the alkali metals.

Because a bare pyrex or quartz surface adsorbs alkali atoms, it is important to have the walls of the optical-pumping cell well coated with at least a film of alkali metal if vapor pressure curves are to be used. Otherwise the density of alkali atoms in a cell at a given temperature can be considerably lower than that predicted on the basis of vapor pressure curves. This is especially true for sodium.

This means that the vapor pressure curves can only be used to determine the density of alkali atoms over the very narrow temperature range in which optical-pumping signals can be observed with the alkali metal on the walls of the cell. Yet it is obviously important to observe the energy dependence of the quantity $\langle v\sigma_{SE}\rangle$ in order to provide a critical test to existing phase shift calculations.

The method for optical pumping at high and low temperatures described in Section 1 enables one to measure electron linewidths at fixed density over a wide range of temperatures. In order to measure the density of alkali atoms in the cell under conditions where the density is not determined by the temperature of the cell, one can pass a beam of white light through the cell and monitor the fractional absorption of the white light by the alkali atoms. This fractional absorption can then be compared to the absorption measured under conditions which allow the use of vapor pressure curves. Since the absorption of white light is insensitive to changes in the absorption line shape and line center due to changes in the temperature, equal fractional absorptions implies equal alkali atom densities. Care must be taken, however, to avoid serious errors due to density gradients within the cell. The presence of density gradients can be checked for by masking off various parts of the cell and monitoring the absorption of pumping light through the corresponding portions of the cell volume.

The source of electrons is typically an r.f. discharge or the ionizing radiation from tritium gas in the cell. The tritium produces very strong and quiet electron signals, but it can not be used in quartz cells at high temperatures because of diffusion through the walls.

Experimental values for $\langle v\sigma_{SE}\rangle$ for e-Rb and e-Na collisions at various temperatures are given in Table XII. Theoretical values for the case of sodium are given for comparison.

Additional information of electron–alkali atom scattering phase shifts can be obtained by measuring the spin-exchange frequency shift $\delta\omega_0$. As was shown in Section VI, the ratio of the frequency shift $\delta\omega_0$ to the linewidth $\Delta\omega$ is independent of the density of alkali atoms and is given by

$$\frac{\delta\omega_0}{\Delta\omega}=\frac{1}{2}\frac{\left\langle v\sum\limits_{l=0}(2l+1)\sin^2(\delta_l^3-\delta_l^1)\right\rangle}{\left\langle v\sum\limits_{l=0}^{\infty}(2l+1)\sin^2(\delta^3-\delta_l^1)\right\rangle}\overline{\langle\sigma_z(A)\rangle} \qquad \text{...(D.5)}$$

where $\overline{\langle\sigma_z(A)\rangle}$ is the electronic polarization of the alkali atoms. Since the sign of the shift changes with the sign of the polarization, the frequency shift can be measured by measuring the electron resonance frequency with left and right circularly polarized pumping light.

To relate the ratio $\delta\omega_0/\Delta\omega$ to the scattering phase shifts, one must measure the magnitude and sign of $\overline{\langle\sigma_z(A)\rangle}$. If the hyperfine components of the pumping

TABLE XII. *The temperature dependence of the electron resonance linewidth $\Delta\nu$ at constant alkali density N for e-Na and e-Rb collisions. The ratio $\Delta\nu/N = \langle v\sigma_{SE}\rangle$, where v is the velocity of the electrons, σ_{SE} is the spin-exchange cross section, and the bracket indicates an average over the thermal velocity distribution of the electrons. The density N is determined from the published vapor pressure curves and the measured cell temperature with the cell walls coated with alkali metal*

Na	Expt.* (Davis and Balling, 1973)	Theory (Moores and Norcross, 1972)
	$\frac{\Delta\nu}{N}(403°\text{K}) = (1{\cdot}3 \pm 0{\cdot}3) \times 10^{-7}$ cm³/sec	$\frac{\Delta\nu}{N}(403°\text{K}) = 0{\cdot}45 \times 10^{-7}$ cm³/sec
	$\frac{\Delta\nu/N\,(725°\text{K})}{\Delta\nu/N\,(403°\text{K})} = 1{\cdot}48 \pm 0{\cdot}15$	$\frac{\Delta\nu/N\,(725°\text{K})}{\Delta\nu/N\,(403°\text{K})} = 1{\cdot}73$
Rb	Expt.† (Davis and Balling, 1973)	
	$\frac{\Delta\nu}{N}(298°\text{K}) = (2{\cdot}1 \pm 0{\cdot}5) \times 10^{-7}$ cm³/sec	
	$\frac{\Delta\nu/N\,(660°\text{K})}{\Delta\nu/N\,(298°\text{K})} = 1{\cdot}32 \pm 0{\cdot}15$	

* N(403°K) = 5×10^{10}Na atoms/cm³ (Iol *et al.*, 1971).

† N(300°K) = $1{\cdot}1 \times 10^{10}$Rb atoms/cm³ (Ditchburn, R. W. and Gilmour, J. C., 1941). *Rev. Mod. Phys.* **13**, 310.

light are essentially equal, the fractional absorption of the pumping light by the alkali atoms is proportional to $1 - \overline{\langle\sigma_z(A)\rangle}$. The polarization $\overline{\langle\sigma_z(A)\rangle}$ can be held at zero by applying a continuous saturating r.f. field at the low field Zeeman frequency. The polarization magnitude $|\overline{\langle\sigma_z(A)\rangle}|$ is obtained by comparing the fractional absorption of the pumping light with no r.f. field with the absorption in the presence of the saturating r.f. field.

The sign of $\overline{\langle\sigma_z(A)\rangle}$ can be determined by observing the relative amplitudes of the Zeeman transitions in an intermediate static magnetic field in which they are resolved to determine which states are more heavily populated.

For the case of Rb-e collisions (Davis and Balling, 1972), no temperature dependence of the frequency shift was observed in the temperature range (200–600°K).

E. SPIN-EXCHANGE BETWEEN ALKALI ATOMS

Numerous experiments have been performed to measure the spin-exchange cross section for collisions between atoms. The most recent and reliable of these have taken advantage of the fact that the relaxation of $\overline{\langle \mathbf{I}.\mathbf{J}\rangle}$, in contrast to the relaxation of $\overline{\langle J_z\rangle}$, is not complicated by the effect of nuclear spin. The double exponential decay of $\overline{\langle J_z\rangle}$ due to a single relaxation mechanism makes data analysis more difficult.

Several calculations of spin-exchange cross sections for collisions between alkali atoms have been made, and they are in reasonable agreement with experimental observations. Since the optical-pumping measurements necessarily involve a thermal distribution in the velocities of the atoms, the observed relaxation times have been related to the spin-exchange cross section by assuming that the cross section has little or no temperature dependence. This assumption is consistent with theoretical predictions, and it appears to be born out by comparisons of optical-pumping data near room temperature with data obtained in an E.P.R. experiment (Ressler *et al.*, 1969) in the temperature range (500–700°K).

The relaxation of $\overline{\langle \mathbf{I}.\mathbf{J}\rangle}$ due to collisions with buffer gas atoms or the walls, spin-exchange collisions with another species of alkali atom, and self-exchange collisions under the conditions that $\overline{\langle J_z\rangle} = 0$ for all alkali atoms is given by the simple equation (Grossetete, 1964, 1968).

$$\frac{\mathrm{d}\overline{\langle \mathbf{I}.\mathbf{J}\rangle}}{\mathrm{d}t} = -\left(\frac{1}{T_1} + \frac{1}{T_e} + \frac{1}{T_{e'}}\right)\overline{\langle \mathbf{I}.\mathbf{J}\rangle}, \qquad \text{...(E.1)}$$

where T_e is the self-exchange time, T_e' is the spin-exchange time for collisions with the 2nd species of alkali atoms, and T_1 is the relaxation time due to wall collisions, etc.

The relaxation of $\overline{\langle \mathbf{I}.\mathbf{J}\rangle}$ under the conditions that $\overline{\langle \mathbf{J}_z\rangle}=0$ can be observed in several ways. The difference in hyperfine populations is achieved with optical pumping using D_1 or D_2 lines or both. If the hyperfine components are unequal, hyperfine pumping can be carried out with unpolarized light and no polarization of the alkali atoms will result. On the other hand, if the pumping radiation is circularly polarized and Zeeman pumping is used to achieve differences in the hyperfine population levels, one must apply saturating r.f. fields at the Zeeman frequencies of the alkali atoms to maintain $\overline{\langle J_z\rangle}=0$. The decay of $\overline{\langle \mathbf{I}.\mathbf{J}\rangle}$ in the absence of the pumping light can be observed using Franzen's method or by using a second weak monitoring beam. The control and measurement of the alkali atom densities must be accomplished in essentially the same manner as described in the previous section.

Of course, it is not necessary to measure the spin-exchange time in this manner. One could also observe the relaxation of $\overline{\langle J_z\rangle}$ and fit the data to a double exponential to measure the spin relaxation due to collisions between different species of atoms or make the measurements in a high magnetic field. In a high field, the nuclear and electronic spins decouple and the relaxation of $\overline{\langle J_z\rangle}$ is unaffected by the hyperfine interaction. Alternatively, one could measure the linewidths of resolved Zeeman transitions. None of these methods appear to be as convenient or reliable as the observation of $\overline{\langle \mathbf{I}.\mathbf{J}\rangle}$.

A comparison of experimental measurements of spin-exchange cross sections with theoretical calculations are given in Tables XIII and XIV. In all cases the experimental cross section has been obtained from the observed relaxation time by assuming that the cross section has no energy dependence.

The agreement between experiment and theory is best for the most recent calculations of Chang and Walker and it would appear that spin-exchange collisions between alkali atoms in this energy range are reasonably well understood. It would be interesting to see, however, how well the predicted temperature dependence of the thermally averaged cross sections would agree with optical-pumping measurements over an extended temperature range.

F. SPIN-RELAXATION TIMES

The relaxation of atomic polarization through collisions with inert buffer gas atoms and with the walls of the optical-pumping cell has been investigated experimentally and theoretically for more than a decade. Spin-relaxation times are of intrinsic interest, because they shed light on the weak interatomic forces which produce spin disorientation. Unfortunately, not all of the possible pitfalls in experimental technique and in the theoretical interpretation of the observations were appreciated in early experiments. As a result, many of the early measurements of spin-relaxation times are suspect.

TABLE XIII. *Experimental values for the thermally averaged spin-exchange cross section in units of* 10^{-14} cm^2 *for collisions between alkali atoms. The experiments were performed at the indicated temperatures*

Na–Na	K–K	Rb–Rb	Rb–Cs	Cs–Cs	Temperature (°K)	Reference
$1 \cdot 109 \pm 0 \cdot 005$					390	Moretti and Strumia, 1971
$1 \cdot 03 \pm 0 \cdot 21$	$1 \cdot 45 \pm 0 \cdot 21$	$1 \cdot 9 \pm 0 \cdot 2$		$2 \cdot 06 \pm 0 \cdot 02$	500–700	Ressler *et al.*, 1969*
				$2 \cdot 18 \pm 0 \cdot 12$	300	Beverini *et al.*, 1971
	$2 \cdot 7 \pm 0 \cdot 7$	$2 \cdot 54 \pm 0 \cdot 38$	$2 \cdot 00 \pm 0 \cdot 36$	$2 \cdot 22 \pm 0 \cdot 35$	350	Grossetete and Brossel, 1967
		$1 \cdot 9 \pm 0 \cdot 2$	$2 \cdot 3 \pm 0 \cdot 2$		351	Gibbs and Hull, 1967
		$1 \cdot 85 \pm 0 \cdot 23$			363	Jarrett, 1964
		$2 \cdot 02 \pm 0 \cdot 2$			343	Vanier, 1967†

* The measurements of Ressler *et al.* (1969) were made in an E.P.R. experiment.
† Those of Vanier (1967) were made in a Rb-maser experiment.

TABLE XIV. *Theoretical calculations of spin-exchange cross sections in units of* 10^{-14} cm^2 *for collisions between alkali atoms at specified temperatures*

	H	Li	Na	K	Rb	Cs	Temperature (°K)	References
H	0·2	0·5	0·5	0·6	0·6	0·6	540	Dalgarno and Rudge, 1965
	0·19	0·35	0·36	0·39	0·40	0·41	500	Smirnov and Chibsov, 1965
	0·18	0·33	0·34	0·37	0·37	0·38	1000	Smirnov and Chibsov, 1965
Li		0·9	1·0	1·1	1·1	1·2	540	Dalgarno and Rudge, 1965
		0·89	0·93	1·1	1·1	1·2	500	Smirnov and Chibsov, 1965
		0·85	0·88	1·0	1·1	1·1	1000	Smirnov and Chibsov, 1965
		1·05					300	Chang and Walker, 1968*
		1·0					500	Chang and Walker, 1968*
Na			1·1	1·2	1·3	1·4	540	Dalgarno and Rudge, 1965
			0·98	1·2	1·2	1·0	500	Smirnov and Chibsov, 1965
			0·94	1·1	1·2	1·2	1000	Smirnov and Chibsov, 1965
			1·24				300	Chang and Walker, 1968*
			1·19				500	Chang and Walker, 1968*
K				1·5	1·5	1·6	540	Dalgarno and Rudge, 1965
				1·4	1·5	1·6	500	Smirnov and Chibsov, 1965
				1·3	1·4	1·5	1000	Smirnov and Chibsov, 1965
				2·44			300	Chang and Walker, 1968*
				2·34			500	Chang and Walker, 1968*
Rb					1·6	1·8	540	Dalgarno and Rudge, 1965
					1·6	1·7	500	Smirnov and Chibsov, 1965
					1·5	1·6	1000	Smirnov and Chibsov, 1965
					2·70		300	Chang and Walker, 1968*
					2·59		500	Chang and Walker, 1968*
Cs						1·9	540	Dalgarno and Rudge, 1965
						1·8	500	Smirnov and Chibsov, 1965
						1·7	1000	Smirnov and Chibsov, 1965
						2·82	300	Chang and Walker, 1968*
						2·73	500	Chang and Walker, 1968*

* The cross sections of Chang and Walker (1968) are thermally averaged.

(1) *Experimental Considerations*

Most spin-relaxation experiments involve the observation of transient optical-pumping signals. The atoms are prepared in an initial state of polarization by optical pumping. The pumping light is switched off, and the polarization decays with time. The time dependence of the polarization can be monitored by allowing the polarization to decay in the dark (Franzen, 1959) for various lengths of time, turning the pumping light back on and recording the initial light absorption. It is becoming more common to continuously monitor the polarization with a second much weaker beam (Bouchiat and Grossetete, 1966; Minguzzi *et al.*, 1966a).

As Bouchiat (1965) has pointed out, it is important to understand the spectral profile of the monitoring beam in order to know the physical significance of the measured absorption. As was discussed in Section V, the absorption of D_1 light can be proportional to $(1 \pm 2\overline{\langle J_z \rangle})$, $(1 \pm 2\overline{\langle J_x \rangle})$, $\overline{\langle \mathbf{I}.\mathbf{J} \rangle}$ or a linear combination of the above depending upon the direction of the monitoring beam relative to the pumping beam, the polarization, and the relative intensities of the hyperfine components in the monitoring beam. The degree of optical thickness of the sample can be important if the signals are to be linear functions of these observables.

It is crucial to avoid contamination of the cell by traces of highly disorienting gases or volatile substances in measurements of buffer gas relaxation time (Franz, 1965). The relaxation due to trace amounts of certain contaminants (pump oil is a particular example) can be significantly large.

Measured relaxation times must be extrapolated to zero alkali density to avoid error due to spin-exchange collisions between alkali atoms.

(2) *Theoretical Interpretation of Transient Signals*

In the earliest relaxation experiments, it was assumed that the pumping beam was monitoring $\overline{\langle J_z \rangle}$ and that the decay of $\overline{\langle J_z \rangle}$ was exponential. This will not, however, be true in most experimental situations. Masnou-Seeuws and Bouchiat (1967) have written a definitive paper on alkali atom spin relaxation under various cell conditions. They have derived the form of the transient signals to be expected for cells with paraffin coated walls with and without buffer gas and for cells with bare walls and a buffer gas.

For very high buffer gases, relaxation at the walls of the cell can be neglected and relaxation occurs simply through collisions with the inert buffer gas atoms. The principal spin disorientating interaction is the spin-orbit coupling between the alkali valence electron spin and the relative orbital angular momentum of the two particles in an alkali atom-buffer gas atom collision (Herman, 1964, 1965). In such collisions the valence electron spin is randomized. If there were no hyperfine interaction in the ground state of the alkali atom, the relaxation of $\overline{\langle J_z \rangle}$ would be a single exponential characterized by a relaxation time

$$\frac{1}{T_g} = \langle v N_g \sigma_g \rangle, \qquad \text{...(F.1)}$$

where v is the velocity of the alkali atoms relative to the buffer gas atoms, N_g is density of the buffer gas, σ_g is the cross section for spin disorientation, and the bracket indicates an average over the atomic velocity distribution. As was discussed in Sections V and VI, the coupled equations which describe the relaxation of $\overline{\langle J_z \rangle}$ and $\overline{\langle \mathbf{I}_z \rangle}$ due to spin-randomizing collisions are

$$\frac{d\overline{\langle J_z \rangle}}{dt} = -\frac{\overline{\langle J_z \rangle}}{T_g} + \frac{2}{(2I+1)^2} \frac{\overline{\langle I_z \rangle}}{T_g}, \qquad \text{...(F.2)}$$

$$\frac{d\overline{\langle J_z \rangle}}{dt} = -\frac{2}{(2I+1)^2} \frac{\overline{\langle I_z \rangle}}{T_g}. \qquad \text{...(F.3)}$$

The relaxation of $\overline{\langle J_z \rangle}$ in the dark is therefore described by

$$\frac{\overline{\langle J_z(t) \rangle}}{\overline{\langle J_z(0) \rangle}} = (1-a)\exp(-t/T_g) + a \exp\left(\frac{-2t}{(2I+1)^2 T_g}\right), \qquad \text{...(F.3)}$$

where

$$a = \frac{2\overline{\langle I_z(0) \rangle}}{\overline{\langle J_z(0) \rangle}} \frac{1}{(4I^2+4I-1)}. \qquad \text{...(F.4)}$$

The form of equations (F.2) and (F.3) are identical to those obtained in Section VI for the relaxation of $\overline{\langle J_z \rangle}$ due to spin-exchange collisions with another species of alkali atoms having zero polarization. As was the case for spin-exchange relaxation, the relaxation of $\overline{\langle \mathbf{I}.\mathbf{J} \rangle}$ is a simple exponential (Masnou-Seeuws and Bouchiat, 1967) described by the equation

$$\frac{d\overline{\langle \mathbf{I}.\mathbf{J} \rangle}}{dt} = \frac{-\overline{\langle \mathbf{I}.\mathbf{J} \rangle}}{T_g}. \qquad \text{...(F.5)}$$

It should be noted that self-exchange collisions between the alkali atoms will slow down the relaxation of $\overline{\langle \mathbf{I}.\mathbf{J} \rangle}$ if $\overline{\langle J_z \rangle} \neq 0$.

In a strong magnetic field (Paschen–Back limit), **I** and **J** are decoupled and precess separately about the steady magnetic field. In this case, the nuclear spin will not affect the relaxation of $\overline{\langle J_z \rangle}$ and it will decay exponentially with time constant T_g (Franz *et al.*, 1971).

A second simple situation is the case of an evacuated optical-pumping cell with paraffin-coated walls. The relaxation occurs again through the interaction of the electron spin with the time varying fields in the vicinity of the wall. The disorientation collisions are weak and randomize the electron spin without directly affecting the nuclear spin. The equations governing the relaxation of $\overline{\langle J_z \rangle}$, $\overline{\langle I_z \rangle}$, and $\overline{\langle \mathbf{I}.\mathbf{J} \rangle}$ are of the same form (Bouchiat and Brossel, 1966) as equations (F.2), (F.3) and (F.5).

For cells with moderate buffer gas densities and bare cell walls, the diffusion of the alkali atoms to the walls becomes an important mode of relaxation. Experiment has shown that the collisions with a bare glass wall are "strong" and produce "uniform" relaxation. After a single wall collision, an alkali atom can be found in any ground state sublevel with equal probability. The observables $\overline{\langle \mathbf{I}.\mathbf{J} \rangle}$, $\overline{\langle J_z \rangle}$, and $\overline{\langle I_z \rangle}$ are all zero at the cell wall. The time derivatives of the observables are given by (Masnou-Seeuws and Bouchiat, 1967; Minguzzi *et al.*, 1966a; Franz, 1972)

$$\frac{\mathrm{d}\overline{\langle J_z \rangle}}{\mathrm{d}t} = -D\nabla^2\overline{\langle J_z \rangle} - \frac{\overline{\langle J_z \rangle}}{T_g} + \frac{2\overline{\langle I_z \rangle}}{(2I+1)^2 T_g}, \qquad \text{...(F.6)}$$

$$\frac{\mathrm{d}\overline{\langle I_z \rangle}}{\mathrm{d}t} = -D\nabla^2\overline{\langle I_z \rangle} - \frac{2\overline{\langle I_z \rangle}}{(2I+1)^2 T_g}, \qquad \text{...(F.7)}$$

and

$$\frac{\mathrm{d}\overline{\langle \mathbf{I}.\mathbf{J} \rangle}}{\mathrm{d}t} = -D\nabla^2\overline{\langle \mathbf{I}.\mathbf{J} \rangle} - \frac{\overline{\langle \mathbf{I}.\mathbf{J} \rangle}}{T_g}, \qquad \text{...(F.8)}$$

where D is the diffusion constant for a given temperature and buffer gas pressure p. The solution of these diffusion equations subject to the boundary conditions

$$\overline{\langle J_z \rangle} = \overline{\langle I_z \rangle} = \overline{\langle \mathbf{I}.\mathbf{J} \rangle} = 0 \qquad \text{...(F.9)}$$

yields in general an infinite denumberable set of time constants corresponding to the various diffusion modes.

The relaxation of $\overline{\langle \mathbf{I}.\mathbf{J} \rangle}$ in a cylindrical cell is described by the general solution (Beverini *et al.*, 1971)

$$\overline{\langle I.J(t) \rangle} = \overline{\langle \mathbf{I}.\mathbf{J} \rangle}_0 \sum_{\substack{v=0 \\ i=1}}^{\infty} B_{iv}\, \mathrm{e}^{-t/\tau_{iv}} \qquad \text{...(F.10)}$$

where

$$\frac{1}{\tau_{iv}} = [\pi^2(2v+1)^2/L^2 + \mu_i^2/r^2]\, D + \frac{1}{T_g}, \qquad \text{...(F.11)}$$

and

$$B_{iv} = \tau_{iv} J_1\left(\frac{\mu_i r_p}{r}\right) \cdot J_1\left(\frac{\mu_i r_a}{r}\right) \Big/ [\pi\mu_i(2v+1) J_1(\mu_i)]^2. \qquad \text{...(F.12)}$$

In these equations μ_i is the ith zero of $J_0(x)$, L is the length of the cell, r is the radius of the cell, r_p is the radius of the pumping beam and r_a is the radius of the detection beam. A traditional approach to equation (F.10) is to neglect diffusion modes higher than the first (Franzen's approximation) which gives the simple result

$$\frac{\mathrm{d}\overline{\langle \mathbf{I}.\mathbf{J}\rangle}}{\mathrm{d}t} = -\frac{\overline{\langle \mathbf{I}.\mathbf{J}\rangle}}{\tau_1}, \qquad \text{...(F.13)}$$

where

$$\frac{1}{\tau_1} = \frac{1}{T_g} + \left[\frac{\pi^2}{L^2} + \frac{\mu_1^2}{r^2}\right] D. \qquad \text{...(F.14)}$$

Keeping only the first diffusion mode, we have for the relaxation of $\overline{\langle J_z\rangle}$ and $\overline{\langle I_z\rangle}$

$$\frac{\mathrm{d}\overline{\langle J_z\rangle}}{\mathrm{d}t} = -\frac{\overline{\langle J_z\rangle}}{\tau_1} + \frac{2}{(2I+1)^2}\frac{\overline{\langle I_z\rangle}}{T_1}, \qquad \text{...(F.15)}$$

and

$$\frac{\mathrm{d}\overline{\langle I_z\rangle}}{\mathrm{d}t} = -\frac{\overline{\langle I_z\rangle}}{\tau_1'}, \qquad \text{...(F.16)}$$

where

$$\frac{1}{\tau_1'} = \frac{2}{(2I+1)^2 T_g} + \left[\frac{\pi^2}{L^2} + \frac{\mu_1^2}{r^2}\right] \mathrm{D}. \qquad \text{...(F.17)}$$

These equations have the same form for a cell with spherical geometry but with different values for the relaxation time describing the diffusion to the walls (Minguzzi *et al.*, 1966a).

Masnou-Seeuws and Bouchiat (1967) have looked at the effect of including the second diffusion mode, and the full series solution has been fitted to the data for the relaxation of Na and Cs by Strumia and co-workers (Minguzzi *et al.*, 1966a; Moretti and Strumia, 1971; Strumia and Moretti, 1972). As is pointed out in these last references, anomalously large temperature dependences observed in earlier relaxation experiments may be attributed to the neglect of higher diffusion modes, especially in samples which are not optically thin. Franz (1972) has recently investigated by computer simulation the validity of Franzen's approximation. He finds that the approximation is a good one for optically thin samples and low polarizations.

Masnou-Seeuws and Bouchiat have also investigated in detail the situation of well-coated cell walls and a buffer gas as a function of the buffer gas density and the quality of the wall coating. When the buffer gas pressure is low and the coating is of high quality, the relaxation is dominated by buffer gas relaxation and only the first diffusion mode need be considered.

(3) *Results of Relaxation Experiments*

Perhaps the most dramatic discovery which has come out of relaxation experiments is evidence of the existence of short lived Rb-Kr molecules (Aymar *et al.*, 1967; Bouchiat *et al.*, 1967, 1969). Bouchiat *et al.* observed the relaxation of $\overline{\langle J_z \rangle}$ and $\overline{\langle \mathbf{I}.\mathbf{J} \rangle}$ as a function of Kr density and of the static magnetic field. They carefully coated the cell walls to avoid wall effects. At high fields the relaxation of $\overline{\langle J_z \rangle}$ and $\overline{\langle \mathbf{I}.\mathbf{J} \rangle}$ depended linearly on buffer gas density and was consistent with straightforward spin randomization (Herman, 1964). At lower fields, however, the relaxation time for $\overline{\langle J_z \rangle}$ exhibited an unexpected field dependence and pressure dependence which would be explained by the formation of Rb–Kr molecules.

More prosaic information, which has been extracted from studies of spin relaxation in cells containing a buffer gas, is given in Tables XV and XVI. Table XV gives values for the diffusion constant D_0 at standard temperature and pressure for Na and Cs atoms in various buffer gases. Table XVI contains a list of spin-disorientation cross sections. These tables only include the results of recent experiments in which the subtleties of the theoretical interpretation of transient signals was clearly understood.

The pressure dependence of the spin-relaxation time of Cs atoms in a cell with uncoated walls containing a buffer gas for various buffer gases (Beverini *et al.*, 1971) is shown in Fig. 21. Those curves are typical of what to expect with the other alkali atoms as well. The relaxation time increases as the pressure increases until the spin relaxation due to the buffer gas becomes comparable to the relaxation arising from diffusion to the walls. The optimum pressure for the longest possible relaxation time is of course a function of the spin disorientation cross section of the gas.

The relatively high spin-disorientation cross section exhibited in Table XVI for N_2 is not indicative of its quality as a buffer gas. An important relaxation mechanism is the trapping of radiation in the optical-pumping cell. Atoms excited by the pumping light emit partially depolarized resonance radiation in all directions. These re-radiated photons relax the ground state polarization without contributing to the pumping rate. Franz (1968) has observed large polarizations in optically pumped Cs in an N_2 buffer gas which he attributes to quenching of the excited state. The non-radiative transitions to the ground state reduce the number of re-radiated photons and decrease this contribution to the spin relaxation.

Transverse spin-relaxation times have not been as thoroughly studied as longitudinal relaxation times. The transverse relaxation can be measured by measuring the width of resonance signals, or by observing the decay of the transverse polarization (Cohen-Tannoudji, 1962) following a 90° pulse of resonant r.f. Successive 90° and 180° pulses can also be used to observe spin echoes (Novikov, 1965; Ruff, 1966). The spin-echo technique can be used to avoid the necessity of an ultra homogeneous field which would be necessary for the two methods mentioned above. An inhomogeneous field is an important transverse relaxation mechanism in that it causes the precessing spins to get out of phase with each other.

TABLE XV. *The diffusion constant D_0 (760 Torr; 0°C) of Na and Cs atoms measured in various buffer gases in units of cm²/sec*

	Na	Cs
He	0·745 ± 0·02 (Strumia and Moretti, 1972)	0·204 ± 0·04 (Beverini *et al.*, 1971)
Ne	0·364 ± 0·01 (Strumia and Moretti, 1972)	0·153 ± 0·14 (Beverini *et al.*, 1971) 0·4 ± 0·1 (Franz *et al.*, 1971)
A	0·283 ± 0·017 (Strumia and Moretti, 1972)	0·134 ± 0·02 (Beverini *et al.*, 1971)
N_2		0·087 ± 0·015 (Beverini *et al.*, 1971)

TABLE XVI. *Spin-disorientation cross sections for Na, Rb and Cs in various buffer gases*

	Na(10^{-25} cm²)	Rb(10^{-21} cm²)	Cs(10^{-23} cm²)
He	0·115 ± 0·035 (Strumia and Moretti, 1972)		2·8 ± 0·3 (Beverini *et al.*, 1971)
Ne	15·52 ± 0·4 (Strumia and Moretti, 1972)		9·27 ± 0·9 (Beverini *et al.*, 1971) 22 ± 4 (Franz *et al.*, 1971)
A	422 ± 20 (Strumia and Moretti, 1972)		104 ± 10 (Beverini *et al.*, 1971)
Kr		27 ± 3 (Bouchiat, 1967)	
N_2			60·0 ± 4·4 (Beverini *et al.*, 1971)

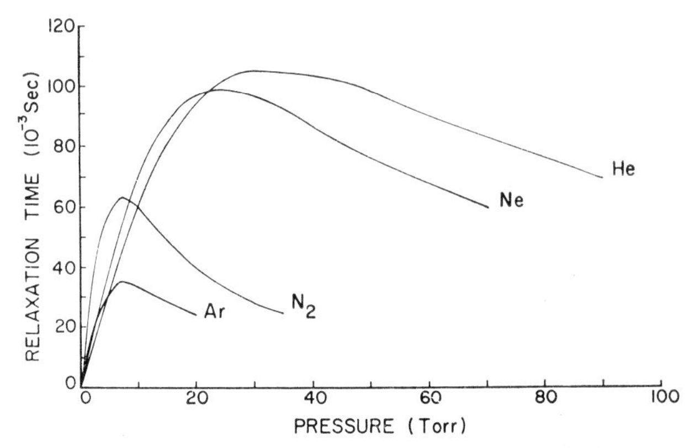

FIG. 21. The pressure dependence of the spin-relaxation time T_g for Cs atoms in various buffer gases. These curves were obtained from the data of Beverini *et al.* (1971).

(4) *Relaxation in the Excited State*

Throughout this chapter, we have assumed for the sake of expediency that the optical-pumping of alkali atoms was accomplished with D_1 radiation and that there was complete mixing in the excited state. Complete excited-state mixing is a simplifying idealization which is more applicable to the lighter alkali atoms than to the heavier alkalis. Since the equilibrium populations of the alkali atom ground state sublevels depend upon the collisional processes which affect the polarization of the excited state during the optical-pumping cycle, optical pumping can be used to study relaxation effects in the excited state (Franz and Franz, 1966).

Several authors have independently (Franz and Franz, 1966; Fricke *et al.*, 1967; Elbel and Naumann, 1967; Violino, 1968) introduced the idea of using D_2 optical pumping as a means for studying relaxation in the $^2P_{3/2}$ state. Early work neglected the effects of nuclear spin, but more recent studies have corrected this deficiency (Papp and Franz, 1972; Elbel *et al.*, 1972; Zhitnikov *et al.*, 1969, 1970). As yet the results of such studies are not completely satisfying. Although it appears that the role of the nuclear spin is well understood (Papp and Franz, 1972), there is still some discrepancy between the results of different workers (Elbel *et al.*, 1972); and excited disorientation cross sections cannot be extracted from the results without assuming a particular model for the excited state relaxation.

G. OPTICAL-PUMPING ORIENTATION OF IONS

In terms of precision, the most successful application of optical-pumping technique to the orientation of ions has been the orientation of the ground state of He^+ in a series of experiments by Dehmelt and co-workers culminating in a precision measurement of the hyperfine splitting of $^3He^+$ (Schuessler, 1969). In these experiments the He^+ ions are stored in an r.f. ion trap and polarized by spin-exchange collisions with an incoming beam of optically pumped Cs atoms. The change in polarization of the He^+ ions produced by a resonant r.f. field is monitored via charge-exchange collisions with the Cs atoms. This detection scheme hinges on the spin dependence of the charge-exchange cross section. Ions with spins parallel to the spins of neutral Cs atoms have a longer lifetime than ions with antiparallel spins. A decrease in the polarization of the He^+ ions increases the rate at which they are lost from the trap. The signal is obtained by monitoring the number of ions in the trap. Because of the long lifetime of an ion in the trap, ultra precise hyperfine resonance measurements can be made.

The ground states of Sr^+ and Ba^+ ions have been directly optically pumped (Ackermann *et al.*, 1967; Sichart *et al.*, 1970) in cells containing an inert buffer gas. The technique is completely analogous to the optical pumping of alkali atoms. The light sources were hollow cathode discharge tubes. The ions are produced in a D.C. discharge maintained in a sidearm and diffuse into the absorption cell. The canonical density of $10^{10} - 10^{11}$ atoms/cm^3 is maintained with a background density of neutrals three orders of magnitude larger. Relaxation times of the order of milliseconds were observed which results in low polarizations and broad resonance linewidths. One of the limitations on the length of the relaxation time is the large spin disorientation cross section for ions colliding with buffer gas atoms (Sichart *et al.*, 1970). Despite these difficulties it proved possible to measure the hyperfine structure of $^{135}Ba^+$ and $^{137}Ba^+$ to 0·1 ppm (Sichart *et al.*, 1970).

The ground states of Ca^+, Sr^+ (Gibbs and Churchill, 1971) and Hg^+ (Hoverson and Schuessler, 1972) have been polarized by spin-exchange collisions with optically pumped Rb atoms and Zeeman resonances have been observed. In the experiment by Gibbs and Hull, the ions were produced by a weak D.C. discharge in a sidearm attached to the cell containing a few Torr buffer gas. The ion densities in the cell were of the order of 10^{10}/cm^3. They concluded that the spin-exchange coupling of the ion-Rb atom spin systems was dominated by e-ion and e-Rb collisions with the free electrons acting as messengers. The signal-to-noise ratio was poor and the linewidths large since the spin-exchange experiment suffered from essentially the same limitations as the optical-pumping studies of Sr^+ and Ba^+.

Schearer has introduced a novel method for polarizing the excited and ground states of atomic and molecular ions by Penning collisions with optically pumped metastable He atoms (Schearer, 1969c; Schearer and Holton, 1970; Schearer and Riseberg, 1971a, b). Ions in excited states are produced by Penning collisions with the oriented metastable He atoms. The polarization of the He atoms is transferred to the ion. Changes in the polarization of the ion excited states can be monitored by observing the fluorescence of the ions as they radiate down to the ground states. In this way, Schearer has observed r.f.

transitions in the $5^2D_{5/2}$ state of Cd^+ and the $4^2D_{3/2}$ and $4^2D_{5/2}$ states of Zn^+ (Schearer and Holton, 1970a). By measuring the linewidth of the resonance, he obtains the lifetimes of the states. He reports the observation of excited state polarization in Cd^+, Zn^+, Mg^+, Ca^+, Sr^+ and Ba^+. He has also observed the polarization of the ground states of Ca^+, Sr^+, and Ba^+ by monitoring the absorption of resonance radiation. An appreciable amount of the polarization in the excited state of the ions produced by Penning collisions is conserved when the ions radiate down to the ground state.

Finally, nuclear resonance signals have been observed in the free ions Rb^+ (Mitchell and Fortson, 1968) and Cs^+ (Nienstadt *et al.*, 1972). In the original experiment by Mitchell and Fortson, Rb^+ ions were produced in a discharge in a sidearm, diffused into an absorption cell and underwent charge-exchange collisions with optically pumped Rb atoms. The charge-exchange collision leaves the nuclear polarization achieved by optical pumping intact and produces a Rb^+ ion with an oriented nucleus. NMR disorients the nuclear spins and charge-exchange collisions depolarize the Rb atoms resulting in a decrease in the transmitted light intensity. Because relaxation times for the nuclear spins are long, the charge-exchange collisions are the principal source of linewidth. In an analogous experiment, Nienstadt *et al.* (1972) have observed NMR in Cs^+. The ions were produced by photoionization rather than a discharge.

H. OPTICAL PUMPING OF ATOMIC P STATES

In recent years, the taboo against optically pumping the P states of atoms because of large disorientation cross sections has been broken in several experiments.

Schearer (1968c, 1969a, b) has succeeded in polarizing the $^3P_2$ metastable states of Ne, Ar, and Xe by an optical-pumping scheme analogous to the optical pumping of metastable He. By measuring the widths of observed Zeeman transitions as a function of buffer gas pressure, he obtained values for the transverse relaxation time. The disorientation cross sections he obtains are surprisingly small and are shown in Table XVII.

The selection rules $M_J \nleftrightarrow -M_J$ (Gallagher, 1967) and $\Delta M_I = 0$ result in low electron randomization cross section for the $^2P_{1/2}$ ground state of Tl. These cross sections have been measured in transient and optical-pumping experiments (Gibbs *et al.*, 1970) for various buffer gases as shown in Table XVII.

The optical pumping of the $^3P_0$ diamagnetic ground state of Pb^{207} and the $^3P_1$, $^3P_2$, and $^3D_2$ metastable states of lead is described in an interesting series of papers by Gibbs *et al.* (1969a, b; Gibbs and White, 1969; Gibbs, 1972). The pumping scheme for the orientation spin-$\frac{1}{2}$ nucleus in the $^3P_0$ ground state is analogous to that for the pumping of the $^1S_0$ state in Hg, Cd, and Zn with the additional complication that atoms excited to the $^3P_1$, excited state can

TABLE XVII. *Disorientation cross sections for P-state atoms in various buffer gases. The values of* Pb^{207} *are upper limits to the cross sections and are for nuclear disorientation*

	$Ne(^3P_2)$ (Schearer, 1969b) (10^{-16} cm^2)	$Ar(^3P_2)$ (Schearer, 1969b) (10^{-16} cm^2)	$Xe(^3P_2)$ (Schearer, 1969b) (10^{-16} cm^2)	$Tl(^2P_{1/2})$ (Gibbs *et al.*, 1970) (10^{-20} cm^2)	$Pb^{207}(^3P_0)$ (Gibbs *et al.*, 1969c) Upper limit (10^{-24} cm^2)
He	0·43	6·2	15	60	19
Ne	16·6	26	38	14	20
Ar		100	61	110	29
Kr				220	34
Xe		127	190	620	12
H_2				380	18
N_2				920	18

decay down to the $^3P_1$, $^3P_2$, and $^3D_2$ metastable states as well as to the ground state. Considerable information has been extracted from these experiments. Upper limits for the disorientation cross section for $^3P_0$ Pb atoms in various buffer gases are given in Table XVII. The g_J factors for the metastable states have been measured, and a value of $\mu_I = 0{\cdot}57235(2)\ \mu_N$ for Pb^{207} has been obtained which differs by 2% from the NMR value. In addition, cross sections for the destruction of alignment in the metastable states were measured. In studies of the relaxation of the nuclear spin due to wall collisions in an evacuated quartz cell an interesting effect was observed which was also seen in the thallium optical-pumping experiment. The relaxation time was considerably lengthened by exposing the quartz wall to several Torr of H_2 for several hours.

I. g-FACTOR SHIFTS DUE TO RESONANT AND NONRESONANT R.F. FIELDS

In our treatment of the interaction of polarized atoms with an applied r.f. field, we replaced a linearly oscillating field with a rotating field and neglected the effect of the counterrotating component. A more rigorous analysis shows that the presence of the counterrotating component shifts the observed g-factor slightly. This is the well known Bloch–Siegert shift (Bloch and Siegert, 1940). If the observed g-factor is g_{obs} and the true g factor is g, then the first-order correction to g is given by

$$g \simeq g_{\text{obs}}\left(1 - \frac{H_1^2}{4H_0^2}\right).$$

where $2H_1$ is the amplitude of the linearly oscillating field and H_0 is the static magnetic field.

Nonresonant r.f. fields can also shift the effective g-factors of polarized atoms. Because the ratio H_1/H_0 can be made relatively large in optical-pumping experiments, the application of resonant and nonresonant r.f. fields can be used to produce dramatic changes in the effective g-factors of polarized atoms. Recently these effects have been studied theoretically and experimentally (Haroche and Cohen-Tannoudji, 1970; Harouche *et al.*, 1970; Novikov and Malyshev, 1972), and it can be said that the effects are well understood.

VIII. THE CONSTRUCTION AND OPERATION OF AN ALKALI OPTICAL-PUMPING APPARATUS

This final section is intended to assist the investigator, in setting up an alkali optical-pumping apparatus. Although we will deal with the optical pumping of alkali atoms, much of the discussion applies to the optical pumping of other atoms as well.

A. LIGHT SOURCES

The signal-to-noise ratio in optical-pumping experiments is largely determined by the intensity and stability of the pumping light source. Cesium, rubidium, and potassium are the easiest atoms to optically pump, because it is a simple matter to construct a stable, intense light source and to filter out the D_2 resonance line. The wavelengths of the D_1 lines for the Cs, Rb, K, and Na are 8944, 7948, 7699, and 5896 Å respectively.

The typical light source is an electrodeless r.f. discharge lamp. There are many possible design variations, but a completely adequate light source can be constructed along the following lines. A 25 cm^3 pyrex flask containing a few droplets of the alkali metal and filled with 2 Torr of argon as a buffer gas is situated in the tank coil of an r.f. amplifier. The amplifier power should be 50 watts or more. The r.f. amplifier is driven by a crystal controlled oscillator. The filaments of all tubes should be D.C. powered, and the plate voltages should come from a regulated power supply to avoid 60 cycle noise in the light output. The r.f. frequency is not critical, but a convenient range is 10–50 MHz. At higher frequencies it is difficult to build a resonant tank circuit with a coil large enough to contain the lamp. (One can also capacitively couple the power into the lamp or use microwave excitation.) Perhaps the simplest approach is to borrow a transmitter design from the VHF section of the Radio Amateur's Handbook.

It is usually necessary to ignite the lamp with a Tesla coil. The tank circuit should be tuned to dip the amplifier power tube plate current. At first, the discharge is purely argon. As the lamp warms up the alkali atoms take over, and the lamp turns color (purple). At this point, the tank circuit will usually require retuning.

A simple way to keep the alkali metal from condensing on the front face of the lamp is to direct a gentle stream of air onto the rear face of the lamp. The alkali metal will condense on the coolest part of the lamp. The temperature of the lamp can be controlled with the air or by varying the plate voltage of the r.f. amplifier or both. The difficulty with this scheme is that the alkali droplets move around on the wall of the lamp causing the light intensity to occasionally jump.

A more sophisticated means of controlling the alkali density in the lamp is necessary for an ultra quiet light source. For example, the alkali metal can be distilled into a sidearm attached to the lamp. The sidearm is electrically heated and a controlled amount of alkali vapor is driven into the discharge region. The lamp is enclosed in a box so that its temperature is stabilized. By adjusting the plate voltage of the r.f. amplifier and the heating of the sidearm one can maintain sufficient alkali vapor in the discharge region and optimize the light intensity without allowing any alkali metal to condense on the walls. The basic setup is shown in Fig. 22.

The alkali vapor reacts with the pyrex walls of the lamp, and over a period of time produces a brown coating on the walls. Although the coating may appear opaque to the eye it will transmit the D_1 lines of Cs, Rb, and K which lie in the infrared. The intensity of the pumping light is not seriously reduced.

Argon is preferable to neon or helium as a buffer gas. The lamp is more difficult to ignite when helium is used as a buffer gas. Neon, on the other hand, breaks down so easily that the lamp may oscillate between the mode in which the discharge is carried by the neon and the desired mode of operation in which the discharge is carried by the alkali atoms.

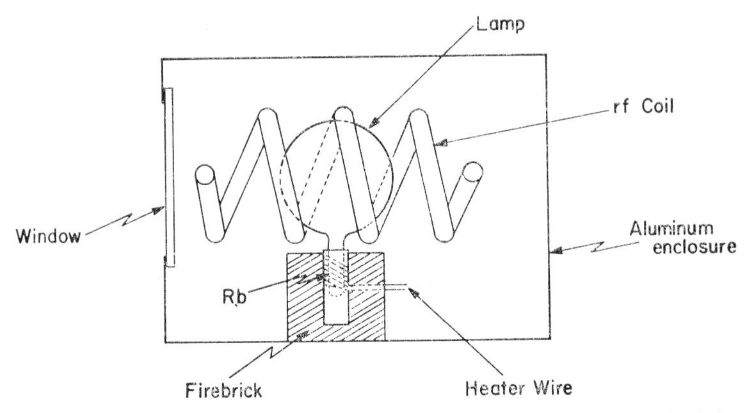

FIG. 22. A lamp housing for an alkali r.f. light source which allows control of the alkali density in the discharge region of the lamp.

Interference filters which select the D_1 radiation are readily purchased. One should request the maximum possible transmission of the D_1 line.

The pumping light is circularly polarized by passing the light through a linear polarizer and then through a quarter-wave plate. Light which is linearly polarized along a principal axis of the quarter wave plate is transmitted 90° out of phase with respect to light polarized along the other principal axis. If the linear polarization of the light incident on the quarter wave plate makes an angle of 45° with respect to one axis of the quarter wave plate, the light transmitted by the plate will be circularly polarized.

Plastic polarizing sheets and quarter-wave plate for the appropriate wavelength can be purchased from Polaroid Corporation. They should be mounted in a holder so that the linear polarizer can be rotated to reverse the sign of the polarization. A simple method for adjusting the orientation of the linear polarizer relative to the quarter-wave plate to achieve circular polarization is the following.

The linear polarizer and quarter-wave plate are placed between the eye and a mirror with the linear polarizer nearest the eye and the quarter-wave plate in back of it. Room light passing through the polarizer and quarter-wave plate is reflected from the mirror and passes back through the two elements to the eye. The polarizer and quarter-wave plate are rotated with respect to each other until there is maximum extinction of the reflected light. Maximum extinction corresponds to maximum circular polarization.

The construction of the light source for the optical-pumping of sodium is a bit more troublesome. The sodium reacts more violently with pyrex than do the heavier alkalis, and the yellow resonance radiation does not penetrate the resulting wall coating. In addition, the D_1 and D_2 lines are too close together for the use of a simple interference filter. Successful electrodeless microwave-excited discharge lamps have been constructed (Moretti and Strumia, 1971; Ioli *et al.*, 1970) with the aid of Na resistant glass. A Lyot polarizing filter can be used to filter the D_1 or D_2 lines or pass them with opposite circular polarizations so that they both pump in the same direction.

A quick and dirty approach, which is adequate for most purposes, is to purchase a General Electric NA-1 commercial sodium lamp and power it with a simple one-transistor current regulated D.C. supply. The lamp operates well with 20–25 volts across the electrodes and a current of 3·5–4·0 amperes. The lamp should be situated in an oven to maintain the correct operating temperature and the lamp's internal heaters should not be used except to start up the lamp. A warm lamp can be started with a Tesla coil, but it is sometimes difficult. The temperature of the lamp oven can be adjusted so that self-reversal in the lamp results in unequal intensities in the D_1 and D_2 lines. Thus an adequate optical-pumping signal can be obtained without the use of a Lyot filter.

Although it is possible to construct a light source for the optical-pumping of lithium (Minguzzi *et al.*, 1966b), the hollow cathode source is not nearly effective enough to justify the trouble. It is much easier to polarize lithium by spin-exchange collisions with another alkali e.g. rubidium (Wright *et al.*, 1969).

B. SIGNAL DETECTION

The most popular method for detecting the optical-pumping signal is by monitoring the intensity of the pumping light transmitted by the optical-pumping cell. The photo-detector to be used is a matter of taste. A simple photo-cell or other low gain photosensitive device coupled to the input of a low noise preamplifier (e.g. Tektronix 122) will do. A possible arrangement is shown in Fig. 23. The microammeter is useful in monitoring long term changes in the light transmitted by the optical-pumping cell due to changes in the light source intensity or due to changes in the density of alkali atoms in the cell. It is important to avoid 60 cycle pickup by the photodetector circuit and the photodetector should be well shielded. The use of a photo-multiplier tube instead of a photocell eliminates the need for a preamplifier.

The lens system which transmits the pumping light through the optical-pumping cell and onto the photodetector need not be particularly sophisticated. It should be noted that most interference filters are intended to be used with light incident normal to their surface. With proper shielding and avoidance of ground loops, the principal source of noise should be the light source. Under

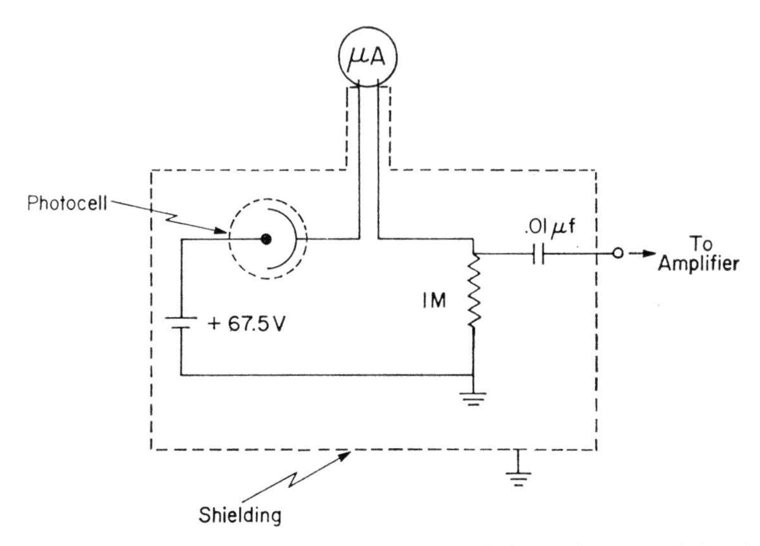

FIG. 23. A possible photodetection circuit for an optical-pumping transmission signal. The shielding is mandatory to prevent 60 Hz pick-up.

these conditions, the intensity of light reaching the photodetector is not critical. One should, however, try to maximize the intensity of pumping light entering the cell so as to maximize the polarization of the alkali atoms.

The optical-pumping signal can be displayed on an oscilloscope or phase-sensitive detector. The reference frequency of the phase-sensitive detector is the frequency with which the resonant r.f. field is chopped (or the frequency at which the resonant r.f. or steady field is modulated).

C. THE MAGNETIC FIELD

By far the most desirable method for producing the steady magnetic field is with a magnetically shielded solenoid. The introduction of the shielded solenoid by Pipkin (Hanson and Pipkin, 1965) was a tremendous improvement in conventional optical-pumping technique. The basic idea is to produce a homogeneous magnetic field by means of a solenoid equipped with field shimming coils. The solenoid is enclosed in a set of concentric cylindrical magnetic shields which virtually eliminate the perturbing effects of random external fields. At one stroke, problems due to the earth's field, magnetic storms, and 60 cycle fields, which plagued early optical-pumping experiments, are eliminated by the magnetic shields. With the solenoid and field correction coils, it is easy to produce fields up to 50 G which are homogeneous to a few ppm. The solenoid-shield system is ideal for ultra precise magnetic moment measurements, and is a real convenience for almost every type of optical-pumping experiment.

The design and construction details of the solenoid used in Pipkin's original experiments are described in the paper by Hanson and Pipkin (1965). This solenoid was 12 in. in diameter and 36 in. long and was surrounded by three concentric cylindrical magnetic shields. The end caps of the shields had 3 in. diameter holes in the center to pass the optical-pumping beam.

The principal source of field inhomogeneity over a small volume at the center of the solenoid is the variation of the axial magnetic field as a function of z, the distance from the center of the solenoid. The functional dependence of the field upon z can be expanded in a power series in z with all odd powers vanishing because of symmetry. The field shimming coils are designed to cancel the 2nd and 4th order terms separately. For example, in free space, a Helmholtz pair produces a field with a vanishing 2nd order dependence on z and thus can be used to cancel the 4th order z dependence of the main solenoid field. The solenoid described by Hanson and Pipkin was designed to operate in free space with appropriate correction coils. The presence of the shields modified the field inside the solenoid so that the performance of the correction coils was not optimum. The correction coils can be designed for operation inside the shield system, however, by a computer solution of the appropriate boundary value problem as described by Hanson and Pipkin.

Recently, Lambert (1973) has recalculated the magnetic field of a solenoid inside a cylindrical shield using a Green's function technique. He discovered that the expression for the axial magnetic field given in equation (32) of the paper by Hanson and Pipkin is missing a constant term. This equation should be corrected to read

$$B_z(0, z) = \frac{8\pi NIa}{10h}\left[\frac{C}{2a} + \sum_{m=1}^{\infty} \frac{R_1(m, b, a)}{I_0\left(\dfrac{m\pi b}{h}\right)} \sin\left(\frac{m\pi c}{h}\right)\cos\left(\frac{m\pi z}{h}\right)\right],$$

where the special functions and constants are as defined by Hanson and Pipkin.

Hanson and Pipkin used two types of high permeability shielding metals. The innermost shield was constructed of Armco iron. The outer shields were made of "Mu-metal". The idea was that the inner shield, which would return the flux of the solenoid, would have a reasonably high permeability at high fields and that the outer shields would have a high permeability at low fields. The shields were equipped with demagnetization coils which allowed the experimenter to produce strong A.C. fields in various directions within the shields themselves. After a large change in the field of the solenoid had been made, 60 cycle current was passed through the demagnetizing coils and slowly decreased to zero. This process shook up the domains in the shields and allowed them to adjust to the new field. This process was necessary to avoid large inhomogeneities due to magnetization of the shields themselves.

The solenoid was constructed on a large lathe. An aluminum cylinder was coated with epoxy. Grooves were cut in the epoxy and the coils of the solenoid were wound onto the grooves. Each winding layer was then coated with epoxy and the grooves for the next layer were cut. The cylinder was slotted lengthwise so that eddy currents would not reduce the effect of the A.C. demagnetizing of the shields. The quoted tolerances on the dimensions and uniformity of the solenoid windings described by Hanson and Pipkin should be taken with a grain of salt. A large slotted aluminum cylinder deforms very easily.

The solenoid-shield system described by Hanson and Pipkin can obviously be improved upon. Correction coils should be designed to operate within a particular shield cylinder of known dimensions. Moly-permalloy cylinders can be effectively used for both inner and outer shields. Moly-permalloy appears to be superior to Armco iron or "Mu-metal". Molly-permalloy shields are constructed and annealed to specification by Alleghany Ludlum Corp. The dimensions of a solenoid are necessarily determined by the swing of the machine shop lathe. Robinson (White *et al.*, 1968) has obtained very good results with a 10-in. diameter solenoid. If the budget is tight, two shields are probably as good as one, and end caps are really only necessary on the innermost shield. The demagnetizing windings are only needed on the inner shield.

If a solenoid similar to the one described by Hanson and Pipkin (20 G/amp) is to be operated in the neighborhood of 50 G, a special power supply must be constructed to deliver $\sim$2 amp with a short term stability of 10^{-7} or better. Otherwise the drift of the resonance frequencies will be too great for precision work. Commercial current regulated supplies are not stable enough. Hanson and Pipkin describe a current regulating circuit utilizing an operational amplifier which senses and reduces to zero the difference in voltage between a mercury battery and the voltage drop across a water cooled 10 Ω resistor in series with the solenoid. A more sophisticated field stabilization scheme (White *et al.*, 1968) is to use an optical-pumping signal to monitor field drift inside the solenoid.

If experiments are to be conducted only at low magnetic fields, a solenoid is not a necessity. A simple Helmholtz pair inside the shields is adequate. The shields themselves are the most important part of the system.

If one absolutely cannot afford to purchase magnetic shields and is willing to work with magnetic fields of the order of the earth's field, an article by Lambert and Wright (1967) will be of interest. The pumping light is directed along the horizontal component of the earth's field and the vertical component of the earth's field is cancelled by a Helmholtz pair. The field along the direction of the pumping light is varied by means of a second Helmholtz pair. A high degree of field homogeneity over the sample volume is achieved by the clever use of pairs of bar magnets which are adjusted to shim the field in a systematic way.

There are two major difficulties with such a setup. It is only suitable for low field operation, and it is subject to the perturbing magnetic fields which are always present in the laboratory. In particular, some way must be found to eliminate 60 cycle modulation of the magnetic field. One way is to use a feedback system in which the 60 cycle field is picked up in a sensing coil, amplified, and fed back out of phase to correction coils which cancel the field. Another approach is to simply apply a 60 cycle current of variable amplitude and phase to correction coils and manually adjust the amplitude and phase of the current to cancel the 60 cycle field.

Once the optical-pumping apparatus is in operation, the width of the alkali atom Zeeman resonance at low r.f. fields in a cell containing buffer gas (little motional field averaging) is a good indicator of field inhomogeneity. At steady magnetic fields between zero and 50 G, the experimenter should not settle for linewidths in excess of 100 Hz.

Finally, the utmost care should be taken to avoid the use of materials which may be slightly magnetic in parts of the apparatus which will be near the optical-pumping cell. Asbestos, certain types of firebrick, bakelite, Chromel-alumel thermocouples, brass plated steel screws, and coaxial cable with steel in the braided shield are all examples of materials which will ruin the homogeneity of the field if placed near the cell.

D. R.F. GENERATION AND MEASUREMENT

The resonant r.f. field is usually produced by a Helmholtz pair surrounding the cell with the plane of the coil pair perpendicular to the steady magnetic field. The coils are driven by a signal generator or other suitable r.f. sources. In a typical optical-pumping experiment, the r.f. is chopped at a low frequency ($\sim$10 Hz) by a switch driven by an audio oscillator. The chopping rate is the reference frequency for the phase-sensitive detector.

At relatively low frequencies, an ordinary mercury relay switch will serve as a chopper. At higher frequencies, a more sophisticated switch such as a coaxial relay or solid state switching device is necessary to prevent leakage of r.f. through the switch when it is in the off position. Such leakage can cause serious distortions of the resonance line shape. The on and off periods of the switch should be as equal as possible. For frequencies up to 50 MHz, a frequency synthesizer or signal generator which is capable of delivering 1.0 volt r.m.s. into a 50 Ω impedence is a quite adequate r.f. source. A frequency synthesizer is a very desirable piece of equipment for serious optical-pumping work. It provides ultra stable frequencies with sensitive variable frequency tuning, adequate power, and distortion-free wave forms. Without modification, ordinary signal generators do not have sufficiently stable frequency output or sensitive tuning for work at frequencies above 10 MHz.

A simple and inexpensive way to extend the range of the frequency synthesizer up to 1000 MHz is to amplify the output of the frequency synthesizer to a power of 1 watt or more in a commercial wideband low distortion amplifier designed for use with a synthesizer and frequency multiply with a varactor diode. Because of the very large power available to the input of the frequency multiplier, one can easily filter the final frequency as many times as necessary to eliminate spurious harmonics, and the transmission of r.f. power into the sample region is not too critical. When possible, the waveform should be observed directly on an oscilloscope to minimize distortion. At higher frequencies, it is probably best to take the plunge and invest in commercial phase-locked microwave equipment.

The output of the signal generator can be directly coupled to a Helmholtz pair to produce the r.f. field. The field strength required to saturate a transition depends upon the relaxation times T_1 and T_2 for the polarized atoms. The requirement for saturation is that $\omega_1^2 \gg T_1 T_2$. A Helmholtz pair with one or two turns will usually suffice, and the small number of turns allows one to use the same coil at higher frequencies without a serious loss of field strength due to the increase in the inductive impedance of the coils. The size of the Helmholtz pair should be as large as possible relative to the size of the cell to avoid inhomogeneities in the r.f. field over the sample volume. The combination of inhomogeneity in the steady magnetic field and in the r.f. field can produce an asymmetric line shape.

At microwave frequencies, a microwave horn or a crude cavity may be necessary to produce the r.f. field in the cell. Field strengths of the order of 1 mG will saturate a transition with a basic linewidth of 100 Hz or less.

Frequency measurements are easily made with commercial solid state frequency counters. Plug-ins allow one to cover a wide range of frequencies. If no frequency standard is available in the laboratory, a phase comparator should be purchased so that the counter's internal time base can be checked against WWVB.

In the course of an experiment it will be necessary to vary the amplitude of the r.f. field. In terms of reproducibility, it is worth the expense to use precision attenuators. These attenuators should be between the r.f. chopper and the input to the frequency counter. A few dB attenuation in the line is usually necessary as a pad. If the input to the counter sees the chopper directly, it usually counts incorrectly. In general, chopping the r.f. is the best way to obtain the optical-pumping signal. If one is searching for a resonance and the exact frequency is not known, it may be more convenient to modulate the steady field as in an N.M.R. experiment. There is less chance of sweeping through the resonance too quickly and missing it.

E. SAMPLE PREPARATION

A typical vacuum-gas handling system for the preparation of optical-pumping cells is shown in Fig. 24. The cell is evacuated and filled with the metals or gases involved in the experiment and then removed by sealing it off from the system with a torch. For most experiments the preparation of the optical-pumping cell does not require ultra-high vacuum conditions, and the vacuum-gas handling system can be a modest one with a small diffusion pump and stopcocks for valves. Such a system will usually not produce a vacuum better than 10^{-6} Torr.

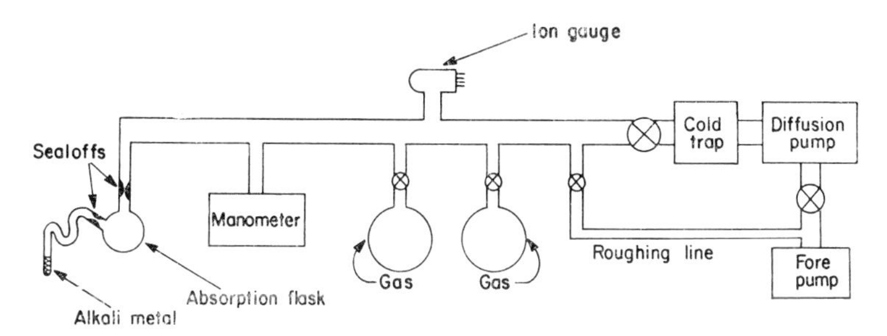

FIG. 24. A typical vacuum-gas handling system for the preparation of optical-pumping cells.

If one is interested in measuring buffer gas relaxation times, however, care must be taken to avoid contamination of the optical-pumping cell by pump oil, and an ultra clean system should be used. Probably the only way to be really safe is to use an ion pump and an all metal-glass system which can be thoroughly baked out. A thoroughly bakeable system with bakeable valves is superior to the system with stopcocks even when ultra clean conditions are not required. The system will pump down much more quickly and optical-pumping cells can be prepared that much faster.

Spherical cells of pyrex or quartz can be easily constructed from commercial spherical flasks. Quartz cells require a graded seal in the tubulation which attaches to the pyrex of the vacuum system. Spherical cells are quite adequate in most cases, but when the ultimate in signal-to-noise is required, cylindrical cells with flat windows are preferable. This is especially true if the cells are to be very small. The size of the cell depends on the nature of the experiment. Cell volumes ranging from 25 to 500 cm^3 have been reported in the literature. If magnetic field inhomogeneity is a problem, the smallest possible cell should be used.

If an r.f. discharge is to be run in the optical-pumping cell, electrodes must be introduced into the cell. A convenient cell design is shown in Fig. 25. The discharge runs between two electrodes located in a turret above the cell. The products of the discharge must diffuse through the buffer gas down into the optical-pumping region. This arrangement minimizes the perturbing effects of the discharge upon the optically-pumped atoms and allows greater control of the density of atoms or ions, created by the discharge. Glass-covered tungsten electrodes can be sealed into the turret of a pyrex cell. This can now be done with quartz as well. Another approach is to have small diameter pyrex (or quartz)

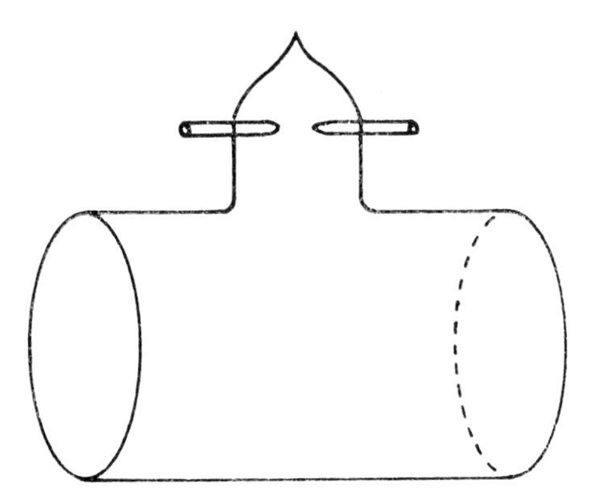

FIG. 25. A possible optical-pumping cell design when a discharge is needed. The discharge products diffuse down into the optical-pumping region from the turret in which the electrodes are placed.

tubes which are closed off on one end sealed into the turret. When the optical-pumping cell is in place in the apparatus, metal wire electrodes connected to the r.f. discharge unit are inserted into the tubes in the turret. The closed ends of these tubes should be almost touching inside the turret so that very weak continuous discharges can be maintained. If a continuous discharge is too strong, the discharge may be pulsed. Pulsed operation is not usually necessary when the electrodes are not in the optical-pumping region.

The cells can be outgassed under vacuum in a number of ways. The simplest is to flame the cell with a Bunsen burner. The heating of the cell should be reasonably uniform to avoid cracking it, but there is usually no problem with a well annealed cell. If sodium yellow appears in the flame, one is close to the softening point of pyrex. At this temperature, vapor dissolved in the pyrex will start coming out of the walls and registering on the ion gauge. This is not gas adsorbed on the surface, which the outgassing procedure is intended to drive off, and it will keep coming out of the glass indefinitely at this temperature. With a quartz cell this will not happen at temperatures below 900°C. There is no danger of straining and cracking a quartz cell with a Bunsen burner. The cells can also be outgassed by heating the cell in an oven or by filling the cell with a little inert gas and running a discharge in the cell, evacuating the cell, and repeating the process a few times.

The cell shown in Fig. 24 has a sidearm attached to it containing the alkali metal to be distilled into the cell. The metal is heated with a Bunsen burner and driven into the cell.

Gases to be introduced into the cell can be purchased in reagent grade in 1-liter pyrex flasks with breakseals. After they are attached to the system and the tubulation is evacuated, the valve to the system is closed and the breakseal can be broken by a glass-covered iron slug raised by a magnet and dropped.

The preparation of optical-pumping cells with wall coatings, which eliminate the need for a buffer gas, is described in detail by Bouchiat (Bouchiat and Brossel, 1966; White *et al.*, 1968); Since the author has no first-hand experience with coated cells, the reader is referred to this article. Coated cells are especially desirable in precision measurements of magnetic moments. The absence of a buffer gas means that there is a dramatic reduction in the linewidth due to field inhomogeneity because of motional narrowing. The coatings can only be used, however, at cell temperatures below their melting point ($< 100°C$).

F. OPTICAL PUMPING AT HIGH AND LOW TEMPERATURES

In order to perform alkali optical-pumping experiments over an extended temperature range, the alkali atom density in the optical-pumping cell must be controlled independently of the cell temperature. The basic cell design which allows this is shown in Fig. 26. While the optical-pumping cell is on the vacuum system, the alkali metal is distilled into the small (6 mm diam.) tubulation ("tip") at the bottom of the cell. For work at high temperatures, the cell is baked out in an oven for an hour at 550°C for pyrex and 900°C for quartz.

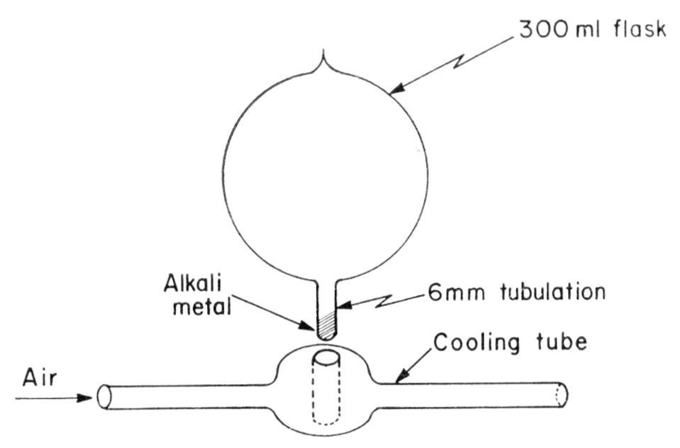

FIG. 26. A cell design which allows one to control the alkali density in the optical-pumping region independently of the temperature of the cell.

This drives out the last of the alkali metal which may be adsorbed on the cell wall. During the bakeout, the tip is inside a pryex socket similar to the one shown in Fig. 26. The socket is cooled by a flow of water. After the bakeout, the cell is filled with buffer gas and removed from the system. If the cell is to be used for low temperature work, no extensive bakeout is necessary.

During an experiment, the cell is situated in a firebrick oven equipped with pyrex windows to allow the passage of the optical-pumping light beam. The "tip" of the cell is again inside a pyrex socket. To work at high temperatures, the oven is heated and the tip is kept cool by a regulated flow of air through the socket. At high temperatures the flow of air regulates the temperature of the tip and hence the density of alkali atoms in the cell.

To work at low temperatures, the oven can be cooled by a flow of dry nitrogen passing through liquid nitrogen. In order to drive alkali atoms up into the cell, hot air is flowed through the socket to heat the tip. It is helpful to start the cooling of the oven slowly so that the nitrogen flushes out the water vapor in the oven before frost can build up on the cell or on the inside of the oven windows. Warm air can be directed at the outside face of each oven window to prevent frost from forming there.

The choice of firebrick to be used in the oven is important. The lowest density brick available should be used. This firebrick is white due to the low iron content. The low density brick is the best insulator and the low iron content is necessary to prevent distortion of the steady magnetic field. Surprisingly, some types of asbestos cement can be used in the construction of the oven without affecting the field homogeneity. This should be checked out, however. Because the principle source of heat loss at high temperatures is radiation through the windows, double windows do not appreciably improve the high temperature performance of the oven. Quartz windows should not be used at very high temperatures because they cloud up very quickly with "bloom".

Pyrex cells should be used for work below 500°C. They are more cheaply and easily constructed. At temperatures above the pyrex strain point (~515°C) water vapor will start to come out of the walls and at ~600°C the cell will begin to collapse. For higher temperature work, quartz cells can be used. A major disadvantage with quartz is the rapid diffusion of helium through the walls at high temperatures. It is not absolutely necessary to drive the alkali metal into the tip while the cell is on the vacuum system. This can be done in the optical-pumping oven, but because of the presence of the buffer gas the process is much slower.

G. OBTAINING THE SIGNAL

With a newly constructed apparatus, there is often some difficulty in obtaining an optical-pumping signal for the first time. A calculation of the steady magnetic field to be expected at a given current passing through the Helmholtz pair or solenoid should be the starting point. The spin resonance frequency of a free electron is 2·8 MHz/G and the low field Zeeman resonance of an alkali atom with nuclear spin I is $(2{\cdot}8/2I+1)$ MHz/G.

The temperature of the optical-pumping cell should be adjusted so that the alkali vapor pressure is $10^{-6}-10^{-5}$ Torr. If a signal is not observed with maximum r.f. power, one should check to make sure r.f. is actually being applied to the cell and that it is being chopped. Next, one should check to see if the alkali atoms are being polarized by the pumping light. Switch the magnetic field on and off a few times. If a transient blip appears on the oscilloscope and it does not appear when the circular polarizer is set to a position midway between left and right-circular polarization (linear polarization) the atoms are being optically pumped. If a transient signal does not appear when the field is turned on or off, try applying a 60 Hz field in place of the static field. If a 60 Hz signal appears on the oscilloscope, and the signal decreases or goes away when the polarizer is set to linear polarization, the atoms are being optically pumped.

If the atoms are not being optically-pumped, the light source is probably not optimized and its operating conditions should be changed. One should, however, check to make sure the circular polarizer has not been put into place backwards. Once an alkali signal has been obtained, the operation of the light source with respect to noise and signal amplitude should be optimized.

The magnetic field inhomogeneity across the cell can be determined by measuring the linewidth of the alkali Zeeman resonance at lower and lower r.f. powers until r.f. broadening is no longer a factor. The field should be shimmed to achieved the desired homogeneity.

One can expect to observe spin-exchange optical-pumping of a second species of paramagnetic atom when the density of the second species is somewhat greater than the optimum density of the optically-pumped alkali atoms. If the density of the second species is determined by the vapor pressure of a metal, one should aim for a vapor pressure $\sim 10^{-5}$ Torr. To spin-exchange optically-pumped lithium for example, a cell temperature of $\sim 400°$C is necessary. This rule of thumb is possible because of the small variation of the spin-exchange cross section from atom to atom.

If the atoms to be spin-exchange polarized are in the form of a molecular gas, a few Torr of the gas and a weak r.f. discharge in a turret above the cell should provide sufficient atoms for a decent signal. Whenever a discharge is run, it is highly probable that one will observe a free electron signal and/or a hydrogen signal. Hydrogen is very often present as a contaminant.

If more than one species of atom is present with the same low field Zeeman resonance frequency, they can be distinguished by increasing the field strength until the various Zeeman transitions are resolved. The splitting of the transitions gives a rough measure of the atoms hyperfine frequency. For example, the difference in frequency between two adjacent Zeeman transitions for a hydrogen-like atom is $\sim 2\nu^2/\Delta\nu$, where ν is the Zeeman frequency and $\Delta\nu$ is the hyperfine frequency.

Acknowledgements

I wish to express my appreciation to R. H. Lambert whose comments and constructive criticism of the manuscript were extremely helpful. I also wish to thank S. J. Davis who prepared the oscilloscope photographs and chart recordings of Rb optical-pumping signals which I have used as illustrations.

Review Articles and Books

Bernheim, R. A. "Optical Pumping." Benjamin, New York, 1965.
Brossel, J. *In:* "Quantum Optics and Electronics." (Ed. C. DeWitt and A. Blandin), p. 189. Gordon and Breach, New York, 1965.
Cohen-Tannoudji, C. and Kastler, A. *In:* "Progress in Optics", (Ed. E. Wolf), p. 3. North-Holland, Amsterdam, 1966.
DeZafra, R. L. (1960). *Am. J. Phys.* **28**, 646.
Happer, W. *In:* "Progress in Quantum Electronics" (Ed. K. W. H. Stevens and J. H. Sanders) Vol. 1, Part 2, p. 51 Pergamon Press, Oxford, 1970.
Happer, W. (1972). *Rev. Mod. Phys.* **44**, 169.
Kastler, A. (1963). *J. Opt. Soc. Am.* **53**, 902.
Major, F. G. *In:* "Methods of Experimental Physics" (Ed. B. Bederson and W. L. Fite) Vol. 7, Part B, p. 1. Academic Press, New York and London 1968.
Skrotskii, G. V. and Izyumova, T. G. (1961). *Soviet Phys. Usp.* **4**, 177.

Bibliography

Ackermann, H., zu Putlitz, G. and Weber, E. W. (1967). *Phys. Lett.* **24A**, 567.
Anderson, L. W. and Ramsey, A. T. (1963). *Phys. Rev.* **132**, 712.
Anderson, L. W., Pipkin, F. M. and Baird, J. C. Jr. (1958). *Phys. Rev. Lett.* **1**, 229.
Anderson, L. W., Pipkin, F. M. and Baird, J. C. Jr. (1960) *Phys. Rev.* **120**, 1279; (1961) ibid, **122**, 1962.
Arditi, M. and Carver, T. R. (1958). *Phys. Rev.* **112**, 449.
Arditi, M. and Carver, T. R. (1961). *Phys. Rev.* **124**, 800.
Aymar, M., Bouchiat, M. A. and Brossel, J. (1967). *Phys. Lett.* **24A**, 753.
Balling, L. C. (1966). *Phys. Rev.* **151**, A1.
Balling, L. C. and Pipkin, F. M. (1964). *Phys. Rev.* **136**, A46.
Balling, L. C. and Pipkin, F. M. (1965). *Phys. Rev.* **139**, A19.
Balling, L. C., Hanson, R. J. and Pipkin, F. M. (1964), *Phys. Rev.* **133**, A607; ibid., **133** AB1.
Balling, L. C., Lambert, R. H., Wright, J. J. and Weiss, R. E. (1969). *Phys. Rev. Lett.* **22**, 161
Barrat, J. and Cohen-Tannoudji, C. (1961). *Compt. Rend.* **252**, 93; ibid., 255.
Beahn, T. J. and Bedard, F. D. (1972). *Bull. Am. Phys. Soc.* **17**, 1127.
Beaty, E. C., Bender, P. L. and Chi, A. R. (1958). *Phys. Rev.* **112**, 450.
Beckmann, A., Boeklen, K. D. and Elke, D. (1974). *Z. Physik* **270**, 173.

Bell, W. E. and Bloom, A. L. (1957). *Phys. Rev.* **107**, 1559.
Bender, P. L. (1963). *Phys. Rev.* **132**, 2154.
Bender, P. L. (1964). *Phys. Rev.* **134**, A1174.
Bender, P. L., Beaty, E. C. and Chi, A. R. (1958). *Phys. Rev. Lett.* **1**, 311.
Beverini, N., Minguzzi, P. and Strumia, F. (1971) *Phys. Rev.* **A4**, 550; (1972) ibid., **A5**, 993.
Bloch, F. and Siegert, A. (1940). *Phys. Rev.* **57**, 522.
Bloom, A. L. and Carr, J. B. (1960). *Phys. Rev.* **119**, 1946.
Böklen, K. D., Dankwort, W., Pitz, E. and Penselin, S. (1967). *Z. Physik* **200**, 467.
Bouchiat, C. C., Bouchiat, M. A. and Pottier, L. (1969). *Phys. Rev.* **181**, 144.
Bouchiat, M. A. (1963). *J. Phys. Radium* **24**, 379; ibid., 611.
Bouchiat, M. A. (1965). *J. Phys. Radium* **26**, 415.
Bouchiat, M. A. and Brossel, J. (1962). *Compt. Rend.* **254**, 3650.
Bouchiat, M. A. and Brossel, J. (1966). *Phys. Rev.* **147**, 41.
Bouchiat, M. A. and Grossetete, F. (1966). *J. Phys. Radium* **27**, 353.
Bouchiat, M. A., Carver, T. R. and Varnum, C. M. (1960). *Phys. Rev. Lett.* **5**, 373.
Bouchiat, M. A., Brossel, J. and Pottier, L. (1967). *Phys. Rev. Lett.* **19**, 817.
Brown, R. A. and Pipkin, F. M. (1968). *Phys. Rev.* **174**, 48.
Cagnac, B. (1961). *Ann. Phys.* (*Paris*) **6**, 467.
Chaney, R. L. and McDermott, M. N. (1969). *Phys. Lett.* **29A**, 103.
Chang, C. K. and Walker, R. H. (1968). *Phys. Rev.* **178**, 198.
Chase, W. and Lambert, R. H. (1973). *Bull. Am. Phys. Soc.* **18**, 120.
Childs, W. J., Goodman, L. S. and Kieffer, L. J. (1961). *Phys. Rev.* **122**, 891.
Clarke, G. A. (1962). *J. Chem. Phys.* **36**, 2211.
Close, F. E. and Osborn, H. (1971). *Phys. Lett.* **34B**, 400.
Cohen-Tannoudji, C. (1962). *Ann. Phys.* (*Paris*) **7**, 423; ibid., 469.
Cohen-Tannoudji, C. (1963). *J. Phys. Radium* **24**, 653.
Colegrove, F. D. and Franken, P. A. (1960). *Phys. Rev.* **119**, 680.
Colegrove, F. D., Schearer, L. D. and Walters, G. K. (1964). *Phys. Rev.* **135**, A353.
Condon, E. U. and Shortley, G. H. "Theory of Atomic Spectra." Cambridge University Press, New York, 1935.
Crampton, S. B. (1971). *Phys. Rev.* **A3**, 515.
Crampton, S. B., Kleppner, D. and Ramsey, N. F. (1963). *Phys. Rev. Lett.* **11**, 338.
Crampton, S. B., Robinson, H. G., Kleppner, D. and Ramsey, N. F. (1966). *Phys. Rev.* **141**, 55.
Crampton, S. B., Berg, H. C., Robinson, H. G. and Ramsey, N. F. (1970). *Phys. Rev. Lett.* **24**, 195.
Crampton, S. B., Duvivier, J. A., Read, G. S. and Williams, E. R. (1972). *Phys. Rev.* **A5**, 1752.
Dahmen, H. and Penselin, S. (1967). *Z. Physik* **200**, 456.
Dalgarno, A. and Rudge, M. R. H. (1965). *Proc. Roy. Soc.* (*London*) **286**, 519.
Das, G. and Ray, S. (1970). *Phys. Rev. Lett.* **24**, 1391.
Davis, S. J. and Balling, L. C. (1972). *Phys. Rev.* **A6**, 1479.
Davis, S. J. and Balling, L. C. *Phys. Rev.* (to be published).
Davis, S. J., Wright, J. J. and Balling, L. C (1971). *Phys. Rev.* **3**, 1220.
Davydov, A. S. "Quantum Mechanics." Pergamon Press, Oxford, 1965.
Dehmelt, H. G. (1957). *Phys. Rev.* **105**, 1924.
Dehmelt, H. G. (1958). *Phys. Rev.* **109**, 381.
Duffy, Jr., W. (1959). *Phys. Rev.* **115**, 1012.
Dutta, C. M., Dutta, N. C. and Das, T. P. (1973). *Phys. Rev.* **A7**, 60.

Edmonds, A. R. "Angular Momentum in Quantum Mechanics." Princeton University Press, Princeton, N.J., 1957.
Elbel, M. and Naumann, F. (1967). *Z. Physik*. **204**, 501; ibid., **208**, 104.
Elbel, M., Koch, A. and Schneider, W. (1972). *Z. Physik*. **255**, 14.
Ensberg, E. S. and Morgan, C. L. (1968). *Phys. Lett*. **28A**, 106.
Ensberg, E. S. and Morgan, C. L. (1971). *Phys. Rev*. **A3**, 2143.
Ensberg, E. S. and zu Putlitz, G. (1969). *Phys. Rev. Lett*. **22**, 1349.
Ernst, K. and Strumia, F. (1968). *Phys. Rev*. **170**, 48.
Faustov, R. (1970). *Phys. Letters* **33B**, 422.
Fernando, P. C. B., Rochester, G. K., Spaulding, I. J. and Smith, K. F. (1960). *Phil. Mag*. **5**, 1291.
Franz, F. A. (1965). *Phys. Rev*. **139**, A603.
Franz, F. A. (1966). *Phys. Rev*. **141**, 105.
Franz, F. A. (1968). *Phys. Lett*. **27A**, 457.
Franz, F. A. (1969). *Phys. Lett*. **29A**, 326.
Franz, F. A. (1970). *App. Phys. Lett*. **16**, 391.
Franz, F. A. (1972). *Phys. Rev*. **A6**, 1921.
Franz, F. A. and Franz, J. R. (1966). *Phys. Rev*. **148**, 82.
Franz, F. A. and Lüscher, E. *Phys. Rev*. **135**, A582.
Franz, F. A., Marshall, T. R. and Munarin, J. A. (1971). *Phys. Lett*. **36A**, 31.
Franzen, W. (1959). *Phys. Rev*. **115**, 850.
Franzen, W. and Emslie, A. G. (1957). *Phys. Rev*. **108**, 1453.
Fricke, J., Haas, J., Lüscher, E. and Franz, F. A. (1967). *Phys. Rev*. **163**, 45.
Gallagher, A. (1967). *Phys. Rev*. **157**, 68; ibid. **163**, 206.
Gibbs, H. M. (1965). *Phys. Rev*. **139**, A1374.
Gibbs, H. M. (1971). *Phys. Rev*. **A3**, 500.
Gibbs, H. M. (1972). *Phys. Rev*. **A5**, 2408.
Gibbs, H. M., Churchill, G. G., Marshall, T. R., Papp, J. F. and Franz, F. A. (1970). *Phys. Rev. Lett*. **25**, 263.
Gibbs, H. M. and Churchill, G. G. (1971). *Phys. Rev*. **A3**, 1617.
Gibbs, H. M. and Hull, R. J. (1967). *Phys. Rev*. **153**, 132.
Gibbs, H. M. and White, C. M. (1969). *Phys. Rev*. **188**, 180.
Gibbs, H. M., Chang, B. and Greenhow, R. C. (1969). *Phys. Rev. Lett*. **22**, 270.
Gibbs, H. M., Chang, B. and Greenhow, R. C. (1969). *Phys. Rev*. **188**, 172.
Grossetete, F. (1964). *J. Phys. Radium* **25**, 383.
Grossetete, F. (1968). *J. Phys. Radium* **29**, 456.
Grossetete, F. and Brossel, J. (1967). *Compt. Rend*. **264**, 381.
Grotch, H. and Hegstrom, R. A. (1971). *Phys. Rev*. **A4**, 59.
Hanson, R. J. and Pipkin, F. M. (1965). *Rev. Sci. Instr*. **36**, 179.
Happer, W. "Light Propagation and Light Shifts in Optical Pumping Experiments—Progress in Quantum Electronics." Pergamon Press, Oxford, 1970.
Happer, W. and Mathur, B. S. (1967). *Phys. Rev. Lett*. **18**, 577; ibid., 727; (1967) ibid., *Phys. Rev*. **163**, 12.
Haroche, S. and Cohen-Tannoudji, C. (1970). *Phys. Rev. Lett*. **24**, 974.
Haroche, S., Cohen-Tannoudji, C., Audouin, C. and Schermann, J. P. (1970). *Phys. Rev. Lett*. **24**, 861.
Heitler, W. "The Quantum Theory of Radiation." University Press, Oxford, 1954.
Herman, R. (1964). *Phys. Rev*. **136**, A1576.
Herman, R. (1965). *Phys. Rev*. **137**, A1062.
Herman, R. M. (1968). *Phys. Rev*. **175**, 10.

Hofmann-Reinecke, H., Haas, J. and Fricke, J. (1969). *Z. Naturforsch.* **24A**, 182.
Hoverson, S. J. and Schuessler, H. A. 3rd International Conference on Atomic Physics, Boulder, Colorado, 1972.
Hughes, V. W. "Recent Research in Molecular Beams" (Ed. I. Estermann). Academic Press, Inc., New York, 1959.
Hughes, W. M. and Robinson, H. G. (1969). *Phys. Rev. Lett.* **23**, 1209.
Hughes, W. M. and Robinson, H. G. (1969). *Bull. Am. Phys. Soc.* **14**, 524.
Hull, R. J. and Brink, G. O. (1970). *Phys. Rev.* **A1**, 685.
Ikenberry, D. and Das, T. P. (1971). *Phys. Rev. Lett.* **27**, 79.
Ioli, N., Minguzzi, P. and Strumia, F. (1970). *J. Opt. Soc. Am.* **60**, 1192.
Ioli, N., Strumia, F. and Moretti, A. (1971). *J. Opt. Soc. Am.* **61**, 1251.
Jarrett, S. M. (1964). *Phys. Rev.* **133**, A111.
Kastler, A. (1950). *J. Phys. Radium* **11**, 255.
Kunik, D. and Kaldor, U. (1971). *J. Chem. Phys.* **55**, 4127.
Lambert, R. H. (1973) private communication.
Lambert, R. H. (1970). *Phys. Rev.* **A1**, 1841.
Lambert, R. H. and Pipkin, F. M. (1962). *Phys. Rev.* **128**, 198.
Lambert, R. H. and Pipkin, F. M. (1963). *Phys. Rev.* **129**, 1233.
Lambert, R. H. and Wright, J. J. (1967). *Rev. Sci. Instr.* **38**, 1385.
Larson, D. J., Valberg, P. A. and Ramsey, N. F. (1969). *Phys. Rev. Lett.* **23**, 1369.
Legowski, S. (1964). *J. Chem. Phys.* **41**, 1313.
Lehmann, J. C. (1964). *J. Phys. Radium* **25**, 809.
Lehmann, J. C. and Brossel, J. (1964). *Compt. Rend.* **258**, 869.
Lehmann, J. C. and Brossel, J. (1966). *Compt. Rend.* **262B**, 624.
Masnou-Seeuws, F. and Bouchiat, M. A. (1967). *J. Phys. Radium* **28**, 406.
Mathur, B. S., Crampton, S. B. and Kleppner, D. (1966). *Bull. Am. Phys. Soc.* **11**, 328.
Mathur, B. S., Tang, H. and Happer, W. (1968). *Phys. Rev.* **171**, 11.
Merzbacher, E. "Quantum Mechanics." John Wiley and Sons, New York, 1961.
Minguzzi, P., Strumia, F., and Violino, P. (1966a). *Nuovo Cimento* **46B**, 145.
Minguzzi, P., Strumia, F. and Violino, P. (1966b). *J. Opt. Soc. Am.* **56**, 707.
Mitchell, J. K. and Fortson, E. N. (1968). *Phys. Rev. Lett.* **21**, 1621.
Moores, D. L. and Norcross, D. W. (1972). *J. Phys.* **B5**, 1482.
Moretti, A. and Strumia, F. (1971). *Phys. Rev.* **A3**, 349.
Nienstadt, H., Schmidt, G., Ullrich, S., Weber, H. G. and zu Putlitz, G. (1972). *Phys. Lett.* **41A**, 249.
Novikov, D. N. (1965). *Opt. Spectry.* (*USSR*) **18**, 419.
Novikov, D. N. and Malyshev, L. G. (1972). *JETP Lett.* **15**, 89.
Olschewski, L. and Otten, E. W. (1966). *Z. Physik* **196**, 77.
Olschewski, L. and Otten, E. W. (1967). *Z. Physik* **200**, 224.
Papp, J. F. and Franz, F. A. (1972). *Phys. Rev.* **A5**, 1763.
Pendlebury, J. M. and Smith, K. F. (1964). *Proc. Phys. Soc.* (*London*) **84**, 849.
Penselin, S., Moran, T., Cohen, V. W. and Winkler, G. W. (1962). *Phys. Rev.* **127**, 524.
Perl, W. (1953). *Phys. Rev.* **91**, 852.
Phillips, M. (1952). *Phys. Rev.* **88**, 202.
Pichanick, F. M. and Woodgate, G. K. (1961). *Proc. Roy. Soc.* (*London*) **A263**, 89.
Pipkin, F. M. and Lambert, R. H. (1962). *Phys. Rev.* **127**, 787.
Ramsey, A. T. and Anderson, L. W. (1964). *Bull. Am. Phys. Soc.* **9**, 625.
Rao, B. K. and Das, T. P. (1969). *Phys. Rev.* **185**, 95.
Rao, B. K., Ikenberry, D. and Das, T. P. (1970). *Phys. Rev.* **A2**, 1411.
Ray, S., Lyons, J. D. and Das, T. P. (1968). *Phys. Rev.* **174**, 104; (1969) ibid., **181**,465.

Ray, S. and Kaufman, S. L. (1972). *Phys. Rev. Lett.* **29**, 895.
Ray, S., Lyons, J. D. and Das, T. P. (1968). *Phys. Rev.* **174**, 112; (1969) ibid., **181**, 465.
Ray, S., Das, G., Maldonado, P. and Wahl, A. C. (1970). *Phys. Rev.* **A2**, 2196.
Ressler, N. W., Sands, R. H. and Stark, T. E. (1969). *Phys. Rev.* **184**, 102.
Robinson, H. G. and Myint, T. (1964). *App. Phys. Lett.* **5**, 116.
Robinson, H. G., Hayne, G. S., Hughes, W. M. and White, C. W. (1968). Proceedings of the International Conference on Optical Pumping and Atomic Line Shape. Poland, 1968.
Rosner, D. and Pipkin, F. M. (1970). *Phys. Rev.* **A1**, 571.
Rozwadski, M. and Lipworth, E. (1965). *J. Chem. Phys.* **43**, 2347.
Ruff, G. A. (1966). *Phys. Rev. Lett.* **16**, 976.
Ruff, G. A. and Carver, T. R. (1965). *Phys. Rev. Lett.* **15**, 282.
Sandars, P. G. H. and Woodgate, G. K. (1960). *Proc. Roy. Soc.* (*London*) **A257**, 269.
Schearer, L. D. (1962). *Phys. Rev.* **127**, 512.
Schearer, L. D. (1967). *Phys. Rev.* **160**, 76.
Schearer, L. D. (1968a). *Phys. Rev.* **171**, 81.
Schearer, L. D. (1968b). *Phys. Lett.* **27A**, 544.
Schearer, L. D. (1968c). *Phys. Rev. Lett.* **21**, 660.
Schearer, L. D. (1969a). *Phys. Rev.* **180**, 83.
Schearer, L. D. (1969b). *Phys. Rev.* **188**, 505.
Schearer, L. D. (1969c). *Phys. Rev. Lett.* **22**, 629.
Schearer, L. D. and Holton, W. C. (1970). *Phys. Rev. Lett.* **24**, 1214.
Schearer, L. D. and Riseberg, L. A. (1970). *Phys. Letters* **33A**, 325.
Schearer, L. D. and Riseberg, L. A. (1971a). *Phys. Lett.* **35A**, 267.
Schearer, L. D. and Riseberg, L. A. (1971b). *Phys. Rev. Lett.* **26**, 599.
Schearer, L. D. and Sinclair, F. D. (1968). *Phys. Rev.* **175**, 36.
Schearer, L. D., Colegrove, F. D. and Walters, G. K. (1963). "Large He^3 Nuclear Polarization," *Phys. Rev. Lett.* **10**, 108.
Schuessler, H. A., Fortson, E. N. and Dehmelt, H. G. (1969). *Phys. Rev.* **187**, 5.
Schwartz, C. (1955). *Phys. Rev.* **97**, 380.
Sinchart, F. v., Stöckman, H. J., Ackermann, H. and zu Putlitz, G. (1970). *Z. Phys.* **236**, 97.
Smirnov, B. M. and Chibsov, M. I. (1965). *Soviet Physics JETP* **21**, 624.
Spence, P. W. and McDermott, M. N. (1967). *Phys. Lett.* **24A**, 430.
Strumia, F. and Moretti, A. 3rd International Conference on Atomic Physics, Boulder, Colorado, 1972.
Tiedeman, J. S. and Robinson, H. G. 3rd International Conference on Atomic Physics, Boulder, Colorado, 1972.
Tilgner, R., Fricke, J. and Haas, J. (1969). *Helv. Phys. Acta* **42**, 740.
Ting, Y. and Lew, H. (1957). *Phys. Rev.* **105**, 581.
Title, R. S. and Smith, K. F. (1960). *Phil. Mag.* **5**, 1281.
Vanden Bout, P. A., Ehlers, V. J., Nierenberg, W. A. and Shugart, H. A. (1967). *Phys. Rev.* **158**, 1078.
Vanden Bout, P. A., Aygun, E., Ehlers, V. J., Incesu, T., Saplakoglu, A. and Shugart, H. A. (1968). *Phys. Rev.* **165**, 88.
Vanier, J. (1967). *Phys. Rev. Lett.* **18**, 33.
Violino, P. (1968). *Nuovo Cimento* **54B**, 61.
Walters, G. K., Colegrove, F. D. and Schearer, L. D. (1962). *Phys. Rev. Lett.* **8**, 439.
Weiss, R. E., Lambert, R. H. and Balling, L. C. (1970). *Phys. Rev.* **A2**, 1745.
Wesley, J. C. and Rich, A. (1971). *Phys. Rev.* **A4**, 1341.

White, C. W., Hughes, W. M., Hayne, G. S. and Robinson, H. G. (1968). *Phys. Rev.* **174**, 23.
Winkler, P. F., Kleppner, D., Myint, T., and Walther, F. G. (1972). *Phys. Rev.* **A5**, 83.
Winter, J. M. (1959). *Ann. Phys. (Paris)* **4**, 745.
Wright, J. J. (1972). *Phys. Rev.* **A6**, 524.
Wright, J. J., Balling, L. C. and Lambert, R. H. (1969). *Phys. Rev.* **183**, 180.
Wright, J. J., Balling, L. C. and Lambert, R. H. (1970). *Phys. Rev.* **A1**, 1018.
Zak, B. D. and Shugart, H. A. (1972). *Phys. Rev.* **A6**, 1715.
Zhitnikov, R. A., Kuleshov, P. P. and Okunevitch, A. I. (1969). *Phys. Letters* **29A**, 239.
Zhitnikov, R. A., Kuleshov, P. P., Okunevitch, A. and Sevastyanov, B. N. (1970). *Soviet Physics JETP* **31**, 445

GASEOUS ION LASERS

C. C. DAVIS and T. A. KING

Physics Department, Schuster Laboratory,
University of Manchester, Manchester, England

I. Introduction

This article aims to provide a comprehensive survey of all the important aspects of laser action in gaseous ions, dealing in particular with the spectroscopy of these ion lasers, the important and interesting physical processes which occur in them, their properties, technology and applications. A number of excellent reviews of some or all of these subjects have appeared previously in the literature. The spectroscopy of ion lasers has been dealt with specifically by Bridges and Chester (1965b) and Willett (1971), whilst Bridges *et al.* (1971) have dealt with ion laser plasmas. Previous comprehensive reviews of noble gas ion lasers have been given by Paananen (1966), Kitaeva *et al.* (1970) and Bridges and Chester (1971).

Because of their primary physical interest and technological importance most attention is given here to the noble gas ion lasers, particularly the argon ion laser; however, close attention is also given to the more recently developed metal vapour ion lasers. These lasers and notably the helium–cadmium laser, although not as thoroughly investigated or technologically developed as the noble gas ion lasers, appear to have considerable potential and are beginning to replace the helium–neon laser in many applications.

A. Historical Background

It was realized very early in the development of the laser that a laser using a gaseous amplifying medium would have many advantages. The homogeneity of such a medium and the narrow spontaneous emission linewidth of most transitions in gases offered the prospect of highly coherent oscillators possessing good beam quality and stability. The operation of the first gas laser, by Javan, Bennett and Herriott, announced in 1961, marked the beginning of a period of intensive research and development aimed at discovering new gas laser transitions, understanding, improving and investigating applications for these devices. The first gas laser operating on the 1·1523 μm transition in neon excited in a gas discharge in a helium–neon mixture was also the first CW laser of any kind. The years between 1961 and 1963 marked the discovery of many new laser transitions in a large number of neutral atomic and molecular gases which extended the range of laser wavelengths both into the visible and infrared.

The first laser action in gaseous ions was obtained by Bell towards the end of 1963 who observed two visible and two infrared transitions between levels of singly ionized mercury (Bell, 1964). This first ion laser was excited in a pulsed mode by discharging capacitors charged to high voltage into a discharge tube filled with a 500:1 mixture of helium:mercury at a pressure of 0·5 torr. The laser was interesting in that it gave visible wavelength operation, (5678 Å,) high peak powers (up to 40 W) and exhibited high gain (more than 0·8 dB cm^{-1}). The possibility of obtaining CW laser action in an ionic system of this kind also seemed likely as the length of the laser pulse could be extended by increasing the length of the discharge excitation pulse. Prior to this first observation of ion laser action, very few visible laser transitions had been observed in neutral atomic or molecular gases, the exceptions being the group of CW red transitions observed in helium–neon lasers by White and Rigden (1962) and Bloom (1963) and the group of yellow and red transitions observed by Mathias and Parker (1963) in pulsed discharges through carbon monoxide.

Early in 1964 a report appeared in the magazine *Electronics* (p. 17, Jan. 24) of a laser transition observed at a wavelength of 5225 Å in a mercury–argon mixture, and laser action was subsequently reported (Heard *et al.*, 1964). The transition was not assigned to either argon or mercury but it may have been the earliest observation of a noble gas ion laser transition. Very shortly after the first announcement of ion laser action by Bell, Bridges (1964) reported the observation of ten laser transitions in the blue and green regions of the visible spectrum between levels of singly ionized argon. These transitions were excited in the pulsed mode using either pure argon or mixtures of argon with helium or neon buffer gases. Independently, and almost simultaneously, similar results were reported by Convert *et al.* (1964a, b) and Bennett *et al.* (1964). Some of the observed transitions exhibited very high gain (several dB m^{-1}), produced large peak output powers (several hundred watts) and would oscillate quasi-CW producing output powers of about 10 W for periods up to a milli-second. [Note: the term "quasi-CW" in what follows indicates that the duration of laser action is limited solely by the length of the exciting current pulse and lasts much longer than the times involved in any collisional or radiative excitation and relaxation processes in the discharge plasma.)

Bridges (1964) soon extended pulsed ion laser operation to cover the whole visible spectrum in singly ionized xenon and krypton. Almost immediately afterwards Gordon *et al.* (1964) reported continuous laser action on many of the previously reported pulsed transitions in singly ionized argon, krypton and xenon. Other authors extended the number of ion laser transitions into the infrared with mercury (Bloom *et al.*, 1964) and with the rare gases, in the visible (Laures *et al.*, 1964a, b) into the ultraviolet (Dana *et al.*, 1965) and into the infrared (Horrigan *et al.*, 1965). Gerritsen and Goedertier (1964) reported the first laser transition between levels of a double ionized atom, mercury. Laser action in gaseous ions was augmented by McFarlane (1964a, b) with laser action in singly and multiply ionized oxygen, carbon, nitrogen and argon and singly ionized transitions in chlorine. The number of ion laser transitions was further extended by Bridges and Chester (1965a) who reported 118 visible and ultraviolet laser transitions in 7 elements and Cheo and Cooper (1965) who reported 55 ultraviolet transitions in 6 elements. The discovery of new ion laser transitions had by the middle of 1965 covered 11 elements with 230 reported transitions (Bridges and Chester, 1965b).

Most of these early ion laser transitions were discovered using pulsed excitation although CW laser action had been obtained for some of the transitions by this time. Nearly all these pulsed laser transitions were observed only during the exciting current pulse indicating that electron impact was the dominant mechanism involved in exciting their upper levels. However there were exceptions; some of the transitions in singly ionized mercury (Bell and Bloom, 1965), krypton and xenon (Laures and Dana, 1964a, b; Laures, 1965) and all the transitions in singly ionized iodine (Fowles and Jensen, 1964a, b; Jensen and Fowles, 1964a) would oscillate only in the discharge afterglow and would operate only in the presence of a suitable buffer gas. Various excitation mechanisms were proposed to explain this behaviour such as excitation transfer from metastable atoms (Laures *et al.*, 1964a, b; Dana and Laures, 1965) and charge transfer (Fowles and Jensen, 1964a, b; Jensen and Fowles, 1964a; Dyson, 1965).

Since 1965 the number of ion laser transitions has been further increased to many more wavelengths in many different elements. In their updated listing of gaseous ion laser transitions Bridges and Chester (1971) list 446 identified and unidentified transitions between singly and multiply ionized levels of 29 elements. A survey of the literature up to the end of 1972 reveals over 530 such transitions in the spectra of 32 elements (Table I). Many of these transitions have been made to operate continuously, and very high power outputs in the visible and ultraviolet have been reported. Advances in the technology of gaseous ion lasers have allowed single frequency operation, longitudinal and transverse mode selection, frequency stabilization and passive and active mode locking. This class of gas laser is, at the present time, the most powerful direct generator of CW visible and ultraviolet coherent radiation. The wide range of available wavelengths, and the high beam quality and stability obtainable from these devices has made them useful in many applications, for example in basic physics experiments in solid state physics, gaseous electronics, plasma diagnostics, optical field-particle interaction studies and photon statistics, ultraviolet, visible and infrared emission and absorption spectroscopy, light scattering (Raman, Brillouin and Rayleigh spectroscopy), the generation of tunable continuous visible and infrared radiation in the pumping of dye lasers and parametric oscillators and for pumping solid-state lasers and quantum counters. They also find application in holography, memory systems, wide-screen displays, ranging, surveying and optical radar, the electronics industry (in the production of microcircuits and resistor trimming), guidance systems, optical data processing, information retrieval and storage systems, computers, communications, cutting, drilling, welding, biology, dentistry and medicine (for example as retinal photocoagulators) and in the entertainment industry (discotheques).

II. Comparison of Gaseous Neutral and Ion Lasers

All gas laser transitions can be excited in one or more of five main types of gas discharge (excluding those neutral atomic and molecular laser transitions which rely for all or part of their excitation on either photodissociative or chemical pumping).

(i) Low current density D.C., A.C., R.F. and microwave excited discharges.
(ii) High current density D.C., A.C., or pulsed low pressure arcs and high electron density R.F. or microwave discharges.
(iii) Discharge afterglows.
(iv) Hollow cathode discharges.
(v) Fast rise-time pulsed discharges.

The plasma in each of these different types of discharge has properties which favour the excitation of neutral or ion laser transition with different dominant excitation mechanisms as will be discussed further in Section IV. Neutral laser transitions, excited by electron impact or energy transfer processes, involve mechanisms (i), (ii) and (v) above whilst ion laser transitions are excited in all five types.

Neutral and ion lasers excited in (i) give CW operation, in many cases excellent beam quality, stability and only moderate cost and difficulty of construction. However, very few CW neutral atomic and no CW neutral molecular gas lasers will operate at short wavelengths, particularly in the visible and ultraviolet regions of the spectrum, and no neutral atomic gas laser has been made to operate CW below 5433 Å (the $5s'[\frac{1}{2}]_1^0 \rightarrow 3p[\frac{1}{2}]_1$ neon transition in a helium–neon laser (Perry, 1971)). The helium–neon laser operating at 6328 Å remains the only important short wavelength neutral gas laser, but it provides only modest output powers. Numerous CW ion laser transitions operate in the short wavelength region down to 3250 Å (in the He–Cd laser).

If an attempt is made to increase the power output of a neutral or ion laser of type (i) by increasing the discharge current, the power output of neutral laser transitions and of ion laser transitions excited by processes involving neutrals generally saturates at low power levels. The excited neutral population becomes almost independent of current at high currents because electron collisions destroy levels both in an upward and downward direction. However, this is not true for ion levels which can increase their population up to much higher currents and high power laser operation can be obtained in discharges of type (ii). Discharge afterglows, where the electrons have thermalized, favour the excitation of laser transitions which operate as a consequence of energy or charge transfer processes in the absence of electron collisional excitation. These types of laser transition are primarily of academic interest and generally have lower energy levels which in a conventional discharge plasma are excessively populated through electron collision.

In a hollow cathode discharge, the plasma consists largely of an extended negative glow where electrons, which have left the cathode surface and dropped through the cathode fall, have acquired high energies. The very non-Maxwellian electron energy distribution in such a discharge contains many electrons with high energies and favours the excitation of high lying (ion rather than neutral) levels. This type of discharge is used almost exclusively for exciting ion laser transitions, particularly those by charge transfer processes (see Section V).

Fast risetime, pulsed discharges occur when the temporal rate of current density is much faster than the radiative rates of decay in the plasma, particularly the decay rate of the upper laser level. This type of discharge, although it can be used to excite laser transitions which operate in other types of discharge, is most useful for exciting transitions at very short wavelengths where a very fast high pump rate is necessary to reach the population inversion necessary for laser oscillation to occur, or transitions for which population inversion can only be maintained on a transient basis and which are self-terminating. Although there are at present no important ion laser transitions which fall into this fast-pulse-excited category, recent developments in the field of fast-pulse excited neutral atomic and molecular lasers suggest that this is an area where future important development of ion lasers may take place (see Section IV, F).

III. Spectroscopy of Ion Lasers

Ion laser oscillation has to date been observed in 32 elements, up to the fifth stage of ionization (quadruply charged) in some ion transitions. The very great number of available wavelengths is listed in Table I and subdivided by elements into Tables I.1 to I.33. This has been done according to the position of the element in the periodic table as has been the practice in previous extensive listings of ion laser transitions (Bridges and Chester, 1971; Willett, 1971). Table I is intended as an updated and corrected version of these previous listings and, as far as is known, contains all ion laser transitions observed up to the end of July, 1973.

The layout of the table follows the same general form as that originally used by Bridges and Chester (1965b) and subsequently by Bridges and Chester (1971).

Column 1 gives the calculated wavelength in air of the laser transitions. This wavelength has been calculated from the most recent energy level values available in the literature for the ion concerned using the wavenumber tables of Coleman *et al.* (1960). In a few cases where very accurate spontaneous emission wavelength measurements for the laser transition are available these are listed instead of a value calculated from energy levels. Where the assignment of the laser transition is not known but spontaneous emission wavelength measurements are available these are listed with the appropriate reference in square brackets. An asterisk (*) indicates a strong or characteristic line in the gas indicated in column 6, (?) indicates that the existence of the laser line is in doubt, (...) indicates that the laser line has not been confirmed in spontaneous emission.

Column 2 gives the most accurately measured value of the laser wavelength available in the literature together with its estimated error, if this is known.

Column 3 indicates the state of ionization in which the laser transition occurs, for example III corresponds to the doubly charged ion. Where the state of ionization is given with a question mark, for example III?, this indicates that this is the likely state of ionization from which the transition arises, deduced for example from its threshold current, but which has not been otherwise confirmed.

Column 4 gives the assignment of the laser transition with the upper level first. LS coupling notation is generally used except for a few transitions. Where the transition assignment is taken from energy levels listed by Moore (1949), her level notation is used, otherwise the level assignments follow those given in the most recent literature on the spectrum of the ion in question. The core configuration is given in parentheses except where the core corresponds to the ground configuration of the next state of ionization. For example the ground state of Ar III is a ^3P configuration so the notation $4p\,^2D^0_{5/2}$ for a particular argon II level implies $(^3P)4p\,^2D^0_{5/2}$. Where the transition assignment is uncertain it appears with a question mark (?); where it is unknown as (...).

Column 5 lists the source (S) from which the transition assignment and calculated wavelength have been taken. The abbreviation CM refers to Moore (1949).

Column 6 indicates whether the transition is observed in a pure gas or a gas mixture discharge (D).

Column 7 gives the dominant excitation mechanism for the transition. EI indicates electron impact, CT charge transfer, PI Penning ionization and ET excitation transfer.

Column 8 gives reference numbers for some papers which have reported the transition; the complete reference list is given on pp. 432–436. The first reference indicated is the earliest report of the observation of the transition, the list of other references is not intended to cover all subsequent observation and discussion of the transition.

Abbreviations: CW = Continuous oscillation reported; E = Error in classification or wavelength; G = Gain measured; P = Power output reported; hfs = Hyperfine structure investigated; M = Microwave excited; RF = Radio frequency excited; S = Super-radiant emission reported; a, b, c, d, e—See further discussion in notes to table.

TABLE I.1. *Neon ion laser transitions*

Calculated wavelength Å (air)	Measured wavelength Å (air)	Ion	Transition assignment	S	D	EM	References
2357·96	2358·00 ± 0·06	IV	$3p\,^4D^0_{7/2} \rightarrow 3s\,^4P_{5/2}$	2	Ne	EI	1
2473·40	2473·50 ± 0·06	III	$a\,^3D^0_3 \rightarrow (^2D^0)3p\,^3P_2$	3	Ne	EI	1
2677·90	2677·98 ± 0·06	III	$3p\,^3P_{2,0} \rightarrow 3s\,^3S^0_1$	CM	Ne	EI	4, 1
2678·64	2678·68 ± 0·06	III	$3p\,^3P_1 \rightarrow 3s\,^3S^0_1$	CM	Ne	EI	4, 1
2777·65	2777·5 ± 0·5	III	$(^2D^0)3p\,^3D_3 \rightarrow (^2D^0)3s\,^3D^0_3$	CM	Ne	EI	4
2866·65 [3]	2866·88 ± 0·06	III ? IV ?	…		Ne	EI	4, 1
3319·72	3319·84 ± 0·06	II	$(^1D)3p\,^2P^0_{1/2} \rightarrow (^1D)3s\,^2D_{3/2}$	5	Ne	EI	6, 1, 44
* 3323·73	3323·74 ± 0·05	II	$3p\,^2P^0_{3/2} \rightarrow 3s\,^2P_{3/2}$	5	Ne	EI	4, 1, 6, 44, 28 (CW, P), 29 (CW), 30 (CW, G, P), 33 (S)
? …	3324·37 ± 0·1	?	…		Ne		6
3327·15	3327·5 ± 0·5	II	$3p\,^4D^0_{3/2} \rightarrow 3s\,^4P_{3/2}$	5	Ne	EI	4, 44
3329·16	3329·02 ± 0·1	II	$3d\,^4D_{7/2} \rightarrow 3p\,^4D^0_{7/2}$	5	Ne	EI	6
? 3330·73	3330·67 ± 0·1	II	$3d\,^2F_{5/2} \rightarrow 3p\,^2D^0_{5/2}$	5	Ne + Xe	EI	6 (E) (a)
3331·14	3331·07 ± 0·1	III	$(^2P^0)3p\,^3D_2 \rightarrow (^2P^0)3s\,^1P^0_1$	CM	Ne + Xe	EI	6
					Ne		
3345·45	3345·50 ± 0·06	II	$(^1D)3p\,^2P^0_{3/2} \rightarrow (^1D)3s\,^2D_{5/2}$	5	Ne	EI	6(E), 1, 30 (CW), 33 (S)
* 3378·22	3378·33 ± 0·06	II	$3p\,^2P^0_{1/2} \rightarrow 3s\,^2P_{1/2}$	5	Ne	EI	4, 1, 6, 30 (CW, G, P), 28 (CW), 33 (S)
* 3392·80	3392·86 ± 0·06	II	$3p\,^2P^0_{3/2} \rightarrow 3s\,^2P_{1/2}$	5	Ne	EI	4, 1, 30 (CW, P)
? 3393·18	3393·40 ± 0·1	II	$3d\,^2D_{3/2} \rightarrow 3p\,^2D^0_{5/2}$	5	Ne	EI	6 (b)
* 3713·08	3713·09 ± ?	II	$3p\,^2D^0_{5/2} \rightarrow 3s\,^2P_{3/2}$	5	Ne	EI	7 (CW, G), 30 (CW, P)

Notes to Table I.1. (a) Dana *et al.* (1965) assign this transition to Ne II although, as they have used xenon in the same discharge tube, it might possibly be the unclassified Xe III or Xe IV transition observed in spontaneous emission by Humphreys (1965), Gallardo *et al.* (1970) and Gallego Lluesma *et al.* (1973). (b) This is probably one and the same as the line listed at 3392·80 Å which is a strong laser line and was not reported in addition to the observed line by Dana *et al.* (1965). A line observed at 3079·83 ± 0·1 Å and previously assigned to Ne II, $(^2P^0)3p\,^3P_0 \rightarrow (^2P^0)3s\,^1P^0_1$ at a calculated wavelength in air of 3079·86 Å (Dana *et al.* 1965; Willett, 1971) is in fact an unclassified xenon III transition observed by Humphreys (1965), Gallardo *et al.* (1970) and Gallego Lluesma *et al.* (1973). The same remarks apply to a line observed at 3644·95 ± 0·5 and tentatively assigned to Ne II, $3d\,^2P_{3/2} \rightarrow 3p\,^2P^0_{1/2}$ at 3644·86 Å (Dana *et al.*, 1965; Willett, 1971), which could be the Xe II transition $(^1D)6p\,^2P^0_{1/2} \rightarrow 5d\,^2P_{3/2}$ at 3644·90 or more likely the very strong persistent unclassified xenon IV transition at 3645·57 Å (Gallego Lluesma *et al.*, 1973). A line reported by Dana *et al.* (1965) at 3378·88 ± 0·1 Å is not listed as it is probably one and the same as the strong line at 3378·22 Å.

TABLE I.2. *Argon ion laser transitions*

Calculated wavelength Å (air)	Measured wavelength Å (air)	Ion	Transition assignment	S	D	EM	References
2624·93	2624·90 ± 0·06	IV	$(^1D)4p\,^2D^0_{5/2} \rightarrow (^1D)4s\,^2D_{5/2}$	CM	Ar	EI	1
2753·92 [9]	2753·91 ± 0·06	III	...	10	Ar	EI	4, 1
2884·16	2884·24 ± 0·06	III	$(^2D^0)4p\,^3P_2 \rightarrow (^2D^0)4s\,^3D^0_3$	CM	Ar	EI	6, 1
* 2913·00	2912·92 ± 0·06	IV	$4p\,^2D^0_{5/2} \rightarrow 4s\,^2P_{3/2}$	CM	Ar	EI	4, 1, 6, 183 (G)
* 2926·27	2926·24 ± 0·06	IV	$4p\,^2D^0_{3/2} \rightarrow 4s\,^2P_{1/2}$	CM	Ar	EI	4, 1, 6
3002·66 [9]	3002·64 ± 0·06	?	...		Ar	EI	4, 1, 6 (E)
3024·05	3024 ± 0·5	III	$(^2P^0)4p\,^3D_3 \rightarrow (^2P^0)4s\,^3P^0_2$	CM	Ar	EI	4
? 3047·05	3047 ± 1	II	$?\ 4f[3]^0_{5/2} \rightarrow (^1D)3d\,^2F_{7/2}$	12	Ar	EI	4 (a)
3054·84	3054·8 ± 0·5	III	$(^2P^0)4p\,^3D_2 \rightarrow (^2P^0)4s\,^3P^0_1$	CM	Ar	EI	4
* 3336·13	3336·21 ± 0·06	III	$(^2D^0)4p\,^3F_4 \rightarrow (^2D^0)4s\,^3D^0_3$	CM	Ar	EI	4, 1, 6, 30 (CW, G, P)
* 3344·72	3344·79 ± 0·06	III	$(^2D^0)4p\,^3F_3 \rightarrow (^2D^0)4s\,^3D^0_2$	CM	Ar	EI	4, 1, 6, 30 (CW, P)
* 3358·49	3358·52 ± 0·06	III	$(^2D^0)4p\,^3F_2 \rightarrow (^2D^0)4s\,^3D^0_1$	CM	Ar	EI	4, 1, 6, 30 (CW, P)
* 3511·12	3511·19 ± 0·04	III	$4p\,^3P_2 \rightarrow 4s\,^3S^0_1$	CM	Ar	EI	11, 1, 4, 6, 28 (CW), 33 (S), 30 (CW, P) (b)
3514·18	3514·22 ± 0·05	III	$4p\,^3P_1 \rightarrow 4s\,^3S^0_1$	CM	Ar	EI	11, 4, 6 (c)
3576·61	3576·9 ± 0·5	II	$4d\,^4F_{7/2} \rightarrow 4p\,^4D^0_{5/2}$	12	Ar	EI	4 (E)
* 3637·89 [9]	3637·86 ± 0·04	III	$?\ (^2D^0)4p\,^1F_3 \rightarrow (^2D^0)4s\,^1D^0_2$	11	Ar	EI	11, 1, 4, 6 (d), 7 (CW, P), 30 (CW, G, P)
? 3704·48 [9]	3705·2 ± 0·5	III ?	...		Ar	EI	4 (d)
3795·32	3795·28 ± 0·06	III	$(^2P^0)4p\,^3D_3 \rightarrow (^2P^0)3d\,^3P^0_2$	CM	Ar	EI	4, 6, 177 (CW)
3858·29	3858·26 ± 0·06	III	$(^2P^0)4p\,^3D_2 \rightarrow (^2P^0)3d\,^3P^0_1$	CM	Ar	EI	4, 177 (CW)
...	4088·6 ± 0·2	?	...		Ar	EI	177 (CW) (e)
4146·71	4146·60 ± 0·04	III	$(^2D^0)4p\,^3P_2 \rightarrow (^2P^0)4s\,^3P^2_0$	CM	Ar	EI	4, 177 (CW)
4182·97 [16]		III ?					
4182·98 [9]	4182·92 ± 0·06	IV ?	...		Ar	EI	4, 30 (CW) (f)
4370·75	4370·73 ± 0·06	II	$(^1D)4p\,^2D^0_{3/2} \rightarrow 3d\,^2D_{3/2}$	12	Ar	EI	4, 31 (CW), 44 (CW)
4383·75	4383·6 ± 0·6	II	$4p\,^4S^0_{3/2} \rightarrow 4s\,^2P_{3/2}$	12	Ar	EI	32
4481·81	4482 ± ?	II	$(^1D)4p\,^2D^0_{5/2} \rightarrow 3d\,^2D_{5/2}$	12	Ar	EI	17 (CW), 44 (CW)
* 4545·05	4545·04 ± 0·1	II	$4p\,^2P^0_{3/2} \rightarrow 4s\,^2P_{3/2}$	12	Ar	EI	18 (G, P), 19, 44 (CW), 20 (CW, P), 4

*	4579·35	4579·36 ± 0·16	II	$4p\,^2S^0_{1/2} \rightarrow 4s\,^2P_{1/2}$	12	Ar	EI	18 (P), 19, 20 (CW, P), 21 (G), 4, 44 (CW)
	4609·56	4609·57 ± 0·1	II	$(^1D)4p\,^2F^0_{7/2} \rightarrow (^1D)4s\,^2D_{5/2}$	12	Ar	EI	4
*	4657·89	4657·95 ± 0·02	II	$4p\,^2P^0_{1/2} \rightarrow 4s\,^2P_{3/2}$	12	Ar	EI	18 (P), 19, 20 (CW, P), 21 (G), 4, 44 (CW)
	4726·86	4657·95 ± 0·02	II	$4p\,^2D^0_{3/2} \rightarrow 4s\,^2P_{3/2}$	12	Ar	EI	18 (P), 20, 4, 44 (CW)
*	4764·86	4764·88 ± 0·04	II	$4p\,^2P^0_{3/2} \rightarrow 4s\,^2P_{1/2}$	12	Ar	EI	18 (P), 23 (E, G), 19 (G), 21 (G, P, S), 20 (CW, P), 4, 22, 44 (CW)
*	4879·86	4879·86 ± 0·04	II	$4p\,^2D^0_{5/2} \rightarrow 4s\,^2P_{3/2}$	12	Ar	EI	18 (E, G, P), 19 (G), 4, 21 (G, P, S), 20 (CW), 22 (E), 44 (CW)
	4889·03	4889·06 ± 0·06	II	$4p\,^2P^0_{1/2} \rightarrow 4s\,^2P_{1/2}$	12	Ar	EI	4, 24, 31 (CW)
*	4965·07	4965·09 ± 0·02	II	$4p\,^2D^0_{3/2} \rightarrow 4s\,^2P_{1/2}$	12	Ar	EI	18 (P), 19, 21 (G, P, S), 44 (CW), 20 (CW, P), 4, 22
?	4992·8 [25]	4992·55 ± 0·05	III ?	…		Ar	EI	24 (g)
*	5017·16	5017·17 ± 0·02	II	$(^1D)4p\,^2F^0_{5/2} \rightarrow 3d\,^2D_{3/2}$	12	Ar	EI	18 (P), 19 (E), 21 (G, P), 44 (CW), 20 (CW, P), 4
	5141·79	5141·8 ± 0·5	II	$(^1D)4p\,^2F^0_{7/2} \rightarrow 3d\,^2D_{5/2}$	12	Ar	EI	4, 24, 31 (CW)
*	5145·32	5145·33 ± 0·02	II	$4p\,^4D^0_{5/2} \rightarrow 4s\,^2P_{3/2}$	12	Ar	EI	18 (G, P), 19 (G), 21 (G, P), 20 (CW, P), 4, 22, 44 (CW)
	5286·90	5287 ± 1	II	$4p\,^4D^0_{3/2} \rightarrow 4s\,^2P_{1/2}$	12	Ar	EI	18 (P), 20 (CW, P), 4, 44 (CW)
	5502·20	5502·2 ± 0·5	III	$(^2D^0)4p\,^3D_3 \rightarrow (^2P^0)4s\,^3P^0_2$	CM	Ar	EI	4, 24
	…	6730 ± 0·5	?	…		Ar		34 (h)
	7348·05	7348·04 ± 0·05	II	$(^1D)3d\,^2D_{5/2} \rightarrow 4p\,^2D^0_{5/2}$	12	Ar	EI	33 (S)
	7505·15	7505·08 ± 0·05	II	$(^1D)3d\,^2P_{3/2} \rightarrow 4p\,^2S^0_{1/2}$	12	Ar	EI	33 (S)
	8771·86	8780 ± 3	II	$4p\,^2P_{3/2} \rightarrow (^1D)4s\,^2D_{5/2}$	12	Ar	EI	26 (j)
	10923·44	10923 ± 1	II	$4p\,^2P_{3/2} \rightarrow 3d\,^2D_{5/2}$	12	Ar	EI	26, 27 (CW, P), 28 (CW)

TABLE I.2—*continued*

Notes to Table I.2. (a) Listed by Bridges and Chester (1965a) but not included by Bridges and Chester (1965b, 1971). The assignment of the line is uncertain, two other possible assignments are (Minnhagen, 1963)

$$4f[3]^0_{7/2} \rightarrow ({}^1D)\,3d\,{}^2F_{7/2} \quad \text{at } 3048{\cdot}02\ \text{Å}$$

$$4f[4]^0_{7/2} \rightarrow ({}^1D)\,3d\,{}^2F_{5/2} \quad \text{at } 3046{\cdot}08\ \text{Å}$$

(b) The measured wavelength of this line is the mean of the three equally accurate measurements given by Cheo and Cooper (1965a), Bridges and Chester (1965a) and McFarlane (1964). (c) The measured wavelength of this line is the mean of the two equally accurate measurements given by Bridges and Chester (1965a) and McFarlane (1964). (d) The assignment of the strong laser transition at 3637·89 Å was originally made by McFarlane (1964) by analogy with the classified isoelectronic laser transition at 5392·16 Å in Cl II. However, in view of the nearby unidentified line at 3705·2 Å this assignment may still be in some doubt. The transition at 3637·89 Å has however been confirmed as an Ar III transition both by its intensity dependence on beam energy in a beam-foil spectroscopy experiment (Denis and Dufay, 1969) and from the behaviour of its quenching current in laser operation (Davis and King, 1972). (e) Observed in a tungsten disc segmented bore tube and is distinct from the silicon laser transition at 4088·85 Å. It could be the Ar IV line $4p^4P^0_{1/2} \rightarrow h$ (calculated wavelength 4089·18 Å) observed in spontaneous emission at 4089·04 Å (Rao, 1938). (f) Bridges and Chester (1965a) suggest that this is an Ar III line, however Striganov and Sventitski (1968) list an unclassified Ar IV line at a wavelength of 4182·97 Å. (g) Not confirmed in the literature after its initial observation. (h) This unclassified line was reported to be very repeatable over several weeks in an argon discharge (Hodges and Tang, 1970). There do not seem to be any likely impurity candidates with lines in this region listed in wavelength tables (Harrison, 1969). (j) Originally assigned by Bridges and Chester (1965b). Although the calculated wavelength is outside the reported measurement error the assignment is made because the transition at 8771·86 Å shares its upper and lower levels respectively with the upper and lower levels of other classified laser transitions.

TABLE I.3. *Krypton ion laser transitions*

Calculated wavelength Å (air)	Measured wavelength Å (air)	Ion	Transition assignment	S	D	EM	References
? 2649·27	2649·41 ± 0·06	II ?	? $({}^3P_2)4f[1]^0_{1/2} \rightarrow ({}^3P)4d\,{}^2P_{1/2}$	35	Kr	EI	1 (a)
2664·41	2664·50 ± 0·06	II	$({}^1D)5d\,{}^2P_{3/2} \rightarrow 5p^4\,D^0_{1/2}$	35	Kr	EI	1
2741·39 [36]	2741·5 ± 0·06	?	...		Kr		1 (b)
3049·70 [36]	3049·74 ± 0·06	?	...		Kr		4, 1 (b)
3124·38	3124·43 ± 0·06	III	$({}^2D^0)5p\,{}^1D_2 \rightarrow ({}^2D^0)5s\,{}^1D^0_2$	CM	Kr	EI	1
3239·51	3239·43 ± 0·06	III	$({}^2P^0)5p\,{}^1D_2 \rightarrow ({}^2P^0)5s\,{}^1P^0_1$	CM	Kr	EI	4, 1
3374·96	3375·0 ± 0·5	III	$({}^2P^0)5p\,{}^3D_3 \rightarrow ({}^2P^0)5s\,{}^3P^0_2$	CM	Kr	EI	4, 30 (CW)
* 3507·42	3507·39 ± 0·05	III	$5p\,{}^3P_2 \rightarrow 5s\,{}^3S^0_1$	CM	Kr	EI	4, 1, 28 (CW, P), 30 (CW, G, P) (b)

*	3564·23	3564·20 ± 0·06	III	$5p\,^3P_1 \to 5s\,^3S^0_1$	CM	Kr	EI	1, 30 (CW, G. P)
	3771·34	3771·34 ± 0·05	II	$(^1D)4d\,^2S_{1/2} \to 5p\,^4P^0_{3/2}$	35	Kr	EI	33 (E, S, P) (C)
*	4067·37	4067·36 ± 0·06	III	$(^2D^0)5p\,^1F_3 \to (^2D^0)5s\,^1D^0_2$	CM	Kr	EI	4, 7 (CW), 30 (CW, P)
*	4131·33	4131·38 ± 0·06	III	$5p\,^5P_2 \to 5s\,^3S^0_1$	CM	Kr	EI	4, 30 (CW, P)
	4154·44	4154·45 ± 0·04	III	$(^2D^0)5p\,^3F_3 \to (^2D^0)5s\,^1D^0_2$	CM	Kr	EI	4
	4171·79	4171·81 ± 0·1	III	$5p\,^5P_1 \to 5s\,^3S^0_1$	CM	Kr	EI	4
	4226·58	4226·51 ± 0·06	III	$(^2D^0)5p\,^3F_2 \to (^2D^0)4d\,^3D^0_1$	CM	Kr	EI	4
	4317·80	4318 ± ?	II	$6s\,^4P_{5/2} \to 5p\,^4P^0_{5/2}$	35	Kr + He	ET	37 (P), 38 (d)
	4386·53	4387 ± ?	II	$6s\,^4P_{5/2} \to 5p\,^4P^0_{3/2}$	35	Kr + He	ET	37 (E), 38 (E) (d)
	4443·29	4443·2 ± 0·04	III	$(^2D^0)5p\,^3D_2 \to (^2D^0)4d\,^3D^0_1$	CM	Kr	EI	4
	4577·20	4577·20 ± 0·1	II	$(^1D)5p\,^2F^0_{7/2} \to (^1D)5s\,^2D_{5/2}$	35	Kr	EI	39 (G), 4, 31 (CW), 52 (CW)
	4582·85	4583 ± ?	II	$6s\,^4P_{3/2} \to 5p\,^4D^0_{5/2}$	35	Kr + He	ET	37, 38 (d)
	4615·28	4615·20 ± 0·1	II	$5p\,^2P^0_{3/2} \to 5s\,^2P_{3/2}$	35	Kr	EI	49
*	4619·15	4619·17 ± 0·1	II	$5p\,^2D^0_{5/2} \to 5s\,^2P_{3/2}$	35	Kr	EI	39 (G), 4, 44 (CW), 52 (CW), 53 (CW)
	4633·88	4633·92 ± 0·06	II	$(^1D)5p\,^2F^0_{5/2} \to (^1D)5s\,^2D_{3/2}$	35	Kr	EI	39 (G), 4, 31 (CW)
?	4650·16	4650·16 ± 0·1	II	$5p\,^2P^0_{1/2} \to 5s\,^4P_{1/2}$	35	Kr	EI	4 (e)
*	4680·41	4680·45 ± 0·06	II	$5p\,^2S^0_{1/2} \to 5s\,^2P_{1/2}$	35	Kr	EI	39 (G), 4, 44 (CW), 52 (CW), 53 (CW)
	4694·43	4695 ± ?	II	$6s\,^4P_{5/2} \to 5p\,^4D^0_{7/2}$	35	Kr + He	ET	37 (P), 38 (d)
	4710·46	4710·3 ± 0·6	III	$(^2D^0)5p\,^3F_4 \to (^2D^0)4d\,^3D^0_3$	CM	Kr	EI	32
	4754·47	4754·5 ± 0·3	III	$(^2D^0)5p\,^1F_3 \to (^2D^0)4d\,^3D^0_3$	CM	Kr	EI	32 (E)
*	4762·43	4762·44 ± 0·06	II	$5p\,^2D^0_{3/2} \to 5s\,^2P_{1/2}$	35	Kr	EI	39 (G), 4, 31 (CW), 51 (RF), 44 (CW), 52 (CW), 53 (CW)
	4765·73	4765·71 ± 0·1	II	$5p\,^4D^0_{5/2} \to 5s\,^4P_{3/2}$	35	Kr	EI	39 (G), 4, 44 (CW), 52 (CW)
	4796·33	4796·3 ± 0·6	II	$5d\,^4D_{1/2} \to 5p\,^4S^0_{3/2}$	35	Kr	EI	32 (E) (c)
*	4825·18	4825·18 ± 0·06	II	$5p\,^4S^0_{3/2} \to 5s\,^2P_{1/2}$	35	Kr	EI	39 (G), 4, 52 (CW), 31 (CW), 44 (CW)
	4846·60	4846·66 ± 0·06	II	$5p\,^2P^0_{1/2} \to 5s\,^2P_{3/2}$	35	Kr	EI	4, 44 (CW), 52 (CW)
	5016·45	5016·4 ± 0·1	III	$(^2D^0)5p\,^1D_2 \to (^2P^0)4d\,^1F^0_3$	CM	Kr	EI	50, 34 (CW)
	5022·39	5022 ± ?	II	$5p\,^4D^0_{3/2} \to 5s\,^2P_{3/2}$	35	Kr	EI	17, 31 (CW)
	5037·47 [43]	5037·5 ± 0·6	II	…		Kr	EI	32 (f)
	5125·72	5126 ± ?	II	$6s\,^4P_{3/2} \to 5p\,^4D^0_{3/2}$	35	Kr + He	ET	37, 38 (d)

TABLE I.3—*continued*

Calculated wavelength Å (air)	Measured wavelength Å (air)	Ion	Transition assignment	S	D	EM	References
5208·32	5208·32 ± 0·04	II	$5p^4P^0_{3/2} \rightarrow 5s^4P_{3/2}$	35	Kr	EI	39 (G), 4, 51 (RF), 44 (CW), 52 (CW), 53 (CW)
5217·93	5218·2 ± 0·4	II	$5p^4D^0_{1/2} \rightarrow 5s^2P_{1/2}$	35	Kr	EI	32
5308·65	5308·68 ± 0·04	II	$5p^4P^0_{5/2} \rightarrow 5s^4P_{3/2}$	35	Kr	EI	39 (G), 4, 44 (CW), 52 (CW), 53 (CW)
5501·43	5501·5 ± 0·5	III	$(^2D^0)5p^3F_3 \rightarrow (^2P^0)4d^3D^0_2$	CM	Kr	EI	32 (E)
5597·32	5597·7 ± 1	III	$(^2D^0)5p^3P_2 \rightarrow (^2P^0)5s^3P^0_2$	CM	Kr	EI	32
* 5681·88	5681·92 ± 0·04	II	$5p^4D^0_{5/2} \rightarrow 5s^2P_{3/2}$	35	Kr	EI	39 (G), 4, 51 (RF), 44 (CW), 52 (CW), 53 (CW)
5752·97	5753 ·4 ± 0·5	II	$5p^4D^0_{3/2} \rightarrow 5s^2P_{1/2}$	35	Kr	EI	17, 182 (CW), 34 (CW)
5935·06 } 5935·29 }	5935·3 ± 0·6	{ III / II	$(^2D^0)5p^3P_2 \rightarrow (^2P^0)5s^1P^0_1$ or $5d^4P_{3/2} \rightarrow (^1D)5p^2P^0_{3/2}$	CM / 35 }	Kr	EI	32 (E) (c)
...	6037·6 ± 0·8	II ? III ?	...		Kr	EI	46 (g)
...	6072 ± 1	?	...		Kr	EI	48
6168·80	6168·80 ± 0·5	II	$(^1D)5p^2F^0_{5/2} \rightarrow 4d^4P_{3/2}$	35	Kr	EI	31 (E, CW) (c)
6310·24	6310·3 ± 0·8	III	$(^2D^0)5p^3P_2 \rightarrow (^2P^0)4d^3D^0_1$	CM	Kr	EI	46
6312·76 [47]	6312·6 ± 0·8	III ?	...		Kr	EI	46, 34 (CW, P) (h)
6414·60	6417 ± 1	II	$(^1D)5p^2P^0_{3/2} \rightarrow 4d^2D_{3/2}$	35	Kr	EI	48 (E) (c)
* 6470·88	6471·0 ± 0·5	II	$5p^4P^0_{5/2} \rightarrow 5s^2P_{3/2}$	35	Kr	EI	39 (G), 4, 28 (CW, P), 31 (CW), 44 (CW), 52 (CW)
6570·07	6570·0 ± 0·5	II	$(^1D)5p^2D^0_{5/2} \rightarrow 4d^2D_{5/2}$	35	Kr	EI	39 (E, G), 4 (E), 31 (CW), 44 (CW) (c)
6602·75	6602·8 ± 0·8	II	$(^1D)5p^2P^0_{1/2} \rightarrow 4d^2D_{5/2}$	35	Kr	EI	46 (E), 34 (CW)
6764·42	6764·57 ± 0·1	II	$5p^4P^0_{1/2} \rightarrow 5s^2P_{1/2}$	35	Kr	EI	39 (G), 4, 31 (CW), 44 (CW), 52 (CW)

6870·85	6870·96 ± 0·1	II	$(^1D)5p\,^2F^0_{5/2} \rightarrow 4d\,^2D_{3/2}$	35	Kr	EI	39 (E, G), 4 (E), 31 (CW), 44 (CW) (c)
7435·78	7435·60 ± ?	II	$(^1D)4d\,^2D_{5/2} \rightarrow 5p\,^4D^0_{5/2}$	35	Kr	EI	184
7525·46	7525·5 ± 0·1	II	$5p\,^4P^0_{3/2} \rightarrow 5s\,^2P_{1/2}$	35	Kr	EI	45 (CW) (j)
7931·41	7933·6 ± ?	II	$(^1D)5p\,^2F^0_{7/2} \rightarrow 4d\,^2D_{5/2}$	35	Kr	EI	44 (E, CW) (c)
7993·22	7993·0 ± 0·5	II	$5p\,^4P^0_{3/2} \rightarrow 4d\,^4D_{1/2}$	35	Kr	EI	39 (G), 4, 26, 44 (CW)
8280·34	8280·3 ± 0·1	II	$(^1D)5p\,^2F^0_{5/2} \rightarrow 4d\,^2D_{5/2}$	35	Kr	EI	44 (CW), 45 (E, CW) (k)
8473·33	8473·00 ± ?	II	$(^1D)4d\,^2D_{3/2} \rightarrow 5p\,^4D^0_{3/2}$	35	Kr	EI	184
8587·78 [36]	8589 ± 3	III ?	...		Kr	EI	26
8690·14	8692·5 ± ?	II	$5p\,^2P^0_{1/2} \rightarrow (^1D)5s\,^2D_{3/2}$	35	Kr	EI	44 (CW)
8978·69	8978·40 ± ?	II	$(^1D)4d\,^2D_{5/2} \rightarrow 5p\,^4D^0_{3/2}$	35	Kr	EI	184 (E)
10659·6	10659·60 ± ?	II	$(^1D)4d\,^2D_{5/2} \rightarrow 5p\,^2P^0_{3/2}$	35	Kr	EI	184
13294·04	13295·00 ± ?	II	$(^1D)4d\,^2D_{5/2} \rightarrow 5p\,^2D^0_{3/2}$	35	Kr	EI	184

Notes to Table I.3. (a) Observed in spontaneous emission at 2649·27 Å and assigned to Kr II (Harrison, 1969). The transition assignment has been obtained from the revised energy level scheme of krypton II of Minnhagen *et al.* (1969) and corrects the assignment given by Bridges and Chester (1971a). In LS coupling notation the transition assignment is

$$(^3P_2)\,4f\,^4P^0_{1/2} \rightarrow (^3P)\,4d\,^2P_{1/2}$$

(b) The measured wavelength value is the mean of the two equally accurate measurements reported by Cheo and Cooper (1965) and Bridges and Chester (1965a). (c) Transition assignment of previous listings (Willett, 1971; Bridges and Chester, 1971a) revised using recent energy level listing for Kr II of Minnhagen *et al.* (1969). (d) Excited in a discharge afterglow in a reaction involving energy transfer between helium metastables and ground state krypton ions

$$He^*(2\,^3S_1) + Kr^+ \rightarrow He(^1S_0) + Kr^{+*}(6s\,^4P_{3/2,\,5/2}) + \Delta E_\infty$$

(e) Assignment in doubt, could be the unclassified xenon (III–IV) transition at 4650·25 Å (Gallego Lluesma *et al.* 1973) or the carbon III transition $3p\,^3P^0_1 \rightarrow 3s\,^3S_1$ at 4650·25 Å (Bockasten, 1955). The krypton II assignment given has been revised using the energy level listing of Minnhagen *et al.* (1969). (f) No likely classification of these lines as Kr II could be obtained from the revised energy level listing of Minnhagen *et al.* (1969). (g) Two possible assignments have been proposed by Neusel (1966)

Kr III $(^2D_0)\;5p\,^3P_1 \rightarrow (^2P^0)\;4d\,^3D^0_1$ at 6037·16 Å

Kr II $(^3P_2)\;4f[2]_{3/2} \rightarrow (^1D)\;4d\,^2P_{3/2}$ at 6038·11 Å

[$5f\,^4F^0_{3/2} \rightarrow (^1D)\;4d\,^2P_{3/2}$ in LS notation]

(h) An unclassified transition has been observed at this wavelength and probably belongs to the spectrum of Kr III (Bridges, unpublished observations; Humphreys, unpublished observations). (j) Evidence for the correctness of the assignment of these transitions comes from its observed competition with 7993·22 Å with which it has a common upper level. (k) Originally assigned by Johnson and Webb (1967), the assignment here has been revised using the energy level listing of Minnhagen *et al.* (1969).

TABLE I.4. *Xenon ion laser transitions*

Calculated wavelength Å (air)	Measured wavelength Å (air)	Ion	Transition assignment	S	D	EM	References
2477·39 [55] [191]	2477·18 ± 0·06	?	...		Xe	EI	1 (a)
2691·84 [55] [191]	2691·82 ± 0·06	III	...		Xe	EI	1 (a)
2983·85 / ? 2983·72 [191]	2983·7 ± 0·5	III or IV	$(^2P^0)6p32_1 \rightarrow (^2P^0)6s\,^3P_0^0$ or ...	CM	Xe	EI	4 (b)
3079·71 [191]	3079·78 ± 0·06	III	...		Xe	EI	4, 1, 6 (E) (a)
? 3246·84 / 3246·97 [191]	3246·94 ± 0·06	III or IV	$(^2P^0)6p\,^3D_3 \rightarrow (^2D^0)5d\,^3D_3^0$ or ...	CM	Xe	EI	1, 192 (b)
3305·85 [191] / 3306·06 [191]	3305·92 ± 0·06	III / IV	... / ...		Xe	EI	4 (E), 1, 6 (E) (c)
* 3330·83 [191]	3330·82 ± 0·06	III–IV	...		Xe	EI	4, 1, 6 (E), 192 (P) (a)
3350·01 [191]	3350·04 ± 0·06	IV	...		Xe	EI	1 (a)
* 3454·24	3454·23 ± 0·06	III	$(^2D^0)6p\,^1D_2 \rightarrow (^2D^0)6s\,^1D_2^0$	CM	Xe	EI	1, 6, 30 (CW, P), 192
3483·26 [191]	3482·92 ± 0·06	IV	...		Xe	EI	4, 1 (a)
3542·33	3542·31 ± 0·05	III	$(^2D^0)6p\,^3P_2 \rightarrow (^2D^0)6s\,^3D_3^0$	CM	Xe	EI	33 (S)
3596·60 [63]	3596 ± 1	III	...		Xe	EI	30 (CW)
* 3645·57 [191]	3645·46 ± 0·06	IV	...		Xe	EI	6 (E), 1, 185 (G, P), 192 (P) (d)
3669·20 [55] [191]	3669·20 ± 0·06	III	...		Xe	EI	1, 177 (CW), 192 (a)
3745·71	3745·73 ± 0·06	III	$(^2D^0)6p\,^1D_2 \rightarrow (^2D^0)5d\,^1D_2^0$	CM	Xe	EI	1, 30 (CW), 192
3759·91 [191]	3760·00 ± 0·04	IV	...		Xe	EI	4 (a)
* 3780·97	3780·99 ± 0·06	III	$6p\,^3P_2 \rightarrow 6s\,^3S_1^0$	CM	Xe	EI	4, 1, 6, 30 (CW, G, P), 192
3803·21 [191]	3803·27 ± 0·06	IV	...		Xe	EI	1, 185, 192 (a)
3841·52 / 3841·86	3841·00 ± 1	III	$(^2D^0)6p\,^3F_2 \rightarrow (^2D^0)6s\,^3D_2^0$ or $(^2D^0)6p\,^3P_1 \rightarrow (^2D^0)5d\,^3D_2^0$	CM	Xe	EI	177 (CW)
3972·93 [191]	3972·93 ± 0·06	IV	...		Xe	EI	1, 185, 192 (a)

	3992·84 3992·55	3993·0 ± 1	III	$(^2D^0)6p\,^3P_1 \rightarrow (^2D^0)6s\,^1D_2^0$ or $6p\,^5P_2 \rightarrow 5d\,^3D_1^0$	CM	Xe	EI	32, 177 (CW) (e)
	4050·05	4049·9 ± 0·2	III	$6p\,^3P_1 \rightarrow 6s\,^3S_1^0$	CM	Xe	EI	46
*	4060·41	4060·48 ± 0·06	III	$(^2P^0)6p32_1 \rightarrow (^2P^0)5d25_1^0$	CM	Xe	EI	4, 6, 30 (CW), 42, 185 (P), 192
	4145·72	4145·3 ± 0·6	III	$(^2D^0)6p\,^3D_2 \rightarrow (^2D^0)5d\,^3D_1^0$	CM	Xe	EI	32, 177 (CW)
	4214·01	4214·05 ± 0·06	III	$(^2D^0)6p\,^3P_2 \rightarrow (^2D^0)5d\,^3D_3^0$	CM	Xe	EI	4, 30 (CW, G, P)
	4240·24	4240·26 ± 0·1	III	$(^2D^0)6p\,^1D_2 \rightarrow (^2P^0)5d17_3^0$	CM	Xe	EI	4, 30 (CW, P)
	4272·59	4272·60 ± 0·06	III	$(^2D^0)6p\,^3F_4 \rightarrow (^2D^0)5d\,^3D_3^0$	CM	Xe	EI	4, 30 (CW, P)
	4285·88	4285·92 ± 0·06	III	$(^2D^0)6p\,^3D_3 \rightarrow (^2D^0)6s\,^1D_2^0$	CM	Xe	EI	4
	4296·39	4296·33 ± 0·05	II	$7s\,^4P_{1/2} \rightarrow 6p\,^4P_{3/2}^0$	CM	Xe	EI	33 (S, P)
	4305·71 [191]	4305·75 ± 0·03	IV	...		Xe	EI	4 (E), 55, 42 (E), 185 (G, P), 192 (P) (f)
	4413·08 [36]	4413·0 ± 0·6	II ?	...		Xe	EI	32
	4434·15	4434·15 ± 0·1	III	$(^2D^0)6p\,^3F_2 \rightarrow (^2D^0)5d\,^3D_1^0$	CM	Xe	EI	4, 34 (CW)
	4503·45	4503·5 ± 0·6	III	$(^2P^0)6p32_1 \rightarrow (^2P^0)5d27\,_2^0$	CM	Xe	EI	32 (E)
	4558·80 [191]	4558·74 ± 0·06	IV	...		Xe	EI	68 (g)
*	4603·03	4603·02 ± 0·04	II	$6p\,^4D_{3/2}^0 \rightarrow 65\,^4P_{3/2}$	CM	Xe	EI	39 (G), 17, 4, 53 (CW), 44 (CW), 65 (CW, M)
	4647·43 [191]	4647·40 ± 0·04	IV	...		Xe	EI	4 (h)
	4650·25 [191] 4650·83 [191]	4650·4 ± 0·1	III–IV	...		Xe	EI	4 (j)
	4673·68	4673·73 ± 0·06	III	$(^2D^0)6p\,^1F_3 \rightarrow (^2D^0)6s\,^1D_2^0$	CM	Xe	EI	4, 44 (CW), 59(CW), 66 (G, P), 31 (CW)
	4683·54	4683·57 ± 0·06	III	$6p\,^5P_2 \rightarrow 6s\,^3S_1^0$	CM	Xe	EI	4, 59 (CW)
	4723·14 [191] 4723·57	4723·1 ± 0·6	?	...	CM	Xe	EI	49 (a)
			III	or $6p\,^5P_1 \rightarrow 6s\,^3S_1^0$	CM	Xe		
	[55] 4748·94 [191]	4748·7 ± 0·6	III	...		Xe	EI	31 (CW), 49 (a)
	4794·48	4794·5 ± 0·6	III	$(^2D^0)6p\,^3D_1 \rightarrow (^2D^0)5d\,^3D_1^0$	CM	Xe	EI	32
	4862·49	4862 ± ?	II	$7s\,^4P_{5/2} \rightarrow 6p\,^4P_{5/2}^0$	CM	Xe + Ne	ET	54, 38 (k)
	4869·46	4869·48 ± 0·06	III	$(^2D^0)6p\,^3F_3 \rightarrow (^2D^0)5d\,^3D_2^0$	CM	Xe	EI	4, 44 (CW), 31 (CW)
	4887·30	4887 ± 1	II	$6p\,^2P_{3/2}^0 \rightarrow 6s\,^2P_{3/2}$	CM	Xe	EI	17 (CW) (l)
?	4921·48	4921·5 ± ?	II	$6p\,^2D_{5/2}^0 \rightarrow 5s\,^2P_{3/2}$	CM	Xe	EI	58 (m)

TABLE I.4—*continued*

Calculated wavelength Å (air)	Measured wavelength Å (air)	Ion	Transition assignment	S	D	EM	References
* 4954·13 [191]	4954·10 ± 0·06	IV	...		Xe	EI	4, 55, 42, 59 (CW), 66 (P), 185 (GP), 186 (G, P), 192 (G, P) (d)
4965·08	4965·00 ± 0·06	II	$(^1D)7s\,^2D_{3/2} \rightarrow (^1D)6p\,^2P^0_{3/2}$	CM	Xe	EI	4, 182 (CW)
4972·69	4972·71 ± 0·05	II	$(^1D)6p\,^2P^0_{3/2} \rightarrow 5d\,^2D_{5/2}$	CM	Xe	EI	58, 42, 33 (S) (n)
5007·74 [191] } 5007·80 [36] }	5007·80 ± 0·03	IV	...		Xe	EI	4, 55, 42, 59 (CW), 66 (G, P), 185 (G, P), 186 (G, P) (d)
? 5012·79	5012·8 ± ?	II	$10_{5/2} \rightarrow 6p\,^4D^0_{3/2}$	CM	Xe	EI	58 (p)
5044·92	5044·89 ± 0·04	II	$(^1D)6p\,^2P^0_{1/2} \rightarrow (^1D)6s\,^2D_{3/2}$	CM	Xe	EI	39 (G), 17, 4, 58, 53 (CW), 65 (CW, M), 44 (CW)
5157·08 [191]	5157·04 ± 0·06	IV	...		Xe	EI	68 (a)
* 5159·02 [191]	5159·08 ± 0·03	IV	...		Xe	EI	4, 55, 42, 59 (CW), 66 (G,P), 185 (G,P), 186 (G,P), 192 (a)
? 5188·05	5188·1 ± ?	II	$7s\,^2P_{3/2} \rightarrow 6p\,^2D^0_{5/2}$	CM	Xe	EI	58 (q)
5223·64	5223·4 ± 0·6	III	$(^2D^0)6p\,^1F_3 \rightarrow (^2D^0)5d\,^1D^0_2$	CM	Xe	EI	32
5238·93	5238·89 ± 0·06	III	$(^2D^0)6p\,^3P_2 \rightarrow (^2P^0)5d\,13^0_1$	CM	Xe	EI	4, 31 (CW), 59 (CW), 186
5256·30 [36]	5256·5 ± 0·6	III ?	...		Xe	EI	32
[55] } * 5260·17 [191] }	5260·17 ± 0·03	IV	...		Xe	EI	4, 42, 55, 31 (CW), 66 (G, P), 68 (P) (r) 185 (E, G, P), 186 (GP)
[55] } 5260·43 [191] }	5260·43 ± 0·03	II	$6p\,^2P^0_{3/2} \rightarrow 6s\,^2P_{1/2}$	CM	Xe	EI	58, 31 (CW), 65 (CW, M) (r)
5261·95	5261·5 ± 1	II	$(^1D)6p\,^2D^0_{3/2} \rightarrow (^1D)6s\,^2D_{3/2}$	CM	Xe	EI	39 (G), 17, 4, 58, 53 (CW), 44 (CW)
? 5292·21	5292·2 ± ?	II	$6p\,^4P^0_{5/2} \rightarrow 6s\,^4P_{5/2}$	CM	Xe	EI	58 (s)

	5313·87	5314 ± ?	II	$7s\,^4P_{5/2} \rightarrow 6p\,^4D_{7/2}$	CM	Xe + Ne	EI	34 (G), 38 (k)
	...	5343·34 ± 0·05	IV ? V ?	...		Xe	EI	186, 68
?	5339·35	5339·4 ± ?	II	$6p\,^4P^0_{3/2} \rightarrow 6s\,^4P_{5/2}$	CM	Xe	EI	58 (t)
?	5341·31 [191]		?	...				
*	[55] 5352·90 [191]	5352·90 ± 0·03	IV	...		Xe	EI	4, 42, 59 (CW) (d) 61, 55, 66 (G. P), 68 (P), 185 (GP), 186 (G, P)
	5367·06	5367·0 ± 0·06	III	$(^2D^0)6p\,^3F_2 \rightarrow (^2D^0)5d\,^3D^0_2$	CM	Xe	EI	32 (E)
*	[55] 5394·60 [191]	5394·60 ± 0·03	IV	...		Xe	EI	4, 55, 66 (P), 68 (P), 42, 59 (CW), 185 (G, P), 186 (GP) (d)
	5401·00	5400·9 ± 0·3	III	$(^2D^0)6p\,^3P_2 \rightarrow (^2P^0)5d\,15^0_2$	CM	Xe	EI	67
	5413·52	5413·5 ± 0·6	III	$(^2D^0)6p\,^3P_2 \rightarrow (^2P^0)5d\,17^0_3$	CM	Xe	EI	32
*	5419·15	5419·16 ± 0·06	II	$6p\,^4D^0_{5/2} \rightarrow 6s\,^4P_{3/2}$	CM	Xe	EI	39 (G), 44 (CW), 17, 4, 58, 53 (CW), 65 (CW, M)
	5454·33	5454·6 ± 0·6	III	$6d\,^5D^0_0 \rightarrow (^2D^0)6p\,4_1$	CM	Xe	EI	32 (E), 66 (G, P)
	5499·42 [191]	5499·31 ± 0·04	?	...		Xe	EI	186, 68 (a)
	5524·39	5524·50 ± 0·5	III	$(^2D^0)6p\,^1D_2 \rightarrow (^2P^0)6s\,^3P^0_2$	CM	Xe	EI	31 (CW), 59 (CW)
	5592·27 [191]	5592·35 ± 0·05	IV ? V ?	...		Xe	EI	186, 68 (u)
	5659·37	5659 ± 1	II	$6p\,^2P^0_{1/2} \rightarrow 5d\,^4P_{1/2}$	CM	Xe	EI	17 (CW) (a)
	5659·92 [191]		II	or ...				
	5726·91	5727 ± 1	II	$(^1D)6p\,^2D^0_{5/2} \rightarrow (^1D)5d\,^2F_{5/2}$	CM	Xe + Ne	ET	17 (CW), 38 (E) (v)
	5751·02	5751 ± 1	II	$6p\,^2D^0_{3/2} \rightarrow 5d\,^4P_{1/2}$	CM	Xe	EI	17 (CW) (l)
	5893·28	5893·30 ± 0·03	II	$(^1D)6p\,^2P^0_{3/2} \rightarrow (^1D)5d\,^2D_{5/2}$	CM	Xe	EI	33 (S), 55
*	[55] 5955·67 [191]	5955·67 ± 0·03	IV	...		Xe	EI	4, 42, 55, 66 (P), 68 (P), 186 (P) (d)
*	5971·11	5971·12 ± 0·06	II	$(^1D)6p\,^2P^0_{3/2} \rightarrow (^1D)6s\,^2D_{3/2}$	CM	Xe	EI	39 (G), 17, 4, 58, 44 (CW), 53 (CW), 65 (CW, M) (w)
?	5976·44	5976·5 ± ?	II	$6p\,^4P^0_{3/2} \rightarrow 6s\,^4P_{3/2}$	CM	Xe	EI	58 (w)
	6093·61	6094 ± ?	II	$7s\,^4P_{3/2} \rightarrow 6p\,^4D^0_{3/2}$	CM	Xe + Ne	ET	38 (k)
	6176·15 [191]	6176·2 ± 0·8	III	...		Xe	EI	46, 34 (CW) (a)
	6238·25	6238·9 ± 0·8	III	$(^2D^0)6p\,^1F_3 \rightarrow (^2P^0)5d\,17^0_3$	CM	Xe	EI	46, 34 (CW)
	6270·81	6270·90 ± 0·1	II	$(^1D)6p\,^2F^0_{5/2} \rightarrow (^1D)6s\,^2D_{3/2}$	CM	Xe	EI	39 (G), 17, 4, 53 (CW), 44 (CW), 65 (CW, M)

TABLE I.4—*continued*

Calculated wavelength Å (air)	Measured wavelength Å (air)	Ion	Transition assignment	S	D	EM	References
6286·41 [36]	6286·6 ± 0·6	II	...		Xe	EI	32, 66
6343·43 [191]	6343·18 ± 0·3	?	...		Xe	EI	66 (P) (a)
6528·62	6528·5 ± 0·5	II	$(^1D)6p\,^2F^0_{7/2} \rightarrow (^1D)5d\,^2F_{5/2}$	CM	Xe	EI	31 (CW), 44 (CW), 65 (CW, M)
6694·31	6696·2 ± ?	II	$6p\,^4P^0_{3/2} \rightarrow 5d\,^4D_{1/2}$	CM	Xe	EI	44 (CW)
...	6699·50 ± 0·3	?	...		Xe	EI	68
6702·23	6702 ± 1	II	$(^1D)6p\,^2P^0_{3/2} \rightarrow (^1D)5d\,^2F_{5/2}$	CM	Xe + He	?	65 (CW, M) (x)
7072·34	7074·3 ± ?	II	$37^0_{5/2} \rightarrow 6d\,^4D_{5/2}$	CM	Xe	EI	44 (CW)
7149·03	7148·94 ± 0·6	II	$6p\,^4D^0_{3/2} \rightarrow 65\,^2P_{3/2}$	CM	Xe	EI	49 (E, G), 44 (CW), 65 (CW, M)
* 7827·63	7828 ± 3	II	$35^0_{5/2} \rightarrow 16_{3/2}$	CM	Xe	EI	26
* 7988·00	7989 ± 3	II	$6p\,^4P^0_{1/2} \rightarrow 6s\,^4P_{1/2}$	CM	Xe	EI	26
8332·71	8330 ± 3	II	$27^0_{5/2} \rightarrow 6d\,^4D_{5/2}$	CM	Xe	EI	26
? 8446·19	8443 ± 3	II	$27^0_{5/2} \rightarrow 6d\,^4D_{3/2}$	CM	Xe	EI	26 (y)
8566·94 [191]	8569 ± 3	?	...		Xe	EI	26 (a)
8582·51	8582 ± 3	II	$31^0_{3/2} \rightarrow 10_{5/2}$	CM	Xe	EI	26
* 8716·17	8714 ± 3	II	$6p\,^4D^0_{3/2} \rightarrow 5d\,^2P_{3/2}$	CM	Xe	EI	26, 44 (CW)
* 9059·30	9063 ± 4	II	$27^0_{5/2} \rightarrow 16_{3/2}$	CM	Xe	EI	26
9265·39	9265 ± 4	II	$(^1S)\,5d\,^2D_{3/2} \rightarrow 6p\,^4D^0_{5/2}$	CM	Xe	EI	26
9288·54	9287 ± 4	II	$13^0_{1/2} \rightarrow (^1D)\,5d\,^2S_{1/2}$	CM	Xe	EI	26
* 9698·60	9697 ± 2	II	$6p\,^4D^0_{3/2} \rightarrow 5d\,^4P_{5/2}$	CM	Xe	EI	26, 64 (CW, P)
10633·85	10634 ± 6	II	$6p\,^4D^0_{3/2} \rightarrow 5d\,^4P_{3/2}$	CM	Xe	EI	26
* ...	10950 ± 6	?	...		Xe	EI	26

Notes to Table I.4. (a) Ionic assignment investigated in spontaneous emission experiments by Gallego Lluesma *et al.* (1973). (b) The first assignment has appeared in previous listings (Willett, 1971; Bridges and Chester, 1971a) but now appears questionable in view of the recent spontaneous emission observations of Gallego Lluesma *et al.* (1973). (c) Gallego Lluesma *et al.* (1973) observed four spontaneous transitions near the wavelength of this line, the two closest to the measured laser wavelength are listed. It is possible that there may be two distinct laser transitions with closely spaced wavelengths which oscillate near the measured value of 3305·92 Å. Willett (1971) lists the line at 3305·92 Å as an Ar II transition; this error arises because of the doubtful classification given for this line by Dana *et al.* (1965). (d) Ionic assignment investigated by studies of threshold and quenching current in laser emission (Davis and King, 1972) and in spontaneous emission by Gallego Lluesma *et al.* (1973). (e) The second assignment suggested by Bridges and Chester (1971a). More accurate measurement of the laser wavelength is required for positive assignment of this transition. (f) The wavelength of this transition has been accurately measured by Gallardo *et al.* (1970) and Gallego Lluesma *et al.* (1973). These measurements almost certainly preclude the original assignment of this transition by Bridges and Chester (1965a, b) to Xe III, $(^2D^0)6p\,^3D_3 \rightarrow 5d\,^3D_3$ at 4305·86 Å. In addition this latter transition is a very weak spontaneous line (Humphreys, 1936) whilst the line observed by Gallardo *et al.* (1970) and Gallego Lluesma *et al.* (1973) was rather strong. (g) Laser oscillation at this wavelength is strongly anticorrelated to laser oscillation at 5394·6 Å suggesting these two transitions may have an upper or lower level in common. (h) Originally assigned to the carbon III transition $3p\,^3P_2^0 \rightarrow 3s\,^3S_1$ at 4647·42 Å (Bockasten, 1955) by Bridges and Chester (1965a, b, 1971a) who believed that carbon contaminant entered the discharge through cathode decomposition. However this transition has been confirmed in spontaneous emission in a carbon-free discharge by Gallego Lluesma *et al.* (1973). This transition has also been observed in laser oscillation from a krypton discharge but this would have been due to the presence of some xenon which is usually the dominant impurity in "spectroscopically" pure krypton. (j) Originally assigned to the carbon III transition $3p\,^3P_1^0 \rightarrow 3s\,^3S_1$ at 4650·25 Å (Bockasten, 1955) by Bridges and Chester (1965a, b, 1971a) and believed to be due to sputtered carbon contaminant but has since been confirmed in spontaneous emission in a carbon-free discharge by Gallego Lluesma *et al.* (1973). (k) Excited in a discharge afterglow in a reaction involving energy transfer between mean metastables and ground state xenon ions $Ne^*(3s) + Xe^+ \rightarrow Ne(^1S_0) + Xe^{+*}(7s\,^4P_{3/2,\,5/2}) + \Delta E_\infty$. (l) Originally assigned by Bridges and Chester (1965b). (m) Originally assigned by Heard and Peterson (1964) to the transition in Xe II $6p\,^2P_{1/2}^0 \rightarrow 6s\,^2P_{1/2}$ at 4919·65 Å. The assignment given here is suggested by the present authors. Heard and Peterson may have originally assigned the observed line to the transition indicated because of "sudden" perturbation considerations (see Section IV); however the assignment listed in the tables satisfies the condition $\Delta J = \Delta L = 1$ found for many ion laser transitions. The existence of this laser transitions remains extremely doubtful in view of the unknown accuracy of the original wavelength measurement of Heard and Peterson, particularly since these authors did not apparently observe some of the very strong characteristic laser transitions marked * in the table, for example in this case 4954·13 Å. (n) Assignment originally suggested by Dahlquist (1965). (p) The spontaneous emission wavelength and assignment given here are the nearest to the measured laser wavelength. It seems likely however because of the probable large inaccuracy of the wavelength measurement of Heard and Peterson (1964) that this ion line is the strong unclassified transition at 5007·80 Å. Willett (1971) incorrectly assigns the transition listed in the table at 5012·79 Å to Xe III. (q) Similar remarks apply as to (p) above. The observed line may be the strong unclassified transition at 5159·02 Å. (r) These two lines were originally thought to be a single one but high resolution spectroscopy has revealed the two lines and their differing ionic character, see Gallardo *et al.* (1970); Bridges and Chester (1971b) and Gallego Lluesma *et al.* (1973). (s) Similar remarks apply as under (p) above. The observed line may be the strong unclassifield transition at 5260·17 Å. (t) Originally assigned by Heard and Peterson (1964) to the weak spontaneous transition Xe II $6d\,^4D_{5/2} \rightarrow 6p\,^4D_{3/2}^0$ at 5327·90 Å. The assignment listed here is nearer to the measured wavelength and is a very strong line in spontaneous emission (Humphreys, 1939). However, as pointed out by Bridges and Chester (1965b), it is likely that the transition observed by Heard and Peterson was an unclassified xenon transition, possibly the one at 5341·31 Å (Gallego Lluesma *et al.*, 1973) or more probably the strong transition at 5352·90 Å. (u) Hoffman and Toschek (1970) originally tentatively assigned this line to the Xe II transition $(^1D)7s\,^2D_{3/2} \rightarrow (^1D)6p\,^2D_{3/2}^0$ at 5591·71 Å, a wavelength which lay considerably outside their experimental error. This laser line must correspond to the unclassified xenon transition observed by Gallego Lluesma *et al.* (1973). (v) Incorrectly assigned by Dana and Laures (1965) and repeated by Willett (1971), the error arising from an error in the paper by Humphreys (1939) which was later corrected (Humphreys *et al.*, 1939). This transition is excited in an excitation transfer process as under (k). (w) Heard and Peterson (1964) originally assigned their laser line observed at 5976·5 Å to the transition listed here at 5971·11 Å. The second assignment listed here is the closest match in spontaneous emission to their measured wavelength and is included here in case it was in fact the transition observed by Heard and Peterson. (x) A Penning ionization process involving helium and xenon metastables has been suggested (Willett, 1970) as a possible excitation mechanism for this transition which operated in a microwave excited xenon discharge in which helium appeared necessary for oscillation $He^*(2\,^3S_1) + Xe(6s) \rightarrow He(^1S_0) + Xe^{+*}(^1D)6p\,^2P_{3/2}^0 + \Delta E_\infty$. (y) Could be the oxygen I line $3p\,^3P_{0,\,2,\,1} \rightarrow 3s\,^3S_1^0$ which oscillates at the four closely spaced wavelengths 8446·28 ,8446·38, 8446·72 and 8446·86 Å (see Willett, 1971 for references). However the assignment list is plausible as the upper level $27_{5/2}^0$ is known to be the upper level of two other laser transitions at 8332·71 and 9059·30 Å. The transition observed by Sinclair (1965) at 8408 ± 3 Å and listed by Willett (1971) is an unclassified neutral xenon transition.

TABLE I.5. *Magnesium ion laser transitions*

Calculated wavelength Å (air)	Measured wavelength Å (air)	Ion	Transition assignment	S	D	EM	References
9218·25	9218 ± 1·5	II	$4p\,^2P_{3/2} \to 4s\,^2S_{1/2}$	101	Mg + He	PI	99 (CW)
9244·27	9244 ± 1·5	II	$4p\,^2P_{1/2} \to 4s\,^2S_{1/2}$	101	Mg + He	PI	99 (CW)
10914·23, 10915·27	10915 ± 1·5	II	$4p\,^2P_{3/2} \to 3d\,^2D_{5/2}$ or $4p\,^2P_{3/2} \to 3d\,^2D_{3/2}$	101	Mg + He	PI	99 (CW, P)
10951·78	10952 ± 1·5	II	$4p\,^2P_{1/2} \to 3d\,^2D_{3/2}$	101	Mg + He	PI	99 (CW, P)
24041·5	24041·6 ± 3	II	$5p\,^2P_{3/2} \to 4d\,^2D_{5/2}$	101	Mg + He, Ne or Ar	EI	100
24124·6	24125·2 ± 3	II	$5p\,^2P_{1/2} \to 4d\,^2D_{3/2}$	101	Mg + He, Ne or Ar	EI	100

TABLE I.6. *Calcium ion laser transitions*

Calculated wavelength Å (air)	Measured wavelength Å (air)	Ion	Transition assignment	S	D	EM	References
8542·09	8541·8 ± 0·6	II	$4p\,^2P^0_{3/2} \to 3d\,^2D_{5/2}$	159	Ca + He	EI	159 (G, P, S)
8662·14	8662·0 ± 0·6	II	$4p\,^2P^0_{1/2} \to 3d\,^2D_{3/2}$	159	Ca + He	EI	158 (G, P, S)

TABLE I.7. *Strontium ion laser transitions*

Calculated wavelength Å (air)	Measured wavelength Å (air)	Ion	Transition assignment	S	D	EM	References
10327·31	10327·15 ± 0·4	II	$5p\,^2P^0_{3/2} \to 4d\,^2D_{5/2}$	CM	Sr + He or Ne	EI	113 (G, S), 139
10914·89	10914·5 ± 0·5	II	$5p\,^2P^0_{1/2} \to 4d\,^2D_{3/2}$	CM	Sr + He or Ne	EI	113 (G, S)

Calculated wavelength Å (air)	Measured wavelength Å (air)	Ion	Transition assignment	S	D	EM	References
25924·3	25923 ± 1·5	II	$7p\,^2P^0_{3/2} \to 6d\,^2D_{5/2}$	CM	Ba + He, Ne, Ar or H_2	EI	143, 139
29057·2	29059 ± 2	II	$7p\,^2P^0_{1/2} \to 6d\,^2D_{3/2}$	CM	Ba + He, Ne, Ar or H_2	EI	143, 139

TABLE I.9. *Zinc ion laser transitions*

Calculated wavelength Å (air)	Measured wavelength Å (air)	Ion	Transition assignment	S	D	EM	References
4911·62	4912 ± ?	II	$4f\,^2F^0_{5/2} \to 4d\,^2D_{3/2}$	144	Zn + He	CT + EI	178, 147, 149 (CW), 151 (CW), 152 (CW)
4924·01 / 4923·86	4924 ± ?	II	$4f\,^2F^0_{7/2} \to 4d\,^2D_{5/2}$ or $4f\,^2F^0_{5/2} \to 4d\,^2D_{5/2}$	145	Zn + He	CT + EI	146, 175 (CW, P), 148 (CW), 149 (CW), 152 (CW) (a)
* 5894·33	5894 ± ?	II	$4s^2\,^2D_{3/2} \to 4p\,^2P^0_{1/2}$	144	Zn + He	PI	147 (CW, P), 150 (CW)
6021·18	6021 ± ?	II	$5d\,^2D_{3/2} \to 5p\,^2P^0_{1/2}$	144	Zn + He	CT	152 (CW)
* 6102·49	6102·8 ± 0·7	II	$5d\,^2D_{5/2} \to 5p\,^2P^0_{3/2}$	144	Zn + He	CT	97, 149 (CW), 152 (CW)
* 7478·79 [13]	7478·3 ± 1·6	II	$4s^2\,^2D_{5/2} \to 4p\,^2P^0_{3/2}$	144	Zn + He	PI	97, 98 (CW, G), 147 (CW, P)
7588·48 [13]	7587·5 ± 1·6	II	$5p\,^2P^0_{3/2} \to 5s\,^2S_{1/2}$	144	Zn + He	CT	97, 175 (CW, P), 98 (CW), 147 (CW, P)
7612·90 [13]	7611·8 ± 1·6	II	$6s\,^2S_{1/2} \to 5p\,^2P^0_{1/2}$	144	Zn + He	CT	97
7757·86 [13]	7757 ± ?	II	$6s\,^2S_{1/2} \to 5p\,^2P^0_{3/2}$	144	Zn + He	CT	146 (E)

Notes to Table I.9. (a) The first assignment is the only one considered by previous authors; however, in the light of the recent revised Zn II energy level data of Martin and Kaufman (1970), the present authors have also included a second possible assignment. More accurate measurement of the wavelength of this laser transition is necessary to confirm which assignment is correct.

TABLE I.10. *Cadmium ion laser transitions*

Calculated wavelength Å (air)	Measured wavelength Å (air)	Ion	Transition assignment	S	D	EM	References
* 3250 ± 29	3250 ± ?	II	$5s^{2}\,{}^{2}D_{3/2} \rightarrow 5p\,{}^{2}P^{0}_{1/2}$	CM	Cd + He	PI	161 (CW, P), 98 (CW, G, P)
* 4415·65	4415·6 ± 0·7	II	$5s^{2}\,{}^{2}D_{5/2} \rightarrow 5p\,{}^{2}P^{0}_{3/2}$	CM	Cd + He, Cd + Ne	PI, EI	97, 160 (CW, P), 162 (CW, G, P), 149 (CW), 166 (CW), 167 (CW)
4881·72	4882·0 ± ?	II	$({}^{3}D)\,5p\,{}^{4}F^{0}_{5/2} \rightarrow 5d\,{}^{2}D_{3/2}$	CM	Cd + He		148 (CW)
5025·48	5025·9 ± ?	II	$({}^{3}D)\,5p\,{}^{4}F^{0}_{7/2} \rightarrow 5d\,{}^{2}D_{5/2}$	CM	Cd + He		148 (CW, P)
5337·48	5337 ± ?	II	$4f\,{}^{2}F^{0}_{5/2} \rightarrow 5d\,{}^{2}D_{3/2}$	CM	Cd + He, Cd + Ne	CT	146, 163, 175 (CW, P), 164 (CW), 149 (CW), 166 (CW)
5378·13	5378 ± ?	II	$4f\,{}^{2}F^{0}_{7/2} \rightarrow 5d\,{}^{2}D_{5/2}$	CM	Cd + He, Cd + Ne	CT	146, 163, 175 (CW, P), 164 (CW), 149 (CW), 166 (CW)
6354·80	6355 ± ?	II	$6g\,{}^{2}G_{7/2} \rightarrow 4f\,{}^{2}F^{0}_{5/2}$	CM	Cd + He	CT	164 (CW), 166 (CW)
6360·04	6360 ± ?	II	$6g\,{}^{2}G_{9/2} \rightarrow 4f\,{}^{2}F^{0}_{7/2}$	CM	Cd + He	CT	164 (CW), 166 (CW)
7236·89	7237 ± ?	II	$6f\,{}^{2}F^{0}_{5/2} \rightarrow 6d\,{}^{2}D_{3/2}$	CM	Cd + He	CT	164 (CW), 166 (CW)
7284·32	7284 ± ?	II	$6f\,{}^{2}F^{0}_{7/2} \rightarrow 6d\,{}^{2}D_{5/2}$	CM	Cd + He	CT	164 (CW), 166 (CW)
8066·87	8066·9 ± ?	II	$6p\,{}^{2}P^{0}_{3/2} \rightarrow 6s\,{}^{2}S_{1/2}$	CM	Cd + He	CT	175 (CW, P), 151 C(W)
8390·02 / 8388·89	8390 ± ?	II ?	$?11p\,{}^{2}P^{0}_{1/2} \rightarrow 8s\,{}^{2}S_{1/2}$ or $7s\,{}^{2}S_{1/2} \rightarrow 6p\,{}^{2}P^{0}_{3/2}$	CM	Cd + He		165 (a)
8530·33	8530·9 ± ?	II	$6p\,{}^{2}P^{0}_{1/2} \rightarrow 6s\,{}^{2}S_{1/2}$	CM	Cd + He	CT	151 (CW), 150 (CW)
8877·70	8877·8 ± ?	II	$9s\,{}^{2}S_{1/2} \rightarrow 7p\,{}^{2}P^{0}_{3/2}$	CM	Cd + He	CT	150 (CW)
11863·57	11869 ± ?	II ?	$?10d\,{}^{2}D_{3/2} \rightarrow ({}^{3}D)\,5p\,{}^{2}D^{0}_{5/2}$	CM	Cd + He, Cd + Ne		165 (b)

Notes to Table I.10. (a) The first assignment was suggested by Bridges and Chester (1971a), the second by Willett (1971). Neither transition has been observed in spontaneous emission. (b) Assignment tentatively suggested by Bridges and Chester (1971a).

TABLE I.11. *Mercury ion laser transitions*

Calculated wavelength Å (air)	Measured wavelength Å (air)	Ion	Transition assignment	S	D	EM	References
4797·01	4797 ± 0·1	III	$5d^8 6s^2\,(J=4) \rightarrow 5d^9(^2D_{5/2})6p_{1/2}\,(J=3)$	CM	Hg + He	EI	130 (G, P), 131
5677·17 [13]	5678 ± ?	II	$5f\,^2F^0_{7/2} \rightarrow 6d\,^2D_{5/2}$	CM	Hg + He, Hg + Ar		132 (G, P), 23, 133, 130 (G, P), 22
6149·50 [13]	6150(a)	II	$7p\,^2P^0_{3/2} \rightarrow 7s\,^2S_{1/2}$	CM	Hg + He, Hg + Ar	CT, PI	132 (G, P), 23 (G), 133, 130 (G), 22, 134, 135 (hfs), 138 (CW) (a)
7346·37 [13]	7346 ± ?	II	$7d\,^2D_{5/2} \rightarrow 7p\,^2P^0_{3/2}$	CM	Hg + He		132, 133
7418·1 [13]	7418 ± ?	II	$5d^9_{3/2}6s_{1/2}6p_{3/2}\,^2P^0_{3/2} \rightarrow 7s\,^2S_{1/2}$	CM	Hg, Hg + He		137, 180 (CW)
7944·66 [13]	7945 ± ?	II	$7p\,^2P^0_{1/2} \rightarrow 7s\,^2S_{1/2}$	CM	Hg + He	CT, PI	135, 137 (G, P), 138 (CW), 179
8548·2 [13]	8547 ± ?	II	$5g\,^2G_{7/2} \rightarrow$ "C" $^2F^0_{5/2}$	CM	Hg + He		133
8622·0	8628 ± ?	II ?	? $8p\,^2P^0_{3/2} \rightarrow$ "$^4D^2_{5/2}$"	CM	Hg + He		133
...	8677 ± ?	II ?	...		Hg + He		133
9396·8	9396 ± ?	II	$10s\,^2S_{1/2} \rightarrow 8p\,^2P^0_{3/2}$	CM	Hg + He		133
10583·6	10586 ± ?	II	$8s\,^2S_{1/2} \rightarrow 7p\,^2P^0_{3/2}$	CM	Hg + He		132, 133
...	12545	II ?	...		Hg + He		133
...	12981	II ?	...		Hg + He		133
...	13655	II ?	...		Hg + He		133 (b)
15555	15550	II	$7p\,^2P^0_{3/2} \rightarrow 6d\,^2D_{5/2}$		Hg + He	CT, PI	133

Notes to Table I.11. (a) The wavelength of this transition has been measured very accurately for four isotopes of mercury by Byer *et al.* (1965). The values obtained were

$$^{198}\text{Hg} - \lambda_{\text{vac.}} = 6151{\cdot}1851 \pm 0{\cdot}0005\ \text{Å}\ (\lambda_{\text{air}} = 6149{\cdot}48\ \text{Å})$$
$$^{200}\text{Hg} - \lambda_{\text{vac.}} = 6151{\cdot}1750 \pm 0{\cdot}0005\ \text{Å}\ (\lambda_{\text{air}} = 6149{\cdot}47\ \text{Å})$$
$$^{202}\text{Hg} - \lambda_{\text{vac.}} = 6151{\cdot}1650 \pm 0{\cdot}0005\ \text{Å}\ (\lambda_{\text{air}} = 6149{\cdot}46\ \text{Å})$$
$$^{204}\text{Hg} - \lambda_{\text{vac.}} = 6151{\cdot}153\ \text{Å}\quad (\lambda_{\text{air}} = 6149{\cdot}45\ \text{Å})$$

(b) Could be the Hg I laser transition $7p\,^3P^0_1 \rightarrow 7s\,^3S_1$ at 13673·47 Å originally observed by Bockasten *et al.* (1965). Because of the fairly inaccurate energy levels of Hg II available in the Atomic Energy Levels (Moore, 1958), spontaneous emission wavelengths from Harrison (1969) have been given where possible as these are more accurate than calculated wavelengths. A transition observed by Bloom *et al.* (1964) at 11,181 Å and listed by Bridges and Chester (1965b) as the transition in Hg II, $7g\,^2G_{7/2} \rightarrow 6f\,^2F^0_{5/2}$ at 11,179 Å is in fact the Hg I transition $7p\,^1P^0_1 \rightarrow 7s\,^3S_1$ at 11,176·76 Å originally observed by Bockasten *et al.* (1965).

TABLE I.12. *Boron ion laser transitions*

Calculated wavelength Å (air)	Measured wavelength Å (air)	Ion	Transition assignment	S	D	EM	References
3451·29	3451·32	II	$2p2p\,^1D_2 \rightarrow 2s2p\,^1P_1^0$	95	BCl_3	EI	90

TABLE I.13. *Indium ion laser transitions*

Calculated wavelength Å (air)	Measured wavelength Å (air)	Ion	Transition assignment	S	D	EM	References
4681·11	4680·5 ± 0·7	II	$4f\,^3F_4^0 \rightarrow 5d\,^3D_3$	CM	In + He or Ne	?	97

TABLE I.14. *Carbon ion laser transitions*

Calculated wavelength Å (air)	Measured wavelength Å (air)	Ion	Transition assignment	S	D	EM	References
1548·19	1548·2 (vac)	IV	$2p\,^2P_{3/2}^0 \rightarrow 2s\,^2S_{1/2}$	157		EI	190
1550·77	1550·8 (vac)	IV	$2p\,^2P_{1/2}^0 \rightarrow 2s\,^2S_{1/2}$	157		EI	190
4647·42	4647·40 ± 0·04	III	$3p\,^3P_2^0 \rightarrow 3s\,^3S_1$	125	CO_2	EI	57, 129 (a)
4650·25	4650·21 ± 0·04	III	$3p\,^3P_1^0 \rightarrow 3s\,^3S_1$	125	CO_2	EI	57, 129 (a)

Notes to Table I.14. (a) Also reported by Bridges and Chester (1965a, b; 1971a) as appearing in xenon and krypton (4647 Å) discharges but the transition in these cases was almost certainly one of the unclassified transitions near these carbon III wavelengths in xenon III–IV (see Table I.4).

TABLE I.15. *Silicon ion laser transitions*

Calculated wavelength Å (air)	Measured wavelength Å (air)	Ion	Transition assignment	S	D	EM	References
4088·85	4088·90 ± 0·1	IV	$4p\,^2P_{3/2} \rightarrow 4s\,^2S_{1/2}$	107	Ar	EI	4 (E) (a)
4552·62	4552·59 ± 0·06	III	$4p\,^3P_2 \rightarrow 4s\,^3S_1$	105	PF_5, SF_6, $SiCl_4$	EI	74 (E), 77 (b)
4567·82	4567·84 ± 0·06	III	$4p\,^3P_1 \rightarrow 4s\,^3S_1$	105	PF_5, SF_6, $SiCl_4$	EI	74 (E), 77 (b)
* 6347·10	6347·24 ± 0·06	II	$4p\,^2P^0_{3/2} \rightarrow 4s\,^2S_{1/2}$	106	PF_5, SF_6, $SiCl_4$	EI	74 (E), 77 (b)
* 6371·36	6371·48 ± 0·06	II	$4p\,^2P^0_{1/2} \rightarrow 4s\,^2S_{1/2}$	106	PF_5, SF_6	EI	74 (E) (b)
6671·88	6671·93 ± 0·06	II	$4p\,^4D_{7/2} \rightarrow 4s\,^4P^0_{5/2}$	106	PF_5, $SiCl_4$	EI	74 (E), 77 (c)

Notes to Table I.15. (a) This is the strongest Si IV spontaneous emission line in the visible spectrum; it was only observed in laser oscillation on one occasion but under conditions favouring the production of free silicon by bore erosion in an argon discharge in a quartz tube at low pressure. Was originally listed as appearing in a xenon discharge but was in fact observed in argon, however it is distinct from the argon laser line at 4088·6 Å (see Table I.2). (b) Originally assigned to fluorine (Cheo and Cooper, 1965b); the correct assignment was pointed out by Palenius (1966). (c) Originally assigned to phosphorous (Cheo and Cooper, 1965b); the correct assignment was pointed out by Palenius (1966).

TABLE I.16. *Germanium ion laser transitions*

Calculated wavelength Å (air)	Measured wavelength Å (air)	Ion	Transition assignment	S	D	EM	References
5131·75	5131·5 ± 0·7	II	$4f\,^2F^0_{5/2} \rightarrow 4d\,^2D_{3/2}$	111	Ge + He or Ne	?	97
* 5178·65	5178·4 ± 0·7	II	$4f\,^2F^0_{7/2} \rightarrow 4d\,^2D_{5/2}$	111	Ge + He or Ne	?	97

TABLE I.17. *Tin ion laser transitions*

Calculated wavelength Å (air)	Measured wavelength Å (air)	Ion	Transition assignment	S	D	EM	References
* 5799·18 [13]	5798·7 ± 0·7	II	$4f\,^2F^0_{7/2} \rightarrow 5d\,^2D_{5/2}$	CM	Sn + He or Ne	?	97
* 6453·58 [13]	6453·0 ± 0·7	II	$6p\,^2P^0_{3/2} \rightarrow 6s\,^2S_{1/2}$	CM	Sn + He	PI	97, 98 (CW, G)
...	6579·03 ± 0·06	?	...		$SnCl_4$	EI ?	90, 96 (E) (a)
6844·20 [13]	6844·0 ± 0·7	II	$6p\,^2P^0_{1/2} \rightarrow 6s\,^2S_{1/2}$	CM	Sn + He	PI	97, 98 (CW)

Notes to Table I.17. (a) Might possibly be the same transition as that listed in M.I.T. Wavelength tables at 6579·26 Å which may be the neutral tin transition $10d\,^3D^0_{1,2} \rightarrow 6p\,^3P_1$ at a calculated wavelength of 6580·85 Å. However there is a rather large discrepancy between the measured and calculated wavelengths in this case. It remains uncertain whether the observed laser line corresponds to an unclassified transition in neutral or ionized tin or to the above transition.

TABLE I.18. *Lead ion laser transitions*

Calculated wavelength Å (air)	Measured wavelength Å (air)	Ion	Transition assignment	S	D	EM	References
5372·25	5372·1 ± 0·7	II	$5f\,^2F^0_{7/2} \rightarrow 6s6p^2\,^4P_{5/2}$	CM	Pb + He	EI ?	97
5608·9	5608·6 ± 0·5	II	$7p\,^2P^0_{3/2} \rightarrow 7s\,^2S_{1/2}$	CM	Pb + He	PI	181 (CW)
6659·9	6660·1 ± 0·5	II	$7p\,^2P^0_{1/2} \rightarrow 7s\,^2S_{1/2}$	CM	Pb + He	PI	181 (CW)

TABLE I.19. *Nitrogen ion laser transitions*

Calculated wavelength Å (air)	Measured wavelength Å (air)	Ion	Transition assignment	S	D	EM	References
3367·35	3367·32 ± 0·06	III	$(^3P^0)3p\,^4P_{5/2} \rightarrow (^3P^0)3s\,^4P^0_{5/2}$	CM	N_2	EI	1
3478·71	3478·76 ± 0·05	IV	$3p\,^3P^0_2 \rightarrow 3s\,^3S_1$	127	N_2	EI	57, 1, 128
3482·99	3483·02 ± 0·06	IV	$3p\,^3P^0_1 \rightarrow 3s\,^3S_1$	127	N_2	EI	1
3995·00	3994·99 ± 0·02	II	$3p\,^1D_2 \rightarrow 3s\,^1P^0_1$	126	N_2	EI	129
4097·32	4097·29 ± 0·06	III	$3p\,^2P^0_{3/2} \rightarrow 3s\,^2S_{1/2}$	CM	N_2	EI	57, 128
4103·38	4103·36 ± 0·02	III	$3p\,^2P^0_{1/2} \rightarrow 3s\,^2S_{1/2}$	CM	N_2	EI	57, 129
4510·88	4510·45 ± 0·23	III	$(^3P^0)3p\,^4D_{5/2} \rightarrow (^3P^0)3s\,^4P^0_{3/2}$	CM	N_2	EI	57
4514·87	4514·86 ± 0·03	III	$(^3P^0)3p\,^4D_{7/2} \rightarrow (^3P^0)3S\,^4P^0_{5/2}$	CM	N_2	EI	57, 129
4621·39	4621·0 ± 0·8	II	$3p\,^3P_0 \rightarrow 3s\,^3P^0_1$	126	N_2	EI	32 (a)
4630·54	4630·51 ± 0·02	II	$3p\,^3P_2 \rightarrow 3s\,^3P^0_2$	126	N_2	EI	57, 129
4643·09	4643·9 ± 0·8	II	$3p\,^3P_1 \rightarrow 3s\,^3P^0_2$	126	N_2	EI	32 (a)
5016·39	5016·39	II	$3d\,^3F_2 \rightarrow 3p\,^3D_2$	126	N_2	EI	124 (E) (b)
5666·63	5666·62 ± 0·03	II	$3p\,^3D_2 \rightarrow 3s\,^3P^0_1$	126	N_2	EI	71, 124 (E) (b)
5676·02	5676·03 ± 0·03	II	$3p\,^3D_1 \rightarrow 3s\,^3P^0_0$	126	N_2	EI	71
* 5679·56	5679·53 ± 0·03	II	$3p\,^3D_3 \rightarrow 3s\,^3P^0_2$	126	N_2, air	EI	4, 71, 122, 124 (E) (b)
5686·21	5686·9 ± 0·8	II	$3p\,^3D_1 \rightarrow 3s\,^3P^0_1$	126	N_2	EI	32 (a)
6482·09	6482·6 ± 0·6	II	$3p\,^1P_1 \rightarrow 3s\,^1P^0_1$	CM	N_2	EI	32, 123

Notes to Table I.19. (a) Originally tentatively assigned by Bridges and Chester (1965b). (b) In the mode of excitation of these lines used by Gadetskii *et al.* (1971) a modified form of electron impact excitation occurred. Heard and Peterson (1964c) report a number of laser lines observed in pulsed discharges through nitrogen and mercury–nitrogen mixtures, however their wavelength measurements are not sufficiently accurate to determine whether they correspond to transitions listed in the table or are separate lines.

TABLE I.20. *Phosphorus ion laser transitions*

Calculated wavelength Å (air)	Measured wavelength Å (air)	Ion	Transition assignment	S	D	EM	References
3347·70	3347·76 ± 0·06	IV	$4p\,^3P_2^0 \rightarrow 4s\,^3S_1$	CM	PF_5	EI	74
4222·09	4222·25 ± 0·06	III	$4p\,^2P_{3/2}^0 \rightarrow 4s\,^2S_{1/2}$	CM	PF_5	EI	74
* 6024·18	6024·27 ± 0·06	II	$4p\,^3D_2 \rightarrow 4s\,^3P_1^0$	110	PF_5, P + He or Ne	EI, CT ?	115, 74, 182 (CW)
6034·04	6034·19 ± 0·06	II	$4p\,^3D_1 \rightarrow 4s\,^3P_0^0$	110	PF_5	EI	74
* 6043·12	6043·22 ± 0·06	II	$4p\,^3D_3 \rightarrow 4s\,^3P_2^0$	110	PF_5, P + He or Ne	EI, CT ?	115, 74, 182 (CW)
6087.82	6088·04 ± 0·06	II	$4p\,^3D_1 \rightarrow 4s\,^3P_1^0$	110	PF_5	EI	74
6165·59	6165·74 ± 0·06	II	$4p\,^3D_2 \rightarrow 4s\,^3P_2^0$	110	PF_5	EI	74
* 7845·63	7845·63 ± 1	II	$4p\,^1P_1 \rightarrow 4s\,^1P_1^0$	110	P + He or Ne	EI ?, CT ?	115

TABLE I.21. *Arsenic ion laser transitions*

Calculated wavelength Å (air)	Measured wavelength Å (air)	Ion	Transition assignment	S	D	EM	References
5385·20	5385·1 ± 0·4	II	$6s(\frac{3}{2},\frac{1}{2})^0_1 \to 5p\frac{3}{2}[\frac{3}{2}]_1$ $(6s\,^1P^0_1 \to 5p\,^3P_1)$	102	As + He	CT	188 (CW)
* 5496·95	5496·95 ± 0·4	II	$6s(\frac{3}{2},\frac{1}{2})^0_2 \to 5p\frac{3}{2}[\frac{5}{2}]_3$ $(6s\,^3P^0_2 \to 5p\,^3D_3)$	102	As + He	CT	188 (CW) (a)
5497·73	5497·6 ± 0·4	II	$5p\frac{1}{2}[\frac{1}{2}]_1 \to 5s(\frac{1}{2},\frac{1}{2})^0_0$ $(5p\,^3D_1 \to 5s\,^3P^0_0)$	102	As + He or Ne	CT	188 (CW) (a)
5558·09	5558·2 ± 0·4	II	$5p\frac{1}{2}[\frac{3}{2}]_2 \to 5s(\frac{1}{2},\frac{1}{2})^0_1$ $(5p\,^3D_2 \to 5s\,^3P^0_1)$	102	As + He or Ne	CT	73, 188 (CW)
5651·32	5652 ± 1	II	$5p\frac{3}{2}[\frac{5}{2}]_3 \to 5s(\frac{3}{2},\frac{1}{2})^0_2$ $(5p\,^3D_3 \to 5s\,^3P^0_{2})$	102	As + Ne	?	73
5837·90	5838·0 ± 0·4	II	$6s(\frac{3}{2},\frac{1}{2})^0_2 \to 5p\frac{3}{2}[\frac{3}{2}]_2$ $(6s\,^3P^0_2 \to 5p\,^3P_2)$	102	As + He	CT	188 (CW)
6170·27	6170·2 ± 0·4	II	$5p\frac{1}{2}[\frac{3}{2}]_1 \to 5s(\frac{1}{2},\frac{1}{2})^0_1$ $(5p\,^1P_1 \to 5s\,^3P^0_1)$	102	As + He or Ne	CT	73, 188 (CW)
* 6511·74	6511·8 ± 0·4	II	$6s(\frac{3}{2},\frac{1}{2})^0_1 \to 5p\frac{3}{2}[\frac{5}{2}]_2$ $(6s\,^1P^0_1 \to 5p\,^1D_2)$	102	As + He	CT	188 (CW)
7102·72	7102·5 ± 0·4	II	$5p\frac{3}{2}[\frac{3}{2}]_1 \to 4p^{3}\,^3P^0_2$ $(5p\,^3P_1 \to 4p^{3}\,^3P^0_2)$	102	As + He		188 (CW)

Notes to Table I.21. (a) The first observation of one of these two lines was by Bell *et al.* (1965), however from the wavelength measurement reported therein, 5498 ± 1 Å, it is uncertain which one. As well as the mixed LK, jj, LS coupling notation of Li and Andrew (1971), normal LS coupling notation for the transitions is given in parentheses.

TABLE I.22. *Antimony ion laser transitions*

Calculated wavelength Å (air)	Measured wavelength Å (air)	Ion	Transition assignment	S	D	EM	References
6130·00	6130 ± ?	II	$6p\,^3D_3 \to 6s\,^3P^0_2$	CM	Sb + Ne	EI ?	112

TABLE I.23. *Bismuth ion laser transitions*

Calculated wavelength Å (air)	Measured wavelength Å (air)	Ion	Transition assignment	S	D	EM	References
4561·18	4560·70 ± 0·1	III	$7p\,^2P^0_{1/2} \rightarrow 7s\,^2S_{1/2}$	CM	Bi + He or Ne	EI ?	176 (G)
5719·24	5719·20 ± 0·1	II	$6p_{1/2}7p_{1/2}\,^3P_0 \rightarrow 6p_{1/2}7s\,^3P^0_1$	CM	Bi + He or Ne	EI ?	176 (G)
7599·00	7598·70 ± 0·5	III	$6f\,^2F^0_{5/2} \rightarrow 7d\,^2D_{3/2}$	CM	Bi + He or Ne	EI ?	176 (G)
8068·81	8069·20 ± 0·5	III	$6f\,^2F^0_{7/2} \rightarrow 7d\,^2D_{5/2}$	CM	Bi + He or Ne	EI ?	176 (G)

TABLE I.24. *Oxygen ion laser transitions*

Calculated wavelength Å (air)	Measured wavelength Å (air)	Ion	Transition assignment	S	D	EM	References
2781·04	2781·50 ± 0·50	V	$3p\,^3P^0_2 \rightarrow 3s\,^3S_1$	CM	O_2	EI	189
2983·78	2983·86 ± 0·06	III	$3p\,^1D_2 \rightarrow 3s\,^1P^0_1$	CM	O_2	EI	4 (E), 1, 189 (E)
* 3047·13	3047·15 ± 0·06	III	$3p\,^3P_2 \rightarrow 3s\,^3P^0_2$	CM	O_2	EI	4 (E), 1, 189
3063·42 [121]	3063·46 ± 0·06	IV	$3p\,^2P^0_{3/2} \rightarrow 3s\,^2S_{1/2}$	121	O_2	EI	1, 189
3381·21 / 3381·30	3381·34 ± 0·06	IV	$({}^3P^0)3p\,^4D_{5/2} \rightarrow ({}^3P^0)3s\,^4P^0_{3/2}$ or $({}^3P^0)3p\,^4D_{3/2} \rightarrow ({}^3P^0)3s\,^4P^0_{1/2}$	121	O_2	EI	1
3385·52	3385·54 ± 0·06	IV	$({}^3P^0)3p\,^4D_{7/2} \rightarrow ({}^3P^0)3s\,^4P^0_{5/2}$	121	O_2	EI	1, 189
3727·33	3727·11 ± 0·50	II	$3p\,^4S^0_{3/2} \rightarrow 3s\,^4P_{3/2}$	CM	O_2	EI	189
* 3749·49	3749·48 ± 0·03	II	$3p\,^4S^0_{3/2} \rightarrow 3s\,^4P_{5/2}$	CM	O_2	EI	57, 1, 4, 189 (a)
3754·67	3754·69 ± 0·03	III	$3p\,^3D_2 \rightarrow 3s\,^3P^0_1$	CM	O_2	EI	57, 1, 4 (a)
3759·88	3759·85 ± 0·04	III	$3p\,^3D_3 \rightarrow 3s\,^3P^0_2$	CM	O_2	EI	57, 1, 4, 189
* 4347·38	4347·38 ± 0·04	II	$({}^1D)3p\,^2D^0_{3/2} \rightarrow ({}^1D)3s\,^2D_{3/2}$	CM	O_2	EI	57, 4, 189
* 4351·28	4351·25 ± 0·03	II	$({}^1D)3p\,^2D^0_{5/2} \rightarrow ({}^1D)3s\,^2D_{5/2}$	CM	O_2	EI	57, 4 (a)
* 4414·88	4414·92 ± 0·03	II	$3p\,^2D^0_{5/2} \rightarrow 3s\,^2P_{3/2}$	CM	O_2	EI	57, 4 (a)

Calculated wavelength Å (air)	Measured wavelength Å (air)	Ion	Transition assignment	S	D	EM	References
* 4416·97	4416·97 ± 0·04	II	$3p^2D^0_{3/2} \rightarrow 3s^2P_{1/2}$	CM	O_2	EI	57, 4
...	4605·52 ± 0·09	?	...		O_2	EI	57
4649·14	4649·08 ± 0·1	II	$3p^4D^0_{7/2} \rightarrow 3s^4P_{5/2}$	CM	O_2	EI	4 (b)
5592·37	5592·48 ± 0·05	III	$3p^1P_1 \rightarrow 3s^1P^0_1$	CM	O_2	EI	57, 4, 122 (E), 123 (P), 124
6640·99	6640·2 ± 1	II	$3p^2S^0_{1/2} \rightarrow 3s^2P_{1/2}$	CM	O_2, air	EI	123 (P)
6666·94	6666·94	II	$4p^2P^0_{1/2} \rightarrow 3d^2P_{3/2}$	CM	O_2	EI	124 (c)
6721·36	6721·38 ± 0·04	II	$3p^2S^0_{1/2} \rightarrow 3s^2P_{3/2}$	CM	O_2	EI	57 (E), 124

Notes to Table I.24. (a) The mean value of the two equally accurate wavelength measurements of McFarlane (1964) and Bridges and Chester (1965) are given. (b) This transition may have been reported first by Heard and Peterson (1964c), although the transition they report in CO at 4650 Å could equally well have been the C III transition at 4650·25. Willett (1971) lists an additional oxygen ion laser transition at 2984·61 Å, this arises because of the original incorrect assignment of this transition in Bridges and Chester (1964a). The transitions observed by Heard and Peterson (1964c) at 4120, 4525, 4750, 5440, 5500, 5540, 5679 and 5690 Å in Hg–N_2 mixtures and in CO are listed as oxygen ion laser transitions by Willett (1971). The above transitions were probably measured with large error and almost certainly correspond to transitions in oxygen, nitrogen or carbon which are already listed in the table.

TABLE I.25. *Sulphur ion laser transitions*

Calculated wavelength Å (air)	Measured wavelength Å (air)	Ion	Transition assignment	S	D	EM	References
2638·90 [119] * 2639·14 [117]	2638·98 ± 0·06	V ? II ?	$4s \rightarrow 4p_1$ or ...	119	SO_2	EI	114 (a)
3324·86	3324·86 ± 0·06	III	$4p^3P_2 \rightarrow 3d^3P^0_2$	CM	SO_2	EI	114
3497·34 [118]	3497·37 ± 0·06	III	...	120 116	SO_2, H_2S	EI	114 (b)
* 3709·37	3709·41 ± 0·06	III	$4p^3D_2 \rightarrow 3d^3P^0_1$	CM	SO_2, SF_6, H_2S	EI	114
4925·32	4925·60 ± 0·06	II	$4p^4P^0_{3/2} \rightarrow 4s^4P_{1/2}$	CM	SO_2	EI	114
5014·00	5014·24 ± 0·06	II	$4p^2P^0_{3/2} \rightarrow 4s^2P_{3/2}$	CM	SO_2	EI	114
5032·39	5032·62 ± 0·06	II	$4p^4P^0_{5/2} \rightarrow 4s^4P_{5/2}$	CM	SO_2, SF_6	EI	114
* 5160·11 [116]	5160·32 ± 0·06	III	...	116	SO_2, SF_6	EI	114
* 5219·37 [116]	5219·62 ± 0·06	II ? III ?	...		SO_2, SF_6	EI	114 (c)

TABLE I.25—*continued*

Calculated wavelength Å (air)	Measured wavelength Å (air)	Ion	Transition assignment	S	D	EM	References
* 5320·70	5320·88 ± 0·06	II	$(^1D)4p\,^2F^0_{7/2} \rightarrow (^1D)4s\,^2D_{5/2}$		SO_2, SF_6, H_2S	EI	114, 115, 182 (CW) (d)
* 5345·66	5345·83 ± 0·06	II	$(^1D)4p\,^2F^0_{5/2} \rightarrow (^1D)4s\,^2D_{3/2}$		SO_2, SF_6, H_2S	EI	114, 115, 182 (CW) (d)
5428·64	5428·74 ± 0·06	II	$4p\,^4D^0_{3/2} \rightarrow 4s\,^4P_{1/2}$		SO_2, SF_6 H_2S	EI	114, 115 (d)
* 5432·74	5432·87 ± 0·06	II	$4p\,^4D^0_{5/2} \rightarrow 4s\,^4P_{3/2}$		SO_2, SF_6, H_2S	EI	114, 115, 182 (CW) (d)
* 5453·79	5453·88 ± 0·06	II	$4p\,^4D^0_{7/2} \rightarrow 4s\,^4P_{5/2}$		SO_2, SF_6, H_2S	EI	114, 115, 182 (CW) (d)
5473·60	5473·74 ± 0·06	II	$4p\,^4D^0_{1/2} \rightarrow 4s\,^4P_{1/2}$		SO_2, SF_6, H_2S	EI	114, 115 (d)
5509·65	5509·90 ± 0·06	II	$4p\,^4D^0_{3/2} \rightarrow 4s\,^4P_{3/2}$		SO_2, SF_6, H_2S	EI	114
5564·91	5565·11 ± 0·06	II	$4p\,^4D^0_{5/2} \rightarrow 4s\,^4P_{5/2}$		SO_2	EI	114
* 5639·98	5640·12 ± 0·06	II	$4p\,^2D^0_{5/2} \rightarrow 4s\,^2P_{3/2}$		SO_2, SF_6, H_2S	EI	114, 115, 182 (CW) (d)
* 5646·98	5647·16 ± 0·06	II	$4p\,^2D^0_{3/2} \rightarrow 4s\,^2P_{1/2}$		SO_2, SF_6, H_2S	EI	114, 115 (d)
5819·18	5819·35 ± 0·06	II	$4p\,^2D^0_{3/2} \rightarrow 4s\,^2P_{3/2}$		SO_2	EI	114

Notes to Table I.25. (a) Listed as an SV transition by Cooper and Cheo (1966) on the basis of an assignment of a spontaneous transition at 2638·90 Å by Bowen and Millikan (1925) to the SV transition $4s \rightarrow 4p_1$. However, this seems an unlikely assignment, particularly since no other sulphur ion laser transitions above SIII have been observed. Although the wavelength measurement accuracy of Cooper and Cheo (1966) is stated as ± 0·06 Å, comparison of their measured wavelength with calculated or observed spontaneous emission wavelengths listed in the table indicates their real error may in certain cases be as large as ± 0·28 Å. The laser line at 2638·98 Å may therefore correspond to the spontaneous transition at 2639·14 Å assigned to SII by Bloch and Bloch (1929). (b) Gilles (1931) has suggested that this is the transition SIII, $4d\,^1F^0_3 \rightarrow 4p\,^1D_2$ but no evidence exists for this assignment. (c) Assigned to SII by Gilles (1931) and SIII by Bloch and Bloch (1929). (d) Laser oscillation also observed in discharges through sulphur plus any one of a number of carrier gases: helium, neon, argon, hydrogen, oxygen, nitrogen and air (Fowles and Silfvast, 1965).

TABLE 1.2[illegible]. Selenium ion laser transitions

Calculated wavelength Å (air)	Measured wavelength Å (air)	Ion	Transition assignment	S	D	EM	References
4467·60	4468·0 ± 0·5	II	$5p\,^2P^0_{1/2} \rightarrow 5s\,^2P_{1/2}$	CM	He + Se	CT	169 (CW, G) (a)
4604·31	4604·6 ± 0·5	II	$5p\,^2D^0_{5/2} \rightarrow 5s\,^4P_{5/2}$	CM	He + Se	CT	168 (CW, P), 173 (CW, P)
4618·77	4619·1 ± 0·5	II	$5p\,^4P^0_{5/2} \rightarrow 5s\,^4P_{3/2}$	CM	He + Se	CT	169 (CW, G)
4648·43	4648·6 ± 0·5	II	$5p\,^4P^0_{3/2} \rightarrow 5s\,^4P_{1/2}$	CM	He + Se	CT	168 (CW, P), 173 (CW, P)
4718·23 [170] 4717·82	4718·5 ± 0·5	II	$5p\,^4S^0_{3/2} \rightarrow 3_{1/2}$	CM	He + Se	CT	169 (CW, G) (b)
4740·98	4740·6 ± 0·5	II	$5p\,^2P^0_{3/2} \rightarrow 4p^4\,^2P_{3/2}$	CM	He + Se	CT	169 (CW, G)
4763·68	4764·1 ± 0·5	II	$5p\,^2D^0_{3/2} \rightarrow 5s\,^4P_{3/2}$	CM	He + Se	CT	168 (CW), 173 (CW, P)
4765·54	4765·1 ± 0·5	II	$5p\,^2P^0_{1/2} \rightarrow 4p^4\,^2P_{3/2}$	CM	He + Se	CT	169 (E, CW, G)
4840·63	4840·6 ± 0·5	II	$5p\,^2S^0_{1/2} \rightarrow 5s\,^4P_{3/2}$	CM	He + Se	CT	168 (CW)
4844·99	4845·0 ± 0·5	II	$5p\,^4S^0_{3/2} \rightarrow 5s\,^4P_{5/2}$	CM	He + Se	CT	168 (CW, G), 173 (CW, P)
* 4975·72	4976·1 ± 0·5	II	$5p\,^2D^0_{5/2} \rightarrow 4p^4\,^2P_{3/2}$	CM	He + Se	CT	168 (CW, G, P), 173 (CW, P)
* 4992·79	4992·9 ± 0·5	II	$5p\,^4P^0_{3/2} \rightarrow 5s\,^4P_{3/2}$	CM	He + Se	CT	168 (CW, G, P), 173 (CW, P)
* 5068·66	5068·7 ± 0·5	II	$5p\,^4P^0_{5/2} \rightarrow 5s\,^4P_{5/2}$	CM	He + Se	CT	168 (CW, G. P), 173 (CW, P)
5096·50 [171] 5096·30	5096·1 ± 0·5	II	$5p\,^4D^0_{7/2} \rightarrow 4d\,^4F_{9/2}$	CM	He + Se, Ne + Se	CT ?	73, 168 (CW), 173 (CW) (c)
5142·15	5141·9 ± 0·5	II	$5p\,^4D^0_{3/2} \rightarrow 5s\,^4P_{1/2}$	CM	He + Se	CT	168 (CW, G)
* 5175·94	5176·0 ± 0·5	II	$5p\,^4D^0_{5/2} \rightarrow 5s\,^4P_{3/2}$	CM	He + Se	CT	168 (CW, G, P), 173 (CW, P), 174 (CW, P)
* 5227·49	5227·6 ± 0·5	II	$5p\,^4D^0_{7/2} \rightarrow 5s\,^4P_{5/2}$	CM	He + Se, Ne + Se	CT	73, 168 (CW, G, P), 173 (CW, P), 174 (CW, P)
5253·10	5252·6 ± 0·5	II	$5p\,^2D^0_{5/2} \rightarrow 5s\,^2P_{3/2}$	CM	He + Se	CT	168 (CW, G)
5253·67	5253·2 ± 0·5	II	$5p\,^4D^0_{1/2} \rightarrow 5s\,^4P_{1/2}$	CM	He + Se	CT	168 (CW, G)
5271·15 5271·27	5271·3 ± 0·5	II	$(1D)\,5p\,^2D^0_{5/2} \rightarrow (^1D)\,5s\,^2D_{3/2}$ or $5p\,^4D^0_{5/2} \rightarrow 4d\,^4F_{7/2}$	CM 171	He + Se	CT ?	168 (CW) (d)

TABLE I.26—*continued*

Calculated wavelength Å (air)	Measured wavelength Å (air)	Ion	Transition assignemnt	S	D	EM	References
* 5305·44	5305·5 ± 0·5	II	$5p^2D^0_{3/2} \rightarrow 5s^2P_{1/2}$	CM	He + Se	CT	168 (CW, G, P), 173 (E, CW, P), 174 (CW, P)
5522·44 5522·66	5522·8 ± 0·5	II	$5p^4P^0_{5/2} \rightarrow 4p^{4\,2}P_{3/2}$ or $5p^4P^0_{3/2} \rightarrow 5s^4P_{5/2}$	CM	He + Se	CT	168 (CW), 173 CW, P)
5566·88	5567·1 ± 0·5	II	$5p^4D^0_{3/2} \rightarrow 5s^4P_{3/2}$	CM	He + Se	CT	169 (CW, G)
5591·13	5591·6 ± 0·5	II	$5p^4P^0_{3/2} \rightarrow 5s^2P_{1/2}$	CM	He + Se	CT	168 (CW), 173 (CW, P)
5623·15	5622·8 ± 0·5	II	$5p^4P^0_{1/2} \rightarrow 5s^4P_{1/2}$	CM	He + Se	CT	169 (CW, G)
5697·82	5697·9 ± 0·5	II	$5p^4D^0_{1/2} \rightarrow 5s^4P_{3/2}$	CM	He + Se	CT	168 (CW), 173 (CW, P)
5747·61	5747·9 ± 0·5	II	$5p^4D^0_{5/2} \rightarrow 5s^4P_{5/2}$	CM	He + Se	CT	168 (CW)
5842·61	5842·8 ± 0·5	II	$5p^2S^0_{1/2} \rightarrow 4p^{4\,2}P_{3/2}$	CM	He + Se	CT	169 (E, CW, G)
5866·23	5866·7 ± 0·5	II	$5p^4P^0_{5/2} \rightarrow 5s^2P_{3/2}$	CM	He + Se	CT	169 (CW, G), 173 (E, CW, P)
6055·92	6056·3 ± 0·5	II	$5p^2P^0_{3/2} \rightarrow 7_{3/2}$	CM	He + Se	CT	168 (CW, G), 173 (CW, P)
6065·73	6066·1 ± 0·5	II	$5p^4P^0_{3/2} \rightarrow 4p^{4\,2}P_{3/2}$	CM	He + Se	CT	169 (E, W, G)
6101·98	6102·1 ± 0·5	II	$5p^2D^0_{3/2} \rightarrow 5s^2P_{3/2}$	CM	He + Se	CT	169 (CW, G)
6444·26	6443·9 ± 0·5	II	$5p^2D^0_{5/2} \rightarrow 7_{3/2}$	CM	He + Se	CT	168 (CW), 173 (CW, P)
6490·49	6490·1 ± 0·5	II	$5p^4D^0_{1/2} \rightarrow 5s^2P_{1/2}$	CM	He + Se	CT	168 (CW), 173 (CW, P)
6534·40	6534·6 ± 0·5	II	$5p^2P^0_{1/2} \rightarrow 9_{1/2}$	CM	He + Se	CT	168 (CW)
7063·89	7064·2 ± 0·6	II	$5p^4P^0_{1/2} \rightarrow 5s^2P_{1/2}$	CM	He + Se	CT	169 (CW, G) (e)
7392·06	7392·4 ± 0·6	II	$5p^4P^0_{5/2} \rightarrow 7_{3/2}$	CM	He + Se	CT	169 (CW, G)
7674·72	7674·9 ± 0·7	II	$5p^2P^0_{3/2} \rightarrow (^1D)\,5s^2D_{5/2}$	CM	He + Se	CT	169 (CW, G)
7724·06	7723·6 ± 0·7	II	$5p^4D^0_{1/2} \rightarrow 5s^2P_{3/2}$	CM	He + Se	CT	169 (CW, G)
7796·10	7796·2 ± 0·7	II	$5p^2P^0_{1/2} \rightarrow (^1D)\,5s^2D_{3/2}$	CM	He + Se	CT	169 (CW, G)
7838·77	7839·3 ± 0·7	II	$5p^4P^0_{1/2} \rightarrow 4p^{4\,2}P_{3/2}$	CM	He + Se	CT	169 (CW, G)
8309·30	8308·9 ± 0·8	II	$5p^2D^0_{5/2} \rightarrow (^1D)\,5s^2D_{5/2}$	CM	He + Se	CT	169 (E, CW, G)
...	9249·3 ± 0·9	?	...		He + Se		169 (CW, G) (b)
9955·15	9954·7 ± 1	II	$5p^4P^0_{5/2} \rightarrow (^1D)\,5s^2D_{5/2}$	CM	He + Se	CT	169 (CW, G) (g)
* 10408·83	10409·4 ± 1	II	$5p^4D^0_{1/2} \rightarrow 6_{3/2}$	CM	He + Se	CT	169 (CW, G) (g)
* 12586·73	12587·9 ± 1	II	$5p^4P^0_{3/2} \rightarrow 10_{5/2}$	CM	He + Se	CT	169 (CW, G) (g)

Notes to Table I.26. (a) An alternative assignment to Se II, $14^{o}_{3/2} \rightarrow 5s^{4}P_{5/2}$ at 4467·60 Å is unlikely to be correct. The upper level of the assignment listed in the table is known to be the upper level of other laser transitions. (b) There is quite a large discrepancy between the calculated and measured spontaneous emission wavelengths of this transition. Note that the assignment given by Martin (1935) for this line is incorrect. (c) There is quite a large discrepancy between the calculated and measured spontaneous emission wavelengths of this transition. The assignment may therefore be incorrect. (d) The first assignment given to a spontaneous transition observed at 5271·22 Å by Martin (1935), the second assignment given to a spontaneous transition observed at 5271·11 Å by Krishnamurty and Rao (1935). (e) Not observed in spontaneous emission either by Bartlett (1934), Krishnamurty and Rao (1935) or Martin (1935). (f) No allowed transition between energy levels of Se II listed in Atomic Energy Levels (Moore, 1949) gives a wavelength near the observed wavelength of this laser transition. The assignment $5p^{4}D_{5/2} \rightarrow 8_{1/2}$ lies at 9248·07 Å but offends the strict $\Delta J = 0, \pm 1$ selection rule. (g) No spontaneous emission observations for Se II have been reported in the literature in this wavelength region. Hernqvist and Pultorak (1972) report the observation of Se II laser lines at the approximate wavelengths 4622, 4667, 4781, 5609, 5888 and 6465 Å. Transitions near these wavelengths were not observed by Silfvast and Klein (1970) or Klein and Silfvast (1971) who used long discharge tubes which allowed low gain transitions to oscillate. The lines reported by Hernqvist and Pultorak certainly correspond to other lines already listed in the table, either because their wavelengths were measured only approximately or through grating ghost effects.

TABLE I.27. *Tellurium ion laser transitions*

Calculated wavelength Å (air)	Measured wavelength Å (air)	Ion	Transition assignment	S	D	EM	References
4842·90	4843·3 ± 0·4	II	$122'^{o}_{5/2} \rightarrow 102_{3/2}$	153	Te + He	CT	155 (CW, G)
5020·39 [153]	5020·0 ± 0·4	II	...		Te + He	CT ?	155 (CW, G)
5256·41 [153]	5256·4 ± 0·4	II	...		Te + He	CT ?	155 (CW, G)
* 5449·84	5449·8 ± 0·4	II	$103^{o}_{3/2} \rightarrow 85'_{5/2}$	153	Te + Ne	CT	155 (CW, G)
...	5454·0 ± 0·5	II ?	...		Te + Ne	EI	154
* 5479·08	5479·3 ± 0·4	II	$105^{o}_{3/2} \rightarrow 86_{3/2}$	153	Te + Ne	CT	155 (CW, G)
* 5576·35	5576·5 ± 0·4	II	$112^{o}_{7/2} \rightarrow 94_{5/2}$	153	Te + He, Te + Ne	CT, EI	112, 154 (a) 155 (CW, G)
...	5640·5 ± 0·5	II ?	...		Te + Ne	EI	154
* 5666·20	5666·1 ± 0·4	II	$101^{o}_{3/2} \rightarrow 83_{1/2}$	153	Te + Ne	CT	155 (CW,G)
* 5708·12	5707·9 ± 0·4	II	$103^{o}_{7/2} \rightarrow 85'_{5/2}$	153	Te + He, Te + Ne	CT, EI	112, 154, 155 (CW, G), 156 (CW, P) (a)
5741·64	5741·5 ± 0·4	II	$112^{o}_{3/2} \rightarrow 94_{5/2}$	153	Te + Ne	CT	155 (CW, G)
5755·85	5755·7 ± 0·4	II	$100^{o}_{5/2} \rightarrow 82_{3/2}$	153	Te + He, Te + Ne	CT	155 (CW, G)
5765·25	5764·9 ± 0·4	II	$112^{o}_{5/2} \rightarrow 95_{3/2}$	153	Te + He	CT	155 (CW, G)

TABLE I.27—*continued*

	Calculated wavelength Å (air)	Measured wavelength Å (air)	Ion	Transition assignment	S	D	EM	References
	5851·12	5851·0 ± 0·4	II	$111^0_{5/2} \rightarrow 94_{5/2}$	153	Te + He, Te + Ne	CT	155 (CW, G)
*	5936·15	5935·9 ± 0·4	II	$99^0_{3/2} \rightarrow 82_{3/2}$	153	Te + Ne	CT	154, 155 (CW, G)
	5972·63	5972·3 ± 0·4	II	$111^0_{5/2} \rightarrow 95_{3/2}$	153	Te − He, Te + Ne	CT	155 (CW, G)
*	5974·68	5974·3 ± 0·4	II	$102^0_{5/2} \rightarrow 85_{5/2}$	153	Te + He	CT	155 (CW, G)
	6014·46	6014·7 ± 0·4	II	$105^0_{5/2} \rightarrow 88_{3/2}$	153	Te + Ne	CT	155 (CW, G)
	6082·26	6082·4 ± 0·4	II	$\ldots J = \frac{1}{2} \rightarrow J = \frac{1}{2}$	153	Te + He	CT	155 (CW, G)
*	6230·73	6230·4 ± 0·4	II	$105^0_{3/2} \rightarrow 88_{3/2}$	153	Te + Ne	CT	155 (CW, G)
	6245·45	6245·4 ± 0·4	II	$99^0_{3/2} \rightarrow 83_{1/2}$	153	Te + He, Te + Ne	CT	154, 155 (CW, G)
	6585·12 [153]	6585·0 ± 0·5	II	...		Te + He	CT	155 (CW, G)
	6648·58	6648·2 ± 0·5	II	$97^0_{1/2} \rightarrow 82_{3/2}$	153	Te + Ne	CT	155 (CW, G)
	6676·07	6676·5 ± 0·5	II	$103^0_{3/2} \rightarrow 88_{3/2}$	153	Te + Ne	CT	155 (CW, G)
	6885·14	6885·3 ± 0·5	II	$100^0_{5/2} \rightarrow 85_{5/2}$	153	Te + Ne	CT	155 (CW, G)
*	7039·13	7039·2 ± 0·5	II	$97^0_{1/2} \rightarrow 83_{1/2}$	153	Te + He, Te + Ne	CT	154, 155 (CW, G) (b)
	7801·68	7801·6 ± 0·5	II	$105^0_{3/2} \rightarrow 92_{3/2}$	153	Te + Ne	CT	155 (CW, G)
	7921·69	7921·4 ± 0·6	II	$97^0_{1/2} \rightarrow 85_{3/2}$	153	Te + Ne	CT	155 (CW, G)
*	8604·63 [153]	8604·4 ± 0·6	II	...		Te + He	CT	155 (CW, G)
	8733·81 [153]	8734·3 ± 0·6	II	...	153	Te + He	CT	155 (CW, G)
	8972·10	8971·9 ± 0·6	II	$103^0_{3/2} \rightarrow 92_{5/2}$	153	Te + Ne	CT	155 (CW, G)
	...	8998·2 ± 0·6	II	...		Te + Ne	CT ?	155 (CW, G)
	9378·48	9377·9 ± 0·6	II	$99^0_{3/2} \rightarrow 88_{5/2}$	153	Te + Ne	CT	155 (CW, G) (c)

Notes to Table I.27. (a) There is probably a significant contribution to the pumping of these transitions, when operated in the pulsed mode at low pressure, from electron impact. (b) Webb (1969) obtained oscillation at this wavelength in a Te + He discharge, however Silfvast and Klein (1972) only obtained oscillation in a Te + Ne discharge. (c) Not observed in spontaneous emission by Handrup and Mack (1964). The level notation for Te II is taken from Handrup and Mark (1964). The designations are incomplete and levels are represented by their approximate energy and J value, for example $103^0_{3/2}$ indicates a level with energy $\gtrapprox$ $1{\cdot}03 \times 10^5$ cm^{-1} and $J = 3/2$. A prime indicates a different core configuration. A transition observed by Bell *et al.* (1966) at 6350 Å in a pulsed Te + Ne discharge is an unclassified Te I transition at 6349·7 Å (Harrison, 1969).

TABLE I.28. *Fluorine ion laser transitions*

Calculated wavelength Å (air)	Measured wavelength Å (air)	Ion	Transition assignment	S	D	EM	References
2759·63	2759·59 ± 0·06	III	$3p\,^2D^0_{5/2} \rightarrow 3s\,^2D_{5/2}$	108	PF_5, SF_6	EI	74
2826·13	2826·08 ± 0·06	IV	$3p\,^3D_3 \rightarrow 3s\,^3P^0_2$	CM	PF_5, SF_6	EI	74
3121·54	3121·56 ± 0·06	III	$3p\,^4D^0_{7/2} \rightarrow 3s\,^4P_{5/2}$	108	PF_5, SF_6	EI	74
3174·17	3174·18 ± 0·06	III	$3p\,^2D^0_{5/2} \rightarrow 3s\,^2P_{3/2}$	108	PF_5, SF_6	EI	74 (E)
3202·76	3202·74 ± 0·06	II	$(^2D^0)3p\,^1D_2 \rightarrow (^2D^0\,3s\,^1D^0_2$	109	PF_5, SF_6	EI	74
4024·73	4024·78 ± 0·06	III	$3p\,^3P_2 \rightarrow 3s\,^3S^0_1$	109	PF_5, SF_6	EI	74 (E)

TABLE I.29. *Chlorine ion laser transitions*

Calculated wavelength Å (air)	Measured wavelength Å (air)	Ion	Transition assignment	S	D	EM	References
2632·67 [104]	2632·70 ± 0·06	III	$(^1D)4d\,^2D_{5/2} \rightarrow (^1D)4p\,^2F^0_{7/2}$	CM	Cl_2	EI	74 (a)
3191·46	3191·43 ± 0·06	III	$4p\,^4S^0_{3/2} \rightarrow 4s\,^4P_{5/2}$	CM	Cl_2	EI	74 (E)
* 3392·88	3392·87 ± 0·06	III	$(^1D)4p\,^2D^0_{3/2} \rightarrow (^1D)4s\,^2D_{3/2}$	CM	Cl_2	EI	74 (E)
* 3393·45	3393·45 ± 0·06	III	$(^1D)4p\,^2D^0_{5/2} \rightarrow (^1D)4s\,^2D_{5/2}$	CM	Cl_2	EI	74 (E)
* 3530·03	3530·03 ± 0·06	III	$(^1D)4p\,^2F^0_{7/2} \rightarrow (^1D)4s\,^2D_{5/2}$	CM	Cl_2	EI	74 (E)
* 3560·68	3560·69 ± 0·06	III	$(^1D)4p\,^2F^0_{5/2} \rightarrow (^1D)4s\,^2D_{3/2}$	CM	Cl_2	EI	74 (E)
* 3602·10	3602·10 ± 0·06	III	$4p\,^4D^0_{7/2} \rightarrow 4s\,^4P_{5/2}$	CM	Cl_2	EI	74 (E)
* 3612·82	3612·10 ± 0·06	III	$4p\,^4D^0_{5/2} \rightarrow 4s\,^4P_{3/2}$	CM	Cl_2	EI	74 (E) (b)
3622·68	3622·69 ± 0·06	III	$4p\,^4D^0_{3/2} \rightarrow 4s\,^4P_{1/2}$	CM	Cl_2	EI	74 (E)
* 3720·45	3720·46 ± 0·06	III	$4p\,^2D^0_{5/2} \rightarrow 4s\,^2P_{3/2}$	CM	Cl_2	EI	74 (E)
* 3748·80	3748·78 ± 0·06	III	$4p\,^2D^0_{3/2} \rightarrow 4s\,^2P_{1/2}$	CM	Cl_2	EI	74 (E)
4132·50	4132·5 ± 0·1	II	$(^2D^0)4p\,^1D_2 \rightarrow (^2D^0)4s\,^1D^0_2$	CM	Cl_2	EI	75 (CW, RF)
4740·42	4740·4 ± 0·1	II	$(^2P^0)4p\,^1P_1 \rightarrow (^2P^0)3d\,^1D^0_2$	CM	Cl_2	EI	75 (CW, RF)
4768·70	4768·74 ± 0·06	II	$(^2P^0)4p\,^3D_2 \rightarrow (^2P^0)4s\,^3P^0_1$	CM	Cl_2	EI	74, 75 (CW, RF)

TABLE I.29—*continued*

Calculated wavelength Å (air)	Measured wavelength Å (air)	Ion	Transition assignment	S	D	EM	References
4781·34	4781·34 ± 0·03	II	$(^2P^0)4p\,^3D_3 \rightarrow (^2P^0)4s\,^3P_2^0$	CM	Cl_2, MCl	EI	70, 71, 73 (CW, RF), 74, 75 (CW, RF), 115 (c)
4896·85	4896·88 ± 0·03	II	$(^2D^0)4p\,^3F_4 \rightarrow (^2D^0)4s\,^3D_3^0$	CM	Cl_2, MCl	EI	70, 71, 73 (CW, RF), 74, 75 (CW, RF), 115 (c)
4904·76 [72] 4904·82	4904·73 ± 0·03	II	$(^2D^0)4p\,^3F_3 \rightarrow (^2D^0)4s\,^3D_2^0$	CM	Cl_2, MCl	EI	70, 71, 73 (CW, RF), 74, 75 (CW, RF), 115 (c) (d)
4917·72 [72] 4917·81	4917·66 ± 0·03	II	$(^2D^0)4p\,^3F_2 \rightarrow (^2D^0)4s\,^3D_1^0$	CM	Cl_2, MCl	EI	70, 71, 73 (CW, RF), 74, 75 (CW, RF), 115 (c) (d)
5078·28	5078·30 ± 0·03	II	$(^2D^0)4p\,^3D_3 \rightarrow (^2D^0)4s\,^3D_3^0$	CM	Cl_2, MCl	EI	70, 71, 73 (CW, RF), 74, 75 (CW, RF), 115 (c)
5103·09	5103·1 ± 0·1	II	$(^2D^0)4p\,^3D_2 \rightarrow (^2D^0)4s\,^3D_2^0$	CM	Cl_2	EI	75 (CW, RF)
* 5217·92	5217·90 ± 0·03	II	$4p\,^3P_2 \rightarrow 4s\,^3S_1^0$	CM	Cl_2, $SiCl_4$, MCl	EI	70, 71 (CW), 73 (CW, RF), 74, 76 (CW, RF, P), 77, 75 (CW, RF, P), 115 (c) (e)
5221·35	5221·30 ± 0·03	II	$4p\,^3P_1 \rightarrow 4s\,^3S_1^0$	CM	Cl_2, MCl	EI	70, 71, 73 (CW, RF), 74, 75 (CW, RF), 115 (c)
* 5392·16	5392·15 ± 0·03	II	$(^2D^0)4p\,^1F_3 \rightarrow (^2D^0)4s\,^1D_2^0$	CM	Cl_2, $SiCl_4$, MCl	EI	70, 71, 73 (CW, RF), 74, 76 (CW, RF, P), 75 (CW, RF, P), 115 (c) (e)
6094·72	6094·74 ± 0·03	II	$(^2D^0)4p\,^1P_1 \rightarrow (^2D^0)4s\,^1D_2^0$	CM	Cl_2, $SiCl_4$, MCl	EI	70, 71, 73 (CW, RF), 74, 75 (CW, RF), 115 (c) (e)

Notes to Table I.29. (a) See also Palenius (1966). (b) The measured wavelength of this line which appears in Cheo and Cooper (1965) may be a misprint. (c) Observed in discharges where the inside of the discharge tube was coated with an alkali metal chloride (Fowles *et al.*, 1965), abbreviated MCl in column 6. (d) The measured wavelengths of these lines differ from those calculated by more than the stated error, however the measured spontaneous emission wavelengths listed in the original paper from which the analysis of Cl II in Atomic Energy Levels (Moore, 1949) was taken also differ from those calculated in the same way. (e) Observed in discharges through $SiCl_4$ by Carr and Grow (1967).

TABLE I.30. *Bromine ion laser transitions*

Calculated wavelength Å (air)	Measured wavelength Å (air)	Ion	Transition assignment	S	D	EM	References
4742·69	4742·66 ± 0·03	II	$(^2P^0)5p\,^3D_3 \rightarrow (^2P^0)5s\,^3P_2^0$	91	Br_2	EI	93
5054·66	5054·63 ± 0·05	II	$(^2D^0)5p\,^3F_3 \rightarrow 4d\,^3D_2^0$	91	Br_2	EI	93
* 5182·27	5182·38 ± 0·02	II	$5p\,^3P_2 \rightarrow 5s\,^3S_1^0$	91	Br_2, $Br_2 + Ne$	EI ?	93, 73, 182 (CW) (a)
5238·23	5238·26 ± 0·04	II	$5p\,^3P_1 \rightarrow 5s\,^3S_1^0$	91	Br_2, $Br_2 + Ne$	EI ?	93, 182 (CW)
* 5332·08	5332·03 ± 0·03	II	$(^2D^0)5p\,^1F_3 \rightarrow (^2D^0)5s\,^1D_2^0$	91	Br_2, $Br_2 + Ne$	EI ?	93, 73, 182 (CW) (a)
6117·62	6117·56 ± 0·06	II	$5p\,^5P_2 \rightarrow 5s\,^3S_1^0$	91	Br_2, HBr	EI	94, 90
6168·70	6168·78 ± 0·06	II	$5p\,^5P_1 \rightarrow 5s\,^3S_1^0$	91	Br_2	EI	94

Notes to Table I.30. (a) The two laser lines observed by Bell *et al.* (1965) and measured by them at 5185 ± 1 Å and 5334 ± 1 Å are almost certainly the same transitions as these two listed lines accurately measured by Keefe and Graham (1965). Further weight is lent to the assignment of these lines as they are the analogues of the two strongest transitions in Cl II. The energy levels listed by Rao (1958) were used to calculate the wavelengths listed in the table. [Note: All the energy level values listed in Rao (1958) and Atomic Energy Levels (Moore, 1949) are 5·9 ± 0·5 cm^{-1} too high for all levels above $4p^4$ (Martin and Tech, 1961)]. The fact that some of the measured and calculated wavelengths listed in the table differ by more than the listed experimental error probably reflects deficiencies in the energy level values used to calculate these wavelengths.

TABLE I.31. *Iodine ion laser transitions*

Calculated wavelength Å (air)	Measured wavelength Å (air)	Ion	Transition assignment	S	D	EM	References
...	4533·79 ± 0·03	?	...		I_2	EI	82 (P) (a)
...	4674·40 ± 0·03	?	...		I_2	EI	82 (P) (a)
...	4934·67 ± 0·03	?	...		I_2	EI	82 (P) (a)
4986·92	4986 ± ?	II	$(^2D^0)6p\,^3D_2 \rightarrow 5d\,^3D_1^0$	78	I_2 + He	CT	79, 188 (CW)
5216·27	5216 ± ?	II	$(^2D^0)6p\,^3F_2 \rightarrow 5d\,^3D_1^0$	78	I_2 + He	CT	79, 88 (hfs), 188 (CW)
* 5407·36	5407 ± ?	II	$(^2D^0)6p\,^3D_2 \rightarrow (^2D^0)6s\,^3D_2^0$	78	I_2 + He	CT	81, 79, 83 (hfs), 85 (CW), 88 (hfs)
5625·69	5625 ± ?	II	$6p\,^3P_2 \rightarrow 6s\,^3S_1^0$	78	I_2 + He, I_2 + Ne	EI ?	79
5678·08	5678 ± ?	II	$(^2D^0)6p\,^3F_2 \rightarrow (^2D^0)6s\,^3D_2^0$	78	I_2 + He	CT	81, 79, 83 (hfs), 85 (CW), 88 (hfs)
* 5760·72	5760 ± ?	II	$(^2D^0)6p\,^3D_2 \rightarrow (^2D^0)6s\,^3D_1^0$	78	I_2 + He	CT	80, 81, 79, 83 (hfs), 85 (CW), 88 (hfs)
6068·93	6068·9 ± ?	II	$(^2D^0)6p\,^3F_2 \rightarrow (^2D^0)6s\,^3D_1^0$	78	I_2 + He	CT	87 (G), 88 (hfs)
* 6127·49	6127 ± ?	II	$(^2D^0)6p\,^3D_1 \rightarrow (^2D^0)6s\,^3D_2^0$	78	I_2 + He	CT	80, 81, 79, 83 (hfs), 84 (CW) (b), 85 (CW), 88 (hfs)
6516·18	6516 ± ?	II	$(^2D^0)6p\,^3F_2 \rightarrow 5p^5\,^1P_1^0$	78	I_2 + He	CT	86, 85 (CW), 88 (hfs)
* 6585·21	6585 ± ?	II	$(^2D^0)6p\,^3D_1 \rightarrow (^2D^0)6s\,^3D_1^0$	78	I_2 + He	CT	81, 79 (E), 83 (hfs), 85 (CW, P), 88 (hfs)
6825·23	6825·23	II	$(^2D^0)6p\,^3F_2 \rightarrow (^2D^0)6s\,^3D_3$	78	I_2 + He	?	87
6904·77	6904 ± ?	II	$(^2D^0)6p\,^3D_2 \rightarrow (^2D^0)6s\,^1D_2^0$	78	I_2 + He, Ne or Kr	CT ? EI ?	79 (E)
7032·99	7032 ± ?	II	$(^2D^0)6p\,^3F_2 \rightarrow (^2D^0)5d\,^3D_2^0$	78	I_2 + He	CT	81, 79, 85 (CW), 83 (hfs)

7138·97	7138·97 ± ?	II	$(^2D^0)6p\,^3D_2 \rightarrow (^2D^0)5d\,^3D^0_3$	78	I_2 + He	CT	87, 85 (CW)
7618·50	7618·50 ± ?	II	$(^2D^0)6p\,^3F_2 \rightarrow (^2D^0)5d\,^3D^0_3$	78	I_2 + He	CT	188 (CW)
7735·78	7735·78 ± ?	II	$(^2D^0)6p\,^3D_1 \rightarrow (^2D^0)5d\,^3D^0_2$	78	I_2 + He	CT	188 (CW)
? 8253·84	8250 ± 100	II	$(^2D^0)6p\,^3D_1 \rightarrow (^2D^0)5d\,^3P^0_0$	78	I_2 + He	CT	79, 85 (CW) (c)
8170·07	8170·07 ± ?	II	$(^2D^0)6p\,^3D_2 \rightarrow (^2D^0)5d\,^3F^0_3$	78	I_2 + He	CT	188 (CW) (c)
8804·23	8800 ± 100	II	$(^2D^0)6p\,^3F_2 \rightarrow (^2D^0)5d\,^3F^0_3$	78	I_2 + He	CT	79, 85 (CW)
8877·61	8877·61 ± ?	II	$(^2D^0)6p\,^3F_3 \rightarrow (^2D^0)5d\,^3G^0_4$	78	I_2 + He	CT	188 (CW)
...	10417·2 ± 0·06	?	...		I_2	EI	82 (a)

Notes to Table I.31. (a) There are no spontaneous emission lines of either iodine I or iodine II at these wavelengths and these laser lines almost certainly originate from more highly ionized levels. (b) Also observed in CW operation in a Cd I_2 + He discharge (Collins *et al.*, 1972). (c) These two transitions are probably one and the same; the early wavelength measurement of the laser line at 8250 Å was really insufficiently accurate for the assignment to the transition at 8253·84 Å to be made. The recent observations of Piper *et al.* (1972) have confirmed the existence of a laser transition at 8170·07 Å, but not at 8253·84 Å.

TABLE I.32. *Ytterbium ion laser transitions*

Calculated wavelength Å (air)	Measured wavelength Å (air)	Ion	Transition assignment	S	D	EM	References
16498·42	16498 ± 2	II	$6p\,^2P^0_{3/2} \rightarrow 5d\,^2D_{5/2}$	140	Yb + He, Ne, Ar or H_2	EI	141, 142, 143, 139
24377·26	24377 ± 2	II	$6p\,^2P^0_{1/2} \rightarrow 5d\,^2D_{3/2}$	140	Yb + He, Ne, Ar or H_2	EI	142, 143, 139

TABLE I.33. *Miscellaneous ion laser transitions*

Calculated wavelength Å (air)	Measured wavelength Å (air)	Ion	Transition assignment	S	D	EM	References
...	3545 ± ?				Hg + Ar		22 (a)
...	4645·3 ± 0·5				Ar		4 (b)
...	5225 ± ?				Hg + Ar		180, 134 (c)
...	5610 ± ?	N or O			Air		187
...	5620 ± ?	N or O			Air		187
...	6577·45 ± 0·12				He, Ne		4 (d)
...	7065 ± ?				Hg + Ar		22 (e)

Notes to Table I.33. (a) Probably measured with a large error and may be the Ar II transition at 3511·12 Å. If the wavelength measurement was in fact accurate a possible assignment would be

$$\text{Ar II } 4d\,^2F_{5/2} \rightarrow 4p\,^2D^0_{3/2} \qquad \text{at } 3545{\cdot}60 \text{ Å}.$$

(b) No argon line observed in spontaneous emission lies near this wavelength, it may be the CIV transition, $6f\,^2F^0 \rightarrow 5d\,^2D$ at 4646 Å (Bockasten, 1956) operating through carbon contamination of the discharge. (c) Probably measured with a large error and may be the Ar II transition at 5145·32 Å. If this is the case then this was the first ionized noble gas transition to exhibit laser oscillation (see ref. 180). (d) A very doubtful assignment is to O III, $3d\,^5F_2 \rightarrow 4d\,^3D^0_3$ at 6577·50 Å, a line not observed in spontaneous emission. (e) Quite likely to be the self-terminating Ar I transition observed by Ericson and Lidholt (1967),

$$(^2P^0_{1/2})4p[3/2]_2 \rightarrow 4s[3/2]^0_2 \qquad \text{at } 7067{\cdot}21 \text{ Å}.$$

IV. Excitation and Quenching Mechanisms in Noble Gas Ion Lasers

With very few exceptions the dominant excitation mechanism of the laser transitions in the ionized noble gases is electron impact. However, depending on the particular laser involved, the exact nature of the electron impact processes by which excitation reaches the laser levels may be different. A number of these processes have been proposed for the noble gas ion lasers and a considerable amount of experimental evidence has been obtained which helps to elucidate what is the dominant electron impact process involved in noble gas ion lasers operating in different regimes. Because of its primary importance most experimental and theoretical effort has been aimed at the argon ion laser. However the results obtained for this system are almost certainly valid in general for all ion lasers where electron impact excitation of pure atomic species is involved.

The various electron impact processes which have been considered in connection with the excitation of ion lasers are

(i) Direct electron impact excitation of the upper laser levels from the neutral atom ground state (Fig. 1).
(ii) Electron impact excitation from neutral metastable states (Fig. 2).
(iii) Electron impact excitation from the ion ground state (Fig. 3).
(iv) Electron impact excitation from metastable states of the ion (Fig. 4).
(v) Excitation following radiative cascade from higher levels which are themselves populated by one or more of processes (i)–(iv) (Fig. 5).
(vi) Recombination into excited states (Fig. 6).

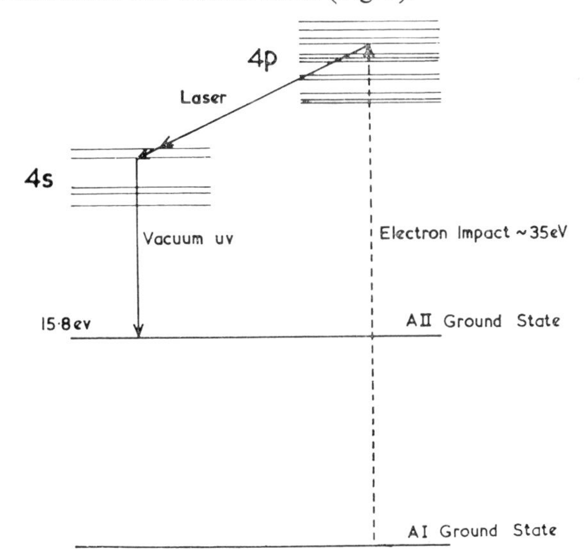

Fig. 1. Single-step excitation from the neutral atom ground state.

Although a number of specific plasma diagnostic experiments have been performed which indicate the relative importance of these different processes under different excitation conditions (see Section VII), much information

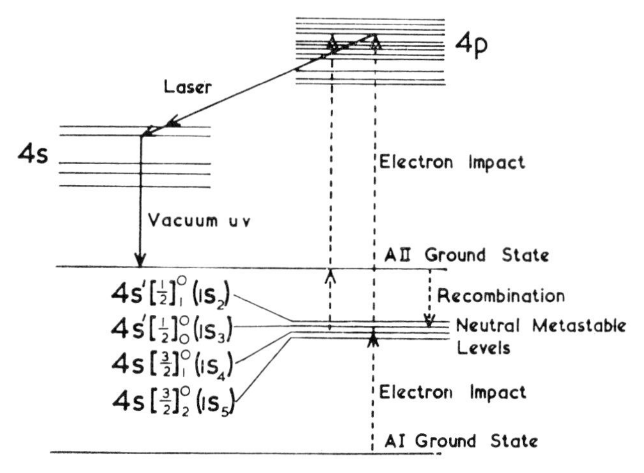

FIG. 2. One- and two-step excitation processes from neutral metastable levels.

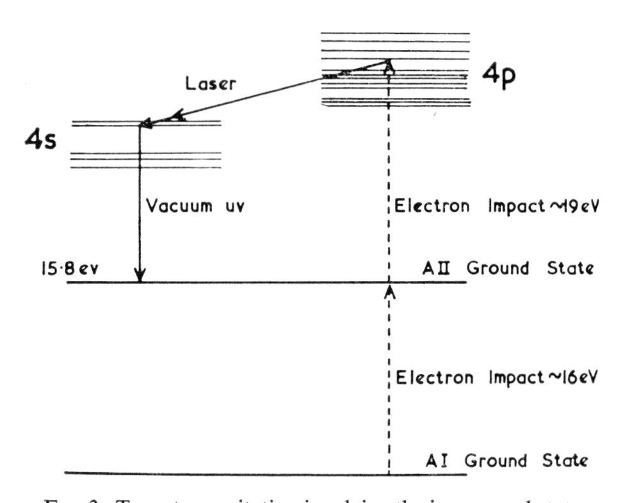

FIG. 3. Two-step excitation involving the ion ground state.

regarding the excitation mechanism in ion lasers has been drawn from observations of the current dependence of output power, or gain, or the intensity of spontaneous emission from the upper laser level and levels cascading into the upper laser level.

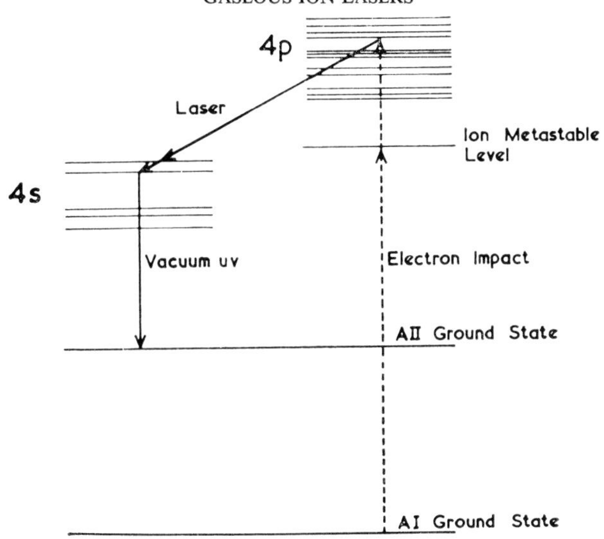

FIG. 4. Two-step excitation involving ion metastable levels.

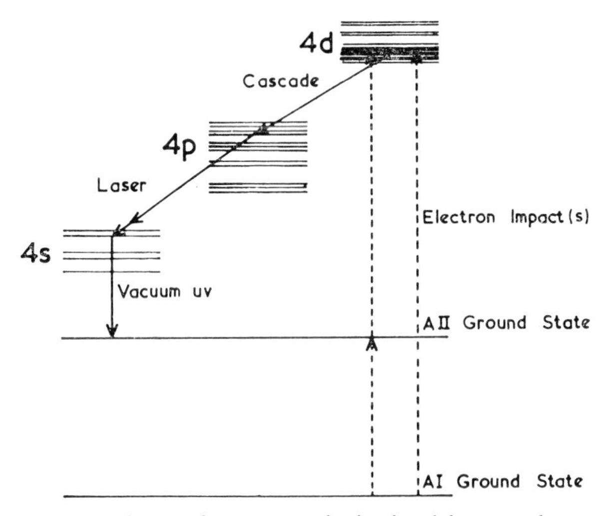

FIG. 5. One- and two-step excitation involving cascades.

The current dependence of these parameters can be predicted from simple rate equations which describe the excitation and relaxation of levels involved in the laser excitation process.

A. RATE EQUATIONS AND EXCITATION MECHANISM

Consider the schematic energy level diagram of an ion laser shown in Fig. 7 where the population and natural lifetime of each level are N_i and τ_i respectively.

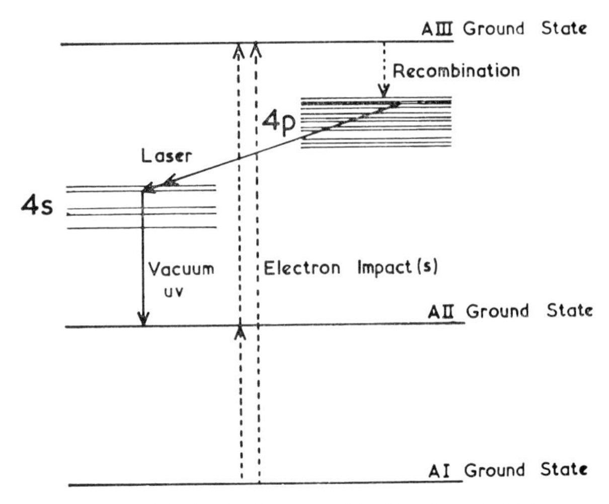

FIG. 6. Excitation by recombination from Ar III.

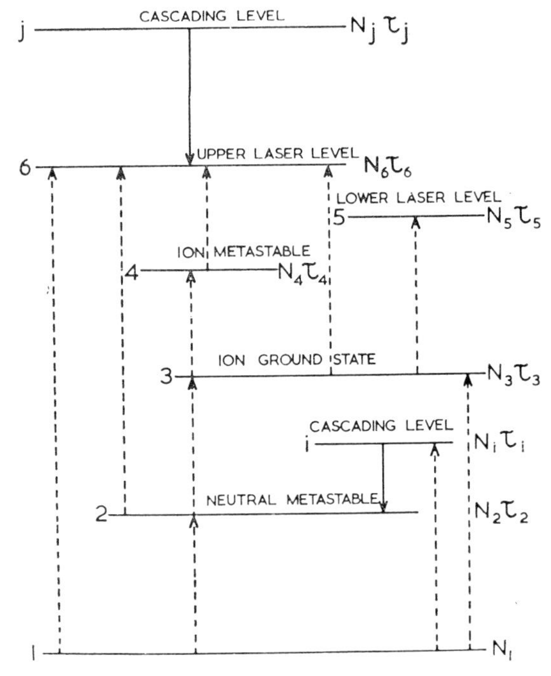

FIG. 7. Generalized energy level diagram of an ion laser showing various excitation processes.

The rate at which each of the various excitation processes occurs in a plasma where the electron density and energy distribution are known can be calculated.

(i) The rate of excitation of the upper laser level 6 direct from the neutral atom ground state is

$$\frac{dN_6}{dt} = N_1 N_e \langle \sigma v \rangle_{61} \qquad ...(4.1)$$

where N_e is the electron density and $\langle \sigma v \rangle_{61}$ is a cross-section for the excitation of level 6 from level 1 averaged over the electron velocity distribution.

If the relaxation of level 6 is dominated by spontaneous emission then its relaxation rate is

$$-\frac{dN_6}{dt} = \frac{N_6}{\tau_6} \qquad ...(4.2)$$

In the absence of any other significant excitation process, in equilibrium,

$$N_6 = N_1 N_e \tau_6 \langle \sigma v \rangle_{61} \qquad ...(4.3)$$

Since in general electron density in the plasma is proportional to current, if the electron temperature in the plasma does not change significantly

$$N_6 \propto I,$$

(i.e. in a laser where the single-step excitation process from the ground state dominates) the population of the upper laser level and hence the spontaneous emission from it will be proportional to current. If $N_6 \gg N_5$ then the laser gain will also be proportional to current. If the laser linewidth is homogeneously broadened the power will be proportional to current but will be proportional to the square of the current if it is inhomogeneously broadened.

(ii) The rate of excitation of the neutral metastable level will be

$$\frac{dN_2}{dt} = N_1 N_e \langle \sigma v \rangle_{21} + \sum_i N_i A_{i2} \qquad ...(4.4)$$

where the summation over i runs over all neutral levels which cascade into the metastable level 2, and A_{i2} is the Einstein coefficient for each of these cascades.

In equilibrium the population of the non-metastable neutral levels i is

$$N_i = N_1 N_e \tau_i \langle \sigma v \rangle_{i1} \qquad ...(4.5)$$

That is

$$\frac{dN_2}{dt} = N_1 N_e \langle \sigma v \rangle_{21} + \sum_i N_1 N_e \tau_i A_{i2} \langle \sigma v \rangle_{i1} \qquad ...(4.6)$$

Since level 2 cannot decay radiatively its relaxation will be mainly caused by electron collisions which either destroy it in an upward or downward direction

$$-\frac{dN_2}{dt} = N_2 N_e \langle \sigma v \rangle_2 \qquad ...(4.7)$$

where $\langle \sigma v \rangle_2$ is a velocity averaged destruction cross-section for level 2.

In equilibrium

$$N_2 = \left(N_1 \langle \sigma v \rangle_{21} + \sum_i N_1 \tau_i A_{i2} \langle \sigma v \rangle_{i1} \right) (\langle \sigma v \rangle_2)^{-1} \qquad \ldots(4.8)$$

and the metastable population is independent of current, i.e. saturated [as is well established in the case of low current density argon ion lasers (Webb, 1968)].

The rate of excitation of level 6 from these neutral metastables is

$$\frac{dN_6}{dt} = N_2 N_e \langle \sigma v \rangle_{62} \qquad \ldots(4.9)$$

and in equilibrium, if radiative relaxation of level 6 dominates and no other significant excitation processes occur,

$$N_6 = N_e \langle \sigma v \rangle_{62} \tau_6 \left(N_1 \langle \sigma v \rangle_{21} + \sum_i N_1 \tau_i A_{i2} \langle \sigma v \rangle_{i1} \right) (\langle \sigma v \rangle_2)^{-1} \qquad \ldots(4.10)$$

Once again the population of the upper laser level is proportional to current. Hence processes (i) and (ii) are identical as far as their current dependence of spontaneous emission, gain and output power are concerned.

(iii) The rate of excitation of the ion ground state is

$$\frac{dN_3}{dt} = N_1 N_e \langle \sigma v \rangle_{31} + N_2 N_e \langle \sigma v \rangle_{21} \qquad \ldots(4.11)$$

The second term represents ionization from neutral metastables. The relaxation of ion ground states is mainly governed by excitation to higher states and by loss to the discharge tube walls where they recombine. In practice since the plasma is quasi-neutral and the majority of ions will be in the ground state

$$N_3 = N_e$$

The rate of excitation of the upper laser level from the ion ground state in this case is

$$\frac{dN_6}{dt} = N_e^2 \langle \sigma v \rangle_{63} \qquad \ldots(4.12)$$

and in equilibrium providing relaxation of level 6 is dominated by radiative processes

$$N_6 = N_e^2 \tau_6 \langle \sigma v \rangle_{63} \qquad \ldots(4.13)$$

This two-step excitation process will therefore lead to a current-squared dependence of spontaneous emission from the upper laser level and of the laser gain.

However, at high current densities, electron collisional destruction of the upper laser level may also occur, in which case

$$-\frac{dN_6}{dt} = \frac{N_6}{\tau_6} + N_6 N_e \langle \sigma v \rangle_6 \qquad \ldots(4.14)$$

where $\langle\sigma v\rangle_6$ is an average destruction cross-section for level 6. Consequently in equilibrium

$$N_6 = \frac{N_e^2\langle\sigma v\rangle_{63}}{1/\tau_2 + N_e\langle\sigma v\rangle_6} \qquad \ldots(4.15)$$

and if electron collisional destruction dominates over radiative decay

$$N_6 = N_e\langle\sigma v\rangle_{63}(\langle\sigma v\rangle_6)^{-1} \qquad \ldots(4.16)$$

and a linear dependence on current results (providing of course the electron energy distribution does not change with current).

(iv) The rate of excitation of neutral metastable levels from the ion ground state and by cascades is

$$\frac{dN_4}{dt} = N_e^2\langle\sigma v\rangle_{43} + \sum_j N_j A_{j4} \qquad \ldots(4.17)$$

where the summation over j runs over all levels populating the ion metastable level. By analogy with process (ii) the population of non-metastable levels j is

$$N_j = N_e^2\tau_j\langle\sigma v\rangle_{j3} \qquad \ldots(4.18)$$

and

$$\frac{dN_4}{dt} = N_e^2\left(\langle\sigma v\rangle_{43} + \sum_j \tau_j A_{j4}\langle\sigma v\rangle_{j3}\right) \qquad \ldots(4.19)$$

If the destruction of the ion metastables occurs mainly by electron collisions

$$-\frac{dN_4}{dt} = N_4 N_e\langle\sigma v\rangle_4 \qquad \ldots(4.20)$$

and in equilibrium

$$N_4 = N_e\left(\langle\sigma v\rangle_{43} + \sum_j \tau_j A_{j4}\langle\sigma v\rangle_{j3}\right)(\langle\sigma v\rangle_4)^{-1} \qquad \ldots(4.21)$$

where $\langle\sigma v\rangle_4$ is an average cross-section for destruction of ion metastables. In this case the excitation of the upper laser level from these metastables is

$$\frac{dN_6}{dt} = N_e N_4\langle\sigma v\rangle_{64} \qquad \ldots(4.22)$$

and once again if radiative decay is the dominant relaxation mechanism of level 6, in equilibrium

$$N_6 = \tau_6 N_e^2\langle\sigma v\rangle_{64}\left(\langle\sigma v\rangle_{43} + \sum_j \tau_j A_{j4}\langle\sigma v\rangle_{j3}\right)(\langle\sigma v\rangle_4)^{-1} \qquad \ldots(4.23)$$

and the population of the upper laser level is once more proportional to the square of the current.

However if the ion metastables are not saturated their destruction may follow an equation of the form

$$-\frac{dN_4}{dt} = k_4 N_4 \qquad \ldots(4.24)$$

In this case

$$N_4 = \frac{N_e^2}{k}\left(\langle\sigma v\rangle_{43} + \sum_j \tau_j A_{j4}\langle\sigma v\rangle_{j3}\right) \qquad \ldots(4.25)$$

and

$$N_6 = \frac{\tau_6 N_e^3\langle\sigma v\rangle_{64}}{k}\left(\langle\sigma v\rangle_{43} + \sum_j \tau_i A_{j4}\langle\sigma v\rangle_{j3}\right) \qquad \ldots(4.26)$$

and the population of the upper laser level will be proportional to the third power of the current.

(v) If an ion level lying above the upper laser level is populated by process (i) its population will be

$$N_j = N_1 N_e \tau_j \langle\sigma v\rangle_{j1} \qquad \ldots(4.27)$$

and the rate at which these levels populate the upper laser level by cascade is

$$\frac{\mathrm{d}N_6}{\mathrm{d}t} = \sum_j N_j A_{j6} \qquad \ldots(4.28)$$

and in equilibrium

$$N_6 = \tau_6 N_1 N_e \sum_j A_{j6} \tau_j \langle\sigma v\rangle_{j1} \qquad \ldots(4.29)$$

If however the cascading levels are populated by two-step processes from the ion ground state then

$$N_j = N_e^2 \tau_j \langle\sigma v\rangle_{j3} \qquad \ldots(4.30)$$

and

$$N_6 = \tau_6 N_e^2 \sum_j A_{j6} \tau_j \langle\sigma v\rangle_{j3} \qquad \ldots(4.31)$$

(vi) Recombination processes into excited ion levels can probably be neglected and will not be considered further here.

In a real situation this rate equation approach should consider all the processes (i)–(v) occurring together, in which case the population of the upper laser level might be expected to follow an equation of the form

$$N_6 = aI + bI^2 + cI^3. \qquad \ldots(4.32)$$

The population of the lower laser level will also follow a general equation of this form

$$N_5 = a' I + b' I^2 + c' I^3 \qquad \ldots(4.33)$$

So the population inversion and gain of the system will follow a similar equation

$$\Delta N = \left(N_6 - \frac{g_6}{g_5} N_5\right) = AI + BI^2 + CI^3 \qquad \ldots(4.34)$$

where now A, B and C may be negative.

Clearly, the degree of complexity of these rate equations is determined by the number of simultaneously occurring excitation processes; the exact contribution of each individual process in the discharge plasma of an ion laser being dependent on the following factors:

(i) The electron energy distribution in the plasma.
(ii) The densities of atoms, ions and electrons in the plasma.
(iii) The cross-sections of different electron impact processes.
(iv) The populations of various excited neutral and ionic states of the particles in the plasma.

In addition the overall observed behaviour of the laser will be further complicated by other parameters of the discharge plasma, such as the temperatures of neutral and ionized atoms in the plasma and the spatial distributions and bulk drift velocities of charged and neutral particles. These are discussed in Section VII.

B. SINGLY-IONIZED ION LASERS

The succeeding sections discuss in detail each of the individual processes which have been proposed in order to explain the excitation of singly-ionized electron impact excited ion lasers. Although all these processes occur to a greater or lesser extent simultaneously, under certain conditions one process may dominate. Before embarking on a detailed discussion it is convenient to summarize the broad conclusions which have been reached about excitation mechanisms in singly-ionized argon lasers:

(i) Single-step excitation from the neutral atom ground state is only dominant in the excitation of fast pulsed argon ion lasers.
(ii) The contribution which neutral metastables make to the pumping remains uncertain, although they probably contribute significantly to the formation of ground state ions.
(iii) Excitation from ground and metastable states of the ion dominates the pumping of low current density CW argon ion lasers.
(iv) Cascade from higher levels, themselves excited from ground and metastable states of the ion, contributes a significant proportion of the pumping particularly at high current densities.

(1) *Single-step Excitation in Ion Lasers*

The single-step electron impact excitation process in singly-ionized argon is shown schematically in Fig. 1; energetic electrons with energy > 35 eV preferentially excite levels in the $4p$ configuration and population inversion between the $4p$ and $4s$ configurations results. This single-step process was first suggested by Bennett *et al.* (1964) on the basis of "sudden" perturbation considerations and consists of simultaneous ionization and excitation of a neutral atom by collision with a fast electron

$$\mathrm{Ar} + e \rightarrow \mathrm{Ar}^{+*} + 2e \quad \ldots(4.35)$$

It is considered to occur faster than the time required for the remaining electrons in the ion core to change their wave-functions. Unlike excitation by collision with slow electrons, where the dipole approximation gives only states which are of parity opposite to the ground state, in the "sudden" approximation the excited states produced by electron collision have the same parity as the ion ground state, in argon $3p^5\,{}^2P^0_{3/2}$. Because the single-step excitation process requires high energy electrons it can only be important in discharge plasmas where large numbers of such electrons are present. Its importance in a particular ion laser can be estimated from a knowledge of the plasma parameters of the system and the cross-sections for electron impact excitation to the various laser levels. Koozekanani (1966, 1967, 1968) has calculated the cross-section for the excitation of various Ne II, Ar II and Kr II states by this process in terms of the total cross-section for ionization. When the fast electron strips one of the neutral atoms outer electrons away, the resulting ion exists in an unstable state which can be designated $|(p^5)LSJM\rangle$ (where L, S, J and M are the conventional quantum numbers of the state in LS Russell–Saunders coupling). This state then relaxes to various states of the single ionized atom; which can be designated $|l'^4 L_c S_c, nl'' s, L' S' J' M'\rangle$. l', L_c and S_c are appropriate quantum numbers for the Ar III core electrons, n, l'' and s are the quantum numbers of the excited electron, and $L' S' J'$ and M' are total quantum numbers for the excited ion state. The expansion of $|(p^5)LSJM\rangle$ in terms of a linear combination of these states gives a series of coefficients which are related to the cross-section for excitation to the particular state in the linear combination (Koozekanani, 1966). The largest cross-sections for argon $4p$ states are typically 1/200 the total ionization cross-section; (see also Vainshtein and Vinogradov, 1967).

Specifically, the largest cross-sections calculated for excitation of np states in the singly-ionized noble gases are for ${}^2P^0_{1/2}$ and ${}^2P^0_{3/2}$ states; intermediate coupling calculations give non-zero cross-sections for excitation to other np states.

Koozekanani (1967) predicts that for the excitation of $np\,{}^2P_{1/2}$ and $np\,{}^2P_{3/2}$ levels the level cross-section divided by the degeneracy should be a constant, the experimental data of Latimer and St. John (1970) bear out this prediction in the case of argon. Some of the excitation functions observed by Latimer and St. John are shown in Fig. 8; the largest cross-section observed is for excitation of the $4p\,{}^2P_{3/2}$ state, the upper level of the 4765 Å laser transition. The maximum cross-section observed for this state is 10^{-18} cm^2, observed at an electron energy of 90 eV. The maximum cross-section occurs at lower electron energies for some other states, for example $4p\,{}^2D_{5/2}$ the upper level of the 4880 Å laser

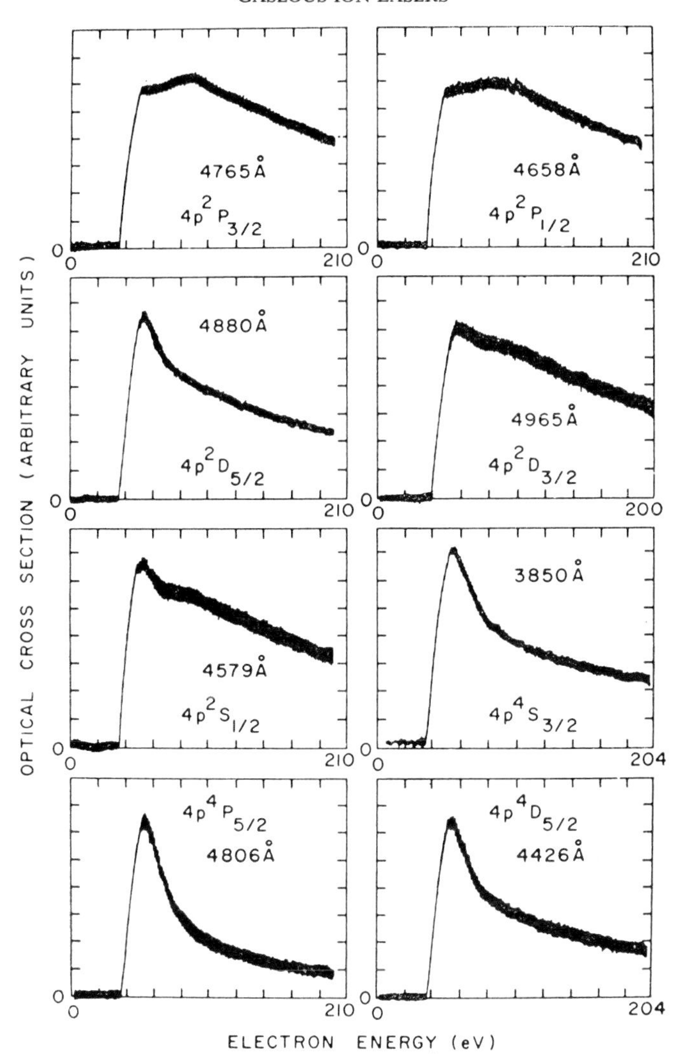

FIG. 8. Optical excitation cross-sections for the Ar II 4*p* levels (Latimer and St. John, 1970).

transition. The measurements of Latimer and St. John indicate that the gain of the 4765 Å laser transition should exceed that of the 4880 Å, under conditions where single electron impact is the dominant excitation mechanism, in a plasma with a Maxwellian temperature distribution and electron temperatures greater than 20 eV. The 4765 Å line is observed to have the higher gain in fast pulsed argon ion lasers (Bennett *et al.*, 1964; Davis and King, 1971) and it is in this type of excitation where high electron temperatures are expected.

Additional measurements of the single-step excitation cross-section for argon II states which are in agreement with Latimer and St. John (1970) have been made by Clout and Heddle (1969, 1971) and Felt'san and Povch (1970). The general shape of the excitation functions observed by these investigators are in good agreement although the maximum excitation cross-sections for some levels differ by up to a factor of 5. The excitation function and cross-section measurements reported by Bennett *et al.* (1966) and Hammer and Wen (1967) are not in as good agreement with other investigators because their measurements were made at relatively high pressures at which multiple collision effects were probably significant. The maximum cross-sections for single electron collisional excitation observed for different argon II levels by various investigators are listed in Table II.

TABLE II. *Maximum cross-sections for simultaneous ionization and excitation of argon atoms to various upper laser levels by electron impact*

Upper level	Important laser transition from level (Å)	Electron energy at peak (eV)	Level cross-section (10^{-20} cm^2)
$4p\,^2S$	4579	54[a]	26[a]
		51[b]	15·8[b]
		60[c]	8·6[d]
$4p\,^2P$	4658	90[a]	54[a]
		80[b]	24·2[b]
		97–102[c]	36[c]
$4p\,^2P$	4765	90[a]	101[a]
		85[b]	59·5[b]
		97–105[c]	102[c]
$4p\,^2D$	4880	54[a]	79·5[a]
		52[b]	55·2[b]
		60[c]	115[c]
$4p\,^2D$	4965	54[a]	48·5[a]
		55[b]	36·8[b]
		60[c]	6·8[c]

[a] Latimer and St. John (1970).
[b] Clout and Heddle (1971).
[c] Felt'san and Povch (1970).
[d] Deduced from line cross-section measured by Felt'san and Povch using branching ratio measured by Schumaker and Popenoe (1969).

The excitation functions and cross-sections for single electron excitation of some singly-ionized neon levels have been determined by Walker and St. John (1972). The largest values again occur for levels $3p\,^2P_{3/2}$ and $3p\,^2P_{1/2}$ as predicted by sudden perturbation theory; these are the upper levels of the strongest neon II laser transitions. However, the shape of the excitation functions is quite different from the shapes observed for corresponding argon II levels, see for example Fig. 9. The maximum cross-sections, of the order of 10^{-19} cm^2, occur at higher energies and the excitation functions are much broader, and generally similar to one another in shape. This would suggest little relative change in the gains of laser transitions from different levels in discharge plasmas with different electron temperatures.

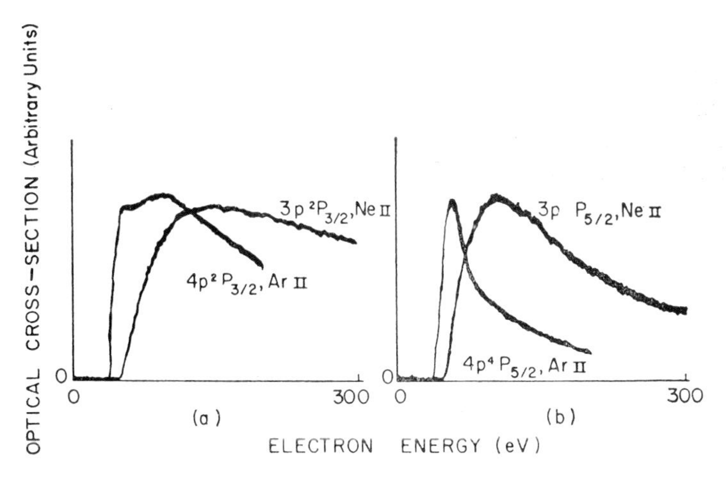

FIG. 9. Comparison of the optical excitation cross-sections for Ne II and Ar II levels (Walker and St. John, 1972).

Excitation functions and single electron collisional excitation cross-sections for some krypton II levels have been obtained by Rozgachev and Yaroslavtseva (1970). The largest cross-sections occur for $5p$ levels but the largest among these levels no longer occurs for $^2P_{3/2}$ and $^2P_{1/2}$ levels. The largest cross-section observed, 5.10^{-17} cm^2, occurs for the $5p\,^4D^0_{5/2}$ level, the upper level of the strong 5682 Å laser transition. However the very large measured value for this level has not been corrected for cascade contributions and is probably much larger than the true value. The larger deviation between the predictions of simple sudden perturbation theory and experiment in the case of krypton is probably accountable for by the considerable mixing of wavefunctions which causes the levels of krypton to depart from a pure simple LS coupled form (Koozekanani, 1968). Little is known experimentally about the cross-section for direct electron impact excitation of the lower (mainly ns)levels of the singly-ionized noble gas lasers. However, since these laser levels probably obey pure LS coupling rather better than the upper levels the predictions of sudden perturbation theory should hold and these lower levels should have very small cross-sections for excitation from the neutral atom ground state.

Because of the availability of reliable excitation cross-sections for the single-step excitation process from the neutral atom ground state in the noble gases, the importance of this process in a particular discharge plasma can be ascertained on the basis of electron temperature and density measurements.

The total excitation rate of an upper laser level j by the single-step electron collisional process from the neutral atom ground state is

$$R_j = NN_e \int_{E_j}^{\infty} f(E)\,\sigma_j(E)\sqrt{\frac{2E}{m}}\,\mathrm{d}E \qquad \ldots(4.36)$$

where N is the density of ground state atoms, N_e is the electron density, m is the mass of the electron, E_j is the energy of state j, $\sigma_j(E)$ is the excitation cross-section of this state as a function of energy and $f(E).\mathrm{d}E$ is the fraction of electrons with energy between E and $E + \mathrm{d}E$. If $f(E)$ is Maxwellian, then

$$f(E) = 2\left(\frac{E}{\pi k^3 T_e^3}\right)^{1/2} \mathrm{e}^{-E/kT_e} \qquad \ldots(4.37)$$

Since it is probably reasonable to assume that the electron energy distribution in most ion laser plasmas is nearly Maxwellian, then if T_e, N, N_e and $\sigma_j(E)$ are known the pumping rate due to the single step process can be calculated. Rudko (1967) performed this calculation for excitation of the upper level ($4p\,^2D_{5/2}^0$) of the strong 4880 Å argon laser transition; he assumed an electron temperature of 5 eV (1 eV ≡ 11,600 °K) as measured by Kitaeva *et al.* (1966b) and used the cross-section measured for the single-step process by Hammer and Wen (1967). The calculated pumping rate for the $4p\,^2D_{5/2}^0$ was

$$R_j = \frac{2NN_e}{(\pi m k^3 T_e^3/2)^{1/2}} \times 3{\cdot}49 \times 10^{-20}\,\mathrm{s}^{-1} \qquad \ldots(4.38)$$

With values of $N = 10^{15}\,\mathrm{cm}^{-3}$ and $N_e = 1{\cdot}5.10^{13}\,\mathrm{cm}^{-3}$, the calculated pumping rate was only 1% of that measured.

Mercer (1967) performed a similar calculation, using cross-section measurements of Bennett *et al.* (1966) and an electron temperature of 3·9 eV. He concluded that, under his experimental conditions (current density 190 A cm^{-2} in 0·3 torr of argon), the contribution of single step pumping to the total excitation of the $4p\,^2D_{5/2}^0$ state was 8%.

These calculations are somewhat unreliable as the cross-section data, particularly those of Bennett, are considered inaccurate and the pumping rates calculated, particularly by Mercer, are too high; for example, Mercer's calculations indicated a contribution to the total pumping of the upper level of the 4765 Å laser transition ($4p\,^2P_{3/2}^0$) of 100% for $T_e > 4{\cdot}6$ eV.

(a) *CW ion lasers*. In view of the electron temperatures and electron densities measured in CW argon ion lasers (see Tables IV and VI) the contribution of single step pumping from the neutral atom ground state is small. Its percentage contribution has been estimated by Mercer *et al.* (1967) from line-shape studies and is largest for the $4p\,^2P^0_{3/2}$ state, with a contribution of about 20% to the total pumping at the maximum E/p_0 (axial field strength/filling pressure) values used (110 V cm^{-1} $torr^{-1}$). The contribution of the single step process to the pumping of all states increases with E/p_0 (Ballik *et al.*, 1966) as does the electron temperature.

Although the contribution of single step processes from the neutral atom ground state can generally be neglected in low current density CW lasers (where most careful diagnostic measurements of plasma parameters have been made) the uncertainty in the values of electron density and temperature in very high current density CW ion lasers [up to 3500 A cm^{-2} (Donin, 1969)] prevents precise assessment of its importance in these devices.

(b) *Pulsed ion lasers.* In pulsed ion lasers where much higher E/p_0 values can occur than in CW ion lasers, the contribution of the single step process is much more important. The prompt pulsed laser action which occurs in fast-risetime high current pulsed discharges and in the rising edge of more conventional high current pulsed discharges is dominantly excited by single electron collisional processes. This is admirably illustrated in the pulsed argon ion laser where the order of time-appearance in the rising edge is generally 4765, 4880, 4965, 5145 Å (Smith and Dunn, 1968; Davis, 1970) following the same sequence as the cross-section values for the upper levels of these transitions. The behaviour of the strong neon–ion laser transitions under conditions of fast pulsed excitation again fits into a pattern consistent with single step excitation from the neutral atom ground state. The strongest observed transition, at 3324 Å has the largest cross-section for electron impact excitation from the ground state (Walker and St. John, 1972). The 3378 Å transition is less strong, as is the 3393 Å transition, probably because of competition effects which occur because of its common upper level with 3324 Å. Bearing in mind the very favourable lifetime ratios that exist for the transitions in the singly-ionized noble gas lasers (see later Section IX, G and Table XII) it is clear that under conditions where sufficient high energy electrons are available single-step excitation can provide large population inversions.

The role of the single-step excitation mechanism in exciting many of the laser transitions observed in high current pulsed discharges in other pure gases or vapours (see spectroscopy Table I) is unclear. Although it is certain that the excitation mechanism of these transitions involves electron impact (EI) the lack of definite cross-section information on the single-step process precludes the conclusion that single-step electron collision is the dominant excitation process involved. However in all pulsed gas lasers which operate in fast-risetime high current pulsed discharges at low pressure the single step process is likely to be dominant.

(2) *Excitation from Neutral Metastables*

In many gas discharges large numbers of neutral metastables build up and provide an easy path for ionization since they present a large cross-section and have a lower threshold for ionization than does direct ionization from the neutral atom ground state.

In so far as neutral metastables are involved in the production of ground state ions they must be considered as an integral part of any excitation scheme involving electron impact on ion ground states. However, simultaneous ionization and excitation of neutral metastables can lead to preferential excitation of upper laser levels. This process is shown schematically in the case of argon in Fig. 2.

The metastable levels of most importance in this process are the two $4s$ and two $4s'$ levels, which are the four lowest lying excited levels of argon above the ground state. Two of these states $4s'[\frac{1}{2}]_0^0$ and $4s[\frac{3}{2}]_2^0$ are true metastables; at neutral atom ground state densities $> 10^{14}\ \mathrm{cm}^{-3}$ the other two levels $4s'[\frac{1}{2}]_1^0$ and $4s[\frac{3}{2}]_1^0$ became quasi-metastable by resonance trapping of their radiation to the ground state. Measurements made by Miller *et al.* (1964–7) indicate that the four levels are in thermal equilibrium. If the total metastable population is N_0 then on the basis of their degeneracies

$$\begin{aligned} N(4s[\tfrac{3}{2}]_2^0) &= 0{\cdot}417 N_0 \\ N(4s[\tfrac{3}{2}]_1^0) = N(4s'[\tfrac{1}{2}]_1^0) &= 0{\cdot}250 N_0 \\ N(4s'[\tfrac{1}{2}]_0^0) &= 0{\cdot}083\, N_0. \end{aligned} \qquad \ldots(4.39)$$

Bennett and Lichten (1965) originally considered the process involving electron impact excitation of these metastables as a possible explanation of the current-squared dependence which was observed for the excitation rate of upper laser levels in argon. However, the population of excited states of neutral argon excited in a discharge saturates at low currents (Webb, 1968) and the current dependence of excitation from these states would show a linear dependence on current and be indistinguishable from excitation from the neutral atom ground state.

The cross-section for the production of upper laser levels in the noble gases by the removal of the metastable electron has been considered by Koozekanani (1967, 1968) in the "sudden" approximation. In the metastable state the atom presents a large cross-sectional area since the electron has a classical orbit of much larger radius than in its unexcited state. For example, in argon the average radius of the orbit of the $4s$ metastable electron is 14·67 a.u. whereas for the unexcited $3p$ electrons it is 4·2 a.u. Koozekanani finds that preferential excitation cross-sections for the process

$$np^5(n+1)\,s + e \rightarrow np^4\,mp + 2e \qquad \ldots(4.40)$$

exist in the noble gases for excitation of $np^4(n+1)p$ states, particularly for $np^4(n+1)\,^2P^0_{3/2}$ and $np^4(n+1)\,^2P^0_{1/2}$ states, that is for the same states which are favoured in direct electron impact excitation from the neutral atom ground state. The calculated maximum cross-sections are typically 10^{-3} of the total cross-section for ionization from the metastable state. As the cross-section for this ionization from the metastable $np^5(n+1)s$ levels is an order of magnitude greater than for ionization from the neutral atom ground state and has a lower threshold (about 4 eV in argon), excitation of these upper laser levels will be much more important than from the ground state, given equal populations of neutral atom ground states and metastables. The importance of the excitation of the metastables in a given situation will depend on their population. Very little direct information is available on the concentration of these metastables under the conditions of laser excitation in pulsed and continuous discharges. Miller *et al.* (1964–7) have reported that the metastable densities observed in low current density argon ion lasers are too low to account for the laser action.

Measurements of metastable densities in low current density argon ion laser discharges have also been made by Cotman and Johnson (1967). The measured metastable densities either decrease slowly or are fairly independent of electron density and increase as the tube bore is reduced.

Even in small bore discharge tubes the metastable densities are insufficiently high for excitation from neutral metastables to make a significant contribution to direct pumping of the upper laser levels. However, in low electron temperature plasmas the rate of direct pumping from metastables may be as great as from the neutral atom ground state, even though under these circumstances both rates are small. Kitaeva *et al.* (1967b) have estimated the metastable densities in a CW argon ion laser at 0·37 torr operating at a current density of 400 A cm^{-2} to be $1{\cdot}7\,.\,10^{14}$ cm^{-3}, although they conclude that in this system the main contribution of the metastables is to provide an easy ionization path for the production of ground state ions.

The exact contribution of neutral metastables to the pumping of both CW and pulsed ion lasers remains uncertain and further work on densities in high current density discharges is needed. Their role in such systems is generally neglected by many authors although they may provide a significant contribution to the population of either ground or excited state ions. Some investigations have pointed indirectly to such a contribution although no definite conclusions can be drawn. Vladimirova *et al.* (1971) found that in a typical argon ion discharge, at current densities up to 500 A cm^{-2}, a substantial contribution to the excitation of the $4p\,^2D^0_{5/2}$ and $4p\,^4D^0_{5/2}$ levels of argon II was made by stepwise processes through states whose population decreased with rising discharge current. Neutral metastable states would show this population decrease at high current densities. Davis and King (1971) observed an apparent linear dependence of gain on discharge current in a pulsed argon ion laser under conditions where direct electron impact excitation from the neutral atom ground state was probably not important. This result could indicate a contribution from some other more easily excited neutral state, although electron collisional destruction of the upper laser levels could have masked a current squared dependence of excitation of these levels.

(3) *Two-step Excitation Involving Ion Ground States*

This process is shown schematically in the case of argon in Fig. 3. The first step of the process produces a ground state ion by electron impact on a ground state neutral atom, or alternatively a neutral metastable. In the second step of the process electron impact excitation of the upper laser levels results. Numerous investigators (Gordon *et al.*, 1964, 1965; Ballik *et al.*, 1966; Kitaeva *et al.*, 1966a, b, 1967, 1968, 1969; Mercer *et al.*, 1967; Mercer, 1967a, b; Rudko, 1967; Rudko and Tang, 1966, 1967; Borisova *et al.*, 1967b; Beigman *et al.*, 1967; Webb, 1968; Lebedeva *et al.*, 1968, 1970; Brandi, 1972) have shown that this process plus contributions from cascade processes and excitation via ion metastables is the dominant excitation process in low current density (< 500 A cm^{-2}) CW argon ion lasers. It is probably also dominant in other CW argon ion lasers at higher current densities and in similar CW noble gas and other ion lasers.

The two step process was originally proposed (Gordon *et al.*, 1964) because of an observed current-squared dependence of spontaneous emission from the upper laser levels as a function of current, and an I^4 dependence of the laser power output. In an inhomogeneously broadened laser system above threshold this I^4 dependence of power output is expected if the gain is proportional to the square of the current. Other investigators however have reported an I^2 dependence of the power output (Gordon *et al.*, 1965; Borisova *et al.*, 1967b; Lebedeva *et al.*, 1968) which is consistent with a current squared dependence of the gain in a homogeneously broadened laser. These observations are not inconsistent, as the available evidence shows that the argon ion laser makes a transition from inhomogeneous broadening just above threshold to complete homogeneous broadening well above threshold. These observations alone however do not prove conclusively that the two-step excitation process is occurring. A number of theoretical arguments can be put forward which would appear to minimize its importance involving the ion ground state. Since the lower laser levels in the noble gas ion lasers are strongly connected, and are of opposite parity to the ion ground state they should have large cross-sections in the electric dipole approximation for low energy electron excitation from the ion ground state. The excitation of the upper *np* levels should be forbidden; however, the cross-section for optically forbidden transitions frequently exhibits a large value for a small range of energies near threshold, whilst the cross-section for the allowed process rises much more slowly as is shown schematically in Fig. 10. Electrons with energies between E_T and E_1 will excite preferentially the optically forbidden transition. The importance of two-step excitation to the upper laser levels via the ion ground state depends critically on the electron temperature in a discharge. If this temperature is too high, preferential excitation of the lower laser levels is to be expected. Increase of the electron temperature does not automatically destroy the inversion, however, because at higher electron energies the amount of upper level excitation by single step processes and cascades can increase.

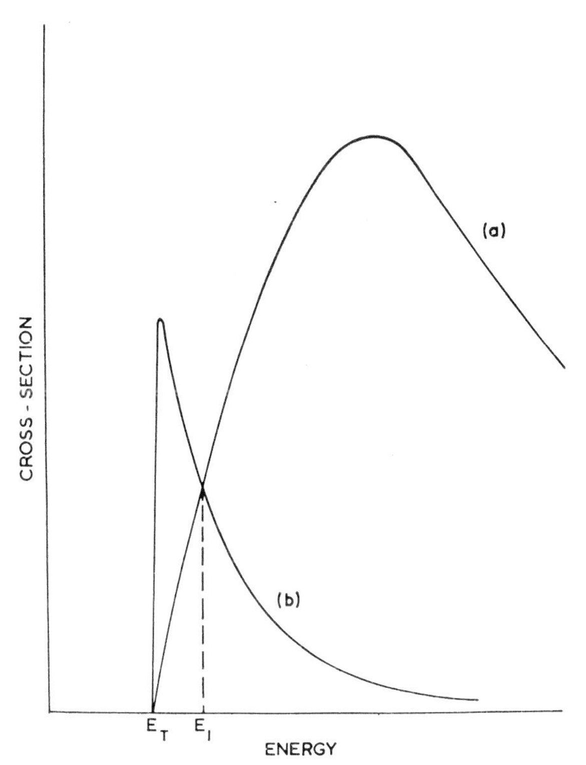

FIG. 10. Qualitative energy dependence of electron excitation cross-sections for (a) optically allowed and (b) optically forbidden transitions.

The only direct evidence that a large cross-section exists for excitation of the upper laser levels from the ion ground state in the noble gas ion lasers has been obtained by Imre *et al.*, (1972) who measured cross sections for the process

$$Ar^+(3p^5) \rightarrow Ar^{+*}(3p^4 4p) + e \qquad \ldots(4.41)$$

in a colliding beam experiment. Their cross-section results are shown in Fig. 11. The maximum cross-sections for the excitation of four of the single ionized

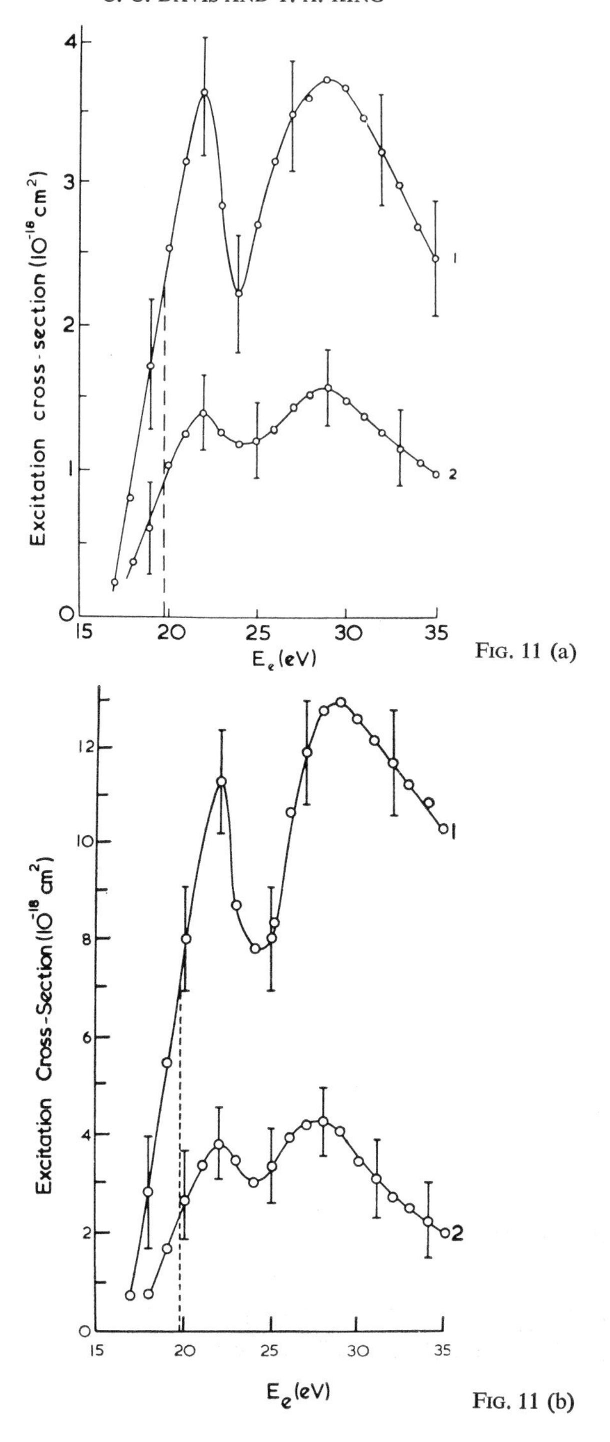

FIG. 11 (a)

FIG. 11 (b)

argon laser transitions are in the order 4880 Å > 4965 Å > 4658 Å > 4545 Å. The maximum cross-sections observed are quite large e.g. $1{\cdot}3 \times 10^{-17}$ cm^2 for the upper level of 4880 Å. However, the absolute values of these cross-sections are not to be relied upon absolutely as the apparatus of Imre *et al.* was calibrated using the data of Felt'san and Povch (1970).

Although an observed I^2 dependence of the spontaneous emission from the upper laser levels in both CW and long pulse argon ion lasers (Gordon *et al.*, 1965; Demtröder, 1966) may indicate the involvement of ion states in the excitation of the upper laser levels, it does not show it conclusively. For example a level excited by a single collision from the neutral atom ground state could show an I^2 population dependence if the electron temperature in the discharge increased with current in such a way as to make $\langle \sigma v \rangle$ (the velocity averaged excitation cross-section) proportional to electron density.

Most of the definitive evidence of the involvement of ion states in the excitation of argon ion lasers has come from spontaneous emission lineshape measurements and, indirectly, from measurements of plasma parameters and cross-sections which show that excitation direct from the neutral atom ground state is negligible under the normal conditions of CW laser excitation.

Kitaeva *et al.* (1966a, b, 1967a, b, 1968, 1969, 1971), Ballik *et al.* (1966), Mercer (1967), Mercer *et al.* (1967), Sze and Bennett (1972) and Sze *et al.* (1972), have shown that the spontaneous emission lines of argon ion laser transitions have lineshapes consistent with excitation from ion states, although they do not show whether the ion states involved are the metastables or the ground state. Further, Webb (1968) has demonstrated that, under low current density CW argon ion laser excitation conditions, the radial intensity profiles of ion transitions correlate well with the square of the theoretical electron density profile. Additional indirect evidence of the two-step excitation process in the CW argon ion laser comes from the success of theoretical models, based on this process, in describing the behaviour of such lasers (Herziger and Seelig, 1968).

(4) *Excitation from Ion Metastable Levels*

As mentioned in the previous section, although the importance of ion states in the excitation of argon ion lasers has been demonstrated, the evidence does not indicate whether the ion ground or metastable levels are involved. The importance of metastable levels in this process was originally suggested because of apparent theoretical drawbacks to the excitation of $4p$ upper laser levels in argon from the ion ground state. As shown previously, excitation via ion

---

FIG. 11. Absolute electron excitation cross-sections for the Ar II lines: a(1) – 4658 Å, $4p\,^2P^0_{1/2}$; a(2) – 4545 Å, $4p\,^2P^0_{3/2}$; b(1) – 4880 Å, $4p\,^2D^0_{5/2}$; b(2) 4965 Å, $4p\,^2D^0_{3/2}$ (Imre *et al.*, 1972).

metastables retains the same current dependence as excitation from the ion ground state under conditions where the ion metastables are in equilibrium with the ion ground state.

Gordon *et al.* (1965) originally came to the conclusion that ion metastables played a significant part in the pumping of the argon ion laser, but there is still little experimental data on exactly what proportion of the pumping comes from these states. Lebedeva *et al.* (1970) observed a term in the current dependence of excitation of the $4p$ configuration of argon II proportional to the third power of the current and concluded that some of the pumping comes from non-metastable states of the ion. Van der Sijde (1972a, d) also observed this third power dependence in a hollow cathode discharge under conditions similar to those of laser excitation.

However, most of the evidence of a large contribution to the pumping of argon II $4p$ levels from ion metastables comes from theoretical considerations. Beigman *et al.* (1967) and Kitaeva *et al.* (1968, 1969) calculated cross-sections $\langle \sigma v \rangle$ averaged over a Maxwellian electron energy distribution for various excitation processes between configurations of argon II. The processes they considered were $3p \rightarrow 4p$; $3p \rightarrow 3d$; $3p \rightarrow 4d$; $3p \rightarrow 5d$; $3p \rightarrow 4s$; $3p \rightarrow 5s$; $3d \rightarrow 4p$ and $4s \rightarrow 4p$. The calculations indicate that under typical CW laser conditions of electron density and temperature, the excitation rates of the $4s$, $4p$ and $4d$ configurations from the ground $3p$ state are very similar, whilst the excitation of $3d$ levels is an order of magnitude larger. The excitation process $4s \rightarrow 4p$ is the most efficient of all, followed closely by the process $3d \rightarrow 4p$. The only drawback to these calculations is that they consider configurations as a whole and do not show which individual levels acquire large populations.

TABLE III. *Low lying ion levels of importance in the excitation of upper laser* 4p *levels in the argon ion laser*

Level	Lifetime (ns)[a]	Level	Lifetime (ns)[a]
$3d\,^4F_{9/2}$	Metastable	$3d\,^4P_{3/2}$	250
$3d\,^4F_{7/2}$	Metastable	$3d\,^4P_{1/2}$	217
$3d\,^4F_{5/2}$	716	$3d\,^2F_{7/2}$	Metastable
$3d\,^4F_{3/2}$	411	$3d\,^2F_{5/2}$	7·07
$3d\,^4D_{7/2}$	Metastable	$3d\,^2D_{5/2}$	0·599
$3d\,^4D_{5/2}$	5870	$3d\,^2D_{3/2}$	0·579
$3d\,^4D_{3/2}$	13,100	$3d\,^2P_{3/2}$	2·88
$3d\,^4D_{1/2}$	68,500	$3d\,^2P_{1/2}$	2·56
$3d\,^4P_{5/2}$	221		
$4s\,^4P$	$67{\cdot}1_{5/2}$		
$4s\,^4P$	$28{\cdot}7_{3/2}$		
$4s\,^4P$	$119_{1/2}$		

[a] The lifetime values are from theoretical calculations of Luyken (1972).

Thus it appears that efficient excitation of the argon II $4p$ configuration can occur via the long-lived $3d$ and $4s\,^4P$ levels which are listed in Table III, together with their calculated lifetimes (Luyken, 1972). The $4s\,^4P$ levels, although not so long lived, are known to have large cascade pump rates ($\sim 10^{18}$ cm^{-3} s^{-1}) from higher levels (Rudko and Tang, 1967).

It should be pointed out that there are no suitable low-lying ion metastable levels in neon II to contribute to the pumping in this laser, for example compare Figs 12 and 13.

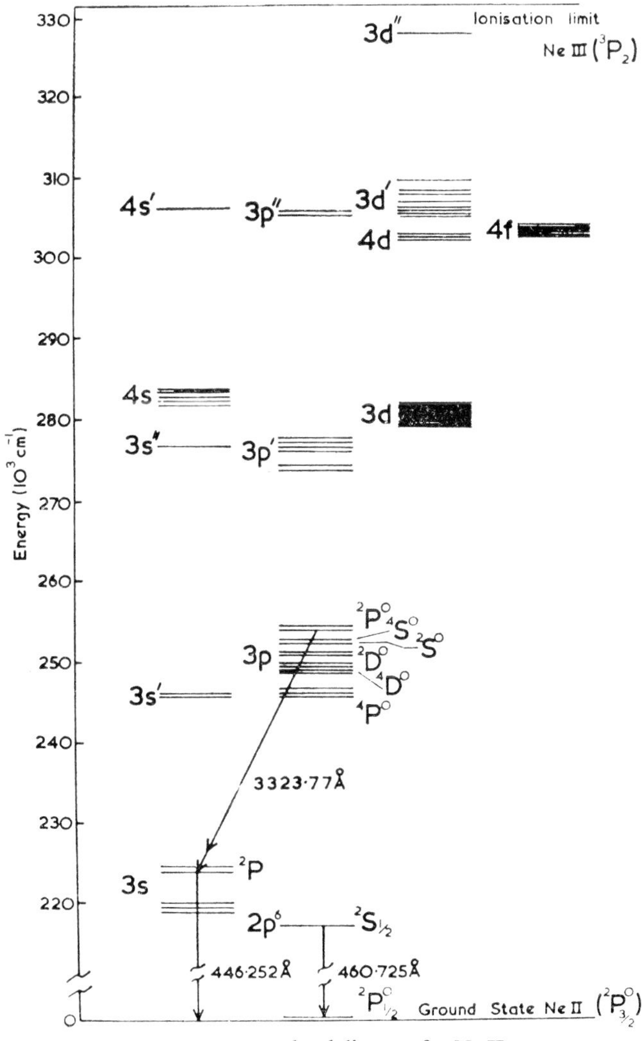

FIG. 12. Energy level diagram for Ne II.

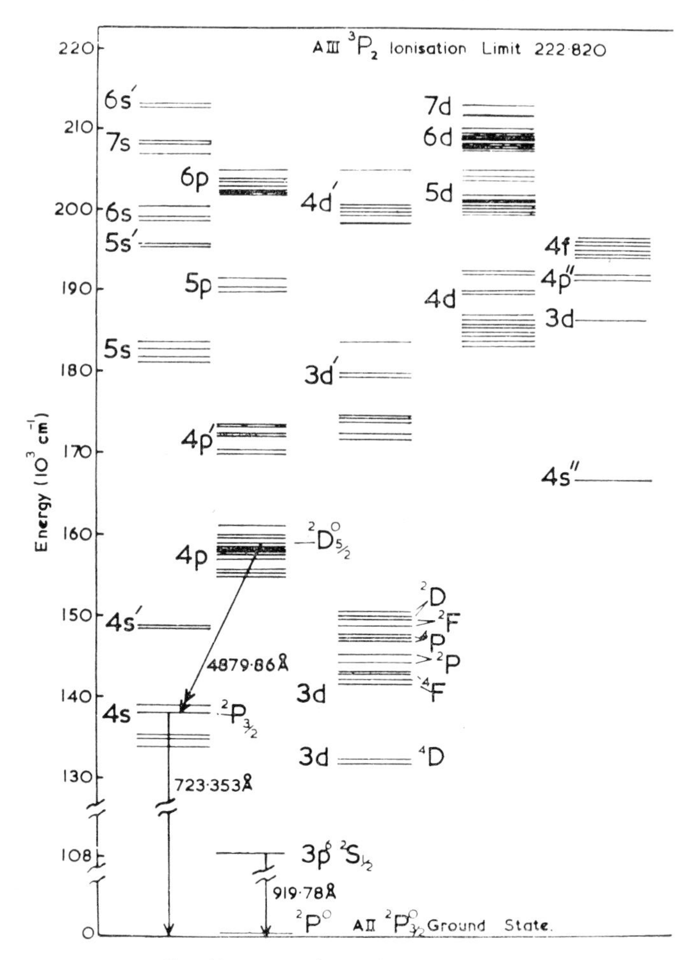

FIG. 13. Energy level diagram for Ar II.

(5) *Excitation by Cascade Processes*

Cascade pumping of the upper level in an ion laser can occur as a non-rate-determining step in conjunction with the single and multi-step electron collisional processes previously discussed. Two typical cascade processes are shown in Fig. 5; here an argon $4d$ state is excited either by one or two step electron collisions and then populates a $4p$ laser level by radiative transfer. The importance of the cascade process has been considered mainly in relation to its contribution to the pumping in CW argon ion lasers. The rate and percentage of cascade pumping of a particular laser upper level can be determined by measuring the total spontaneous emission rates into and out of the state.

Statz *et al.* (1965) first estimated the amount of cascade pumping of the Ar II $4p$ states, using the spectral intensity data of Minnhagen (1963). They added up the intensities of all the lines terminating in individual $4p$ states; this indicated the largest cascade contribution to be from Ar II $4d$ states, with particularly large contributions to the population of the $4p\,^2D^0_{5/2}$ and $4p\,^4D^0_{5/2}$ states, which are the upper levels of the two strongest laser transitions at 4880 and 5145 Å respectively. Rudko and Tang (1966) measured the relative intensity of important transitions into the Ar II $4p$ states under actual CW ion laser conditions and also found a large contribution from $4d$ to $4p$ states. Marantz *et al.* (1966) concluded that, in a 1-mm bore argon laser operating at 6 A, 50% of all the pumping of the $4p\,^2D^0_{5/2}$ level came from cascades. Rudko and Tang (1967)

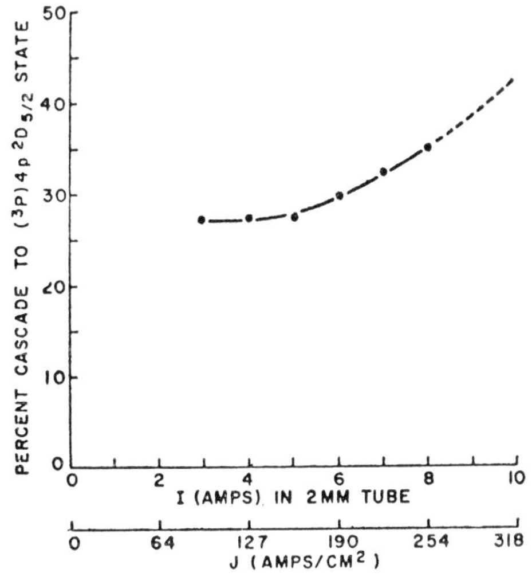

FIG. 14. Current dependence of the cascade contribution to the Ar II 4880 Å upper laser level (Rudko and Tang, 1967).

also found that the percentage cascade contribution to this state increased with current density, as in Fig. 14; they also confirmed an observation, previously made by Miller *et al.* (1967), that if the populations of various levels within a configuration divided by their degeneracies were plotted in a semi-log plot against their energy, levels within a configuration lay on straight lines. This effect and the corresponding level populations are shown in Fig. 15. The straight lines on which levels within a configuration lie are of the form

$$\ln\left(\frac{N_n}{g_n}\right) = -\frac{E_n}{kT_c} + \text{constant} \qquad \ldots(4.42)$$

where T_c can be called the temperature of the configuration. Miller *et al.* concluded that the existence of this configuration temperature implied that rapid thermalization between levels within a configuration was occurring. If this had been so it would have implied that a large cascade contribution to a particular level within a configuration would in fact have helped to populate all the levels within that configuration. However by performing laser perturbation spectroscopy experiments, where laser oscillation at one wavelength was interrupted with an in-cavity chopper and modulation was looked for on transitions coming from different upper levels within the same configuration, Rudko (1967), Rudko and Tang (1967) and Dunn (1968) concluded that

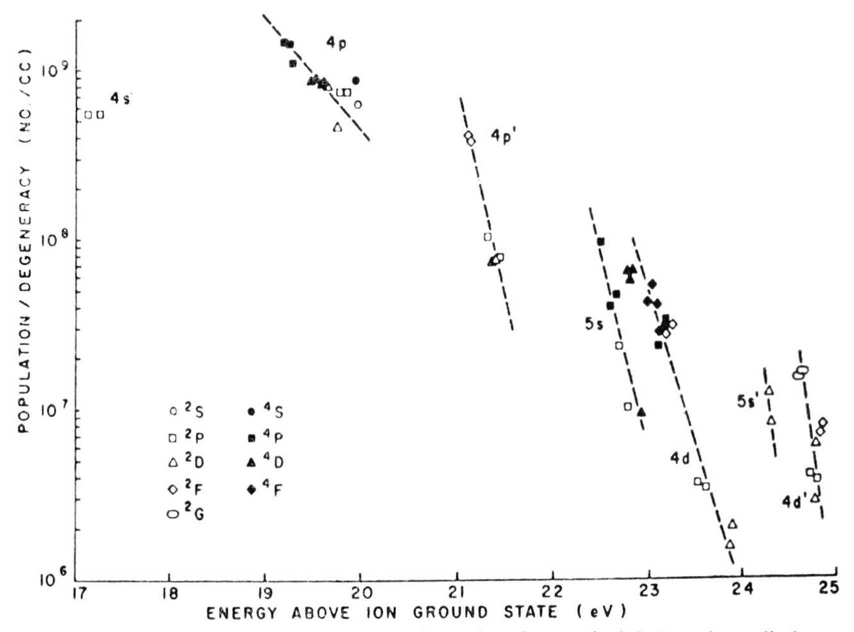

FIG. 15. Population densities of Ar II configurations in a typical CW gas laser discharge (Rudko and Tang, 1967).

thermalization within a configuration occurs only slowly. The existence of a configuration temperature is more of a fortuitous representation of the relative efficiency of excitation of the different levels within a configuration. The existence of an apparent configuration temperature has also been observed by Van der Sijde (1972b, d) in a magnetically confined hollow cathode argon discharge where the conditions approximated to those in a hollow cathode laser.

The percentage contribution that radiative cascades make to the population of the argon ion $4p$ levels has also been investigated by Bridges and Halsted (1967). They found cascade contributions to the $4p\,^2D^0_{5/2}$ and $4p\,^4D^0_{5/2}$ levels of 23 and 22% respectively under lower current density conditions than those used by Rudko and Tang (1967). Similar results have been obtained by Labuda *et al.* (1966). Jennings *et al.* (1970) demonstrated that in a typical hollow cathode, ion laser cascades also made a significant contribution to the population of $4p$ states. Van der Sijde (1972d) reported smaller but significant contributions from cascade to the population of $4p$ levels in a low pressure magnetically confined hollow cathode discharge; largest cascade contributions made in this case were to the levels $^2D^0_{5/2}$ ($\sim 10\%$) and $^4S^0_{3/2}$ ($\sim 15\%$). In addition to this experimental evidence, theoretical considerations (Beigman *et al.*, 1967) also point to a large cascade contribution to the population of the $4p$ configuration. Under CW laser conditions most of the cascades come from the $4d$ configuration with negligible contributions from higher lying configurations such as $5s$, $6s$ and $5d$, these cascading levels themselves being excited via the ion ground or metastable states. The increased importance of cascades in high current density (small bore) CW lasers stems from the higher electron temperatures which obtain in these devices. In the more recent generation of very high power ion lasers pioneered by Boersch and co-workers using high current (but relatively low current density) discharges in wide bore tubes (see Section X G), the cascade contribution to the pumping becomes smaller as these devices operate most efficiently at lower electron temperatures than narrower bore tubes laser (Herziger and Seelig, 1968). At lower electron temperatures the pump rate of higher lying configurations in argon II (e.g. $4d$) falls.

Recombination between ions and electrons may also make a small contribution to the pumping in ion lasers. This process has been considered by Leonov *et al.* (1966); it is shown schematically in Fig. 6. The process, for example

$$\mathrm{Ar}^{++} + e \rightarrow \mathrm{Ar}^{+*}, \qquad \ldots(4.43)$$

would not show any simple current dependence, since the doubly ionized argon population is not simply related to current. This two-body recombination process however appears unlikely to be important in the excitation of any ion laser transitions since the rate at which it occurs would be extremely small except in the presence of a third body, which would imply higher pressure conditions than are common in ion lasers. Most of the recombination processes which occur in ion lasers occur at the discharge tube walls and it is unclear whether excited atoms or ions would be produced to any extent in this manner.

(6) *Excitation in Long-pulse Ion Lasers*

One area of observation where the role of the single-step excitation process is unclear is in long pulse ($> \mu s$) excitation of ion lasers, in particular the argon ion laser. A frequently observed phenomenon in long-pulse excitation of ion lasers is multiple laser pulse outputs in different time regions of the excitation pulse. This effect is illustrated later in Fig. 43 (see p. 367). The excitation of the prompt rising edge pulse (A) can be ascribed to single-step excitation from the neutral atom ground state; however, the excitation in region B has been ascribed to different processes by different authors. Laser action in region B may be quasi-CW as it can often be extended in time duration by extending the length of the exciting current pulse. Demtröder (1966) and Smith and Dunn (1968) ascribe the excitation mechanism in region B of the argon ion laser to the same processes involving the ion ground state as are operative in CW argon ion lasers. However the observations of Davis and King (1971) indicate that a single-step excitation mechanism, perhaps from neutral metastables, may still be involved. These observations and others made in different investigations of pulsed argon ion lasers should not be seen as contradictory since the experimental conditions used by different investigators have varied quite widely. Quantitative analysis of such systems is complicated by the fact that all the plasma parameters which control the operation of the laser vary during the pulse (Hattori and Goto, 1969a; Klein, 1970).

The variation of these plasma parameters with time explains the appearance of double or multiple laser pulses and the existence of a period of no laser oscillation between pulses as in Fig. 43. The cessation of the first pulse has generally been ascribed to the onset of resonance trapping (Bridges and Chester, 1965a; Gordon *et al.*, 1965; Demtröder, 1966; Smith and Dunn, 1968; Klein, 1970). The second pulse is presumed to occur when heating of ions during the pulse increases the Doppler width sufficiently to reduce this resonance trapping. Although this explanation may be valid in certain cases there is also evidence that transient laser action in the rising edge of the current pulse in pulsed argon ion lasers occurs because of the transient high electron temperature associated with the high initial axial field which exists in such a situation (Davis, 1970). Laser action ceases when the axial field can no longer maintain a sufficiently high electron temperature for single-step excitation of the upper laser levels from the neutral atom ground state to occur. When laser action commences again (region B in Fig. 43) a different excitation mechanism is operative (Demtröder, 1966; Smith and Dunn, 1968; Davis and King, 1971) as the relative gains and intensities of different transitions alter. Typically in argon ion lasers the 4765 Å transition is strongest in the rising edge and the 4880 Å transition is strongest in period B. The dead interval between A and B has been shown to be absorbing at high currents (Gordon *et al.*, 1961; Davis and King, 1971) which is consistent either with resonance trapping or electron collisional effects as discussed in Section IV, C.

C. QUENCHING IN SINGLY-IONIZED ION LASERS

Experiments on ion lasers at high current densities described in Section XI, A show that saturation and eventual quenching of the power output occurs as the current density is increased. Similarly, increase or decrease of the operating pressure of an ion laser from optimum will eventually produce the same effect. Saturation with increasing current density has been observed in CW ion lasers (Paananen, 1966; Fendley, 1968; Donin *et al.*, 1969; Boersch *et al.*, 1970) at the highest current densities used, e.g. 900 A cm^{-2} for saturation of Xe III lines (Fendley, 1968). In pulsed ion lasers sufficiently high current densities to cause saturation and quenching are readily obtained. (Bridges and Cheo, 1965a; Cheo and Cooper, 1965b; Demtröder, 1966; Hattori and Goto, 1968a, b, 1969a, b; Jarrett and Barker, 1968; Cottrell, 1968; Klein, 1970; Davis and King, 1971, 1972). A number of possible mechanisms have been suggested for saturation and quenching of ion lasers at high current densities:

(i) Increased excitation of the lower laser levels by electron collisions (Cottrell, 1968; Davis and King, 1971).
(ii) Reduction of the density of neutral atoms by a high degree of ionization (Boersch *et al.*, 1970; Levinson *et al.*, 1968). This particular process would be effective in quenching laser action from levels pumped from neutral atom states.
(iii) Reduction of the density of singly-charged ions by a high degree of multiple ionization (Levinson *et al.*, 1968). This would effectively quench laser action from levels excited from singly-ionized ground or metastable atoms.
(iv) Contamination of the discharge caused by the high current densities used (Boersch *et al.*, 1970). This process may play a small part in high current density CW ion lasers.
(v) Electron destruction of the upper laser levels. (Levinson, 1968; Schacter and Skurnick, 1971; Skurnick and Schacter, 1972).
(vi) Reduction in electron temperature (Boscher *et al.*, 1971).
(vii) Resonance trapping of the vacuum ultra-violet radiation from the lower laser level. (Statz *et al.*, 1965; Bridges and Chester, 1965a; Cheo and Cooper, 1965b; Demtröder, 1966; Gorog and Spong, 1966; Hattori and Goto, 1968a, b, 1969a, b; Preobrazhenski and Shaparev, 1968; Boersch *et al.*, 1970; Klein, 1970).

The quenching of laser oscillation by pressure changes is fairly easily understood. If the pressure in the laser is too high the electron energy distribution will be shifted to lower values of electron temperature which upsets the balance between the excitation of the upper and lower laser levels. At the same time, higher pressures increase resonance trapping of the lower laser level because under equivalent current density conditions the ion (and electron density) are higher at high pressures (because of the lower electron mobility). Reduction of pressure in an ion laser will eventually quench the laser oscillation, although this may not be an easily observed effect in practice. Usually difficulty will be experienced in maintaining the discharge at sufficiently low pressures to cause laser oscillation to cease; under these circumstances the plasma electron temperature will rise and the electron density may fall leading to a change in the relative efficiencies of exciting upper and lower laser levels.

However, saturation and quenching of laser power output with current increase are the important effects to be understood as they clearly have technological implications for the maximum available power densities (W/cm^{-3}) which can be extracted from ion lasers. As far as singly-ionized ion lasers are concerned, the most important of the potential quenching mechanisms mentioned above are increased excitation of the lower laser level (i); electron destruction of the upper laser level (v) and resonance trapping (vii).

The first of these three processes could occur for a number of reasons; in a CW ion laser the excitation depends to a large extent on excitation from the ion ground and metastable levels. These processes predominate over direct excitation of the lower laser levels only because the electron temperature in most low current density ion lasers happens to be in just the right region. The creation of a substantial population inversion is also enhanced by the very favourable lifetime ratios that exist in ion lasers (see Section IX, G). These lifetime ratios τ (upper level)/τ (lower level) are typically of the order of 20 which would imply, neglecting radiation trapping, that the ratio of pumping rates R (lower level)/R (upper level) would have to reach 20 to destroy the inversion. Experiments on CW argon ion lasers show that the electron temperature increases with current density which could shift the balance of excitation to lower laser levels if these are excited on an optically allowed transition from the ion ground state. If the upper laser levels are excited by a single-step process from neutral atom states (or an apparent single-step process involving the ion ground state and electron destruction of the upper laser level) and the lower laser levels are excited from ionic states, then the current density dependence of spontaneous emission, gain and output power (if homogeneous broadening is operative) should be $\propto aj - bj^2$. This kind of dependence has been observed both in the output power (Cottrell, 1968) and the gain (Davis and King, 1971) of pulsed argon ion lasers.

Electron destruction of the upper laser level can cause quenching only if it begins to reduce the effective lifetime of the level appreciably. Statz *et al.* (1965) discounted this process but their calculations are very approximate. Kon'kov *et al.* (1969) and Vladimirova *et al.* (1970a, b) have monitored the change in effective lifetime of the argon $4p\,^2D^0_{5/2}$ and $4p\,^4D^0_{5/2}$ levels (the upper levels of the 4880 and 5145 Å laser transitions) as a function both of current, pressure and magnetic field, finding a monotonic decrease in effective lifetime with both current density and pressure which they attribute to electron collisional deactivation of the upper laser levels. The lifetimes of these levels are reduced by a factor of 2 at current densities of the order of

500 A cm^{-2}. This reduction is probably not sufficient to cause quenching of laser action with increase of current density, particularly since the intensity of emission from the upper levels of the ionized argon laser transitions continues to increase at current densities when saturation and quenching of the laser power output has occurred (Demtröder and Elendt, 1966; Levinson *et al.*, 1968).

(1) *Resonance Trapping*

This process has been generally regarded [see (vii) above] as the main saturation and quenching process operative in CW and pulsed ion lasers. However, further consideration of this process in the light of more recent experimental and theoretical work indicates that it may not be so important as has previously been thought.

In a structure with cylindrical geometry the emission and subsequent reabsorption of a resonance photon many times before it escapes leads to an increase in the effective lifetime of an ion on the tube axis by a factor $1/g$ where g is given by (Holstein, 1947, 1951)

$$g = \frac{1 \cdot 60}{k_0 R(\pi \ln k_0 R)^{1/2}} \quad \text{for a Doppler broadened line}$$

$$k_0 = \frac{g_2 \lambda_0^3 N}{g_i 8\pi^{3/2} \tau_2 v_0} \qquad v_0 = \left(\frac{2kT}{M}\right)^{1/2} \qquad \ldots(4.44)$$

and $g = 1 \cdot 115(\pi k_p R)^{-1/2}$ for a Lorentzian broadened line where

$$k_p = \frac{g_2 \lambda_0^2 N}{g_i 2\pi \tau_2 \gamma_p}$$

R is the radius of the cylinder, N is the (assumed uniform) density of resonance radiation absorbers (in the case of the ion lasers, ground state ions), M is the mass of the emitting species (ions), g_2 and g_1 are the degeneracy factors of the upper and lower levels of the resonance transition, τ_2 is the natural lifetime of the emitting level (lower laser level), λ_0 is the centre wavelength of the resonance (lower laser level draining) transition and γ_p is proportional to the density of electrons and ions in the discharge.

In ion lasers, resonance trapping of the vacuum ultraviolet draining radiation from the lower laser levels can increase their lifetime and destroy the population inversion. A criterion which is usually adopted in deciding whether resonance trapping will quench laser action is to determine whether $\tau_2/g \gg \tau_3$, where τ_3 is the lifetime of the upper laser level (neglecting electron destruction of this level). To determine the magnitude of resonance trapping effects it is useful to consider the 4880 Å laser line in argon, since this transition is more subject to resonance trapping effects because of the degeneracy factor of its lower level ($4s\,^2P_{3/2}$) than lines terminating on the $4s\,^2P_{1/2}$ level. For this transition the resonance draining transition is at 723·353 Å, $\tau_2 = 0{\cdot}36$ ns and $g_2/g_1 = 1$.

In a fast pulsed argon ion laser operating at 4880 Å where the ions may remain cool for a short time, $T_i \simeq 300°K$ and $k_0 = 6{\cdot}68 . 10^{-3}\ N$.

In a typical CW ion laser a value of $T_c = 2700°K$ is appropriate and $k_0 = 2{\cdot}23 . 10^{-3}\ N$.

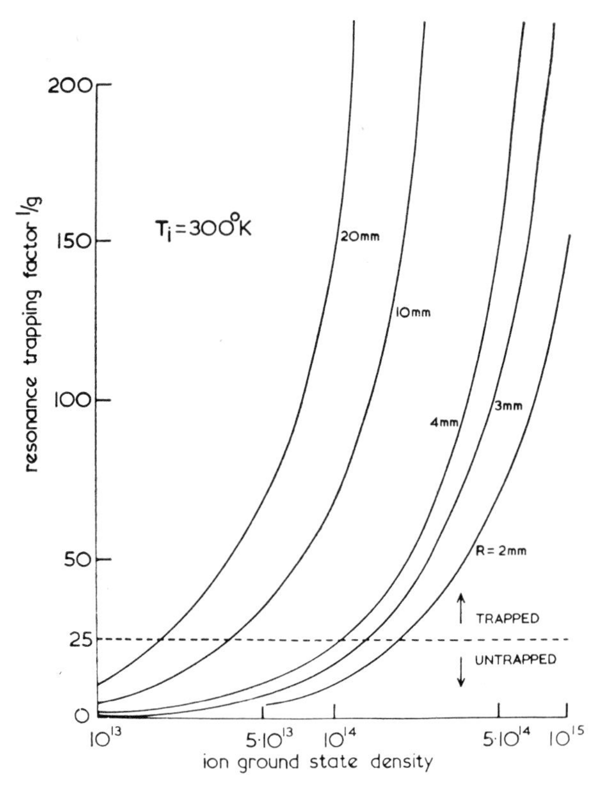

FIG. 16. Variation of the resonance trapping factor $1/g$ with ion ground state density for the lower level of the Ar 4880 Å laser transition for $T_i = 300°K$ and various tube diameters.

Trapping of the lower laser level can be regarded as severe in this case when $1/g = 25$. Figures 16 and 17 show the variation of $1/g$, the resonance trapping factor for various tube diameters at 300 and 2700°K. Clearly, resonance trapping is less severe at the higher ion temperature as the Doppler width of the resonance line is larger. In a small bore (2 mm) CW argon ion laser operating at 4880 Å resonance trapping becomes important at ion ground state densities $\gtrsim 5{\cdot}10^{14}\ \mathrm{cm}^{-3}$, whilst in a larger bore (10 mm) it becomes important at $\gtrsim 5 . 10^{13}\ \mathrm{cm}^{-3}$. Table IV, which will be discussed in more detail in Section VII, A, lists some measured electron (and hence ion) densities in CW argon ion lasers which lead, in many cases, to severe resonance trapping. However, the ability of wide bore argon ion lasers to operate efficiently at current densities where resonance trapping might be expected to be severe (Boersch *et al.*, 1970; Herziger and Seelig, 1968; Wang and Lin, 1972) demonstrates that resonance trapping cannot be as important as the simple considerations above indicate.

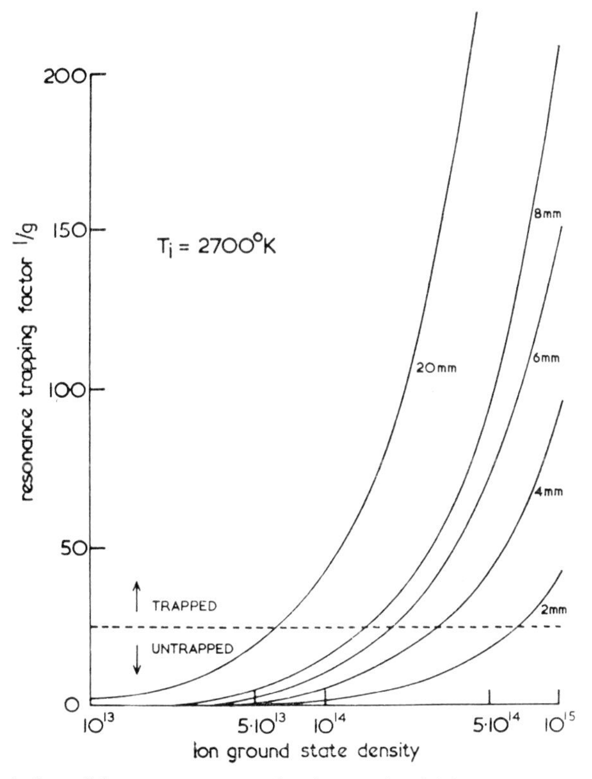

FIG. 17. Variation of the resonance trapping factor $1/g$ with ion ground state density for the lower level of the Ar 4880 Å laser transition for $T_i = 2700$ °K and various tube diameters.

These idealized calculations neglect a number of factors which tend to prevent resonance trapping becoming effective at the sort of ion densities expected:

(i) The ions in the discharge move in a radial field which accelerates an ion from the point where it is produced towards the wall, producing a large spread of radial ion velocities and reducing the effectiveness of trapping.

(ii) There is a radial distribution of ions in the discharge having its maximum on the tube axis which reduces the effective radius of the tube to

TABLE IV. *Electron densities in argon ion lasers*

Current density (A cm^{-2})	Tube bore (mm)	Pressure (torr)	N_e (cm^{-3})	Method of measurement	Author(s)
189	2·6	0·5	$3{\cdot}7.10^{13}$	Stark broadening of H_β line	Kitaeva *et al.* (1966a, 1967)
400	1·6	0·5	$3{\cdot}4.10^{13}$		
150	2·8	0·37	$1{\cdot}8.10^{13}$	Deduced from T_e (calculated from Kagan-Perel lineshape analysis) using Kolesnikov (1962) conductivity formula	Kitaeva *et al.* (1966b)
250	2·8	0·37	$2{\cdot}8.10^{13}$		
350	2·8	0·37	$3{\cdot}2.10^{13}$		
150	2·8	0·62	$2{\cdot}4.10^{13}$		
250	2·8	0·62	$4{\cdot}1.10^{13}$		
350	2·8	0·62	$4{\cdot}8.10^{13}$		
400 (pulsed)	5	0·5	$8{\cdot}10^{12}$	Stark broadening of H_α, H_β lines	Kitaeva *et al.* (1966a, 1967)
560 (pulsed)	5	0·5	1.10^{14}		
820 (pulsed)	5	0·5	2.10^{14}		
630 (pulsed)	4·5	Not stated	1.10^{14}	Stark broadening of H_β lines	Glazunov *et al.* (1967)
1260 (pulsed)	4·5	Not stated	$2{\cdot}6.10^{14}$		
1890 (pulsed)	4·5	Not stated	$4{\cdot}1.10^{14}$		
700 (pulsed)	6	0·037	$8{\cdot}5.10^{14}$	Double probe	Hattori and Goto (1969a)
1400 (pulsed)	6	0·037	$1{\cdot}6.10^{15}$		
2100 (pulsed)	6	0·037	2.10^{15}		
700 (pulsed)	6	0·015	$3{\cdot}9.10^{14}$		
1400 (pulsed)	6	0·015	$6{\cdot}9.10^{14}$		
1750 (pulsed)	6	0·015	$8{\cdot}8.10^{14}$		
250 (pulsed)	10	0·037	3.10^{14}		
500 (pulsed)	10	0·037	$8{\cdot}4.10^{14}$		
750 (pulsed)	10	0·037	$1{\cdot}3.10^{15}$		
250 (pulsed)	10	0·015	$2{\cdot}6.10^{14}$		
500 (pulsed)	10	0·015	$5{\cdot}1.10^{14}$		
750 (pulsed)	10	0·015	$8{\cdot}0.10^{14}$		

25	—	Not stated	6.10^{12}	Theoretical	Zarowin (1969)
120	—	Not stated	2.10^{14}		
250	—	Not stated	6.10^{14}		
500	—	Not stated	$2{\cdot}5.10^{15}$		
15,000 A peak[a]	52	0·012	2.10^{15}	Laser interferometry	Illingworth (1970)
15,000 A peak	52	0·025	1.10^{15}		
470 (pulsed)	3	0·033	$3{\cdot}6.10^{13}$	Calculated from mobility data (Brown, 1966) and measured values of current and axial field	Davis (1970)
1840 (pulsed)	3	0·022	$6{\cdot}9.10^{13}$		
2130 (pulsed)	3	0·050	$5{\cdot}2.10^{14}$		
113–242	7	0·23–0·68	$0{\cdot}7–2{\cdot}2.10^{14}$	Stark broadening of argon I line	Kitaeva *et al.* (1971)
100	3	0·37	6.10^{13}	Laser interferometry	Zory and Lynch (1971)
400	3	0·62	$1{\cdot}7.10^{14}$		
280	1·5	0·14	$4{\cdot}7.10^{13}$	Calculated from measured axial field	Fridrikhov and Fotiady (1971)
280	1·5	0·4	$7{\cdot}8.10^{13}$		
280	1·5	0·66	9.10^{13}		
50	2	1	$6{\cdot}3.10^{13}$	Double probe	Miller and Webb (1971)
100	2	1	$4{\cdot}3.10^{13}$		
150	2	1	$5{\cdot}7.10^{13}$		
200	2	1	$6{\cdot}9.10^{13}$		
250	2	1	$8{\cdot}9.10^{13}$		
530 (pulsed)	11	0·2 (He) 0·004 (Ar)	5.10^{12}[b]	Microwave	Bensimon *et al.* (1971)
250 (pulsed 37 μs)	2	0·040	8.10^{13}	Microwave	Raff and George (1971)
320 (pulsed 37 μs)	2	0·040	$1{\cdot}04.10^{14}$		
640 (pulsed 37 μs)	2	0·040	$2{\cdot}14.10^{14}$		
250 (pulsed 37 μs)	2	0·1	$1{\cdot}32.10^{14}$		
320 (pulsed 37 μs)	2	0·1	$1{\cdot}8.10^{14}$		
640 (pulsed 37 μs)	2	0·1	$3{\cdot}74.10^{14}$		

TABLE IV—*continued*

Current density (A cm^{-2})	Tube bore (mm)	Pressure (torr)	N_e (cm^{-3})	Method of measurement	Author(s)
25	2	0·28	3.10^{12}	Microwave	Pleasance and George (1971); Raff (1971)
120	2	0·28	$4{\cdot}1.10^{12}$		
25	2	0·56	4.10^{12}		
120	2	0·56	$6{\cdot}5.10^{12}$		
250	2	0·56	9.10^{12}	Microwave	Pleasance and George (1971); Raff (1971)
320	2	0·56	$1{\cdot}03.10^{16}$		
25	2	1·17	6.10^{12}		
120	2	1·17	$1{\cdot}06.10^{14}$		
250	2	1·17	$1{\cdot}5.10^{14}$		
320	2	1·17	$1{\cdot}72.10^{14}$		
114	7	0·23	$1{\cdot}5.10^{13}$	Microwave	Kitaeva *et al.* (1972)
164	7	0·6	6.10^{13}		
114	7	0·23	3.10^{13}	Herziger-Seelig (1968) calculation	Kitaeva *et al.* (1972)
164	7	1·6	6.10^{13}		
32	2	0·2	$2{\cdot}05.10^{14}$	Stark broadening of argon I lines	Sze and Bennett (1972)
64	2	0·2	$2{\cdot}2.10^{14}$		
95	2	0·2	$2{\cdot}25.10^{14}$		
127	2	0·2	$2{\cdot}4.10^{14}$		
191	2	0·2	$2{\cdot}7.10^{14}$		
286	2	0·2	$3{\cdot}05.10^{14}$		

[a] In Z-pinch discharge, current density rapidly varying. [b] In afterglow of pulsed discharge.

resonance trapping. Skurnick and Schacter (1972) have shown that for an ion in the lower laser level at radial position r within the tube, the resonance trapping factor is reduced to

$$g(r) \simeq \frac{1}{2k_0 R}\left(1 + \frac{2r^2}{R^2}\right) \quad \ldots(4.45)$$

(iii) Non-radiative destruction of lower laser levels by electron collisions tends to counterbalance the effects of resonance trapping. Some of these non-radiative processes may even help to increase the population of the upper laser level, e.g. in single ionized argon the process $|4s\rangle + e \to |4p\rangle + e$ has a very large cross-section (Beigman *et al.*, 1968).

Merkelo *et al.* (1968) claim that the destruction of lower levels of the argon ion laser is primarily non-radiative at current densities up to 300 A cm^{-2} within the pressure range 0·1–0·5 torr. This was based on an in-cavity perturbation spectroscopy experiment where the laser oscillation was periodically interrupted inside the laser cavity and the simultaneous modulation of the vacuum ultraviolet radiation from the lower laser level studied. None was detected but it cannot be concluded that largely non-radiative destruction of the lower laser levels was responsible for this effect.

For example the population of a lower laser level in this experiment obeys an equation of the form

$$\frac{dN}{dt} = -\frac{N}{\tau} - NN_e\langle\sigma v\rangle + R_s + \sum_j R_j + R_i \quad \ldots(4.46)$$

where R_i is the rate of pumping of this level by electron collisions and R_s is the rate of pumping of the lower level by stimulated emission, R_j is the cascade contribution to the pumping on the spontaneous transition from level j, $\langle\sigma v\rangle$ is a velocity averaged destruction cross-section and τ is the lifetime of the lower level allowing for resonance trapping effects. That is

$$N = \frac{R_s + \sum_j R_j + R_i}{1/\tau + N_e\langle\sigma v\rangle} \quad \ldots(4.47)$$

In the absence of lasing

$$N = \frac{\sum_j R_j' + R_i'}{(1/\tau) + N_e\langle\sigma v\rangle} \quad \ldots(4.48)$$

where R_i' is the modified electron collisional pump rate and R_j' is the modified cascade pump rate from level j in the absence of lasing. The percentage modulation of spontaneous emission from the lower level observed by Merkelo *et al.* is proportional to

$$\frac{R_s + \sum_j (R_j - R_j') + R_i - R_i'}{R_s + \sum_j R_j + R_i}$$

Since $R_j \simeq R_j'$, and $\sum_j R_j \gg R$ only a very small modulation will be observed. A large value of $N_e\langle \sigma v \rangle$ will not affect the magnitude of the unobserved modulation. However further investigation of these effects appear to be desirable.

(2) *Annular Modes and Dark Spaces in High Current Pulsed Ion Lasers*

A number of authors (Cheo and Cooper, 1965; Gorog and Spong, 1966; Cottrell, 1968; Davis, 1970) have observed the appearance of high order transverse modes in pulsed ion lasers at high currents and/or pressures. These modes, shown in Fig. 39 (see p. 357), take the form of a ring with a dark centre or a series of dots distributed annularly around a central dark region. It has been supposed (Cheo and Cooper, 1965; Gorog and Spong, 1966; Preobrazhenski *et al.*, 1968), that these effects are due to resonance trapping of the lower laser levels which preferentially destroys population inversion near the axis of the laser. However, near the walls where the resonance trapping factor is much smaller, population inversion is maintained, allowing high order modes which utilize an annular region near the discharge tube walls to exist. The axial maximum of ion density also encourages resonance trapping near the axis.

Although most experimental observations are consistent with this model, other explanations are also possible. There is considerable evidence that, either through single-step excitation of the upper laser level or two-step excitation of this level plus its electron collisional destruction, the population inversion in a high current laser follows a general equation of the form

$$\Delta N = aj - bj^2 - j_t \qquad \ldots(4.49)$$

where j is current density and a, b and j_t are constants. (j_t can be regarded as the threshold current.)

In ion lasers the radial profile of current density follows that of electron density, $n_e(r)$, which is a maximum on the axis. Thus the radial distribution of population inversion will be

$$\Delta N(r) = a'\, n_e(r) - b'\, n_e^2(r) - j_t \qquad \ldots(4.50)$$

and a' and b' are new constants.

With increasing current density or pressure $n_e(r)$ increases and from the above equation it can be seen that when $b'[n_e(0)]^2 + j_t > a'\, n_e(0)$ axial quenching occurs. With further increase of current density this region of quenching spreads from the axis leaving an annular region of inverted population near the walls.

Davis (1970) has observed the production of annular modes in the output of argon ion laser transitions when operating in a mixed Ar–Xe ion laser at a low partial pressure of argon ($\gtrsim$ 1 mtorr). Resonance trapping cannot be the explanation at these low pressures, although the radial dependence of electron density can still lead to the effect. Thus the existence of a radially dependent deexcitation rate of the laser levels will lead to the production of annular modes. Skurnick and Schacter (1972) have considered a simple model for the radial dependence of population inversion produced in this way and obtain good agreement with experimental observations. It seems that these radially dependent electron collisional effects may dominate over the effect of resonance trapping, even though the latter phenomenon certainly occurs to some extent.

D. EXCITATION AND QUENCHING IN MULTIPLY-IONIZED ION LASERS

Koozekanani (1968c) has considered the possibility of simultaneous double ionization and excitation of argon, neon and krypton by fast electrons (energy about 100 eV) using the "sudden" approximation. The largest cross-sections for this process are to states of the form $|np^3, (n+1)p\rangle$, that is

$$|np^6\rangle + e \rightarrow |np^3, (n+1)p\rangle + 3e \qquad \ldots(4.51)$$

where $n = 3$, 4 or 5 for argon, neon and krypton respectively. The largest cross-sections of all are, in argon for example, for excitation to the states $Ar^{++}4p\,^1D_2^0$ and $4p\,^3P_2^0$; however, even these cross-sections are small on an absolute scale ($\sim 10^{-19}$ cm^2). In practice it is unlikely that this kind of process contributes significantly to the excitation of doubly-ionized laser levels as the cross-sections are so small and the required electron energies high.

Levinson *et al.* (1968) have investigated the excitation processes of Ar III and Ar IV ion laser lines in a high current pulsed laser, finding a correlation between the times of maximum spontaneous emission of ions Ar^{n+} and laser action from states $Ar^{(n+1)+}$. They conclude that multiply-ionized laser transitions are excited from the ground state ion of one lower charge, that is:

$$Ar^{n+} + e \rightarrow Ar^{(n+1)+*} + 2e \qquad \ldots(4.52)$$

This process is analogous to the single-step excitation process of singly-ionized laser levels from states of the neutral atom and might therefore be expected only to be of importance under conditions of high electron temperature, such as can obtain in high current pulsed lasers. The electron temperatures needed are very high, for example the upper level of the strong argon III transition at 3511 Å is about 60 eV above the singly-ionized argon ion ground state. However, continuous laser oscillation on many multiply-ionized laser transitions is observed under conditions where the electron temperature is probably too low for this process to occur. Under these circumstances the excitation process originally proposed by Bridges and Chester is more likely

$$Ar^{n+} + e \rightarrow Ar^{n+*} + e. \qquad ...(4.53)$$

Even this process requires quite a high electron temperature; the upper level of the 3511 Å Ar III transition is 26 eV above the doubly-ionized ion ground state. However, because of the lack of diagnostic information about the high current density plasmas where continuous laser action in multiply-ionized ions occurs, it cannot be estimated with any certainty what percentage of the excitation actually comes from this type of process and what contribution may come from other processes such as cascades.

Quenching of laser action can occur with increase of current even for laser action on transitions in multiply-charged ions, although the very high current densities needed are only readily available in pulsed systems. A similar quenching process can occur as in singly-ionized lasers although in very high current pulsed systems of this kind, where the excitation mechanism is likely to be the one proposed by Levinson *et al.* (1968), a significant contribution to quenching is likely to come from electron excitation of the lower laser levels, for example

$$Ar^{++}(3p^4) + e \rightarrow Ar^{++*}(3p^3, 4s) + e.$$

At the very high current densities where quenching occurs electron collisional destruction of the laser levels is also likely to be significant. In addition, at very high current densities, the plasma may become extensively multiply-ionized and the density of the lasing ions and those of lower charge will become depleted.

E. Z-PINCH PULSED ION LASERS

When a high current discharge is created in a cylindrically symmetric tube the current gives rise to an azimuthal magnetic field which causes charged particles moving parallel to the cylinder axis to experience a force directed inward. This magnetic force is only sufficiently large to dominate the behaviour of the plasma when the current is very large, which makes the effect important only in high current pulsed discharges. The radially directed inward force on the charged particles leads to a reduction in radius of the current carrying region. The discharge is not stable in this pinched condition. The sequence of events which occur in a rapidly pinched discharge can be treated by the Snowplough Model (see, for example, Anderson *et al.*, 1958). It is assumed that when the discharge is becoming ionized the current can be considered to flow in an infinitesimally thin layer on the outside of the cylinder, and the magnetic field does not penetrate the current carrying region. Thus the effect of the magnetic force can be considered as that of a radially inward-moving piston. The resultant pressure force sweeps up all the particles, while increasing their kinetic energy. The particles reach maximum kinetic energy when the discharge is pinched to its minimum radius, usually $\frac{1}{8}$ of the original radius. After reaching the minimum radius, the particles randomize because of collisions with each other and thereby raise the temperature of the gas. The overall current through the discharge decreases slightly as the pinch occurs and sudden changes in the current and voltage across the discharge tube can occur at the instants of maximum contraction. After the first pinching of a high current pulsed discharge, additional expansion and contraction of the current carrying region will occur; these expansions and contractions can be very violent and do not necessarily remain cylindrically and axially symmetric.

Gorog and Singer (1966) have given a simple theoretical description of the dynamic pinch effect which enables its importance under various conditions of current and pressure to be determined. They produce a dimensionless figure of merit K for determining the extent to which the pinch effect occurs. For $K < 1$, maximum compression of the discharge occurs after the first maximum of the magnetic field and the pinch effect becomes less important. If the current pulse in a Z-pinch discharge approximates to a half sinusoid with peak current I_0 reached after a time t then the parameter K (R.M.K.S. units) takes the form

$$K = \frac{\mu_0 t_0^2}{\pi^2 \rho}\left(\frac{I_0}{\pi R^2}\right)^2 \qquad \ldots(4.54)$$

where ρ is the gas density in kg m^{-3}. Since $I_0/\pi R^2$ is the average current density at the peak of the current pulse $= J_0$, K can be re-written in the simple form

$$K = \frac{2{\cdot}4 \times 10^4}{p_0 \text{ (mass number)}} \quad t_0^2 J_0^2$$

where p_0 is the effective discharge pressure at 300°K.

A number of investigations have been made of ion laser action in pinched discharges. The conditions under which the laser action occurs in these are markedly different from those found in conventional pulsed ion lasers with much wider bore discharge tubes (up to 40-mm bore) at much lower pressures (down to 10^{-4} torr) being necessary. Laser action does not occur until the occurrence of the pinch produces the particle densities and electron temperatures necessary for oscillation. In Z-pinch argon ion lasers (Kulagin *et al.*, 1966a, b; Likhachev *et al.*, 1967a, b; Vasil'eva *et al.*, 1969; Illingworth, 1970, 1972; Hashino *et al.*, 1972) the dominant lines are the singly-ionized line at 4765 Å, which is known to be favoured in situations of high electron temperature where direct electron impact of its upper level occurs, and the doubly-ionized lines at 3511 and 3638 Å, which favour high current density operation in conventional ion lasers. Emission on the 4765 Å occurs quite early in the pinch before the maximum compression of the discharge plasma occurs, whilst the doubly-ionized lines appear more nearly at the time of maximum compression. At this time the electron density reaches very high values ($\sim 10^{15}$ cm^{-3}) (Illingworth, 1970) and the discharge is extensively multiply-ionized (neutral particle density before ionization at 10 mtorr $\simeq 3{\cdot}2 . 10^{14}$ cm^{-3}). The shortage of neutral particles under these circumstances probably explains why the 4765 Å terminates before peak compression is reached.

Laser action in Z-pinch discharges has also been observed in neon and nitrogen (Hashino *et al.*, 1972), oxygen (Hashino *et al.*, 1973) where laser action in quadruply-ionized oxygen was observed (OV) and xenon (Papayouanou and Gumeiner, 1970, 1971; Papayouanou *et al.*, 1973) where the group of very strong unclassified visible transitions commonly observed in conventional pulsed xenon lasers were observed.

In addition to the above investigations where experimental conditions were used deliberately to favour the occurrence of Z-pinch effects, the results of other investigations of high current pulsed ion lasers which have been reported in the literature have probably been influenced by pinch effects, e.g. Bennett *et al.* (1964), Hattori and Goto (1969a, b), Simmons and Witte (1970), Klein (1970) and Davis and King (1971, 1972a).

Under conditions where the pinch effect is important ($K > 1$) the time for the first pinch to develop can be written

$$\tau_{\text{pinch}} \simeq 0{\cdot}04 \frac{(p_0(\text{Mass number})\, t_0^2)^{1/4}}{J_0^2} \text{ seconds} \qquad \ldots(4.55)$$

where p_0 is in torr, t_0 in seconds and

$$J_0 = \frac{I_0}{\pi R^2} \text{ A cm}^{-2}$$

and I_0 is the peak current reached in the discharge.

In conventional pulsed ion lasers operating at high currents the time before this pinch occurs is quite short, for example, in operation at 500 A peak current in a 2-mm bore tube operating with argon at a pressure of 10 mtorr and a current pulse rise time of 1 μs, $\tau_{\text{pinch}} \simeq 800$ ns and the pinch occurs in the rising edge of the current pulse. In small bore lasers of this type the occurrence of a pinch is often likely to quench the laser action rather than to enhance it, particularly if the laser transition is one excited from states of the neutral atom as these are likely to be extensively removed by ionization.

F. FAST-PULSE-EXCITED LASERS

These lasers are of current interest because of their ability to produce very short wavelengths, down to 1161 Å in molecular hydrogen (Hodgson and Dreyfus, 1972; Waynant, 1972), 1548 Å in triply-ionized carbon (Waynant,

1973) and 1730 Å in molecular xenon (Johnson and Gerardo, 1973; Bhaumik *et al.*, 1973; Rhodes and Hoff, 1973). These lasers can also provide high peak output powers, for example 3 MW in molecular nitrogen at 3371 Å (Godard, 1973) and relatively high average powers ~ 1·5 W in molecular nitrogen (Targ, 1972), 15 W in atomic copper (Petrash *et al.*, 1973) and can operate with relatively high efficiency, ~ 1% in molecular nitrogen (Godard, 1973) and atomic copper (Petrash *et al.*, 1973). Their importance is likely to increase as the range of transitions and elements which operate in this way is extended. Because of potentially higher output powers and efficiencies and their ability to operate at very short wavelengths, these lasers have stimulated interest in applications such as plasma production and laser produced plasmas for fusion.

Although some of these fast-pulse-excited lasers require fast excitation to achieve inversion at very short wavelengths, many laser transitions require

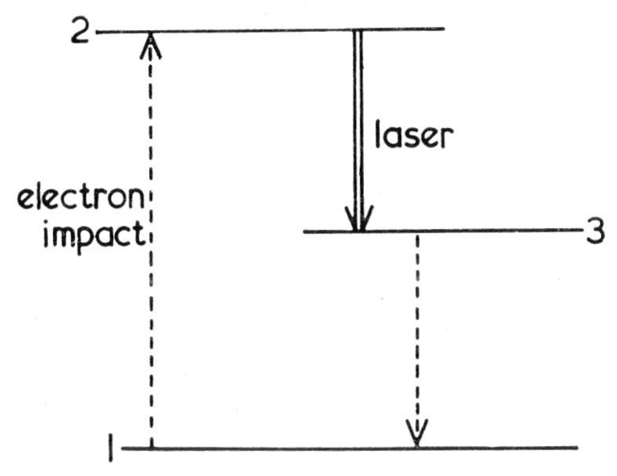

FIG. 18. Schematic energy level diagram of a self-terminating laser system.

excitation on a time scale of the order of or faster than their upper state lifetimes because they are intrinsically self-terminating. Ion laser transitions which fall into this category have been observed in calcium (Walter *et al.*, 1966), strontium (Deech and Sanders, 1968) and carbon (Waynant, 1973), although at present most transitions of this kind have been observed in neutral atoms or molecules (for reviews of this type of pulsed laser see Walter *et al.*, 1966 and Petrash, 1972).

The operation of these "self-terminating" or "cyclic" lasers can be illustrated by reference to Fig. 18. Level 1 is the ground state and the laser transition occurs between levels 2 and 3, the former of which has a large electron impact excitation cross-section from the ground state. This implies that level 2 will be strongly connected radiatively to the ground state and will be short lived. In the simplest kind of self-terminating laser, level 3 is not connected to the ground state and is consequently only slightly populated by electron collisions. Even if level 3 can decay via intermediate states it is likely to be long-lived. Suppose such a three-level system is excited in a fast-rise-time pulsed discharge; for simplicity it is assumed that the electron density in the discharge rises linearly as a function of time. The electron density follows an equation of the kind

$$N_e = \frac{(N_e)_0 \, t}{t_r} \qquad \ldots(4.56)$$

where t_r is a time which characterizes the rate at which the electron density rises.

The rate at which level 2 is excited by electron collisions is

$$\left(\frac{\mathrm{d}N_2}{\mathrm{d}t}\right)_{\mathrm{ec}} = N_e\langle\sigma v\rangle = \frac{N_{e_0}\langle\sigma v\rangle N_1 t}{t_r} \qquad \ldots(4.57)$$

where $\langle\sigma v\rangle$ is an excitation cross-section averaged over the velocity distribution of the electrons and it is assumed that N_1 is independent of time. (In fact $N_1(t) = N_1(t=0) - N_2(t) - N_3(t)$ where $N_2(t) N_3(t) \ll N_1(t=0)$.) Level 2 also decays radiatively at a rate

$$\left(\frac{\mathrm{d}N_2}{\mathrm{d}t}\right)_{\mathrm{rd}} = -\frac{N_2}{\tau_2} \qquad \ldots(4.58)$$

where τ_2 is the lifetime of level 2.

Level 3 in the simplest case does not decay radiatively and is only populated by spontaneous or stimulated emission on the laser transition. In the absence of stimulated emission the rate at which level 3 is populated is

$$\frac{\mathrm{d}N_3}{\mathrm{d}t} = N_2 A_{23} \qquad \ldots(4.59)$$

where A_{23} is the Einstein coefficient for spontaneous emission from level 2 to level 3.

From these considerations the population of level 2 obeys

$$\frac{\mathrm{d}N_2}{\mathrm{d}t} = \frac{N_{e_0}\langle\sigma v\rangle N_1 t}{t_r} - \frac{N_2}{\tau_2} \qquad \ldots(4.60)$$

with the boundary condition that at $t = 0$, $N_2 = 0$, that is

$$N_2 = \frac{N_1 N_{e_0} \tau_2 \langle\sigma v\rangle}{t_r} [t - \tau_2 + \tau_2 \, \mathrm{e}^{-t/\tau_2}] \qquad \ldots(4.61)$$

Consequently

$$N_3 = \frac{N_1 N_{e_0} \tau_2 A_{23} \langle\sigma v\rangle}{t_r} \left[\frac{t^2}{2} - \tau_2 t - \tau^2 \, \mathrm{e}^{-t/\tau_2} + \tau_2^2\right] \qquad \ldots(4.62)$$

where the boundary condition $N_3 = 0$ at $t = 0$ has been used. In the simplest case where there are no competing transitions from level 2, $A_{23} = 1/\tau_2$. Neglecting level degeneracies, the population inversion is given by

$$N_2 - N_3 = \frac{(2N_1 N_{e_0}\langle\sigma v\rangle \tau_2^2}{t_r}[t/\tau_2 + e^{-t/\tau_2} - 1 - \tfrac{1}{4}(t/\tau_2)^2] \quad \ldots(4.63)$$

The behaviour of this function is shown in Fig. 19 showing that a positive population inversion can be maintained only for a time $\sim\tau_2$. Laser oscillation will only occur provided $(N_2 - N_3)_{max} \geqslant (N_2 - N_3)$ threshold.

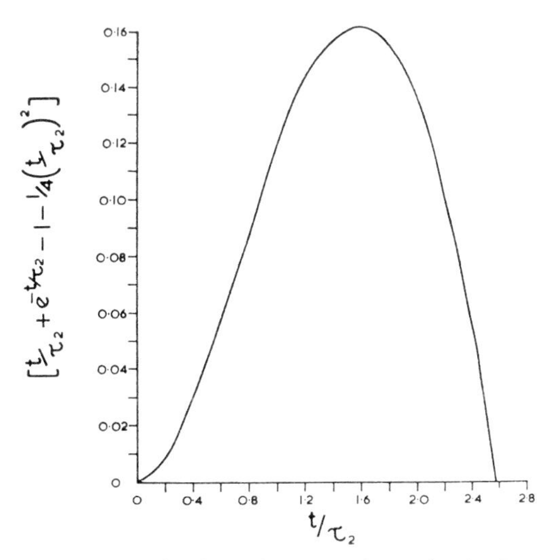

FIG. 19. Population inversion in a self-terminating laser.

Clearly for such a system to exhibit laser oscillation, a sufficient number of atoms must be excited to level 2 to reach the threshold immersion density in times $< \tau_2$. As τ_2 in such a system is likely to be very short (a few ns), this requires t_r to be very short and fast rise-time pulse excitation is essential to achieve laser oscillation. This fast rise-time excitation is usually achieved by using special low inductance discharge tubes, travelling wave (Shipman, 1967) or E-beam excitation (Tkach *et al.*, 1967; Dreyfus and Hodgson, 1972).

V. EXCITATION MECHANISMS IN CW METAL VAPOUR LASERS

As far as this article is concerned the ion lasers which fall into this category are arsenic, cadmium, iodine, lead, magnesium, mercury, phosphorus, selenium, tellurium, tin and zinc. These will be dealt with algebraically and, although not all the transitions from each laser have been observed to operate

CW, the transitions which have only been observed in a pulsed mode will also be discussed. The discussion of these transitions where electron impact excitation may be the dominant excitation mechanism should be considered more closely in connection with the collisional process occurring in noble gas ion lasers discussed in Section IV. The main emphasis of this section is on the CW transitions where the dominant excitation mechanism is either Penning ionization or charge-transfer in metal vapour–helium (or neon) mixtures.

A. FUNDAMENTAL PROCESSES

(1) *Penning Ionization*

This is the excitation transfer process where an excited neutral atom (or molecule) collides with and simultaneously ionizes another neutral atom (or molecule). This reaction is generally only important when the excited atom is metastable as this increases the probability of a collision before the excited atom loses its energy radiatively, although Penning ionization from non-metastable states can occur.

The Penning ionization process can be represented schematically by

$$A_M^* + B \rightarrow A + B^{+*} + e \pm \Delta E_\infty$$

The cross-section for this process is dependent on the particular nature of A_M and B (B may in fact be itself initially in an excited state) but will be very small if the reaction is endothermic with $\Delta E > kT$, where T is the temperature of the colliding particles. The reaction is also likely to have a small cross-section if it is exothermic with $\Delta E > kT_e$ where T_e is the electron temperature in the discharge where the reaction occurs. As far as Penning ionization as a laser excitation mechanism is concerned, the most interesting states A_M are those high excitation energy metastable states in the noble gases, particularly helium, which are listed in Table V.

For the production of slow metastable atoms, electron impact is the most efficient process. Cross-sections are of the same order as for excitation into optically allowed states ($\sim 10^{-16}$ cm^2) if the electron energy is not too high ($\gtrsim$ 100 eV). Under certain conditions in a gas discharge, plasma ion–electron recombination followed by cascading also contributes to the production of metastables as does cascading from higher excited states which have been excited by electron impact. The density of thermal energy metastables which can be obtained in a system, in the absence of any loss through participation in Penning ionization reactions, is governed by the balance between their production and destruction. The various important destruction processes are:

(i) Diffusion to the walls of the container.
(ii) A metastable atom can collide with a neutral atom and be excited into a state which can decay radiatively.

TABLE V. *Characteristics of some important metastable states in the noble gases*

Atom	Ground state	Metastable state	Energy (eV)	Lifetime (s)
He	$1s^{2\,1}S_0$	$2s'\,S_1$	19·82	$2{\cdot}4 \times 10^{-4}$ [c]
		$2s'\,S_0$	20·61	$1{\cdot}97.10^{-2}$
Ne	$2p^{6\,1}S_0$	$3s[3/2]^0_2$	16·61	Long > 0·8[b]
		$3s'[1/2]^0_0$	16·71	Long > 0·8[b]
		$3s[3/2]^0_1$	16·67	*
		$3s'[1/2]^0_1$	16·84	*
Ar	$3p^{6\,1}S_0$	$4s[3/2]^0_2$	11·55	Long > 1·3[b]
		$4s'[1/2]^0_0$	11·72	Long > 1·3[b]
		$4s[3/2]^0_1$	11·62	*
		$4s'[1/2]^0_1$	11·82	*
Kr	$4p^{6\,1}S_0$	$5s[3/2]^0_2$	9·91	Long > 1·0[b]
		$5s'[1/2]^0_0$	10·56	Long > 1·0[b]
		$5s[3/2]^0_1$	10·03	*
		$5s'[1/2]^0_1$	10·64	*
Xe	$5p^{6\,1}S_0$	$6s[3/2]^0_2$	8·31	Long[a]
		$6s'[1/2]^0_0$	9·44	Long[a]
		$6s[3/2]^0_1$	8·43	*
		$6s'[1/2]^0_1$	9·57	*

[a] Van Dyck *et al.* (1971).
[b] Van Dyck *et al.* (1972).
[c] Moos and Woodworth (1973).

States marked with an asterisk are quasi-metastable through resonance trapping of their radiation to the ground state at ground state densities $> 10^{14}$ cm^{-3}.

(iii) Electron collisions destroy metastables in an upward and downward direction.
(iv) Particularly at higher pressures, three body destruction processes involving two other neutral atoms can be important.
(v) Two metastables can collide and one of them can become ionized.

Collisions of metastable atoms with ground state atoms converts them to other metastables in the same configuration. For example, for any of the configurations *ns*, *ns'* of Ne, Ar, Kr and Xe (listed in Table V) collisional mixing will produce a state of thermal equilibrium among the four levels (provided the two states with allowed transitions to the ground state are quasi-metastable). If the total population of the 4 states is N_0, then on the basis of their level degeneracies

$$N(\mathrm{ns}[\tfrac{3}{2}]^0_2) = 0{\cdot}417\ N_0$$
$$N(\mathrm{ns}[\tfrac{3}{2}]^0_1),\ N(\mathrm{ns}'[\tfrac{1}{2}]^0_1) = 0{\cdot}250\ N$$
$$N(\mathrm{ns}'[\tfrac{1}{2}]^0_0) = 0{\cdot}083\ N$$

For noble-gas discharges metastable densities up to 10^{12}–10^{13} cm^{-3}, corresponding to a fraction of about 10^{-4}–10^{-3} metastables have been reported. Of the two helium metastables $2s\,^3S_1$ is the more important as the $2s\,^1S_1$ is very readily destroyed by super-elastic collision with slow electrons

$$e + \mathrm{He}(2s\,^1S_1) \rightarrow \mathrm{He}(2s\,^3S_1) + e + 0{\cdot}79\,\mathrm{eV}$$

The cross-section for this process in the case of thermal electrons is about 10^{-14} cm^2 whilst for the destruction of the $(2s\,^3S_1)$ metastables it is only $\sim 10^{-17}$ cm^2. The He $(2s\,^1S_1)$ is also destroyed by irradiation with light (Fry and Williams, 1969; Hotop *et al.*, 1969). The cross-section for the process where two metastables collide and one becomes ionized has only been determined for the reaction

$$\mathrm{He}(2s\,^3S_1) + \mathrm{He}(2s\,^3S_1) \rightarrow \mathrm{He}^+ + \mathrm{He} + e$$

by Phelps and Molnar (1953) who report a value of 10^{-14} cm^2.

(2) *Charge Transfer*

The charge transfer process is particularly of interest in connection with the excitation of ion laser transitions in thermal energy charge transfer between a ground state atomic (or molecular) ion and a neutral ground state (or metastable) atom, namely

$$\mathrm{A}^+ + \mathrm{B} \rightarrow \mathrm{A} + \mathrm{B}^{+*} \pm \Delta E_\infty$$

Large cross-sections for this reaction are not expected when $\Delta E > kT$. For a given ΔE, much larger cross-sections are likely in the case of the exothermic reaction if one of the particles is molecular and can take up some of the reaction energy in internal vibration.

B. CW "METAL VAPOUR" LASERS

(1) *Arsenic*

The original observation of gas laser action on transitions in ionized arsenic was by Bell *et al.* (1965), who excited low partial pressure arsenic vapour in various noble gas buffers with a pulsed RF discharge. The buffer gas found to be best was neon at a pressure of about 100 mtorr. The upper levels of the arsenic ion laser transitions excited in this manner lie too high for thermal energy charge transfer from neon ion ground states to be their excitation mechanism. In view of the low buffer gas pressures used, the excitation mechanism of the pulsed arsenic ion laser is almost certainly electron impact.

CW laser oscillation on all but one of the previously reported pulsed transitions and on five new transitions has been reported by Piper and Webb (1973). They excited these laser transitions in a helium–arsenic mixture in a hollow cathode structure of the type previously reported by Piper *et al.* (1972).

The dominant excitation mechanism of the transitions observed almost certainly involves charge transfer from ground state helium ions. However, only four of the laser transitions at 5385, 5497, 5838 and 6512 Å from the higher lying upper levels are likely to be excited directly in this manner as,

$$He^+ (^2S_{1/2}) + As \rightarrow As^{+*} (6s) + He (^1S_0) + (1395\text{–}1754 \text{ cm}^{-1})$$

Even so, the relatively large positive energy defect for this reaction (0·17–0·22 eV) suggests that some or most of the excitation occurs via electron collisional deactivation of arsenic levels in closer resonance with the ground state helium ions, particularly the $5d(\frac{1}{2},\frac{3}{2})_2$ level ($5d^3F_2$) (Piper, 1973), the full excitation scheme being

$$He^+ (^2S_{1/2}) + As \rightarrow As^{+*} [5d (\tfrac{1}{2},\tfrac{3}{2})_2] + He (^1S_0) + 248 \text{ cm}^{-1}$$

$$As^{+*} [5d(\tfrac{1}{2},\tfrac{3}{2})_2] + e \rightarrow As^{+*}(6s) + e$$

This process has previously been shown to be of importance in the excitation of some levels in the zinc ion laser (Green *et al.*, 1973). For that group of arsenic ion laser transitions with lower lying $5p$ upper levels, evidence from spontaneous emission intensities suggests that excitation occurs via radiative cascade following charge transfer to arsenic levels in close resonance with the helium ion ground state, once again particularly to $5d(\frac{1}{2},\frac{3}{2})_2$. The optimum helium pressures at which laser action occurs on the two groups of arsenic lines lends weight to this hypothesis. The lines with higher lying upper levels have higher optimum helium pressures, e.g. 55 torr for 5497 and 6512 Å, which is consistent with the electron collisional transfer mechanism as the electron density is higher in the discharge at higher pressures. The optimum pressure for those lines where radiative cascade is important is lower, e.g. 25 torr for the transition at 6170 Å.

For the two strongest CW transitions in the arsenic ion laser at 6512 and 5497 Å power outputs in excess of 10 mW have been obtained (Piper, 1973) and to date saturation of power output with increasing discharge current (up to a maximum of 2·5 Å) has not been observed.

(2) *Cadmium*

The cadmium ion laser has attracted the most interest of all because of its ability to produce short wavelength CW laser emission, notably on the two transitions at 4416 and 3250 Å. The latter of these is the shortest wavelength CW gas laser transition observed to date. CW power outputs of about 200 and 20 mW at 4416 and 3250 Å respectively have been obtained from a cataphoresis type helium–cadmium laser operating with natural cadmium and 3·4 torr of helium at a current of 110 mA in a discharge tube 143 cm long and 2·4 mm bore (Goldsborough, 1969). The technology of such lasers is only a little more complex than that of He–Ne lasers and their efficiency (0·02–0·1 %) compares favourably with typical He–Ne and Ar lasers. The physical characteristics of the various discharge structures used to excite cadmium ion lasers will be dealt with in more detail in Section XII, A. Two types of discharge structure have been found most suitable in providing the conditions appropriate to laser action, namely a uniform distribution of Cd vapour at a pressure between 10^{-3}–10^{-1} torr in a helium discharge at a few torr pressure: the cataphoresis type positive column and the hollow cathode. The different types of discharge used to excite laser action in helium–cadmium lasers (or sometimes neon–cadmium) allow conditions to be achieved which give oscillation on lines which have different dominant excitation mechanisms. In a positive column cataphoresis type helium–cadmium laser, strong CW laser oscillation is obtained only on the 4416 and 3250 Å transitions. It was proposed by Silfvast (1968, 1969) and later demonstrated by Schearer and Padovani (1970), Webb *et al.* (1970) and Silfvast (1971) that the dominant excitation mechanism of both these transitions, which have upper levels $5s^2$ was Penning ionization involving helium $^3S_1$ metastables, $He(^3S_1) + Cd \rightarrow Cd^+(5s^2) + He(^1S_0)$. The total cross-section for this process to all the accessible Cd^+ levels are not accurately known. It has also been suggested (Csillag *et al.*, 1970) that under certain circumstances electron impact was important in populating these $5s^2$ states. However, when the effects of radiative cascade and electron collisional deactivation of the upper and lower laser levels are taken into account, there does not remain any evidence that this is so (Baker, 1973). Although the optimum helium pressures in Cd lasers are generally too high for electron collisional excitation to be important, electron collisions may be more important in promoting laser action observed in Ne–Cd lasers (Csillag *et al.*, 1971). It is interesting to note the general similarity between the operating conditions of He–Cd and He–Ne lasers. Simultaneous laser action at 6328 and 4416 Å has been obtained in a He–Ne–Cd laser (Ahmed and Campillo, 1969; Stefanov and Petrova, 1971). Competition between the 3·39 μm and 6328 Å He–Ne laser lines, and the 4416 Å He–Cd laser line has been observed in such a laser (Giallorenzi and Ahmed, 1971); this indicates that the He $^1S_0$ metastable also helps in populating the 4416 Å upper level. The neon metastables previously mentioned do not have sufficient energy or Penning ionization from them to be a contributing excitation mechanism in these lasers, unless it is via higher lying neon $4s$, $4s'$ states (Racah notation, JK coupling) which become quasi-metastable through resonance trapping. An alternative possible mechanism is by radiative cascade from cadmium levels more nearly in resonance with the neon ion ground state which are excited in a charge transfer process.

The majority of the cadmium ion laser transitions have been observed only in hollow cathode structures where the charge transfer process is enhanced relative to the Penning ionization reaction. This does not however prevent the Penning process continuing to occur and laser action at 4416 Å is still observed in such a structure (Schuebel, 1970c; Sugawara and Tokiwa, 1970; Sugawara *et al.*, 1970a, b). That charge transfer is the dominant excitation mechanism for the majority of the He–Cd ion laser transitions has been demonstrated by Webb *et al.* (1970) and Collins *et al.* (1971b). Webb *et al.* used a flowing afterglow experiment where the contribution to the excitation of various Cd^+ levels

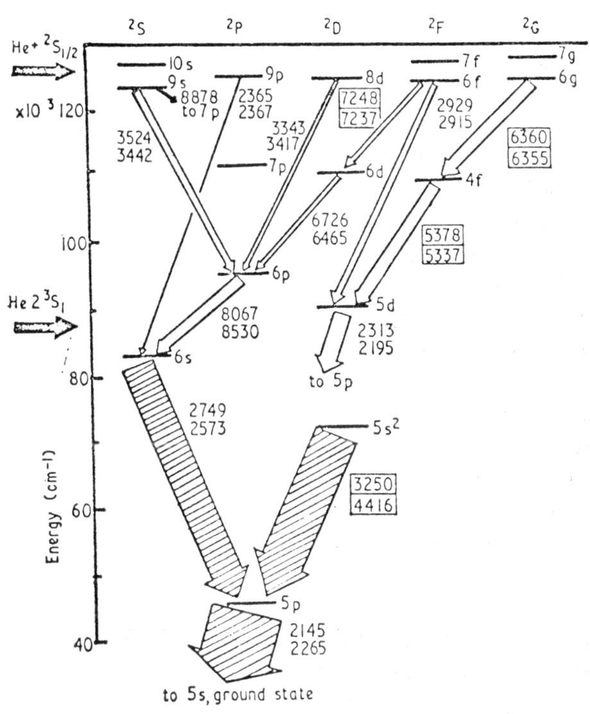

FIG. 20. Energy levels of the cadmium II ion. Transitions excited by charge exchange reactions are indicated by open arrows, transitions excited by Penning reactions are indicated by cross-hatched lines. Known laser transitions CW are boxed. (Webb *et al.*, 1970).

from both Penning ionization and charge transfer could be separated. They concluded that the upper levels of the cadmium ion laser transitions at 7237, 7284, 6360 and 6355 Å are populated directly by charge transfer processes from He^+ ions, whereas the transitions at 5378 and 5337 Å are populated by radiative cascade from higher levels populated directly by charge transfer, particularly the states $6g\,^2G_{7/2}$ and $6g\,^2G_{9/2}$ (Schuebel, 1970a). The upper levels of the 8067 and 8530 Å transitions are populated by a series of radiative cascades which originate in states directly excited by charge transfer.

An energy level diagram of the cadmium ion is shown in Fig. 20 where the various charge transfer, Penning ionization and radiative cascade processes are indicated. One outstanding reason for the efficiency of the He–Cd laser at 4416 Å, considering its short operating wavelength, is the lack of any branching from the upper level of the laser transition; a very high proportion of all the ions which reach the upper laser level can be made to contribute to the laser output. Goldsborough (1969) has obtained a volumetric output of 31 mW cm^{-3} at 4416 Å from a 2·4-mm bore tube [Hodges (1970) incorrectly quotes this volumetric output as 10 mW cm^{-3})].

A very favourable lifetime ratio exists for the two most important CW transitions at 4416 and 3250 Å. The lifetime of the lower level of 4416 Å, is 2·19 ns and of the lower level of 3250 Å, 5·07 ns (Baumann and Smith, 1970). The upper laser levels of the transitions at 4416 and 3250 Å are very long lived [in fact, the transitions from them are forbidden (Hyman, 1971)]. The lifetime of the upper level of the 4416 Å transition has been measured variously as 670 ns (Klein and Maydan, 1970), 830 ns (Geneux and Wanders-Vincenz, 1959), 773 ns (Schearer and Holton, 1970), 783 ns (Barrat and Barrat, 1963) and 994 ns (Schaefer, 1971), and of the 3250 Å transition 310 ns (Schaefer, 1971) and 260 ns (Barrat and Barrat, 1963). In an operating He–Cd laser these lifetimes are shortened by electron collisional deactivation which is one of the mechanisms by which the power output from such a laser saturates with increasing discharge current. The additional and most important mechanism by which saturation and quenching occurs is by electron collisional deactivation of the helium metastables which are required for the Penning ionization pumping process (Giallorenzi and Ahmed, 1971). This can be very efficient even for very low energy electrons.

(3) *Iodine*

Although most of the transitions observed in iodine ion lasers were originally reported using pulsed excitation (Fowles and Jensen, 1964a, b, 1965; Jensen and Fowles, 1964; Kovalchuk and Petrash, 1966; Willett and Heavens, 1966), many have now been observed in CW operation. A single transition at 6127 Å was reported by Collins *et al.* (1972) who used a positive column glow discharge in a CdI_2-He mixture. Subsequently, however, all CW transitions have been observed in hollow cathode lasers (Piper *et al.*, 1972; Piper, 1973).

The dominant excitation mechanism of most of the ionized iodine laser transitions observed in iodine–helium discharges appears to involve thermal energy charge transfer involving ground state helium atoms

$$\mathrm{I}(^2P_{3/2}) + \mathrm{He}^+(^2S_{1/2}) \rightarrow \mathrm{I}((^2D^0)\,6p) + \mathrm{He}\,(^1S_0) + \Delta E.$$

This excitation mechanism was originally suggested by Fowles and Jensen (1964a) and appears to be confirmed by the behaviour of ionized iodine transitions observed in spontaneous emission in a hollow cathode discharge by

Berezin (1969) who observed considerable enhancement of transitions coming from iodine levels with energies in near resonance with ground state helium ions. A further reason why CW iodine ion laser action is easier to obtain in hollow cathode, rather than positive column He–I_2 discharges may be due to the considerable difficulty which can be experienced in running positive columns in such mixtures. The electronegativity of iodine is very high, and in a positive column, unless the electron temperature is high, severe attachment of electrons occurs leading to discharge instability caused by the formation of striations and constrictions (Davis and King, 1972b).

The energy discrepancy ΔE for the charge transfer reaction in the production of the upper laser levels $(^2D^0)6p\,^3F_3$, $(^2D^0)6p\,^3D_2$, $(^2D^0)6p\,^3F_2$ and $(^2D^0)6p\,^3D_1$ is in the range 0·2–0·5 eV. However, this range of energy discrepancy is sufficiently large that some of the population may reach the above four levels following electron collisional transfer from levels such as $(^2D^0)5d\,^1F_3^0$, $(^2D^0)6p\,^1P_1$, $(^2P^0)5d\,^1D_2^0$ or $(^2P^0)5d\,^3D_3^0$.

The excitation mechanism for the three unclassified transitions observed by Kovalchuk and Petrash (1966) must be by electron impact as they were observed in pure iodine vapour only at low pressure (ca. 10^{-3} torr) and not at all when a buffer gas was added. This is consistent with their probable origin as highly ionized transitions which require energetic electrons to excite them.

The upper level of the transition at 5625·69 Å which has been observed in the pulsed mode only in He–I_2 and Ne–I_2 mixtures is accessible by charge transfer only in the case of helium ions. For its excitation in Ne–I_2 mixtures, some other process must be involved, most likely electron impact since the total pressure used when exciting this transition was low ($\sim$ 0·1 torr) as was also the case for the transition at 6905 Å when this transition was excited in Ne–I_2 or Kr–I_2 mixtures (Jensen and Fowles, 1964).

The CW power output from an iodine ion laser is no greater than 3–4 mW on the strongest lines and it does not appear to offer any advantage over the more powerful He–Cd and He–Se "metal vapour" lasers.

(4) *Lead*

Two CW lines have been observed in discharges in He–Pb mixtures: at 5609 and 6660 Å (Silfvast, 1971). The upper levels of both transitions are accessible by Penning ionization.

$$\mathrm{He}(^3S_1) + \mathrm{Pb}(^3P_0) \rightarrow \mathrm{He}(^1S_0) + \mathrm{Pb}^+(7p\,^2P) + \Delta E$$

with energy discrepancies of about 3 eV.

However it is not known to what extent the levels involved receive their excitation following radiative cascade or electron collisional transfer from higher levels. Charge transfer is unlikely to help in populating singly-ionized lead levels as ground state helium ions have sufficient energy to produce doubly-ionized lead ions. The transition at 5372·1 Å which has been observed only in a pulsed mode (Silfvast *et al.*, 1966) has an upper level which is also accessible by Penning ionization with an energy discrepancy of only 0·93 eV.

(5) *Magnesium*

Four CW lines have been observed by Hodges (1971) in He–Mg discharges using a cataphoresis type laser structure. The upper levels of these transitions are all accessible by Penning ionization from helium ^3S or ^1S metastables and the conditions under which they are observed—low vapour pressure of magnesium, a few torr of helium and current densities up to about 7 A cm^{-2}—are appropriate for this excitation mechanism to occur.

The two transitions observed by Cahuzac (1972) in the infrared could be excited only by high current pulses, the noble gas serving only as a buffer and to prevent metal vapour reaching the Brewster windows. Electron impact is most likely the dominant excitation mechanism in this case. As is the practice with pulsed excitation of nobel gas–metal vapour mixtures, Cahuzac maintained the magnesium vapour pressure by enclosing the active part of the discharge tube in an oven.

(6) *Mercury*

Mercury is of interest in being the first material to have exhibited gaseous ion laser action (Bell, 1964); it was also the first element where laser action in doubly-ionized gaseous ions was demonstrated (Gerritsen and Goedertier, 1964). In addition, many of the laser transitions in ionized noble gases appear to have been discovered accidentally in the first instance using discharges in mercury–noble gas mixtures (Convert *et al.*, 1964; Heard *et al.*, 1964).

The laser transitions in ionized mercury are generally observed using Hg–He discharges; however the role of the helium, apart from its obvious role as a carrier gas, in the excitation of the laser upper levels is not quite clear. Dyson (1965) originally suggested that laser action at 6150 Å was due to thermal energy charge transfer

$$\mathrm{He}^{+}\,(^{2}S_{1/2}) + \mathrm{Hg}\,(^{1}S_{0}) \rightarrow \mathrm{He}\,(^{1}S_{0}) + \mathrm{Hg}^{+}\,(7p\,^{2}P^{0}_{3/2}) + 0{\cdot}27\ \mathrm{eV}$$

whereas Suzuki (1965) suggested that a Penning ionization process involving metastable mercury atoms was involved.

$$\mathrm{He}\,(^{3}S_{1}) + \mathrm{Hg}\,(6p\,^{3}P^{0}_{0}) \rightarrow \mathrm{He}\,(^{1}S_{0}) + \mathrm{Hg}^{+}\,(7p\,^{2}P^{0}_{3/2}) + e + 0{\cdot}17\ \mathrm{eV}$$

or

$$\mathrm{He}\,(^{3}S_{1}) + \mathrm{Hg}\,(6p\,^{3}P^{0}_{2}) \rightarrow \mathrm{He}\,(^{1}S_{0}) + \mathrm{Hg}^{+}\,(7p\,^{2}P^{0}_{3/2}) + e + 0{\cdot}96\ \mathrm{eV}.$$

Willett (1970) has pointed out that the results of Dyson and Suzuki are not necessarily contradictory. Dyson investigated discharges in wide bore discharge tubes (2·3 cm i.d.) and Suzuki used narrow bore tubes (6 mm i.d.). The resultant difference in excitation conditions could account for the different apparent principal excitation mechanisms observed. It seems likely that both charge transfer and Penning ionization processes are responsible for excitation of the mercury $7p\,^2P^0_{3/2}$ and $7p\,^2P^0_{1/2}$ levels. That direct electron impact from the neutral mercury atom ground state is the dominant excitation mechanism for these states in Hg–He mixtures, as suggested by Goldsborough and Bloom (1969) is unlikely. Willett (1970) has given a number of reasons why electron impact excitation should be unimportant, for example:

(i) High operating pressure of helium in mercury–helium ion lasers (few torr—tens of torr).
(ii) Laser oscillation occurs at 6150 Å in the afterglow of a pulsed discharge at times longer than electron thermalization times and the decay times of radiative cascade processes into the upper level.
(iii) Preference for oscillation in wide-bore discharge tubes, unlike noble gas lasers in singly-ionized species where electron impact is the dominant excitation mechanism.

In addition CW oscillation at 6150 and 7945 Å has been obtained in a hollow cathode laser (Schuebel, 1971) and it is known that this type of structure favours the excitation of transitions where charge transfer and Penning ionization processes are the dominant excitation mechanisms (Schuebel, 1970a, b, c; Piper, 1972). However, under certain circumstances electron impact excitation may be important; Bennett (1969) has reported that Bell obtained CW oscillation at 6150 Å in pure mercury, by using a special form of discharge in a subsonic gas flow, indicative that electron impact may be the dominant excitation mechanism. Also the observation of pulsed laser action in low pressure Hg–Ar mixtures by Heard (1964) and Convert *et al.* (1964) is consistent with electron impact being the dominant excitation process, either by single electron collision on ground state mercury atoms or multiple collisions involving, for example, ground state mercury ions. The attainment of laser oscillations on the 6150 and 7945 Å transitions is aided, whatever the mechanism by which excitation reaches their upper level, by the expected very favourable lifetime ratio between the upper $7p$ levels and lower $7s$ level. The lower $7s$ level is expected to have a lifetime below 0·2 ns (Goldsborough and Bloom, 1969). The upper levels of the laser transitions in doubly-ionized mercury observed in low pressure ($<$ 0·5 torr) Hg–He mixtures by Gerritsen and Goedertier (1964) are inaccessible even by charge transfer involving mercury metastable $6p\,^3P^0_2$ atoms although this process can produce ground state Hg^{++} ions. These doubly-ionized transitions are almost certainly excited by electron impact.

The upper levels of the classified singly-ionized laser transitions observed only in pulsed Hg–He discharges (see spectroscopy, Table I) are all accessible by charge transfer between He^{+} $(1S_0)$ and Hg $(6p\,^3P^0_{0,2})$. Some of their upper levels, with the exception of $5g\,^2G_{7/2}$, $8p\,^2P^0_{3/2}$ and $10s^2\,S_{1/2}$ are also accessible by Penning ionization involving mercury metastables, for example

$$\mathrm{He}(^1S_1) + \mathrm{Hg}(6p\,^3P^0_2) \rightarrow \mathrm{He}(^1S_0) + \mathrm{Hg}(5f\,^2F^0_{7/2}) + 0{\cdot}37\ \mathrm{eV}.$$

However, it is more likely that electron impact is important in exciting this group of lines, as they are observed in relatively low pressure Hg–He mixtures ($<$ 1·2 torr) (Bloom *et al.*, 1964). Furthermore $He\,^3S_1$ metastables which are likely to be present in much larger densities than the shorter lived $^1S_1$ metastable only have sufficient energy by this reaction to excite the upper level of the transition at 7418 Å. However, Goldsborough and Bloom suggest that this level receives its excitation by a two-step electron collision process.

(7) *Phosphorus*

All the laser transitions observed in phosphorus were originally obtained using high current pulsed excitation of PF_5 vapour (Cheo and Cooper, 1965) or phosphorus vapour with a helium or neon buffer gas (Fowles *et al.*, 1965). The low pressure of PF_5 vapour ($<$ 120 mtorr) and buffer gases (10–1000 m-torr) used suggests that electron impact is the dominant excitation mechanism for all these lines. The behaviour of the 6043 Å transition under pulsed excitation is similar to that observed for many transitions in pulsed noble gas ion lasers, increase of the discharge pressure quenches laser action whilst the total spontaneous emission from the upper level is still increasing with pressure. This is consistent with increasing electron impact excitation of the lower 4*s* level as the electron temperature falls and if resonance trapping of the lower laser level occurs. The lower 4*s* levels of all the singly-ionized phosphorus laser lines are normally very short lived as they have strongly allowed vacuum U.V. transitions to one or more components of the split P II ground state. The lifetimes of the upper levels are not known and only an upper limit can be set to some of them from the data of Miller *et al.* (1971); for example, the lifetime of the P II state $4p\,^3D_1$ is below 3·4 ns, $4p\,^3D_2$ is below 4·5 ns and $4p\,^3D_3$ is below 6·2 ns. The 6043 Å transition is interesting in that it has a very low threshold ($<$ 32 A cm^{-2}) in pulsed excitation (Cheo and Cooper, 1965). CW operation on this and two other transitions has been observed (Bloom *et al.*, 1968) and work on phosphorus lasers is continuing. The upper levels of all these transitions observed in CW operation are accessible by charge transfer from ground state helium ions and this process may play a dominant role in their excitation in He–P discharges.

(8) *Selenium*

Laser action in selenium ions was first observed using pulsed excitation of selenium-noble gas mixtures at low pressures ($\sim$ 100 mtorr) (Bell *et al.*, 1965) where electron impact excitation was probably the dominant mechanism involved in populating the upper laser levels. Subsequently all the transitions in selenium ions have been observed in CW operation using D.C. excitation of He–Se mixtures in cataphoresis type positive column discharges (Silfvast and Klein, 1970; Klein and Silfvast, 1971; Hernqvist and Pultorak, 1972). Charge transfer from ground state helium ions appears to be the dominant excitation mechanism for all the lines as the upper laser levels are inaccessible by Penning ionization and the gas pressures involved ($\sim$ 7 torr He, 5×10^{-3} torr Se) are too high for the discharge electron temperature to be high enough for significant electron impact excitation to occur. The group of 13 $5p$ levels in Se II from which all the laser transitions originate lie from approximately 0·2 eV above to 0·8 eV below the He^+ ground state. However, no preferential excitation of those Se II levels in closed resonance with He^+ is observed. Klein and Silfvast (1971) conclude that mixing collisions with electrons or He atoms is important in distributing the initial charge transfer excitation among the $5p$ levels.

The saturation of laser output power which occurs above 400 mA in a 3 mm bore tube (Klein and Silfvast, 1971) is most likely due to a build up in the lower laser level populations caused by their excitation by electrons or resonance trapping of their draining radiation to the selenium ion ground state.

The very large number of laser transitions in singly-ionized selenium, at present 46, which span the visible and near infrared spectrum and the relatively high output powers and comparatively simple technology of this laser makes it a competitor for noble gas ion lasers for low power ($<$ 100 mW) applications.

(9) *Tin*

The laser transitions in singly-ionized tin were first observed using pulsed discharge excitation of tin–helium or tin–neon mixtures (Silfvast *et al.*, 1966). Vapour pressures of tin in the range $10^{-5} - 2 \times 10^{-4}$ torr were maintained by enclosing the active region of the discharge tube in an oven; carrier gas pressures were in the range 1–2 torr. The exact excitation mechanism under these circumstances is unclear, all the laser upper levels are accessible by charge transfer processes with He^+ or Ne^+ or Penning ionization with helium metastables. The $6p$ levels are also accessible by Penning ionization with neon metastables, for example

$$\mathrm{Sn}\,(^3P_0) + \mathrm{Ne}\,(3s[\tfrac{3}{2}]_2^0) \rightarrow \mathrm{Sn}^+\,(6p\,^2P^0_{3/2}) + \mathrm{Ne}\,(^1S_0) + e + 0{\cdot}3\ \mathrm{eV}.$$

However, at the above carrier gas pressures electron impact excitation of the upper laser levels is probably important

Silfvast (1969) has subsequently observed CW oscillation on the two transitions from $6p$ levels in a He–Sn discharge. He believes that Penning ionization is the dominant excitation mechanism in this case because: (i)

efficient laser action only occurs over a small range of partial pressures of the metal vapour (~ 1·5 mtorr); and (ii) the optimum helium pressure (~ 8 torr) corresponds well with the expected pressure at which maximum helium triplet metastable densities are expected to be reached in a 4-mm bore discharge tube (8–9 torr). The importance of charge transfer processes is probably reduced by the large positive energy defects involved. For example,

$$He^+ (^2S_{1/2}) + Sn (^3P_0) \rightarrow Sn^+ (6p\,^2P^0_{3/2}) + He (^1S_0) + 8{\cdot}3 \text{ eV}.$$

Furthermore, the observations of Silfvast (1969) regarding the importance of Penning ionization in exciting CW laser oscillation in cadmium, tin and zinc have been confirmed independently in the cases of cadmium and zinc [see subsection (2) above and subsection (11) following].

(10) *Tellurium*

Pulsed laser action in tellurium–neon mixtures was first reported by Bell *et al.* (1966) using ring discharge excitation. The low pressures used (0·2 torr neon, 1–2 mtorr tellurium) suggest that electron impact was the dominant excitation mechanism under these conditions. The same excitation mechanism also accounts for the four transitions observed by Webb (1968) which exhibited prompt laser action under low pressure conditions (0·1–0·25 torr). For the rest of the tellurium ion laser transitions observed in pulsed afterglow emission (Webb, 1968) and in CW operation using a cataphoresis type positive column laser (Watanabe *et al.*, 1972; Silfvast and Klein, 1972) the dominant excitation mechanism appears to be charge transfer either from helium ions for those transitions which operate in He–Te discharges or from neon ions for those transitions which operate in Ne–Te discharges. Only five transitions will oscillate CW in both kinds of discharge. Figure 21 shows an energy level diagram for tellurium including metastable and ion ground state levels in helium and neon; the even parity lower laser levels may be populated themselves by charge transfer processes but their populations remain small because they have strong draining transitions to one or more of the tellurium ion ground state levels.

Figure 22 shows a more detailed diagram of the tellurium laser levels showing specific upper laser levels when helium or neon buffer gases are used. For the He–Te laser transitions only one of the 9 upper laser levels ($122'^0_{5/2}$) is energetically near He^+ (0·34 eV below), the others lie in a range 1·5 to 3 eV below He^+. Silfvast and Klein (1972) believe that these levels receive a substantial part of their excitation by radiative cascades from levels excited more nearly in resonance by change transfer. However, electron collisional transfer processes from such levels to the upper laser levels may also occur. A number of the upper laser levels of the transitions observed in Ne–Te mixtures are above the

Ne^+ ground state. Silfvast and Klein believe that charge transfer to those levels only a little above Ne^+ (< 0·54 eV) still occurs, the endothermic energy defect of the reaction being made up by the kinetic energy of the reacting Ne^+

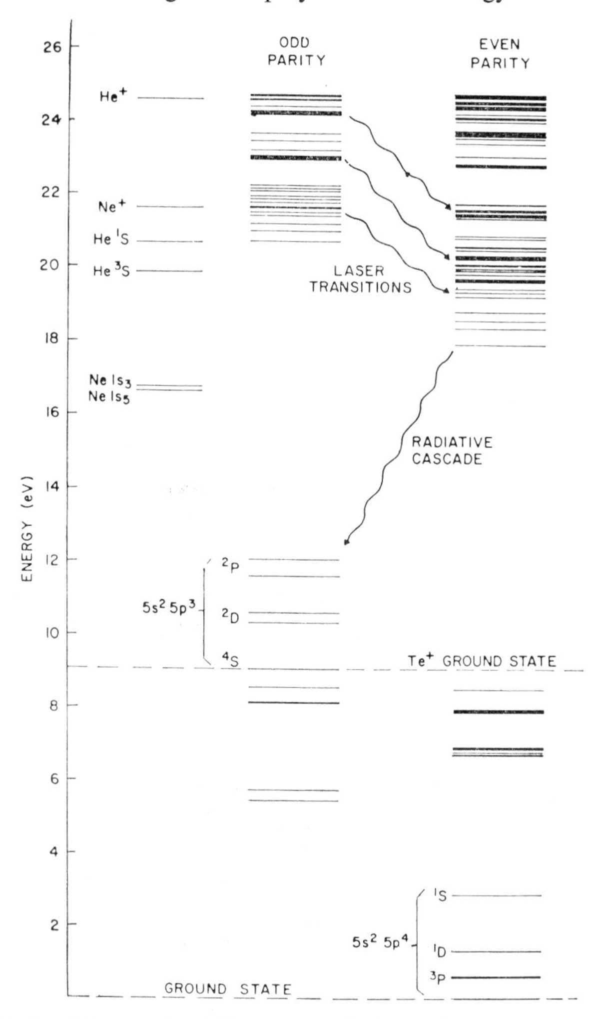

FIG. 21. Partial energy level diagram for tellurium (Silfvast and Klein, 1972).

and Te atoms. However, two of the upper laser levels, $112^0_{3/2}$ and $111^0_{5/2}$ of transitions observed in Ne–Te mixtures are a little too high for the direct charge transfer process to be efficient. For these levels, neutral tellurium metastables are probably involved in reactions of the sort

$$Ne^+ (^2P^0_{3/2}) + Te\,(5p^4\,{}^1D_2) \rightarrow Ne\,(^1S_0) + Te^+\,(112^0_{3/2}) - 0{\cdot}056 \text{ eV}.$$

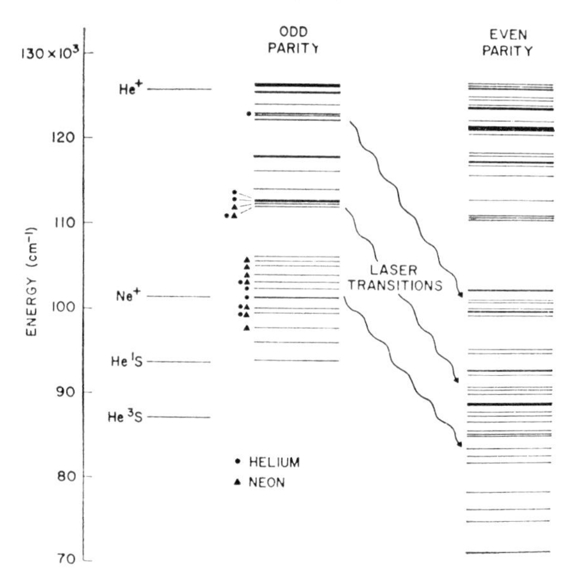

FIG. 22. Detailed diagram of tellurium laser levels. Specific upper laser levels with He and Ne buffer gases are indicated (Silfvast and Klein, 1972).

(11) *Zinc*

Although the zinc ion laser can produce CW power outputs of several milliwatts on its two strongest transitions at 5894 and 7479 Å, its relatively long operating wavelengths and its lower efficiency compared to the helium–neon laser limit its usefulness.

The excitation mechanisms responsible for its operation are similar to those observed in cadmium–helium lasers. For those zinc ion upper laser levels above the energy of helium $^3S_1$ metastables, the excitation mechanism is primarily by charge transfer from ground state helium ions. This same mechanism is probably also dominant in exciting those zinc ion laser transitions observed in pulsed He–Zn discharges (1–2 torr He, 10^{-3}–1 torr Zn) (Fowles and Silfvast, 1965; Silfvast *et al.*, 1966). This excitation mechanism was originally proposed by Jensen and Bennett (1968) as a means for obtaining CW laser oscillation in zinc–helium discharges. The dominance of charge transfer processes in populating the higher lying zinc ion upper laser levels has been demonstrated by fluorescence decay measurements in pulsed afterglow experiments (Jensen *et al.*, 1969; Riseberg and Shearer, 1971) and in flowing afterglow experiments (Webb *et al.*, 1970; Green *et al.*, 1973). The importance of charge transfer processes in populating these higher levels is further demonstrated by their enhancement in hollow cathode zinc-ion lasers (Schuebel, 1970b; Sugawara and Tokiwa, 1970; Sugawara *et al.*, 1970a, b; Jensen *et al.*, 1971; Collins *et al.*, 1971a). Green *et al.* (1973) have shown that direct charge transfer results primarily in the excitation of the $6p\,^2P$ and $5d\,^2D$ levels of zinc ion, which are the upper laser levels most nearly in resonance (see Fig. 23). The $4f\,^2F$ levels are excited by electron collisional de-excitation of $5d\,^2D$ levels, e.g.

$$He^+(^2S_{1/2}) + Zn(^1S_0) \rightarrow He(^1S_0) + Zn^+(5d\,^2D_{3/2}) + 0{\cdot}57 \text{ eV}.$$
$$Zn^+(5d\,^2D_{3/2}) + e \rightarrow Zn^+(4f\,^2F^0_{5/2}) + e + 0{\cdot}09 \text{ eV}.$$

Although the $4f\,^2F$ levels are in quite close resonance with the helium ion ground state the cross-section for their direct excitation is in fact negligibly

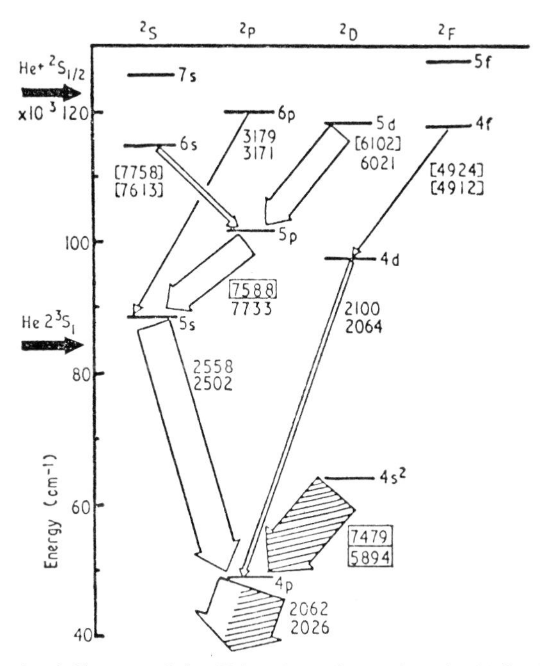

FIG. 23. Energy level diagram of zinc II ion. Notation as for Fig. 20 (Webb *et al.*, 1970). Known pulsed laser transitions are bracketed.

small. The $6s\,^2S_{1/2}$ level which is 1 eV below He^+ receives its excitation primarily by radiative cascade from the $6p\,^2P$ levels.

For those zinc ion upper laser levels below the helium $^3S_1$ metastable level the dominant excitation is by Penning ionization (Webb *et al.*, 1970; Collins *et al.*, 1971; Riseberg and Schearer, 1971). It was originally suggested (Jensen *et al.*, 1969) that He_2^+ molecular ions were responsible for the excitation of the $4s^2\,^2D$ levels. Although these molecular ions have sufficient energy for this reaction to occur they are not in fact responsible for its excitation (Riseberg and Schearer, 1971).

As in the cadmium ion laser, the attainment of population inversion on the zinc ion laser transitions is assisted by the very favourable lifetime ratios between the upper and lower laser levels. For example, the two strong laser transitions from $4s^2\,{}^2D$ levels are strictly forbidden (Hyman, 1971) and have long lifetimes, the measured values are for $4s^2\,{}^2D_{3/2}$, 2·22 μs (Schearer and Holton, 1970), 0·465 μs (Geneux and Wanders-Vincenz, 1959) and for $4s^2\,{}^2D_{5/2}$, 1·61 μs (Schearer and Holton, 1970).

VI. Miscellaneous Ion Lasers

This category includes all the ion lasers other than the noble gases and the CW metal vapour lasers covered in previous sections: once again they will be covered alphabetically.

(1) *Antimony*

A single ion laser transition has been observed by Bell *et al.* (1966) using pulsed ring discharge excitation (Bell, 1965) in a Ne–Sb mixture. Although the Sb^+ upper level involved is accessible by charge transfer, the low neon pressure used (0·2 torr) suggests that electron impact is in fact the dominant excitation mechanism.

(2) *Barium*

As well as a number of neutral barium laser transitions Cahuzac (1970b, 1971) has observed 2 transitions in singly-ionized barium, using fast-risetime high current (up to 250 Å in 5–10 mm bore tubes) pulsed D.C. excitation. The vapour pressure of barium was controlled by placing the discharge tube inside an oven and varied from 2.10^{-4} torr (500°C) to 0·8 torr (850°C). Laser action was observed using a number of different carrier gases at pressures up to a few torr. The dominant excitation of the levels must be by electron impact. The upper laser levels are optically connected to the ground state. However, the lower laser levels are not metastable, their lifetimes are < 0·67 ns ($6d\,{}^2D_{3/2}$) and < 0·56 ns ($6d\,{}^2D_{5/2}$) (Corliss and Bozman, 1962), consequently this laser is not intrinsically of the self-terminating variety.

(3) *Bismuth*

Four ion laser transitions have been reported by Keidan and Mikhalevskii (1968) using pulsed excitation of bismuth (0·05–0·5 torr) in helium or neon carrier gases (0·01–0·1 torr). The laser oscillation does not depend on the nature of the carrier gas and electron impact excitation must be dominant.

(4) *Boron*

A single transition has been observed using a pulsed D.C. discharge in BCl_3 at a pressure in the range 10–50 mtorr. (Cooper and Cheo, 1966). At this low pressure electron impact excitation should be dominant.

(5) *Bromine*

The dominant excitation mechanism for all the transitions observed in pulsed operation in low pressure pure bromine vapour (0·04 torr) (Keefe and Graham, 1965, 1966) or low pressure HBr (10–50 mtorr) (Cooper and Cheo, 1966) must be electron impact. The transitions themselves are isoelectronc with the similar laser transitions observed in the operation of doubly-ionized krypton (see Table I). Two of the transitions have also been observed in Br_2–Ne mixtures (100 mtorr Ne); however, their upper levels are not accessible by charge transfer from Ne^+ ground state ions and electron impact excitation still dominates. CW operation of 3 of the bromine ion laser transitions has also been achieved using RF ring discharge excitation (Bloom, 1968).

(6) *Calcium*

Two transitions of the self-terminating type (see earlier section) have been observed by Walter *et al.*, using high current fast-risetime pulsed D.C. excitation of Ca–He mixtures (1–3 torr He). The upper laser levels are sufficiently low lying ($\sim$ 9 eV) that electron impact excitation is still efficient even in the presence of a few torr of carrier gas.

(7) *Carbon*

Some of the early observations of laser action which were assigned to transitions in ionized carbon did not occur in that element. Notably those laser transitions where the carbon in the discharge was assumed to come from material sputtered from the cathode. Recently (see Waynant, 1973), two self-terminating vacuum ultraviolet transitions in carbon IV have been observed which were excited in a travelling wave low pressure discharge similar to that used to excite vacuum ultraviolet laser action in molecular hydrogen (Waynant, 1972).

(8) *Chlorine*

If it were not for the existence of the argon ion laser the chlorine ion laser might have achieved great commercial importance. Some of the Cl II laser transitions observed in CW operation using R.F. ring discharge excitation (Bell *et al.*, 1965; Goldsborough *et al.*, 1966; Zarowin, 1966) have lower threshold excitation current densities than argon ion lasers at 4880 Å and give output powers of several hundred milliwatts (mostly at 5218 and 5392 Å) with not significantly lower efficiency than in argon ion lasers, and possibly with greater efficiency than krypton ion lasers. However, because the range of available wavelengths from CW chlorine ion lasers largely duplicates those of argon ion lasers and because the chemical reactivity of chlorine limits its suitability for use in arc-discharge excitation systems (except using R.F. excitation) it has not developed as a competitor for the argon ion laser.

The dominant excitation mechanism for all the transitions in chlorine ion lasers is electron impact and a discussion of the various one-, two- and multi-step processes which can occur would be very similar to that previously given in the section on excitation mechanisms in noble gas ion lasers. All the transitions observed in Cl II and Cl III are of the $p \rightarrow s$ type frequently observed in ion lasers, the transitions themselves are often isoelectronic with laser transitions observed in Ar III and Ar IV.

(9) *Fluorine*

Electron impact excitation must be dominant in the production of the fluorine ion laser lines as they were excited in PF_5 or SF_6 vapours at low pressures ($\leqslant$ 20 mtorr) and only oscillated for short times during the rise of the excitation current pulse (Cheo and Cooper, 1965, 1966). This may indicate dominant excitation from the ground state of the neutral fluorine atom or from the ground state of the fluorine ion with one less charge than the lasing ion transition.

(10) *Germanium*

The role of the helium or neon used as a carrier gas in exciting germanium ion laser transitions is uncertain. The upper levels of the two observed laser transitions (Silfvast *et al.*, 1966) are accessible by charge transfer from both helium or neon and by Penning ionization from helium; electron impact excitation may be important also.

(11) *Indium*

Similar considerations apply to the excitation of ion laser action in indium as to germanium (above). However, in this case the upper laser level of the single observed laser transition (Silfvast *et al.*, 1966) is accessible only by charge transfer from helium or neon ions.

(12) *Nitrogen*

Electron impact must be the dominant excitation mechanism of the nitrogen ion laser transitions; but whether the excitation occurs via processes such as

$$N_2 + e \rightarrow N^{+*} + N^* + 2e$$

or

$$\begin{aligned} N_2 + e &\rightarrow 2N + e \\ N + e &\rightarrow N^+ + 2e \end{aligned}$$

or

$$\begin{aligned} N + e &\rightarrow N^+ + 2e \\ N^+ + e &\rightarrow N^{+*} + e \end{aligned}$$

is unclear.

Gadetksii *et al.* (1971) have observed laser oscillation on the transition at 5016 Å under unusual conditions. They use a preparatory pulsed discharge which they propose produces a high concentration of nitrogen metastables. Laser oscillation then occurs in succeeding excitation pulses but with an amplitude that decays exponentially with time as the metastables disappear. They suggest that the nitrogen metastables are formed by dissociative recombination after the preparatory discharge.

$$N_2 + e \rightarrow N_2^+ + 2e$$
$$N_2^+ + e \rightarrow N(^2P) + N(^2D).$$

The metastables then present a large cross-section for electron impact excitation of the upper laser level.

(13) *Oxygen*

Similar remarks apply to the excitation of ion laser action in oxygen as apply in the case of nitrogen: Gadetskii *et al.* (1971) report laser action on a number of transitions where they suggest an excitation mechanism similar to the above.

(14) *Silicon*

The silicon ion laser transitions were originally observed in pulsed discharges in PF_5 or SF_6 vapour (Cheo and Cooper, 1965b). Free fluorine produced by the discharge must have been responsible for the removal of silicon from the walls of the quartz discharge tubes used. One transition at 4089 Å was observed in a very high current pulsed discharge in argon when sputtered silicon from the quartz discharge tube walls must have been produced (Bridges and Chester, 1965a). Most of the transitions have also been observed in pulsed operation in SiCl vapour either with or without added helium or neon (Carr and Grow, 1967). The dominant excitation mechanism for all the transitions must be electron impact. The lower level of the transition at 6672 Å is metastable and the laser transition at this wavelength is consequently self-terminating.

(15) *Strontium*

The two laser transitions observed in singly-ionized strontium (Deech and Sanders, 1968) are analogous to those observed in calcium (Walter *et al.*, 1966). They are self-terminating transitions and in fast-risetime pulsed excitation appear later in the pulse than self-terminating neutral lasers; this suggests that their dominant excitation may be by a 2-step electron impact process involving ion ground states (Walter *et al.*, 1966).

(16) *Sulphur*

Pulsed ion laser action in sulphur has been reported in SO_2, SF_6 and H_2S discharges (Cooper and Cheo, 1966) and in discharges in sulphur plus any one of a number of carrier gases [helium, neon, argon, oxygen, hydrogen, nitrogen or air (Fowles *et al.*, 1965)]. The observation of laser action does not appear to be dependent on the actual discharge mixture and the dominant excitation mechanism responsible for the sulphur ion laser transitions must be electron impact. The most likely mechanism in the case of discharges in sulphur compounds is preliminary dissociation into sulphur atoms or ions by electrons followed by a second electron impact excitation to the upper laser level. Cooper and Cheo (1966) have studied the behaviour of the two strong transitions at 5454 and 5640 Å and have observed similar behaviour to that frequently observed in pulsed noble gas ion lasers, namely, double pulse effects and quenching of laser action with increasing discharge current and at lower currents at higher pressures. They interpret their results in terms of resonance trapping of the lower laser levels, but as discussed previously, an alternative quenching mechanism involving electron impact excitation of the lower laser levels may also contribute.

(17) *Ytterbium*

Two ion laser transitions have been observed (as well as a number of neutral transitions) by Cahuzac (1968, 1970, 1971) and Cahuzac and Brochard (1969). High current pulsed discharge excitation of ytterbium vapour (10^{-4}–0·5 torr, optimum about 0·1 torr) in various carrier gases (0·5–3 torr) was used. The metal vapour pressure was maintained by enclosing the discharge tube in an oven.

Further investigation of the rarer elements in this way may be useful as, at the least, useful spectroscopic information can be obtained.

VII. Plasma Parameters in Ion Lasers

A. NOBLE GAS ION LASERS

The active medium of a noble gas ion laser is a pulsed or CW low pressure plasma with a moderately high electron density ($\lesssim 10^{13}$ cm^{-3}) and electron temperature ($\lesssim 2$ eV), a particle density in the approximate range $3 . 10^{13}$–$3 . 10^{16}$ cm^{-3} (1 mtorr–1 torr equivalent pressure at 273°K) and particle temperatures from ambient (neutral particles in some pulsed ion lasers) to several thousand °K (ions in CW ion lasers). The value of these various quantities in a particular laser help to govern its observed characteristics. However, for a full description of the plasma of an ion laser other phenomena must be considered, for example the spatial distribution of neutral and charged particles within the laser, and their longitudinal and radial drift velocities.

The various parameters of an ion laser plasma control the observed characteristics of the laser in different ways. The particle densities and electron temperature (or their energy distribution, if this is not Maxwellian) determine the rate of excitation of the various excited levels of the laser and the spatial distribution of particles within the laser affects its output transverse mode characteristics. The relative importance of the different mechanisms by which the lineshape of transitions in the laser is determined and their centre frequencies shifted are controlled by the temperatures, densities and drift velocities of the various particles in the plasma, as well as by the spontaneous emission probabilities of these various transitions.

In order to demonstrate the way in which the parameters of an ion laser's plasma affect its characteristics, these various parameters and other phenomena will be discussed separately with particular reference to the argon ion laser where the measured values of these parameters give a fairly comprehensive, but not quite complete, picture of the various processes governing its operation. Very little plasma diagnostic information is available for ion lasers other than argon, however, most of the discussion which follows should also apply to other low pressure arc discharge plasmas used to excite laser action where electron impact is the dominant excitation mechanism.

(1) *Electron Temperatures*

The electron energy distribution of the electrons in most ion lasers approximates to Maxwellian since inter-electron collisions occur at a much greater rate than other processes which perturb the electron energy distribution. [Zarowin (1969) estimates that inter-electron collisions occur at a rate $\sim 10^{24}$ cm^{-3} s^{-1} at an electron density $\sim 10^{14}$ cm^{-3} whilst the electrons undergo inelastic collisions with other particles at a rate of 10^{20} cm^{-3} s^{-1}].

Electron temperatures in argon ion lasers have been measured or calculated by a number of authors whose results are collected together for comparison in Table VI. The difficulty involved in making reliable measurements of the electron temperature reflects itself in the fairly wide range of values which have been measured under approximately identical conditions.

The measurements of Miller and Webb (1971) were made by a double probe method in a narrow bore quartz discharge tube. Difficulties were caused by the probe becoming incandescent and by the fact that a perturbation of the discharge plasma was caused by the probe. The results are also at variance with those of Kitaeva *et al.* (1966b, 1968a) in that no increase of electron temperature with current density was observed. Although the absolute values of electron temperature measured by Kitaeva may be inaccurate, they should reliably indicate an upward trend in electron temperature with current density. The values of T_e are obtained by Kitaeva using a formula derived by Kagan and Perel (1958a, b) which relates the neutral atom temperature T_a (which can be reliably measured from the Doppler width of neutral transitions in spontaneous emission) and transverse ion temperature T_t (measured from the Doppler width observed transversely to the discharge tube axis) by the relation

$$T_t = 0{\cdot}56\, T_a + 0{\cdot}13\, T_e \qquad \ldots(7.1)$$

TABLE VI. *Electron temperatures in argon ion lasers*

Current density (A cm^{-2})	Tube bore (mm)	Pressure (torr)	kT_e (eV)	Method of measurement	Author(s)
150	2·8	0·37	4·3	Deduced from linewidth measurements using Kagan-Perel theory	Kitaeva *et al.* (1966)
250	2·8	0·37	5·9		
350	2·8	0·37	7·8		
150	2·8	0·62	2·8		
250	2·8	0·62	4·6		
350	2·8	0·62	6·6		
[a] (pulsed)		10^{-5}–10^{-2}	90	Deduced from relative line intensities	Tkach *et al.* (1967)
700 (pulsed)	6	0·037	4·5	Double probe	Hattori and Goto (1969a)
1400 (pulsed)	6	0·037	5·3		
2100 (pulsed)	6	0·037	6·3		
700 (pulsed)	6	0·015	7·0		
1400 (pulsed)	6	0·015	8·1		
1950 (pulsed)	6	0·015	10		
250 (pulsed)	10	0·037	3·6		
500 (pulsed)	10	0·037	4·1		
750 (pulsed)	10	0·037	5·8		
250 (pulsed)	10	0·015	4·7		
500 (pulsed)	10	0·015	6·0		
750 (pulsed)	10	0·015	7·8		

25			1·0	Theory	Zarowin (1969)
120			3·0		
250			4·7		
500			7·5		
100[b]	2·05	0·8	2·5	Probe	Burmakin and Evtyunin (1969)
150[b]	2·05	0·8	3·1		
200[b]	2·05	0·8	3·5		
100[b]	2·05	0·6	2·6		
150[b]	2·05	0·6	3·3		
200[b]	2·05	0·6	3·8		
100[b]	2·05	0·2	4·1		
150[b]	2·05	0·2	4·6		
200[b]	2·05	0·2	5·0		
400 (pulsed)	4	0·050	1·7–3·9[c]	Deduced from linewidth measurements using Kagan-Perel theory	Klein (1970)
800 (pulsed)	4	0·050	4·3–6·6[c]		
400 (pulsed)	4	0·015	3·4–5·5[c]		
800 (pulsed)	4	0·015	5·9–8·2[c]		
50	2	1	2·3	Double probe	Miller and Webb (1971)
100	2	1	2		
32–190	2	3	3·5	Deduced from Tonks–Dattner resonances	Prosnitz and George (1971)
32–190	2	3·5	2·7		
32–190	2	4·1	2·6		
32–190	2	4·7	2·2		

TABLE VI—*continued*

Current density (A cm^{-2})	Tube bore (mm)	Pressure (torr)	kT_e (eV)	Method of measurement	Author(s)
280	1·5	0·14	8·8	Calculated from measured axial field	Fridrikhov and Fotiadi (1971)
280	1·5	0·4	5·9		
280	1·5	0·66	5·2		
113	5	0·23	3·9	Deduced from linewidth measurements using Kagan–Perel theory	Kitaeva *et al.* (1971)
163	5	0·23	5·0		
113	5	0·42	3·1		
163	5	0·42	4·1		
242	5	0·42	5·3		
113	5	0·68	2·3		
163	5	0·68	3·1		
242	5	0·68	4·1		
113	7	0·23	4·0	Deduced from linewidth measurements using Kagan–Perel theory	Kitaeva *et al.* (1971)
163	7	0·23	5·1		
113	7	0·42	2·2		
163	7	0·42	3·4		
113	7	0·68	1·3		
163	7	0·68	2·7		
20 A[d]		$1·7.10^3$	3	Deduced from relative line intensities	Van der Sijde (1972a)
40 A[d]		$1·7.10^3$	3·6		
60 A[d]		$1·7–10^3$	3·7		
80 A[d]		$1·7.10^3$	3·8		

114	7	0·23	3·45	Microwave (radio brightness temperature	Kitaeva *et al.* (1972)
114	7	0·6	2·8		
164	7	0·23	4·3		
164	7	0·6	3·8		
114	7	0·23	4·2	Deduced from axial field using Herziger–Seelig (1968) calculation	Kitaeva *et al.* (1972)
114	7	0·6	3·5		
164	7	0·23	4·4		
164	7	0·6	3·6		
35[e]	6	0·97	2·6	Double probe	Maitland and Cornish (1972)
105[e]	6	0·97	2·7		
140[e]	6	0·54	2·15		
210[e]	6	0·54	2·5		
280[e]	6	0·54	2·4		
64	2	0·2	1·1	Deduced from linewidth measurements using modified Kagan–Perel theory	Sze and Bennett (1972)

[a] E-beam discharge, current 10–30 A injected into magnetic field in tube 20–80 mm dia.
[b] In segmented metal tube.
[c] Electron temperature change duringe 50 μs pulse.
[d] In magnetically confined hollow cathode discharge.
[e] In segmented metal tube.

Absolute values of T_e calculated by this method will be in error because of deficiencies in the theory of Kagan and Perel, and because the discharge plasma does not fulfil exactly the conditions required for this theory to be applicable. Closer attention will be given to certain aspects of Kagan–Perel theory in Sections VIII, E and IX, D.

The values of T_e obtained by Kitaeva *et al.* (1972) using a microwave method, where the radio brightness temperature and reflection coefficient of an argon plasma tube were determined, probably reflect accurately the values of T_e to be expected in low current density conventional argon ion lasers using quartz discharge capillaries. Kitaeva also obtains good agreement between his measured values and values calculated using the measured axial field strengths and formulae of Herziger and Seelig (1968). The close agreement may be slightly fortuitous as Herziger and Seelig's formulae were devised using a number of approximate formulae for cross-sections, and make the assumption that the laser plasma can be described by the theory of Tonks and Langmuir (1929) (see also Hernqvist and Fendley, 1967). The measurements of Burmakin and Evtyunin (1969) and Maitland and Cornish (1972) were made in discharge tubes of the segmented type. The method of measurement used by the former authors involved the use of the metal segments of the tube itself as probes. Further work is clearly necessary to explain the fairly large difference in values of T_e and its behaviour with current observed by these two sets of authors.

Most of the measurements of electron temperature in pulsed ion lasers have been made in what can be called the quasi-CW mode, with pulse lengths of several microseconds. The values obtained are consistent with those measured in CW lasers bearing in mind the increased current density obtainable in the pulsed mode. Both Hattori and Goto (1969a) and Klein (1970) have made time resolved measurements of electron density in pulsed argon ion lasers; in the measurements of the former authors the electron temperature followed the current fairly well whilst Klein observed a gradual increase in electron temperature during a constant current 50 μs pulse. However, these slight differences in observed behaviour are attributable to the differences in the methods of excitation and measurement used. No electron temperature measurements have been made in the noble gas ion lasers except argon. However, the predictions of plasma theory (see Section VIII, A on Tonks–Langmuir theory) would suggest that the electron temperatures in these lasers should be in the same sequence as their ionization potentials, that is:

$$T_e\,(\text{neon}) > T_e\,(\text{argon}) > T_e\,(\text{xenon}) > T_e\,(\text{krypton})$$

This tendency may be offset to some extent by differences in optimum filling pressure between these lasers as an increased filling pressure reduces the electron temperature.

(2) *Charged Particle Densities*

To maintain plasma quasi-neutrality in the discharge plasma of an ion laser the charge number densities of electrons and ions must balance, that is

$$N_e(r) = N_i(r) + 2N_{i2}(r) + 3N_{i3}(r) + \cdots + jN_{ij}(r) + \cdots \qquad \ldots(7.2)$$

where $N_i(r)$ is the density of singly charged ions at position r within the plasma and $N_{ij}(r)$ is the density of multiply charged ions of charge j at position r. In typical low current density ion lasers the degree of multiple ionization is very small and $N_e(r) = N_i(r)$. The only place where this will not hold is in the space charge sheath near the walls.

Most quantitative measurements of electron density in ion lasers have been made on low current density argon ion lasers and the results of various investigations are summarized in Table IV. In most of these measurements the spatial distribution of electrons was not taken into account so the measured quantities represent an average over the charged particle density across the tube bore. However, in the recent measurements of Pleasance and George (1971) this was allowed for and the quoted electron density is an average over a Schottky (1924) type electron density profile.

The results obtained by Pleasance and George (1971) are shown in Fig. 24. These measurements should be quite reliable, particularly at the higher pressures used and are in good agreement with the slightly different microwave measurements of Kitaeva *et al.* (1972), the double probe measurements of Miller and Webb (1971) and the values obtained from measurements of changes in plasma refractive index using laser interferometry (Zory and Lynch, 1971). The values of electron density obtained by adding a small quantity of hydrogen to the argon plasma and measuring Stark broadening of the H_α and H_β lines give values of N_e which are probably about a factor of 2 low since the addition of even small quantities of hydrogen affects the mobility of electrons in the discharge (Colli and Facchini, 1952). Measurement of the Stark broadening of neutral argon transitions should give a reliable estimate of the electron density as no perturbation of the discharge occurs in this case. However, the accuracy of such measurements depends on the availability of reliable Stark broadening coefficients. Values of N_e calculated by Sze and Bennett (1972) in this way using the Stark broadening coefficients of Griem (1969) appear to be too large because the calculated Stark broadening coefficients are too small.

It is also possible to estimate the electron density more indirectly from other measured parameters of the discharge such as the axial field and electron mobility or the discharge conductivity and electron temperature. These indirect measurements are only as accurate as the theory on which they are based; care must be taken not to use a theory which does not really apply to plasmas of the sort that occur in ion lasers.

Because of the lack of plasma diagnostic information on noble gas ion laser systems other than argon it is difficult to predict typical electron densities to be found in neon, krypton and xenon ion lasers. However, some indirect

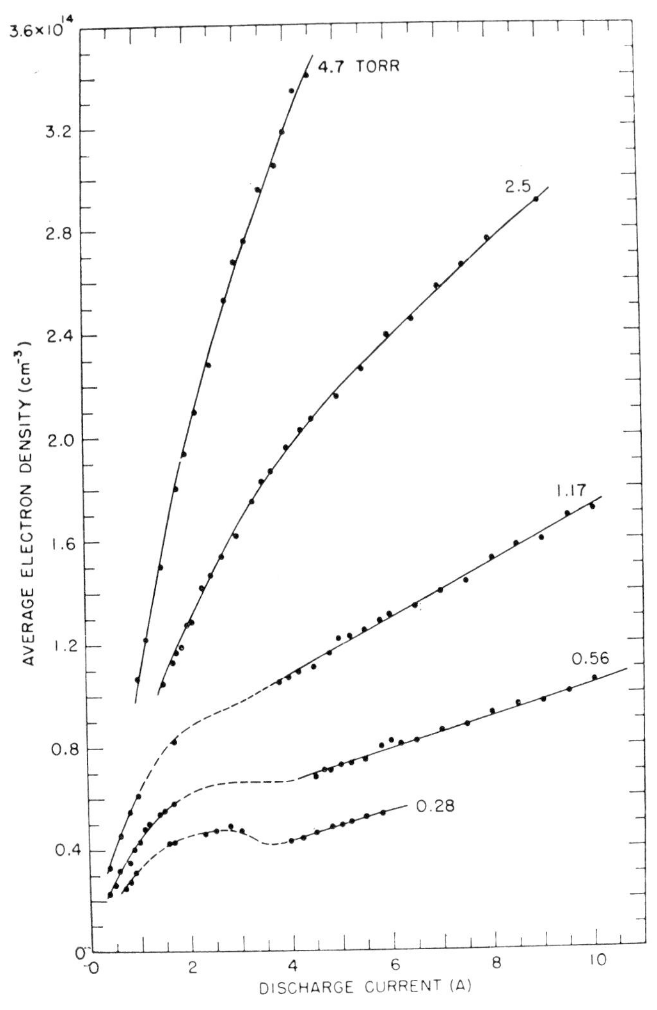

FIG. 24. Electron density in Ar ion laser of 2-mm bore for various filling pressures (Pleasance and George, 1971).

conclusions can be drawn from other measured parameters of these lasers. The current density J flowing in an ion laser is $J = \text{constant}.eN_e\mu_e E$ where N_e is the axial electron density, e the electronic charge, E the axial field in the discharge and the value of the constant is ~ 1 but depends on the precise nature of the radial electron density profile. Under equivalent conditions of E/p_0 in a noble gas ion laser discharge the mobility of the plasma electrons is in the sequence μ_e(neon) > μ_e(argon) > μ_e(krypton) > μ_e(xenon) (Brown, 1966) and so under equivalent conditions of J and E: N_e(neon) < N_e(argon) < N_e (krypton) < N_e(xenon). However, under otherwise identical conditions, the electron temperature in the discharge will tend to follow the sequence of ionization potentials so T_e(neon) > T_e(argon) > T_e(krypton) > T_e(xenon) and this will correspondingly affect the axial field so that E(neon) > E(argon) > E(krypton) > E(neon) and since the mobility increases with field, at constant particle density in the discharge capillary the expected electron densities in ion lasers should be in the sequence N_e(neon) < N_e(argon) < N_e(krypton) < N_e(xenon). However, offsetting this trend in typical noble gas ion lasers is the fact that the particle densities in these lasers under optimum laser conditions may be in the sequence N(neon) > N(argon) > N(krypton) > N(xenon) which could have the effect of helping to maintain similar electron densities in different noble gas ion lasers at the same current density.

It seems likely that the electron densities in long pulse (> few μs) ion lasers are slightly larger under given conditions than in CW lasers. For example Raff and George (1971) measured a value of $N_e = 1{\cdot}04.10^{14}$ cm^{-3} at a current density of 320 A cm^{-2} at 40 mtorr 20 μs after the start of a 37 μs pulse whilst Pleasance and George (1971) measured an almost equal value of 1·03 10^{14} cm^{-3} under identical CW current density conditions but at a filling pressure of 0·56 torr. Now at 0·56 torr the neutral atom temperature in such a laser at a current density of 320 A cm^{-2} is $\approx$ 2500°K (Bennett and Sze, 1972) which implies an equivalent capillary pressure of greater than 61 mtorr. At 40 mtorr this electron density would be reduced because the greater mobility of the plasma electrons at low pressure allows the same current density to flow at lower electron densities than under equivalent conditions at higher pressures.

Hattori and Goto (1969a) also observed a higher electron density than would be expected under equivalent conditions of CW operation and noted that the electron density followed the current pulse quite closely in time. However, direct comparison of the data of these authors with those of other authors is not recommended because of the slightly unusual conditions of inductive excitation used by them and by Goto *et al.* (1971) which are likely to make their plasma parameters different from those which obtain using pulsed D.C. or CW D.C. excitation.

In short duration, fast rise-time and Z-pinch ion laser discharges and in the early stages of the rising edge of the current pulse in a more conventional pulsed D.C. ion laser, measurement of the charged particle density is difficult as it is rapidly time varying. In the rising edge of the current pulse in a pulsed discharge the charged particle density increases rapidly as both the current rises and the axial electric field falls. In the immediate post-breakdown period a large

TABLE VII. *Neutral particle temperatures in argon ion lasers*

Current density (A cm^{-2})	Tube bore (mm)	Pressure (torr)	T_a (°K)	Method of measurement	Author(s)
220	2	2	1750		
220	2	1	2000		
220	2	0·7	2000	Voigt profile analysis of axially emitted lineshape	Ballik *et al.* (1966)
220	2	0·5	1950		
220	2	0·3	1800		
220	2	0·2	1750		
189	2·6	0·5	1650		
200	1·6	0·5	1400		
300	1·6	0·5	1850		
400	1·6	0·5	2350		
500	1·6	0·5	3000	Analysis of emitted lineshape	Kitaeva *et al.* (1966a, 1967a)
400 (pulsed, 4μs)	5	0·06	2000		
560 (pulsed, 4 μs)	5	0·06	4000		
820 (pulsed, 4 μs)	5	0·06	6000		
630 (pulsed, 4·5 ms)	4·5	Not stated	1900		
940 (pulsed, 4·5 ms)	4·5	Not stated	4600	Analysis of emitted lineshape	Glazunov *et al.* (1967)
1260 (pulsed, 4·5 ms)	4·5	Not stated	8700		
305[a]	2·5	Not stated	2900		
410[a]	2·5	but 1 torr	3600	Analysis of emitted lineshape	Donin (1969)
500[a]	2·5		4000		
600[a]	2·5		4200		

410[b]	2·5	1	6000	Voigt profile analysis of axially emitted lineshape	Donin (1970)
800[b]	2·5	1	8500		
95	2	0·2	1300	Voigt profile analysis of axially emitted lineshape	Sze and Bennett (1972)
159	2	0·2	1750		
223	2	0·2	1900		
286	2	0·2	2300		
95	2	0·4	1500		
159	2	0·4	1750		
223	2	0·4	2050		
286	2	0·4	2450		
95	2	0·7	1100		
159	2	0·7	1500		
223	2	0·7	1900		
286	2	0·7	2200		
95	2	1	950		
159	2	1	1300		
223	2	1	1700		
286	2	1	2050		
20 A[c]	—	$1{\cdot}7\text{–}10^{-3}$	4000	Analysis of emitted lineshape	Van der Sijde (1972a)
40 A[c]	—	$1{\cdot}7.10^{-3}$	7000		
60 A[c]	—	$1{\cdot}7.10^{-3}$	11,000		
80 A[c]	—	$1{\cdot}7.10^{-3}$	17,000		

[a] In aluminium disc laser with internal bore layer Al_2O_3.
[b] In segmented metal (molybdenum) tube.
[c] In magnetically confined hollow cathode discharge.

For D.C. discharges in quartz tubes unless stated.

current density may flow at relatively low density because the axial electric field is still very high. Later in the pulse the axial field stabilizes at a lower value and is less rapidly time-varying and the electron density behaviour will approximate to D.C. conditions.

In Z-pinch discharges a different set of phenomena control the behaviour of the plasma. These phenomena and the way they affect the observed laser action in such systems have been discussed in Section IV, E.

(3) *Neutral Atom Temperatures*

Reliable neutral atom temperatures in ion lasers can be deduced directly from measured Doppler broadened linewidths. The neutral temperature is obtained from the Gaussian lineshape from the formula for the full width at half height as,

$$\Delta\nu_D = \frac{2\nu_0}{c}\left(\frac{2kT_a \ln 2}{M}\right)^{1/2} \qquad \ldots(7.3)$$

where ν_0 is the centre frequency of the transition, M is the mass of the emitting species, k is Boltzmann's constant and T_a is the neutral temperature. Dependent on the lifetime values of the upper and lower levels of this transition there will also be a Lorentzian contribution to the lineshape from natural broadening and in addition a contribution from pressure broadening (mainly Stark broadening) which for most neutral lines will dominate over the natural broadening (Kitaeva *et al.*, 1971). In most ion laser plasmas these homogeneous contributions to the linewidths of emitting neutral species will be small, but significant, and should strictly be allowed for by deconvoluting the overall Voigt profile of the transition to obtain the true Gaussian width. Some, but not all, of the measurements of neutral temperatures made in ion lasers, for example those listed for argon ion lasers in Table VII have been made with the Lorentzian contribution to the lineshape allowed for. Very few measurements appear to have been made on neutral atom temperatures in ion lasers other than argon.

The few measurements of ion temperatures in argon, krypton, xenon and mercury ion lasers made by Bloom *et al.* (1965) suggest that, because at equivalent current density the heavier gases have lower ion temperatures, they may also have lower neutral temperatures. However, it is difficult to draw definite conclusions from these measurements because the laser operating conditions under which the measurements were made differ from one laser to another.

Strictly speaking there will be an apparent temperature difference between neutral atom temperatures ($T_{\parallel}^a$) measured from Doppler widths observed along the axis of the tube and temperatures ($T_{\perp}^a$) measured from Doppler widths observed transversely (Kitaeva *et al.*, 1971). These temperature differences arise from the radial motion of ions in the tube. At low values of $p_0 R$ (where p_0 is the effective filling pressure in the discharge capillary and R the tube radius) ions do not make many collisions with neutral atoms on their way to the wall and $T_\perp^a \simeq T_\parallel^a$. At higher values of $p_0 R$ ions make a significant number of charge exchange collisions on their way to the wall. These charge exchange collisions produce neutral atoms with radially directed velocities and a spatial inhomogeneity in the neutral atom velocities results. With increasing tube diameter and/or pressure $T_\perp^i/T_\perp^a$ approaches $T_\parallel^i/T_\parallel^a$ and becomes nearly constant whereas $T_\perp^i/T_\parallel^i$ approaches unity. $T_\parallel^i$ and $T_\perp^i$ are the ion temperatures deduced from measurements of the spontaneous width of ion transitions observed along and transversely to the discharge tube axis [see Section VII, A (5)]. The neutral temperature measured transversely increases with current density at constant pressure and decreases with pressure at constant current density except at low current densities (< 200 A cm^{-2}) in small bore tubes (Kitaeva *et al.*, 1971).

The values of neutral temperature measured in typical CW ion lasers cover a fairly wide range but in a given laser will be lower than the temperature of the ions. The influence that the walls of the discharge tube exert on the neutral temperature is unclear; it may be that in a metal disc or beryllia tube ion lasers (which have a lower internal wall temperature than a quartz tube at comparable current densities) there may be a layer of cooler gas near the walls. However, the measurements of Donin (1969) made on an aluminium disc argon ion laser indicate that the overall bulk neutral temperature remains high. The high neutral temperature in the discharge capillary of an ion laser leads to gas drive-out from the capillary until the pressure exerted by the particles in the capillary matches that exerted by the cooler gas outside the capillary.

(4) *Neutral Particle Densities*

The neutral particle density at the filling pressure p torr at 300°K in an ion laser is $N = 3{\cdot}22.10^{16}\ p$ particles cm^{-3}. As soon as current begins to pass through the capillary the gas temperature in the capillary rises and gas is driven out into the cooler parts of the system (which are at ambient temperature ~ 300 °K). This drive-out takes time on the order of a millisecond to tens of milliseconds to occur (Redaelli and Sona, 1967; Ahmed *et al.*, 1966, 1967). When the system reaches equilibrium the pressure exerted by all the particles in the capillary balances the external pressure. A number of authors (Kitaeva *et al.*, 1966; Ballik *et al.*, 1966; Mercer *et al.*, 1967; Herziger and Seelig, 1968) have used the simple gas law relation to estimate N_0 the particle density in the capillary after drive-out, that is

$$\frac{N_0}{N} = \frac{300}{T_a} \qquad \frac{N_0}{N} = \frac{p_0}{P}$$

where p_0 is the effective filling pressure in the capillary. Measurements of neutral particle densities made by Boguslovskii *et al.* (1967) indicate that this relation holds, at least for the axial neutral particle density.

Considerations of particle density are also complicated by gas pumping effects [see Section X, B (3)] which prevent the simple gas law calculation being entirely valid. Miller *et al.* (1964–7) measured the total atom + ion density in small bore capillary discharges in argon by X-ray absorption and found that the observed particle density, although it fell with increasing capillary gas temperature, was consistently 20–40% higher than predicted by the gas law relationship. In larger bore tubes the gas law relationship probably gives a much more reliable estimate of the particle density in the capillary. In general the ratio of the effective filling pressure at ambient temperature p_0 to the actual filling pressure is given by $p/p_0 = F$ where F varies between $T_a/300$ in wide bore tubes to about $T_a/390$ in small bore tubes.

Values of T_a are available for argon ion lasers (Table VII) so the particle density in the capillary under laser operating conditions can be estimated from the filling pressure. For the other ion lasers very little operating diagnostic information is available and it is difficult to predict exactly how various plasma parameters in these lasers will compare with those measured in argon ion lasers. The optimum filling pressures of CW krypton and xenon lasers are lower than for argon, but for these heavier atoms, their neutral temperatures in the capillary may be lower and gas pumping may not be as severe. The reduction in pressure below the filling pressure may not therefore be as great as in the case of argon so the operating particle density in the capillary may be of the same order.

(5) *Ion Temperatures*

Reliable values of ion temperatures in ion lasers can be obtained from Voigt profile analysis of spontaneous emission lines emitted in the axial direction.

Some typical values of this axial ion temperature $T^i_{\parallel}$ are listed in Table VIII. Not all these measurements have been made with allowance for the Lorentzian contribution to the lineshape. Furthermore these measurements represent an average value over the bore of the discharge tube and do not take into account that the bulk axial drift velocities of the emitting ions varies radially. However, this will only mean that the measured values in Table VIII represent a small overestimation of the actual ion temperature. Bennett *et al.* (1966) have pointed out that the Lorentzian profile emitted in the axial direction by individual ions in the plasma is affected by the fact that the emitting ion is being accelerated in the axial field of the discharge. However, the deviation of the lineshape from Lorentzian is small for the typical axial fields found in ion lasers.

Because the ions in typical noble gas ion lasers move radially with much higher drift velocities than they do along the axis, the widths of spontaneous

TABLE VIII. *Ion temperatures in argon ion lasers*

Current density (A cm^{-2})	Tube bore (mm)	Pressure (torr)	T_i (°K)	Method of measurement	Author(s)
450	Not stated	Not stated	3000	Analysis of axially emitted lineshape	Bloom *et al.* (1965)
220	2	2	1770	Voigt profile analysis of axially emitted lineshape	Ballik *et al.* (1966)
220	2	1	2050		
220	2	0·7	2500		
220	2	0·5	2750		
220	2	0·3	2800		
220	2	0·2	2750		
200	1·6	0·5	2600	Analysis of axially emitted lineshape	Kitaeva *et al.* (1967a)
350	1·6	0·5	3000		
500	1·6	0·5	4600		
315 (pulsed)	4·5	Not stated	4500	Analysis of axially emitted lineshape	Kitaeva *et al.* (1967a)
630 (pulsed)	4·5	Not stated	10,000		
945 (pulsed)	4·5	Not stated	16,000		
400 (pulsed)	4	0·03	1700–3500[a]	Analysis of axially emitted lineshape	Klein (1970)
800 (pulsed)	4	0·03	4600–7300[a]		
400 (pulsed)	4	0·02	2400–4400[a]		
800 (pulsed)	4	0·02	5400·8500[a]		
410	2·5	1	13,000[b]	Voigt profile analysis of axially emitted lineshape	Donin (1970)
800	2·5	1	18,000[b]		
1230	2·5	1	21,000[b]		

TABLE VIII—*continued*

Current density ($A\ cm^{-2}$)	Tube bore (mm)	Pressure (torr)	T_i (°K)	Method of measurement	Author(s)
50 A[c]	—	0·0018	13,000	Analysis of emitted lineshape	Van der Sijde (1972a)
95	2	0·2	1850	Voigt profile analysis of axially emitted lineshape	Sze and Bennett (1972)
159	2	0·2	2300		
223	2	0·2	2700		
286	2	0·2	3200		
95	2	0·4	2600		
159	2	0·4	2200		
223	2	0·4	2500		
286	2	0·4	2900		
95	2	1·0	1000		
159	2	1·0	1500		
223	2	1·0	1850		
286	2	1·0	2300		

[a] Temperature change during 50 μs pulse
[b] In molybdenum disc segmented tube.
[c] Actual current in magnetically confined hollow cathode discharge.

For D.C. currents in quartz tubes unless stated.

emission profiles observed transversely to the tube axis are broader than those observed in the axial direction. The apparent transverse ion temperature $T_{\perp}^{i}$ deduced from the former profiles does not reflect the actual temperature of the ions but is related to other parameters of the discharge. This will be discussed more fully in Sections VIII, E and IX, D. It can be seen by comparison of Tables VII and VIII that ion temperatures are generally higher than the temperatures of neutral atoms in ion lasers. However, the same value of ion temperatures is not always obtained for measurements on transitions from different levels of the ion. For example, measurements made on levels which have a significant contribution to their pumping direct from the neutral atom ground state will tend to be lower and reflect the temperature of the neutral atoms. Evidence for this occurring has been found for the 4765 Å transitions in argon II (Mercer *et al.*, 1967).

(6) *Neutral Particle Drift Velocity and Density Distributions*

Although the charged particles in an ion laser plasma do impart an anode directed force on the neutral particles the resultant mean axially directed velocity of the neutral particles in typical low current density CW lasers is small. Ballik *et al.* (1966) were not able to detect any such axial drift velocity in a CW argon ion laser operating at 160 A cm^{-2}. However a small anode directed drift velocity was detected from Doppler shift measurements by Kitaeva *et al.* (1971). They measured velocities in the range 1.10^3–6.10^3 cm s^{-1} for current densities in the range 100–240 A cm^{-2} in tubes of 5 and 7 mm bore and pressures up to 0·7 torr. At low pressures (below about 0·5 torr) the neutral particle drift velocity increased with current density and tube bore but at higher pressures the dependence of this velocity on the discharge parameters was more complex. The problem of axially directed motion of neutral particles in discharge plasmas will be considered more fully in the section on gas pumping [Section X, B (3)].

Webb (1968) has made extensive measurements on the radial distribution of neutral particles in argon plasmas under conditions of laser excitation from observations of the radial intensity profile of spontaneous transitions. At low current densities < 160 A cm^{-2} and pressures < 2 torr the neutral atom density is independent of radial position. However, at higher currents and pressures the neutral atom density in the centre becomes depleted relative to that near the wall. This is partially due to a radial variation in gas temperature with a maximum temperature on axis and a corresponding minimum in gas density but as the current density becomes very high the radial distribution of electron density will lead to radial variations in neutral atom density.

If the electron and ion densities as a function of radius are $N_e(r)$ and $N_i(r)$ respectively, then for plasma neutrality

$$N_e(r) = N_i(r)$$

This will hold everywhere in the plasma except in the sheath region near the wall. The total particle density (atoms + ions) neglecting multiple ionization is $N_0 = N_e(r) + N_a(r)$. At low current when $N_e(r) \ll N_0$, $N_a(r) = N_0$ and is independent of radial position. However as the degree of ionization of the plasma increases $N_a(r) = N_0 - N_e(r)$ and the neutral atom population becomes radially dependent. Since both the Schottky (ambipolar diffusion) mode of the discharge (Schottky, 1924; see also Cobine, 1958) and the Tonks–Langmuir (free-fall) model of the discharge (Tonk and Langmuir, 1929) predict electron density functions which are a maximum on the tube axis, the onset of significant ionization of the plasma causes the neutral particle density to be larger near the walls than in the centre.

Since the operation of CW and long-pulse argon ion lasers depends on excitation processes involving the ion ground state the radial distribution of neutral particles does not affect directly the spatial variation of population inversion within the tube, which manifests itself in transverse mode effects.

(7) *Charged Particle Density Distributions*

For a discharge in an electropositive gas in a cylindrical tube, the tube wall acquires a negative charge and the resulting radial field controls the motion of the charged particles in the volume of the gas. Except in the space charge sheath which builds up just near the wall, the plasma is quasi-neutral with a cylindrically symmetric electron and ion distribution with $N_e(r) = N_i(r)$ (neglecting multiple ionization). The exact form of $N_e(r)$ will be discussed later in Section VIII on theories of ion laser plasmas, but independent of what discharge regime the plasma falls into—whether it is the Tonks–Langmuir (low pressure) or Schottky (high pressure regime)—$N_e(r)$ is maximum on axis and falls off monotonically towards the wall, at least as far as the edge of the wall sheath. It is generally assumed that there is no radial variation in electron temperature, although just near the wall the electron temperature may be smaller (Bloom *et al.*, 1964).

Provided there is no radial variation in electron temperature then the radial distribution of electron density in an ion laser plasma will control the radial dependence of the various electron collisional processes which populate the laser levels. Webb (1968) has studied the radial dependence of spontaneously emitted argon I and argon II lines under conditions appropriate to laser excitation in a quartz capillary discharge tube. He found very little change in the intensity of argon I lines for changes in current density from 1·6 A cm^{-2} to 160 A cm^{-2} and over quite a wide range of currents the spontaneous emission intensities of argon I lines were independent of radial position. This independence of electron density of the population of the appropriate emitting argon I levels (saturation) implies that any excitation of ionic levels from levels of the neutral atom will exhibit essentially the same current dependence as excitation from the neutral atom ground state.

The radial dependence of the spontaneous emission intensity of argon II lines observed by Webb (1968) gives direct evidence of the mechanism by which these levels are excited. Webb finds excellent proportionality at low pressures between the observed radial intensity profile of argon II levels and the square of the theoretical $N_e(r)$ distribution calculated by Parker (1963), whilst at high pressures he obtains excellent proportionality with the square of the theoretical high pressure $N_e(r)$ distribution (Schottky, 1924). These observations demonstate the dominance of excitation processes of the singly-ionized argon laser levels from states of the ion, at least under the low current density and the relatively low likely electron temperature conditions used by Webb.

In all ion lasers increase of current density (and consequently electron density) eventually leads to quenching of laser action. The various methods by which quenching can occur are discussed in Section IV, C, however they are all related to the radial charged particle distribution in the tube. Knowledge of the charged particle distribution, therefore, helps to explain the various quenching phenomena observed.

(8) *Charged Particle Drift Velocities*

Three types of bulk motion of charged particles are of importance in ion laser plasmas: (i) Longitudinal electron drift—anode directed; (ii) Longitudinal ion drift—cathode directed; (iii) Radial ion drift.

(a) *Longitudinal electron drift velocity.* The average electron drift velocity ($\bar{v}_d$) is related to the current density by

$$\bar{v}_d = J/(\text{constant}\,.\,N_e(0)\,e) \qquad \ldots(7.4)$$

In a plasma at low pressure when Tonks–Langmuir theory applies, the value of the constant is 0·7 (Hernqvist and Fendley, 1967)

$$\text{also } \bar{v}_d = \mu_e\,.\,E$$

where E is the axial field and μ_e the electron mobility in the discharge gas at the appropriate pressure. Some measurements of μ_e in argon at high field strengths have been made by Golant (see Brown, 1966). At high field strength the mobility saturates at $2{\cdot}2\,.\,10^5/p_0$ cm^2 volts^{-1} s^{-1} where p_0 is in torr for values of E/p_0 above 200 volts cm^{-1} torr. Thus the electron drift velocity can be calculated from the electron and current densities or alternatively from the measured axial field and the electron mobility. Some experimental values of longitudinal electron drift velocities are given in Table IX. Various other theoretical relations also exist which connect these quantities with the electron density and are discussed in Section VIII on laser plasma theories. The main

TABLE IX. *Longitudinal electron drift velocities in argon ion lasers*

Current density (A cm^{-2})	Pressure (torr)	V_d^e	Method of measurement	Author(s)
25	Not stated	$2{\cdot}6.10^7$	Theory	Zarowin (1969)
120	Not stated	$4{\cdot}0.10^6$		
250	Not stated	$2{\cdot}6.10^6$		
500	Not stated	$1{\cdot}3.10^6$		
700	0·015	$7{\cdot}2.10^6$	Deduced from current and electron density	Hattori and Goto (1969a)
1400	0·015	$7{\cdot}2.10^6$		
1750	0·015	$7{\cdot}0.10^6$		

effect of axial electron drift in ion lasers, besides the obvious one of current flow, is gas pumping which is discussed separately in Section X, B (3).

(b) *Longitudinal ion drift.* The cathode directed drift of ions, although it is much slower than the corresponding anode directed drift of electrons, is of much more importance in contributing to observed phenomena in ion lasers. The centre emission frequency of an excited ion is shifted to $\nu_0 + \nu_0(v_d^i/c)$ for an observer viewing along the discharge capillary from cathode to anode and to $\nu_0 - \nu_0(v_d^i/c)$ for an observer viewing in the opposite direction where ν_0 is the centre emission frequency of the excited ion at rest and v_d^i is the ion drift velocity. Measurement of the Doppler shift of ions observed from opposite ends of the tube allows the ion drift velocity to be calculated; some typical values obtained in this way are listed in Table X. The Doppler shifts to which these velocities correspond are in the range of several tens to hundreds of MHz. These are typically of the same order or greater than the spacing of longitudinal modes of the laser and lead to interesting mode interaction phenomena which are discussed in Section XI, B (2). The way in which the longitudinal ion drift velocity varies with current density and gas pressure in the discharge capillary is influenced strongly by radial motion of the ions. At low pressures, the ions are accelerated in the longitudinal field for a time governed by their loss at the walls. Since their rate of loss to the wall increases with current density, their axial velocity falls as the current density increases. When the pressure increases, migration to the walls is slowed down and the longitudinal drift velocity of the ions increases. If the pressure is increased still further, the length of time during which an ion can be accelerated by the axial field is mainly determined by collisions. Since this time decreases with increasing pressure the longitudinal ion drift velocity falls again.

(c) *Radial ion drift.* The outward directed mean radial motion of ions which occurs in the plasma of an ion laser leads to a broadening of spectral lines viewed

TABLE X. *Longitudinal ion drift velocities in argon ion lasers*

Current density (A cm^{-2})	Tube bore (mm)	Pressure (torr)	v_d^i (cm s^{-1})	Method of measurement	Author(s)
220	2	1·7	7.10^3	Doppler shift of argon II 4880 Å line	Ballik *et al.* (1966)
220	2	1·0	$1{\cdot}3.10^4$		
220	2	0·5	$1{\cdot}75.10^4$		
220	2	0·4	$1{\cdot}7.10^4$		
220	2	0·3	$1{\cdot}55.10^4$		
220	2	0·2	$1{\cdot}1.10^4$		
200	1·6	0·67	$1{\cdot}10^4$	Doppler shift of argon II 4880 Å line	Kitaeva *et al.* (1967b)
350	1·6	0·67	$1{\cdot}4.10^4$		
500	1·6	0·67	1.10^4		
200	1·6	0·27	$2{\cdot}1.10^4$		
350	1·6	0·27	$1{\cdot}6.10^4$		
400	1·6	0·27	$1{\cdot}3.10^4$		
305[a]	2·5	Not stated but 1 torr	$2{\cdot}6.10^4$	Doppler shift of argon II 4880 Å line	Donin (1969)
410	2·5		$2{\cdot}1.10^4$		
500	2·5		$1{\cdot}9.10^4$		
600	2·5		$2{\cdot}0.10^4$		

TABLE X—*continued*

Current density (A cm^{-2})	Tube bore (mm)	Pressure (torr)	v_d^i (cm s^{-1})	Method of measurement	Author(s)
113	5	0·42	$1{\cdot}15.10^4$	Doppler shift	Kitaeva *et al.* (1971)
163	5	0·42	1.10^4		
239	5	0·42	$8{\cdot}5.10^4$		
113	5	0·55	$1{\cdot}1.10^4$		
163	5	0·55	1.10^4		
239	5	0·55	7.10^3		
113	5	0·68	$1{\cdot}05.10^4$		
163	5	0·68	1.10^4		
239	5	0·68	6.10^3		
113	7	0·3	$1{\cdot}15.10^4$		
163	7	0·3	5.10^3		
239	7	0·3	3.10^3		
113	7	0·42	$8{\cdot}5.10^3$		
163	7	0·42	5.10^3		
213	7	0·42	$4{\cdot}5.10^3$		
113	7	0·55	$7{\cdot}5.10^3$	Doppler shift	Kitaeva *et al.* (1971)
163	7	0·55	5.10^3		
213	7	0·55	5.10^3		
113	7	0·68	7.10^3		
163	7	0·68	$4{\cdot}5.10^3$		
213	7	0·68	5.10^3		

[a] In aluminium disc laser with internal bore layer of Al_2O_3.

transversely to the discharge tube axis. The shape of these lines will be discussed in greater detail in Section IX on lineshapes in ion lasers. Suffice it to say at this point that the theoretical form of the shape is extremely complex except at high pressures where the radial ion motion is controlled by diffusion. However by making certain assumptions about the theoretical form of the lineshape an average radial velocity $v_\perp^i$ can be determined. The most detailed and careful consideration of the problem of radial drift and its effect on the transversely emitted lineshape has been given by Sze and Bennett (1972), some of whose values of $v_\perp^i$ are given in Table XI together with some earlier measurements of

TABLE XI. *Radial ion drift velocities in argon ion lasers*

Current density (A cm^{-2})	Tube bore (mm)	Pressure (torr)	V_r (cm s^{-1})	Method of measurement	Author(s)
85	3	1·0	7·7.10^4	Deduced from axially and transversely emitted lineshapes	Mercer (1967) Mercer *et al.* (1967)
85	3	0·7	1·15.10^5		
85	3	0·5	1·2.10^5		
85	3	0·3	1·17.10^4		
85	3	0·2	1·1.10^4		
64	2	0·2	7·3.10^4		Sze and Bennett (1972)
127	2	0·2	1·21.10^5		
191	2	0·2	1·28.10^5		
286	2	0·2	1·31.10^5		

Mercer (1967) and Mercer *et al.* (1967). It can be noted at this point that in these earlier measurements a significantly smaller average radial velocity was observed for the argon II transition at 4765 Å. Mercer *et al.* (1967) concluded that this arose because of the larger contribution to the pumping of the upper level of the 4765 Å transition direct from states of the neutral atom. Ion states excited in this way do not have as much time to acquire a radially directed drift velocity before they emit as do ion states excited from the ground or metastable states of the ion. Single-step excitation from states of the neutral atom certainly contributes in this way to a lower radial drift velocity but there are other phenomena which would also lead to differences in radial drift velocity for different transitions. Levels which have large excitation cross-sections from the ion ground state may have lower acquired radial drift velocities than levels which are excited predominantly via metastable states of the ion, particularly since charge exchange processes can lead to the production of significant numbers of ground state ions with low acquired radial drift velocities.

B. CW METAL-VAPOUR LASERS

These lasers operate at much lower current density and higher pressure than the noble gas ion lasers. The radial motion of ions is slow, being governed by diffusion, and consequently there are none of the large effects on lineshape and longitudinal drift velocity which occur in noble gas ion lasers. They operate with relatively low current discharges in metal vapour–noble gas mixtures and cataphoresis is the dominant process which governs the distribution of ions longitudinally (Goldsborough, 1969; Sosnowski, 1969), gas pumping being negligible. Goto *et al.* (1971) have measured electron temperatures and densities in He–Cd lasers, and find that the electron temperature changes very little with current but falls with increase of cadmium partial pressure. Ivanov and Sem (1973) also confirm this fall in electron temperature with cadmium density. Some typical measured values are shown in Table XII. Values of T_e measured in a He–Se laser (Goto *et al.*, 1972) fall more slowly with addition of selenium.

Optimum operating pressures in He–Se lasers are higher than for He–Cd so at the same current, electron densities in the former are expected to be higher. This appears to be confirmed by measurements made by a double probe technique by Goto *et al.* (1971b, 1972). It is not yet clear exactly how the electron density varies with cadmium pressure in a He–Cd laser; Goto *et al.* (1971b, 1972) report an initial reduction between 10^{-4}–10^{-3} torr and then an increase with further increase in cadmium pressure. However microwave measurements made by Dunn (1972),and Ivanov and Sem (1973) indicate that the electron density increases monotonically with cadmium pressure.

Further information on the behaviour of N_e and T_e in CW metal vapour lasers will allow the importance of electron destruction of the laser levels to be ascertained as well as the contribution of direct electron impact excitation, for example on ground state metal ions, to be determined.

C. THE EFFECT OF MAGNETIC FIELDS ON ION LASER PLASMA PARAMETERS

In view of the commercial importance of axial magnetic fields in enhancing the power output of ion lasers as described in Section X, F, particularly the noble gas ion lasers, surprisingly little diagnostic work has been performed on ion laser plasmas in such fields.

The primary effect of an axial magnetic field in such plasmas is to reduce charged particle loss to the tube walls, which has the consequence of reducing the axial electric field necessary to maintain the current flow. In theory this reduction in axial electric field should lead to a monotonic reduction in electron temperature with field. This change in electron temperature should however become smaller as the discharge pressure is increased (Forrest and Franklin, 1966a, b). The behaviour of ion laser plasmas in practice sometimes appears to differ slightly from that predicted theoretically. Burmakin and Evtyunin

TABLE XII.

Plasma parameters in He–Cd lasers

Current density (A cm^{-2})	Helium pressure (torr)	Cadmium pressure (torr)	kT (eV)	N_e (cm^{-3})	Method of measurement	Author(s)
2·8	3·2	5.10^{-5}	—	$1·7.10^{12}$	Microwave	Dunn (1972)
2·8	3·2	10^{-3}	—	$2·4.10^{12}$		
2·8	5·0	5.10^{-5}	—	$2·7.10^{12}$		
2·8	5·0	10^{-3}	—	$3·0.10^{12}$		
1·4	3·5	10^{-3}	—	$1·0.10^{12}$	Microwave	Ivanov and Sem (1973)
1·4	3·5	10^{-2}	—	$1·5.10^{12}$		
1·4	3·5	3.10^{-2}	—	$2·5.10^{12}$		
1·4	3·5	3.10^{-3}	6	—	Double probe	Ivanov and Sem (1973)
1·4	3·5	$5·10^{-3}$	4·7	—		
1·4	3·5	10^{-2}	3·9	—		
1·4	3·5	3.10^{-2}	2·5	—		
0·16	2·5	10^{-4}	4·1	$2·2.10^{11}$	Double probe	Goto *et al.* (1971b)
0·16	2·5	10^{-3}	2·7	$1·4.10^{11}$		
0·16	2·5	10^{-2}	1·6	$1·1.10^{11}$		
0·16	2·5	10^{-1}	1·2	$2·0.10^{11}$		
0·49	2·5	10^{-4}	3·2	$6·5.10^{11}$		
0·49	2·5	10^{-3}	2·1	$4·0.10^{11}$		
0·49	2·5	10^{-2}	1·2	$3·7.10^{11}$		
0·49	2·5	10^{-1}	0·9	$6·5.10^{11}$		

Plasma parameters in He–Se lasers

Current density (A cm^{-2})	Helium pressure (torr)	Selenium pressure (torr)	kT (eV)	N_e (cm^{-3})	Method of measurement	Author(s)
0·16	6	10^{-4}	2·9	$1·2.10^{12}$	Double probe	Goto *et al.* (1972)
0·16	6	10^{-3}	2·9	$8·6.10^{11}$		
0·16	6	10^{-2}	2·5	$6·3.10^{11}$		
0·16	6	10^{-1}	2·3	$5·4.10^{11}$		
0·5	6	10^{-4}	2·9	$2·4.10^{12}$		
0·5	6	10^{-3}	2·9	$2·3.10^{12}$		
0·5	6	10^{-2}	2·5	$1·5.10^{12}$		
0·5	6	10^{-1}	2·3	$1·2.10^{12}$		

(1969) observed an initial decrease in axial electric field when an axial magnetic field was applied to an argon ion laser plasma, however further increase of this field at low discharge pressure (for example beyond 500 G at 0·2 torr) leads to an increase in the axial electric field. Fridrikhov and Fotiady (1971) observed an initial increase in axial electric field with the application of a magnetic field which was particularly evident at low pressure (0·14 torr in a 1·5 mm bore tube at a current density of 283 A cm^-). A corresponding increase in electron temperature occurred when the axial field increased. However the results of these two groups of authors should not necessarily be interpreted as evidence of defects in the theory since the measurements themselves might have been subject to other phenomena such as plasma oscillations and gas pumping which are not taken into account in the theory of Forrest and Franklin (1967a, b).

It is generally assumed that the application of an axial magnetic field to an ion laser at constant current causes an increase in electron density in the laser plasma and consequently in many cases an increase in laser output power. Certainly such an increase is expected from a simple consideration of current density and electron drift velocity. As mentioned previously $j = \text{constant} \times N_e e \bar{v}_d$. In an axial magnetic field the mean electron motion now takes the form of a spiral path along the field lines and the axially directed drift velocity is reduced leading to an increase in electron density at constant current (assuming the value of the constant does not change significantly). However the magnitude of the change in average electron density produced by the typical fields used in CW noble gas ion lasers (0–1500 G) is uncertain. Borisova *et al.* (1967) observed a sixfold increase in electron density in a small bore argon ion laser in axial fields of 1000 gauss and a corresponding fall in electron temperature to 2/3 its value in zero field, whilst Fridrikhov and Fotiardy (1971) observed very little increase in N_e for fields up to 800 G in a 1·5-mm bore tube at a current of 5 A. However it is clear that application of an axial magnetic field does produce an enhancement of the charged particle density near the tube axis. In any given situation the question of whether this axial magnetic field will enhance the output of an ion laser will depend on the relative effects on the laser excitation process of the change in electron temperature, electron number density and charged particle spatial distribution. Similar effects can also occur in high current density ion lasers in the absence of any external magnetic field. The self magnetic field of the discharge reduces charged particle loss to the discharge tube walls and enhances the axial electron density.

A transverse magnetic field can also affect the output power of an ion laser; the effects on metal vapour lasers are described in Section XII. The field increases the ion and electron loss rate to the walls leading to subsequent increases in both axial field and electron temperature whilst the electron density falls with increasing field. The greatest perturbation to the plasma parameters occurs at low pressures (Fridrikhov and Fotiady, 1971). Besides affecting the plasma of an ion laser a magnetic field also affects the laser transition itself through Zeeman splitting; the effects of this are discussed separately in Section IX, E.

VIII. Theories of Ion Laser Plasmas

The previous sections have dealt with the various measurable parameters which characterize the plasma of an ion laser. The considerations of these sections are intimately connected with theories of gas discharge plasmas, particularly in so far as the indirect measurement of a number of ion laser parameters depends on the use of an appropriate theory which relates directly measureable plasma quantities with ones that are not as easily measured directly. Further, if the predictions of these theories appear to be valid under those conditions where reliable plasma diagnostic measurements have been made, then it is not unreasonable to use these theoretical methods to predict properties of ion laser plasmas under conditions where diagnostic measurements either have not or cannot easily be made. This section will discuss those theories of ion laser plasmas which describe bulk properties of the plasma at high and low pressure and in axial magnetic fields. Further, some interesting aspects of theories which describe ion motion in plasmas will be considered.

The general classification of discharge plasmas has been considered by Forrest and Franklin (1968). They divide them, for theoretical purposes, into five main types, the classification of a plasma being dependent on the relative sizes of the ion mean free path λ_i, the cylindrical plasma radius R and the axial Debye length λ_{D_0} (which is a measure of the distance over which deviations from charge neutrality can occur in the plasma). The five classifications, with authors who have considered them, are

(a) $\lambda_i \gg R \gg \lambda_{D_0}$ (Tonks and Langmuir, 1929).
(b) $\lambda_i \gg R \simeq \lambda_{D_0}$ (Self, 1963; Parker, 1963).
(c) $\lambda_i \simeq R \gg \lambda_{D_0}$ (Forrest and Franklin, 1966a, b). ...(8.1)
(d) $R \gg \lambda_i \gg \lambda_{D_0}$ (Schottky, 1924).
(e) $R \simeq \lambda_{D_0} \gg \lambda_i$ (Cohen and Kruskal, 1965).

In a consideration of ion laser plasmas, it is not necessary to consider all the five categories separately. Laser plasmas which fall into category (a) will be termed "low" pressure plasmas.

Self (1963) and Parker (1963) treat this low pressure situation more accurately by considering the effect of the sheath on the discharge tube walls. In practice the Debye length in typical ion lasers is $\ll R$ so the plasma does not fall into category (b). However, the calculations made for plasmas which fall into category (b) also apply to those in category (a) where the calculations of Parker (1963) and Forrest and Franklin (1968) provide small corrections to the predictions of Tonks and Langmuir (1929). Laser plasmas which fall into category (c) will be denoted "intermediate" pressure plasmas and those which fall into category (d) "high pressure" laser plasmas. Ion laser plasmas do not generally fall into category (e).

If the complexities introduced by multiple ionization are neglected then the flow of ions towards the walls in the radial field which is built up at all pressures is governed by various processes which control the ion mean free path.

$$\left.\begin{array}{l}\text{(a) Charge-exchange } M^+ + M \rightarrow M + M^+ \\ \text{(b) Momentum transfer } M^+ + M \rightarrow M^+ + M\end{array}\right\} \begin{array}{c}\text{Indistinguishable}\\ \text{processes}\end{array} \quad \ldots(8.2)$$

$$\text{(c) Recombination } M^+ + e(+\text{ third body}) \rightarrow M(+\text{ third body})$$

$$\text{(d) Molecule formation } M^+ + M^{(*)}(+\text{ third body}) \rightarrow M_2^{+(*)}(+\text{ third body})$$

In typical ion laser discharges the dominant processes controlling the motion of the ions are charge exchange and momentum transfer. This has been demonstrated directly in argon laser plasmas from line shape studies by Sze and Bennett (1972) and Kitaeva *et al.* (1971). The charge exchange cross-section σ_{CE} for argon ions in argon is large, $\simeq 5.10^{-15}$ cm^2 at thermal energies (McDaniel, 1964; Hasted, 1964a). Theoretically the total cross-section for processes (a) and (b) together is $\sigma_D = 2\sigma_{CE}$ (Hasted, 1964b). The appropriate value of σ_D in a typical CW argon laser for ions moving in the radial field has been estimated by Webb (1968) to be $= 9{\cdot}3 . 10^{-15}$ cm^2. With this cross-section he calculates the ion mean free path to be

$$\lambda_i = \frac{3{\cdot}3 \times 10^{-3}}{p_0} \text{ cm}$$

where p_0 is the effective filling pressure in torr of the capillary at ambient temperature. As mentioned previously p_0 is related to the filling pressure p by $p_0 = p/F$.

Under typical CW argon ion laser conditions of 300 A cm^{-2} in a 2 mm bore tube and a filling pressure $\simeq 0{\cdot}5$ torr, F is $\simeq 6{\cdot}5$ (from data of Sze and Bennett, 1972) that is $\lambda_i \simeq 4{\cdot}3 \times 10^{-2}$ cm. Because of the spatial distribution of ion generation in the laser tube, Webb (1968) points out that the average distance travelled by an ion before reaching the wall, independent of the pressure, is $\sim 0{\cdot}4\, R$, and in the situation considered above this average distance of $4 . 10^{-2}$ cm is very similar to the ion mean free path, and the laser plasma lies in the "intermediate" pressure region. However, even in this intermediate pressure region where $\lambda_i \sim R$ and the radially directed ions make on average about 1 collision before reaching the wall, it is generally considered that in the absence of magnetic fields Tonks–Langmuir theory gives a better description of the situation than does Schottky theory which probably applies strictly only when $\lambda_i \gtrsim R/10$. Although strictly speaking the plasmas of CW ion lasers do not fall into the Tonks–Langmuir regime, the slightly lower effective filling pressures in pulsed ion lasers make them satisfy the condition $\lambda_i > R$ and Tonks–Langmuir theory will provide an adequate description of the plasma (except in fast rise-time and Z-pinch pulsed lasers).

A. LOW PRESSURES: TONKS–LANGMUIR REGIME

The theory of Tonks and Langmuir (1929) describes a gas discharge plasma where ions are formed in the body of the plasma and then fall radially without collision to the tube wall. Because the electron temperature in such a discharge is much greater than that of the ions, the random electron current to the wall

$$N_e(w) \exp\left(\frac{kT_e}{2\pi m_e}\right)^{1/2}$$

is also much larger. This state of affairs would lead to a loss of plasma quasi-neutrality and so the wall acquires a negative charge which balances the random electron and ion currents at the wall. If the flux of ions to the walls is governed by an average radial drift velocity v_r then the boundary condition at the wall takes the form

$$N_e(w)\left(\frac{kT_e}{2\pi m_e}\right)^{1/2} = N_i(w)\,\bar{v}_r \qquad \text{...(8.3)}$$

where $N_e(w)$ and $N_i(w)$ are the electron and ion densities at the wall. Under normal circumstances this equation is satisfied only by having $N_i(w) > N_e(w)$. The extent of the region where this charge imbalance occurs is typically of the order of a few Debye lengths, where the Debye length in rationalized units is defined by

$$\lambda_{D_0}^2 = \frac{\varepsilon_0 kT_e}{N_{e_0} e^2}$$

and where N_{e_0} is the axial electron density. In typical ion lasers $\lambda_{D_0} \ll R$ and so the plasma is quasi-neutral except very close to the wall.

The theory of Tonks and Langmuir (1929) treats the quasi-neutral plasma and sheath separately, for the region of plasma outside the sheath. They write for the electron density

$$N_e(v) = N_{e_0} \exp\left[\frac{eV(r)}{kT_e}\right] \qquad \text{...(8.4)}$$

where $V(r)$ is the radial distribution of potential. The rate of production of ions $G(r)$ is assumed to be proportional to the electron density.

$$G(r) = ZN_e(r)$$

where Z is the ionization rate. The ions are assumed to be formed at rest and then fall freely towards the walls. The ion density at any radius is then (Parker, 1963),

$$N_i(r) = \int_0^r \frac{G(\rho)\left(\frac{\rho}{r}\right)\mathrm{d}\rho}{\left\{\left(\frac{2e}{M_i}\right)[V(\rho) - V(r)]\right\}^{1/2}}. \qquad \text{...(8.5)}$$

Tonks and Langmuir (1929) working from these assumptions and the assumption of plasma quasi-neutrality were able to calculate the radial distribution of charged particles within the plasma, the radial potential distribution and the parametric variation of the electron temperature. However, they treat the space charge sheath near the discharge tube wall separately from the quasi-neutral plasma and attempt to match their solutions for sheath and plasma at the sheath boundary. To describe the whole region of plasma and sheath more exactly a single solution for $N_e(r)$ and $V(r)$ should be found. This calculation has been carried out by Self (1963) for plane geometry and by Parker (1963) and Forrest and Franklin (1968) for cylindrical geometry.

In these descriptions of plasma and sheath the assumption of quasi-neutrality in the plasma is abandoned, although in practice the actual deviation from quasi-neutrality is negligible outside the sheath region. In the calculations of Forrest and Franklin (1966b, 1968) the plasma is characterized by two parameters α and δ^*

$$\alpha^2 = \frac{\varepsilon_0 Z^2 M_i}{N_{e_0} e^2} = \frac{\lambda_{D0}^2 Z^2 M_i}{kT_e} \qquad \ldots(8.6)$$

α is a measure of the extent to which space charge effects are important in the plasma.

$$\delta^* = \frac{kT_e}{2ZM_i D_e} + \frac{\nu_i}{2Z} \qquad \ldots(8.7)$$

where Z is the ionization rate (electron–ion pairs produced per second per electron), M_i is the ion mass, D_e is the diffusion coefficient of the electrons, ν_i is the collision frequency of the ions for (momentum + charge) transfer and ε_0 is the permittivity of free space.

Parker (1963) characterizes the low pressure plasma by a parameter β, α and β are related by $\beta = \sqrt{2}/\alpha$.

In the Tonks–Langmuir (free-fall) model of the discharge the flow of ions to the walls is collisionless and the parameter δ^* is zero.

In a typical argon ion laser plasma the value of α can be estimated from the ionization rate of about 8.10^6 electrons^{-1} s^{-1} measured by Kitaeva *et al.* (1967b) at 400 A cm^{-2} at pressures in the range 0·27–0·67 torr in a 1·6-mm tube and the electron density at 400 A cm^{-2} and 0·56 torr of $1{\cdot}4.10^{14}$ cm^{-3} [extrapolated from the data of Pleasance and George (1971)]. The value of α thus calculated is $3{\cdot}23.10^{-8}$. With an electron temperature $\sim$ 4 eV and the above electron density the Debye length is $6{\cdot}3.10^{-5}$ cm.

These very small typical calculated values of α and λ_{D_0} demonstrate how little a typical ion laser plasma should differ from quasi-neutrality since at any pressure a small value of α implies that quasi-neutrality is obeyed within the plasma. Complete quasi-neutrality corresponds to $\alpha = 0$; for large values of $\alpha > 10^{-2}$ the boundary between the quasi-neutral plasma and the space charge sheath becomes ill-defined and the assumption of bulk quasi-neutrality within the plasma would no longer apply.

The plasma of an ion laser at "low" pressure should be adequately described by Tonks–Langmuir theory. Because of the very small values of α which apply to these plasmas, any corrections to the predictions of this theory are negligible (Parker, 1963; Forrest and Franklin, 1968). The predictions of Tonks–Langmuir theory have been summarized by Hernqvist and Fendley (1969).

The radial particle density distribution is of the form

$$N_e(r) = N_{e_0} \exp\left[-1{\cdot}155\left(1-\left(1-\left(\frac{r}{R}\right)^2\right)^{1/2}\right)\right] \qquad \ldots(8.8)$$

This particle distribution corresponds to the case $\delta^* = 0$ in Fig. 24. The radial potential distribution within the plasma is

$$V(r) = \frac{1{\cdot}155 \times kT_e}{e}[1-(1-(r/R)^2)^{1/2}]. \qquad \ldots(8.9)$$

The potential at the tube wall is

$$V_w = \frac{kT_e}{c}\ln\left[1{\cdot}045\left(\frac{M_i}{m_e}\right)^{1/2}\right]$$

The difference between $V(R)$ and V_w gives the potential drop in the sheath. The discharge current is

$$I = 0{\cdot}7 N_{e_0} \pi R^2 e\mu_e E$$

and from Forrest and Franklin (1968)

$$\frac{Z}{p}(T_e)\left(\frac{M_i}{kT_e}\right)^{1/2}(pR) = 1{\cdot}109. \qquad \ldots(8.10)$$

The variation in electron temperature with the product pR for the case of argon as calculated by Hernqvist and Fendley (1967) is shown in Fig. 25. The two curves show the predicted behaviour of electron temperatures at low and high pressures.

For the very small values of α which apply in typical ion lasers Tonks–Langmuir theory predicts no variation in electron temperature with current (Forrest and Franklin, 1968). Much of the increase in electron temperature with current that is observed in the argon ion laser is probably due to the reduction in neutral particle density in the capillary caused by drive out and gas pumping which occur as the current is increased.

The effects of magnetic fields on "low" pressure plasmas have been considered by Forrest and Franklin (1966a, b); the effect of the field is best understood in terms of the parameter δ^* which in a plasma where the ions move collisionlessly to the walls is δ^* (free-fall) $= \delta_M$ where

$$\delta_M = \frac{kT_e}{2ZM_i D_e}$$

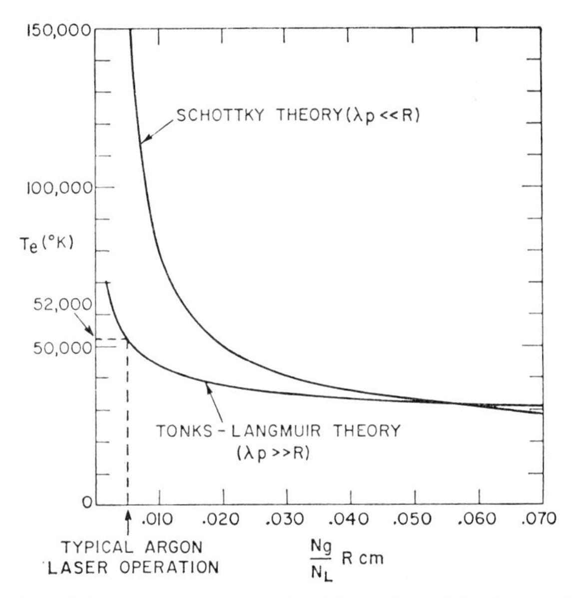

FIG. 25. Variation of electron temperature, T_e, with product of density × radius (Hernqvist and Fendley, 1967).

δ_M is negligibly small in the absence of a magnetic field. When an axial magnetic field is applied the electrons move along the discharge axis in spiral paths of average radius

$$\bar{r}_M = \frac{4}{\pi e B}\left(\frac{2m_e kT_e}{\pi}\right)^{1/2}$$

where B is the magnetic flux.

This radius is $\simeq 2{\cdot}4\,(kT_e)^{1/2}\,B^{-1}$ cm where kT_e is in eV and B in gauss. If the radius of these spiral orbits is smaller than the discharge tube radius, then the electrons will move outward towards the walls by diffusion across the field lines.

In the presence of a magnetic field the electron diffusion coefficient is

$$D_e = \frac{D_{e_0}}{(1+\omega^2\,\tau_e^2)}$$

where D_{e_0} is the electron diffusion coefficient in the absence of the field, $\omega = eB/m_e$ is the cyclotron radial frequency and τ_e is the mean free time between collisions for an electron. Increase of magnetic field increases the value of δ_M and causes the radial potential and charge distribution within the discharge tube to change, curves showing these changes are given by Forrest and Franklin (1966a). The main effect of the magnetic field is to make the electron density distribution adopt the form that it would adopt at higher pressures. This profile is more peaked towards the centre of the discharge tube and corresponds to an increase in axial electron density with both magnetic field and pressure (at the same average electron density). Very high magnetic fields can cause the radial field to reverse direction and this strongly reduces ion loss to the tube walls.

Since $D_e = \mu_e k T_e / e$ where μ_e is the electron mobility, a reduction in D_e caused by magnetic field leads to a subsequent fall in electron temperature. This fall in electron temperature is most pronounced at low pressures.

B. INTERMEDIATE PRESSURES

At intermediate pressures, ions no longer move collisionlessly to the walls and the term $\nu_i/2$ in δ^* must also be considered. Figure 26 which is taken from

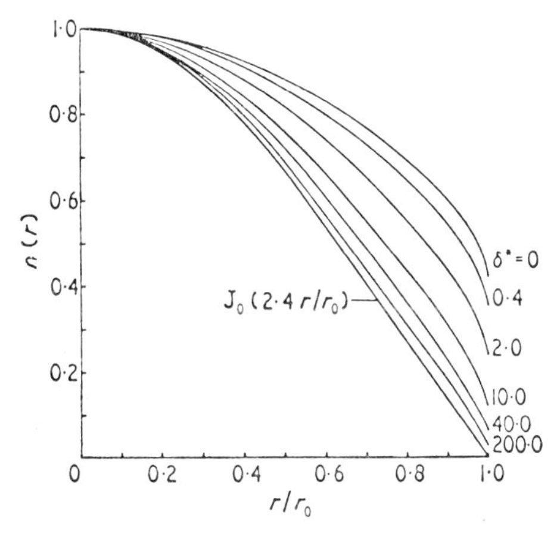

FIG. 26. Variation of the normalized charged particle number density with normalized radius for various values of δ^* (Forrest and Franklin, 1966).

Forrest and Franklin (1966a, b) shows the behaviour of the radial charged particle distribution as a function of δ^*. The gradual change of this profile from the zero magnetic field free-fall case ($\delta^* = 0$) to the diffusion limited behaviour for large δ^* can be interpreted either in terms of the effect of pressure increase on the plasma or the application of an axial magnetic field or both.

Now the collision frequency for the radially moving ions in a plasma is $\nu_i = N\sigma_D \bar{v}_r$.

In a typical argon ion laser plasma with $N \simeq 2{\cdot}7 \times 10^{15}$ particles cm^{-3}, $\sigma_D = 9{\cdot}3 . 10^{-15}$ cm^2 and $\bar{v}_r \simeq 10^5$ cm s^{-1} (Bennett and Sze, 1972)

$$\nu_i = 2{\cdot}5 \times 10^6 \text{ s}^{-1}$$

and

$$\nu_i/2Z \simeq 0{\cdot}3$$

In zero magnetic field the effect on the radial charged particle profile of a term of this order in δ^* is small.

C. HIGH PRESSURES: SCHOTTKY REGIME

At high pressures the motion of the ions in a laser plasma is governed by diffusion and the plasma is adequately described by the theory of Schottky (1924) (see Cobine, 1958).

The radial changed particle distribution becomes

$$N_e(r) = N_{e_0} J\left(r\left(\frac{Z}{D_a}\right)^{1/2}\right) \qquad \text{...(8.11)}$$

where J_0 is the Bessel function of zero order and D_a is the ambipolar diffusion coefficient,

$$D_a = \frac{D_e \mu_e + D_i \mu_i}{\mu_e + \mu_i} \qquad \text{...(8.12)}$$

where D_i is the ion diffusion coefficient, μ_i is the ion mobility and μ_e is the electron mobility.

The first zero of $J_0(x)$ occurs at $x = 2{\cdot}405$. This must correspond to the discharge tube wall; so we can write

$$N_e(r) = N_{e_0} J_0 \frac{(2{\cdot}405r)}{R} \qquad \text{...(8.13)}$$

This charged particle distribution is the asymptotic form for large δ^* as considered by Forrest and Franklin (1966b). Although the application of a magnetic field at high pressure alters the ambipolar diffusion coefficient it does not alter the radial charged particle distributions. The ambipolar diffusion coefficient becomes

$$D_a = D_{a_0}\left(1 + \frac{\mu_i}{\mu_{e_0}}(1 + \omega^2 \tau_e^2)\right)^{-1} \qquad \text{...(8.14)}$$

where D_{a_0} is the value of the ambipolar diffusion coefficient in zero field and μ_{e_0} the electron mobility in zero field.

In the presence of a magnetic field at high pressure when the particle motion is governed by ambipolar diffusion Tonks (1939) gives, for the radial field,

$$E(r) = \left(\frac{D_{e_0}}{1 + w^2 \tau_e^2} - D_i\right)\frac{1}{N}\frac{dN}{dr} \qquad \text{...(8.15)}$$

Since $1/N\mathrm{d}N/\mathrm{d}r$ does not alter with magnetic field when $\omega^2 \tau_e^2$ becomes greater than D_{e_0}/D_i the sign of the field reverses. The magnetic field impedes the motion of the electrons so much that they now appear the more massive particle.

Webb (1968) has studied the radial particle distribution in argon ion laser plasmas and finds that for values of the product $(p_0 R) \gtrsim 0{\cdot}4$ torr mm the radial density profile follows that predicted by Tonks–Langmuir theory closely, whilst for $(P_0 R) \gtrsim 4$ torr mm the radial density follows a Schottky profile. For an ion laser plasma operating in the high pressure region there should be no change in electron temperature either with current or magnetic field.

D. HIGHLY IONIZED PLASMAS

In the above considerations of ion laser plasmas it has been assumed that the degree of ionization is low. Bearing in mind that the typical particle density in a CW argon ion laser is $\simeq 10^{15}$ cm^{-3} or greater and the electron density may reach 10^{14} cm^{-3} (see Table IV) or more this assumption will break down at medium to high current densities. It may, or may not, break down at lower current densities in other noble gas ion lasers. Under these circumstances it may be appropriate to describe certain parameters of the plasma using the theory of fully ionized gases.

Hattori and Goto (1969a) use relations given by Spitzer (1961) to estimate the plasma parameters of a high current inductively pumped argon ion laser. If this laser is highly ionized then the measured discharge conductivity $\sigma = J/E$ is related to the electron temperature and current density by

$$\sigma = 1{\cdot}53 \times 10^{-4}(T_e^{3/2}/\ln \Lambda) \quad \Omega^{-1}\ \text{cm}^{-1} \qquad \ldots(8.16)$$

where

$$\Lambda = \frac{2}{e^3}\left[\frac{(kT_e)^3}{\pi N_e}\right]^{1/2}$$

and the electron drift velocity is given by

$$V_{d_e} = \frac{(kT_e)\,E}{2\pi c e^3 N_e \ln \Lambda}\left(\frac{3kT_e}{m_e}\right)^{1/2}$$

E. ION MOTION IN PLASMAS: KAGAN–PEREL THEORY

As mentioned in previous sections, under normal circumstances the ions in a discharge plasma move both longitudinally under the influence of the axial electric field E and radially under the influence of the radial field E_r.

It is evident from measurements of radial and longitudinal drift velocities in ion lasers that the radial field is larger than the longitudinal field by up to an order of magnitude 5–10. Consequently the overall ion motion is approximately perpendicular to the tube wall. The energy which the ions pick up in this field causes them to strike the wall, where they recombine, with considerable force. This impact of ions upon the discharge tube walls is responsible for the severe sputtering which frequently occurs in ion lasers.

The problem of this ion motion in plasmas has been considered by Kagan and Perel (1957, 1958a, b) Kagan (1958), Zakharov and Pekar (1970) and Sze and Bennett (1972). It is formulated usually in cylindrical coordinates in terms of the Boltzmann equation which describes the motion of the discharge ions in phase space in terms of their distribution function f. If there is no directed motion in the plasma and the ions form a Maxwellian distribution with temperature T_i, then f takes the form

$$f = \rho\left(\frac{M}{2\pi kT_i}\right)^{3/2} \exp\left[\frac{-M_i}{2kT_i}(v_\rho^2 + v_\phi^2 + v^2)\right] \qquad \text{...(8.17)}$$

and the number of ions in the volume of phase space $d\rho\, d\phi\, dz\, dv_\rho\, dv_\phi\, dv$ at the point $(\rho\phi x v_\rho v_\phi v_x)$ is $f d\rho\, d\phi\, dx\, dv_\rho\, dv_\phi\, dv_x$ where ρ is the radial coordinate in the plasma, ϕ the azimuthal angle and x the position of the ion in the axial direction. v_ρ is the radial velocity at radius ρ, v_ϕ is the aximuthal velocity and v_x the longitudinal velocity.

When directed motion of the ions occurs, f changes from the simple Maxwellian form and becomes a function of radial position within the plasma. Kagan and Perel do not solve the Boltzmann equation explicitly but, by making certain assumptions about the form of f, are able to calculate certain average values of parameters of the plasma.

They calculate that the average longitudinal ion drift velocity on axis at low pressures is (Kagan and Perel, 1957),

$$\bar{v}_x = \frac{0{\cdot}69\, eE_\parallel}{M_i Z} \qquad \text{...(8.18)}$$

where $E_\parallel$ is the axial field and Z is the ionization rate in the plasma. The mean radial velocity at low pressures near the tube axis is (Kagan and Perel, 1958a), and near the wall (Kagan and Perel, 1958a),

$$\bar{v}_r^2 = \frac{0{\cdot}67\, kT_a}{M_i} \qquad \text{...(8.19)}$$

and near the wall (Kagan and Perel, 1958a).

$$\bar{v}^2 = \frac{0{\cdot}38\, kT_e}{M_i} \qquad \text{...(8.20)}$$

In general the mean square radial ion velocity at radius ρ is

$$\bar{v}_\rho^2 = A(S)\frac{kT_a}{2M_i} + C(S)\frac{kT_e}{M_i} \qquad \text{...(8.21)}$$

when

$$S = Z\left(\frac{M}{2kT_e}\right)^{1/2} \rho$$

and $A(S)$ and $C(S)$ are functions tabulated by Kagan and Perel (1958a).

Kagan and Perel assume that the one dimensional radial ion velocity distribution at any radius is still Maxwellian, but with mean velocity $\bar{v}$, so that in the radial dimension only

$$f(\rho) \propto \frac{1}{(T_\rho)^{1/2}} \exp\left[\frac{-M_i}{2kT_\rho}(v_\rho - \bar{v}_\rho)\right] \qquad \ldots(8.22)$$

where T_ρ is a radial ion temperature which varies with ρ according to

$$T_\rho = A(S)\frac{T_a}{2} + B(S)\,T_\rho.$$

However, Zakharov and Pekar (1971) have given more careful consideration to the problem of the radial ion velocity distribution at low pressure and have shown that the distribution is not Maxwellian except on the discharge axis and deviates more and more as the tube wall is approached. Figure 27 shows the form of the radial ion velocity distribution at different distances from the tube axis as determined by Zakharov and Pekar.

Kagan and Perel (1958b) calculate that the longitudinal ion drift velocity on the tube axis at any pressure is

$$\bar{v}_x = \frac{eE_\parallel}{M_i\left(Z + \frac{16}{3\lambda_i}\left(\frac{kT_a}{\pi M_i}\right)^{1/2}\right)} \qquad \ldots(8.23)$$

where λ_i is the mean free path. Consequently at high pressures

$$\bar{v}_x = \frac{3eE_\parallel \lambda_i}{16Z}\left(\frac{\pi}{kM_i T_a}\right)^{1/2} \qquad \ldots(8.24)$$

Kagan and Perel assume that the longitudinal distribution function of ion velocities is of the form

$$f(v_x) \propto \frac{1}{(T_x)^{1/2}} \exp\left[-\frac{M_i}{2kT_x}(v_x - \bar{v}_x)^2\right] \qquad \ldots(8.25)$$

where T_x is the longitudinal ion temperature and both $\bar{v}_x$ and T_x are functions of radius.

The consequences of these various assumptions on the exact nature of ion drift in the analysis of lineshapes emitted by ions in ion lasers will be discussed in the section on lineshapes in ion lasers (Section IX).

At high pressures when Schottky theory describes the plasma, the ion motion becomes controlled by diffusion and the radial ion velocity can more easily be determined From Cobine (1958, p. 236) the rate at which ions are moving towards the discharge tube walls at radius r is

$$\left(\frac{\mathrm{d}N}{\mathrm{d}t}\right)_r = -2\pi r D_a \left(\frac{\mathrm{d}N}{\mathrm{d}r}\right)_r \qquad \ldots(8.26)$$

where $N(r)$ is the radial ion density.

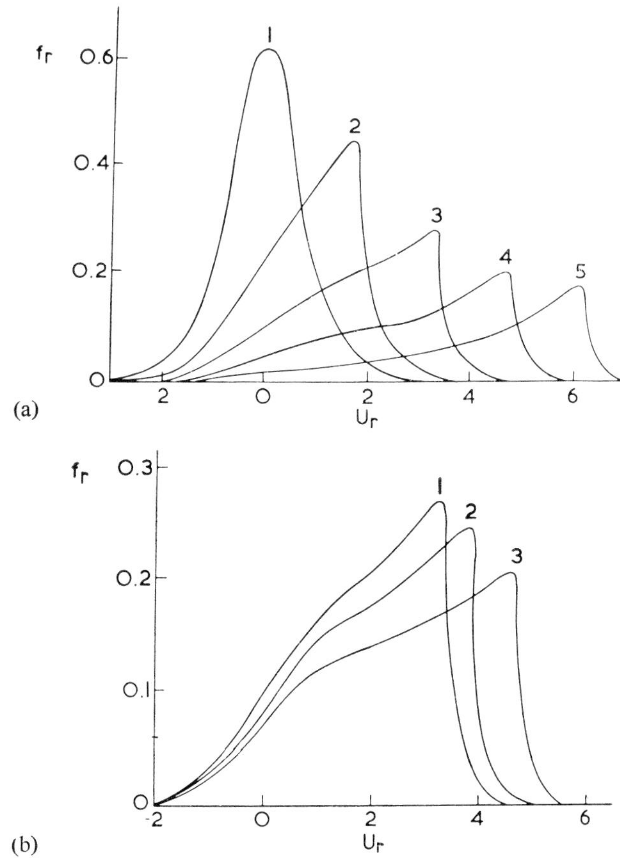

FIG. 27. Ion distribution function as a function of velocity for (a) $\theta = T_e/T = 21{\cdot}667$ and varying $x = r/R$ of (1) 0, (2) 0·324, (3) 0·613, (4) 0·836 and (5) 0·968 and (b) $x = 0{\cdot}613$ and varying θ of (1) 21·667, (2) 28·889 and (3) 43·333.

If the mean outward radial velocity of ions at radius r is V_r then the net flux of ions through the cylindrical surface of radius r is

$$\left(\frac{\mathrm{d}N}{\mathrm{d}t}\right)_r = 2\pi r N_r r. \qquad \ldots(8.27)$$

Then,

$$V_r = -\frac{D_a}{N_r}\left(\frac{\mathrm{d}N}{\mathrm{d}r}\right)_r \qquad \ldots(8.28)$$

and since N_r satisfies

$$N_r = N_0 J_0\left(\frac{2{\cdot}405\, r}{R}\right)$$

$$V_r = -\frac{D_a \dfrac{\mathrm{d}}{\mathrm{d}r}\left[J_0\left(\dfrac{2{\cdot}405\, r}{R}\right)\right]}{J_0\left(\dfrac{2{\cdot}405\, r}{R}\right)} = \frac{D_a J_1\left(\dfrac{2{\cdot}405\, r}{R}\right)}{J_0\left(\dfrac{2{\cdot}405\, r}{R}\right)}$$

where J_1 is the Bessel function of order 1. That is, the mean drift velocity of the ions becomes larger towards the wall. However in the bulk volume of the plasma this drift velocity is too small to cause any significant changes in transversely emitted lineshapes at high pressures.

IX. Spontaneous Emission Line Profiles and Lifetimes in Ion Lasers

The shape of a spontaneoulsy emitted line in an ion laser is controlled by a number of broadening mechanisms: A. Doppler broadening; B. Natural broadening; C. Pressure broadening; D. Apparent broadening caused by ion drift (Kagan–Perel broadening). The relative magnitude of these different mechanisms will determine whether the laser is homogeneously or inhomogeneously broadened and the way in which the broadening of the laser transition controls its output longitudinal mode structure. The spontaneous emission lineshapes of a laser transition along with all other parameters directly influences the gain of the transition (see Section IX, G on lifetimes in ion lasers).

A. Doppler Broadening

The Doppler broadening of both the neutral and ionized transitions is a manifestation of the random velocities of the emitting particles. If these random velocities are Maxwellian distributed then the emitted line shape is a Gaussian with full width at half maximum (FWHM) $\Delta\nu_D$. The normalized form of the frequency distribution lineshape function in this case is

$$g_D(\nu) = \frac{\Delta\nu_D}{2}\left(\frac{\pi}{\ln 2}\right)^{1/2} \exp\{-[2(\nu-\nu_0)/\Delta\nu_D]^2 \ln 2\}$$

where

$$\int_{-\infty}^{\infty} g(\nu) = 1$$

and

$$\Delta\nu_D = \frac{2\nu_0}{c}\left(\frac{2kT\ln 2}{M}\right)^{1/2} \qquad \ldots(9.1)$$

where T is the temperature of the emitting species. At 2500°K, $\Delta\nu_D = [\lambda(\text{mass number})^{1/2}]^{-1}$ 10·75 GHz. Some typical Doppler broadened line-widths calculated from this formula at 300°K and 2500°K are

3324 Å neon II transition: $\Delta\nu_D$ (300°K) = 2·5 GHz, $\Delta\nu_D$ (2500°K) = 7·2 GHz
4480 Å argon II transition: $\Delta\nu_D$ (300°K) = 1·2 HGz, $\Delta\nu_D$ (2500°K) = 3·5 GHz
5682 Å krypton II transition: $\Delta\nu_D$ (300°K) = 0·7 GHz, $\Delta\nu_D$ (2500°K) = 2·1 GHz.

Doppler broadening is generally the dominant broadening mechanism operative in ion lasers, the only exception being perhaps in very high current density and electron density pulsed lasers where Stark broadening may be of comparable magnitude. Since Doppler broadening is an inhomogeneous broadening mechanism, it might be expected that ion lasers would exhibit those features expected of inhomogeneously broadened lasers, particularly

(a) Power output α gain squared.
(b) Multiple longitudinal mode operation possible with adjacent modes spaced by $c/2L$ in frequency (where L is the length of the laser cavity).

In practice, although Doppler broadening is the dominant broadening mechanism in ion lasers, particularly in the noble gas ion lasers, the homogeneous contribution to the linewidth is significant and behaviour more typical of a homogeneously broadened laser is observed.

The main contribution to the homogeneous linewidth comes from natural broadening which is very large in these lasers.

B. NATURAL BROADENING

The natural broadened linewidth (FWH M)$\Delta\nu_N$ of a transition is determined by the lifetimes τ_2 and τ_1 of its upper and lower levels respectively from

$$\Delta\nu_N = \frac{1}{2\pi}\left(\frac{1}{\tau_1} + \frac{1}{\tau_2}\right) \qquad \ldots(9.2)$$

In the noble gases, the lower state lifetimes are very short and the natural linewidths are very large as can be seen from Table XIII. Natural broadening is homogeneous and leads to a Lorentzian line shape whose normalized form is

$$g_N(\nu) = \frac{2}{\pi\Delta\nu_N}\left[\frac{1}{1 + [2(\nu - \nu_0)/\Delta\nu_N]^2}\right] \qquad \ldots(9.3)$$

Although in typical ion lasers this homogeneous contribution to the lineshape is smaller than that of the inhomogeneous contribution, it is sufficiently large that the overall lineshape deviates significantly from a Gaussian and becomes a Voigt profile (see Griem, 1964). The ratio of the inhomogeneous to homogeneous linewidths for some typical transitions in the noble gas lasers are (assuming an ion temperature of 2500°K)

3324 Å NeII—11·2, 4880 Å ArII—7·6, 5682 Å, KrII—6·4.

TABLE XIII. *Lifetimes and natural linewidths in noble gas ion lasers*

Species	Wavelength (Å)	Transition	Upper level lifetime (ns)	Lower level lifetime (ns)	Ratio	Natural linewidth (MHz)
Ne II	3324	$3p\,^2P^0_{3/2} \rightarrow 3s\,^2P_{3/2}$	5[a]	0·26[d]	19	644
Ne II	3378	$3p\,^2P^0_{1/2} \rightarrow 3s\,^2P_{1/2}$	5·9[b]	0·26[d]	23	640
Ne II	3393	$3p\,^2P^0_{3/2} \rightarrow 3s\,^2P_{1/2}$	5[a]	0·26[d]	19	644
Ne II	3713	$3p\,^2D^0_{5/2} \rightarrow 3s\,^2P_{1/2}$	8·3[b,c]	0·26[d]	32	631
Ar III	3511	$4p\,^3P_2 \rightarrow 4s\,^3S^0_1$	3·2[h,i]	—	—	—
Ar III	3638	$(^2D^0)4p\,^1F_3 \rightarrow (^2D^0)4s\,^1D^0_2$	3·6[h,i,j,k]	—	—	—
Ar II	4545	$4p\,^2P^0_{3/2} \rightarrow 4s\,^2P_{3/2}$	9·4[e]	0·36[f]	26	459
Ar II	4579	$4p\,^2S^0_{1/2} \rightarrow 4s\,^2P_{1/2}$	8·8[e]	0·36[f]	24	460
Ar II	4658	$4p\,^2P^0_{1/2} \rightarrow 4s\,^2P_{3/2}$	8·7[e]	0·36[f]	24	460
Ar II	4765	$4p\,^2P^0_{3/2} \rightarrow 4s\,^2P_{1/2}$	9·4[e]	0·36[f]	26	459
Ar II	4880	$4p\,^2D^0_{5/2} \rightarrow 4s\,^2P_{3/2}$	9·1[e]	0·36[f]	25	460
Ar II	4965	$4p\,^2D^0_{3/2} \rightarrow 4s\,^2P_{1/2}$	9·8[e]	0·36[f]	27	458
Ar II	5017	$(^1D)4p\,^2F^0_{5/2} \rightarrow 3d\,^2D_{3/2}$	7·9[e]	0·58[g]	14	295
Ar II	5145	$4p\,^4D^0_{5/2} \rightarrow 4s\,^2P_{3/2}$	7·5[e]	0·36[f]	21	463

TABLE XIII—*continued*

Species	Wavelength (Å)	Transition	Upper level lifetime (ns)	Lower level lifetime (ns)	Ratio	Natural linewidth (MHz)
Kr III	3507	$5p\,^3P_3 \rightarrow 5s\,^3S_1^0$	5·4[h]	—	—	—
Kr III	3564	$5p\,^3P_1 \rightarrow 5s\,^3S_1^0$	5·1[h]	—	—	—
Kr III	4067	$(^2D^0)5p\,^1F_3 \rightarrow (^2D^0)5s\,^1D_2^0$	5·6[h]	—	—	—
Kr III	4131	$5p\,^5P_2 \rightarrow 5s\,^3S_1^0$	7·0[h]	—	—	—
Kr II	4619	$5p\,^2D_{5/2}^0 \rightarrow 5s\,^2P_{3/2}$	7·5[h]	0·52[l, m]	14	327
Kr II	4762	$5p\,^2D_{3/2}^0 \rightarrow 5s\,^2P_{1/2}$	7·8[h]	0·44[l, m]	18	382
Kr II	4825	$5p\,^4S_{3/2}^0 \rightarrow 5s\,^2P_{1/2}$	10·4[h]	0·44[l, m]	24	377
Kr II	5682	$5p\,^4D_{5/2}^0 \rightarrow 5s\,^2P_{3/2}$	7·6[h]	0·52[l, m]	15	327
Kr II	6471	$5p\,^4P_{5/2}^0 \rightarrow 5s\,^2P_{3/2}$	8·5[h]	0·52[l, m]	16	325
Xe II	4603	$6p\,^4D_{3/2}^0 \rightarrow 6s\,^4P_{3/2}$	7[n]	—	—	—
Xe II	5419	$6p\,^4D_{5/2}^0 \rightarrow 6s\,^4P_{3/2}$	11[n]	—	—	—
Xe II	5971	$(^1D)6p\,^2P_{3/2}^0 \rightarrow (^1D)6s\,^2D_{3/2}$	5[n]	—	—	—
Xe II	8716	$6p\,^4D_{3/2}^0 \rightarrow 5d\,^2P_{3/2}$	7[n]	—	—	—

[a] Hesser (1968),
[b] Brand *et al.* (1970),
[c] Denis (1969),
[d] Hodges *et al.* (1970),
[e] Bennett *et al.* (1964),
[f] Statz *et al.* (1965),
[g] Luyken (1972),
[h] Fink *et al.* (1970),
[i] Ceyzeriat *et al.* (1971),
[j] Kernahan *et al.* (1970),
[k] Denis and Dufay (1969),
[l] Marantz (1968)
[m] Marantz *et al.* (1969),
[n] Allen *et al.* (1969).

Analysis of spontaneous emission Voigt profiles for some of the argon ion laser transitions has been carried out by a number of authors (Bennett *et al.*, 1966; Ballik *et al.*, 1966; Korolev *et al.*, 1969; Donin, 1970; Kitaeva *et al.*, 1971; Sze *et al.*, 1972; Sze and Bennett, 1972) using the method originally proposed by Ballik (1966). These investigations have confirmed the existence of a fairly large Lorentzian contribution (of the order of or greater than 500 MHz) to the overall spontaneous emission lineshape of the strong visible transitions. The measured Lorentzian widths are generally slightly larger than the actual linewidth predicted from lifetime values and increase in width with current density (Sze *et al.*, 1972; Sze and Bennett, 1972). The additional contribution to the Lorentzian linewidth almost certainly comes from Stark broadening. The only puzzling fact that remains is that the Lorentzian width of the 5145 Å transition extrapolated to zero current is smaller than its calculated natural linewidth which may indicate that the lower level of this transition has a longer lifetime than predicted.

C. PRESSURE BROADENING

Pressure broadening of the transitions emitted by an atom or ion in a discharge arises through the perturbation caused by other charged and neutral particles in the vicinity of the emitting particle. The dominant pressure broadening mechanism operative in the ion lasers is Stark broadening caused by interaction of the emitting ion with neighbouring electrons. The magnitude of this Stark broadening increases with electron density and is largest for light elements. The magnitude of the Stark-broadening contribution in any ion laser can be estimated from the electron density and temperature in the plasma using the tabulated coefficients of Griem (1964). It appears that in the case of the argon ion transitions at least, these coefficients underestimate the amount of the broadening. In this laser the magnitude of the Stark-broadening contribution to the linewidth can be estimated from the more recent data of Roberts (1968) who concludes that the FWHM of the Stark-broadened Lorentzian line is

$$\Delta\nu_s = 2\Omega(T_e)\,N_e \qquad \ldots(9.4)$$

where $\Omega(T_e)$ is a slowly changing function of T_e.

Roberts gives a value of Ω of 3·1 GHz, for $N_e = 10^{16}$ cm^{-3} and $T_e = 2{\cdot}7$ eV. Thus at an electron density of $5{\cdot}10^{13}$ cm^{-3}, which is fairly typical of CW argon ion lasers, $\Delta\nu_s \simeq 30$ MHz.

In higher current density CW and pulsed lasers the Stark broadening contribution may become comparable with the natural linewidth.

The large overall homogeneous linewidth (natural + Stark broadening) in the noble gas ion lasers makes them behave in many ways as if they were homogeneously broadened overall. In an operating noble gas ion laser the holes burnt in the gain curve by the operating longitudinal modes are generally wider than the spacing between these modes, so the holes overlap, except just above threshold. This can lead to high gain longitudinal modes suppressing neighbouring modes (see Section XI, B). This overlapping of holes also allows over half of the available power in the lineshape to be extracted from the laser in multimode operation, so that the output power of the laser becomes linearly proportional to gain. This has been demonstrated to occur both in CW (Gordon *et al.*, 1965) and pulsed argon ion lasers (Davis and King, 1971). The large homogeneous contribution to the linewidth in the noble gas ion lasers allows a significant proportion of the total available laser power to be extracted from a single mode. Cross-relaxation also contributes to enhance the available power from the whole lineshape in single mode operation. Davis and Lindsay (to be published) have observed a reduction in intensity across the whole spontaneous emission profile of the argon ion spontaneous transition at 6684 Å, viewed in the axial direction, with the onset of lasing at 5145 Å which has the same upper level.

D. KAGAN–PEREL BROADENING

As discussed in previous sections the ions in a discharge plasma have drift velocity components both in the direction of the axis and radially towards the wall. The drift velocity component in the direction of the axis is a function of radial position, particularly at low presusres. Consequently if a spontaneous emission line from such an ion is viewed in the axial direction an average shift in the line centre frequency will be observed. Depending on the viewing direction the observed average centre frequency shift will be

$$\Delta\nu_d \simeq \pm \frac{1}{\lambda_0} \frac{\int_0^R N_i(r)\, v_d(r)\, \mathrm{d}r}{\int_0^R N_i(r)\, \mathrm{d}r}. \qquad \ldots(9.5)$$

providing light emitted by ions at different radial positions is collected uniformly across the tube bore.

This Doppler shift can be ascribed to an average longitudinal ion drift velocity v_d^i where

$$\Delta\nu_d = \pm \frac{v_d^i}{\lambda_0}$$

It is this quantity v_d^i which is generally measured in ion lasers (see Table X).

Bennett *et al.* (1966) have pointed out that because the ions are accelerated along the axis between collisions not only is the centre frequency of the line shifted, but also the lineshape is altered and becomes asymmetrical. However, the amount of asymmetry expected in the lineshapes emitted in the axial direction in typical ion lasers is very small.

A more complicated situation exists when the spontaneous emission is observed in a direction perpendicular to the discharge tube axis. If the observer restricts his attention to a very thin slice of emitting ions taken through the tube axis then he will see the superposition of emitted lineshapes emitted by ions with different radial velocities moving both towards and away from him. Kagan and Perel (1958a) assume that the overall observed profile will still be Gaussian. They take weighted averages of the excited ion velocity across the tube radius and ascribe this average mean square radial velocity to an effective transverse ion temperature $T_{\perp}^{i}$ where

$$\frac{kT_{\perp}^{i}}{M_i} = \frac{\int_0^R N_i(r)\overline{V_r^2}N_e(r)\,\mathrm{d}r}{\int_0^R N_i(r)\,N_e(r)\,\mathrm{d}r} \qquad \ldots(9.6)$$

and obtain a relation of the form

$$T_{\perp}^{i} = 0{\cdot}56T_a + 0{\cdot}13T_e \qquad \ldots(9.7)$$

This relation has been used by a number of investigators (Kitaeva *et al.*, 1969) to obtain values for the electron temperature in ion lasers from measured values of the neutral atom and transverse ion temperatures.

However, Sze and Bennett (1972) have pointed out that the spatial average used by Kagan and Perel should be replaced by

$$\frac{kT_{\perp}^{i}}{M_i} = \frac{\int_0^R N_i(r)\overline{v_r^2}N_e(r)\,r\,\mathrm{d}r}{\int_0^R N_i(r)\,N_e(r)\,r\,\mathrm{d}r} \qquad \ldots(9.8)$$

to take into account the cylindrical geometry of the situation. This leads to a modified expression

$$T_{\perp}^{i} = 0{\cdot}71\,T_a + 0{\cdot}21\,T_e \qquad \ldots(9.9)$$

Also, Sze and Bennett (1972) have considered the situation more carefully and calculate the mean square radial velocity in terms of the number of ions which have acquired a particular velocity v_r by the time they reach the radial position r within the tube. This analysis gives quite close agreement with a modified Kagan–Perel relation of the form

$$T_{\perp}^{i} = 1{\cdot}42\,T_{\parallel}^{i} + 0{\cdot}21\,T_e \qquad \ldots(9.10)$$

where $T_{\parallel}^{i}$ is the measured axial temperature of the ions. In their analysis Sze and Bennett consider the effect of charge exchange collisions on the observed lineshape, since such a collision produces an ion with zero acquired radial velocity. Zakharov and Pekar (1971) neglect such effects but their calculated radial ion velocity distribution should be very reliable at low pressures. They produce a modified Kagan–Perel expression of the form

$$T_{\perp}^{i} = 0{\cdot}64\ T_a + 0{\cdot}118\ T_e \qquad \ldots(9.11)$$

which has been used by Kitaeva *et al.* (1971) to analyse lineshapes in argon ion lasers. Unfortunately, it appears to be the same kind of incorrect spatial average as that of Kagan and Perel.

Further work on the effects of radial motion on the lineshape of ion laser transitions at low pressures appears necessary, for example, taking into account the velocity and density distribution of the neutral atoms within the discharge tube. At high pressures when the laser plasma is controlled by diffusion, the radially directed motion of ions is much slower and very little deviation from a Gaussian lineshape would be expected for ion emission viewed transversely to the discharge tube axis.

E. ZEEMAN SPLITTING OF ION LASER TRANSITIONS IN MAGNETIC FIELDS

Noble gas ion lasers are frequently operated in axial magnetic fields which can increase their power output and efficiency (see Section X, F) by modifying certain fundamental parameters of the discharge plasma (see Section VII, C). However, such fields also modify the output frequency and polarization characteristics of the laser transition because of the Zeeman effect. An axial magnetic field splits the transition into its σ_+ and σ_- components which are right- and left-hand circularly polarized respectively observed in the direction of the field. Thus the gain of the laser transition at a particular frequency becomes dependent on whether the light is right- or left-hand circularly polarized. Further, because the real part of the plasma refractive index will, in general, be different for right- and left-hand circularly polarized light, the laser medium will exhibit Faraday rotation. As a consequence the axial magnetic field causes the light from a laser employing Brewster angle windows to become elliptically polarized, and the light from an internal mirror laser to break up into a set of right- and left-hand circularly polarized modes. In magnetic fields which are high enough to split the gain profiles of the σ_+ and σ_- components by more than the homogeneous linewidth (or the cavity longitudinal mode spacing, if this is greater) threshold operation occurs at two longitudinal mode frequencies near the peaks of the σ_+ and σ_- profiles.

Although Brewster windows define a direction for a linearly polarized output, light which would be linearly polarized in the right plane at one Brewster window suffers a reflection loss at the other because of the Faraday rotation which occurs as the light passes along the laser tube. This leads to overall ellipticity of the output beam. Sinclair (1966) has studied these phenomena for the 4880 Å transition in an argon ion laser, for which the splitting of the σ components is 3·5 MHz gauss^{-1}. At fields of 1000 G this splitting is comparable to the Doppler width of the transition. Sinclair found that the ellipticity of the laser beam was small even at quite high magnetic fields ($\sim$ 0·2 at 1000 G). Even with no magnetic field Sinclair observed that the output was slightly elliptically polarized. This was attributed to stress-induced birefringence in the fused silica windows. Similar effects have been thoroughly investigated in the He–Ne laser (de Lang, 1967). The ellipticity in the output beam, of course, can be converted into a linear polarization by using a linear polarizer. Sinclair concludes that there is very little difference in the output power behaviour as a function of magnetic field in lasers with Brewster windows or internal mirrors.

Labuda *et al.* (1965) and Gorog and Spong (1967) also report equal outputs well above threshold from both types of laser. However, at high fields, Gordon *et al.* (1965) observed the opposite effect. Higher output power is expected in theory from a laser with internal mirrors in an axial magnetic field, particularly for transitions whose Zeeman splitting is large. Certainly, near threshold Brewster window losses are greater than for anti-reflection coated windows (Gorog and Spong, 1967).

In an ion laser, degenerate magnetic sub-levels can be assumed to be equally populated by level mixing collisions. Any asymmetry introduced by laser action with polarized beams is countered by cross-relaxation between these degenerate sub-levels, by for example ion–ion collisions. Bloom *et al.* (1966) and Dunn (1968) have confirmed this in the case of the argon ion laser by magnetic field experiments.

In very high current density ion lasers the self magnetic field of the discharge current produces a small Zeeman splitting ($\simeq$ 200 MHz at 850 A cm^{-2} for the visible transitions in argon ion lasers) but this splitting is sufficiently small as not to influence the observed properties of the laser to a noticeable extent.

F. ISOTOPE EFFECTS IN ION LASERS

The effect of the different isotopes of an ion laser material on its operation can generally be neglected, particularly in the noble gas ion lasers where the isotope shifts of emission centre frequency are much smaller than the line broadening effects in the laser. However, in the helium–cadmium laser the cadmium isotope shifts are sufficiently large to introduce significant asymmetry into the laser gain profile and produce effects which are noticeable in the output mode structure of the laser. Such effects have been observed by Hopkins and Fowles (1968). Figure 28 shows the theoretical gain profile of a natural isotopic abundance inhomogeneously broadened He–Cd laser at 4416 Å. The Gaussian lineshapes for each isotope are shown as well as their sum. By using special isotope mixtures in such a laser the gain profile can be made very broad and the width of mode-locked pulses from the laser reduced (Silfvast and Smith,

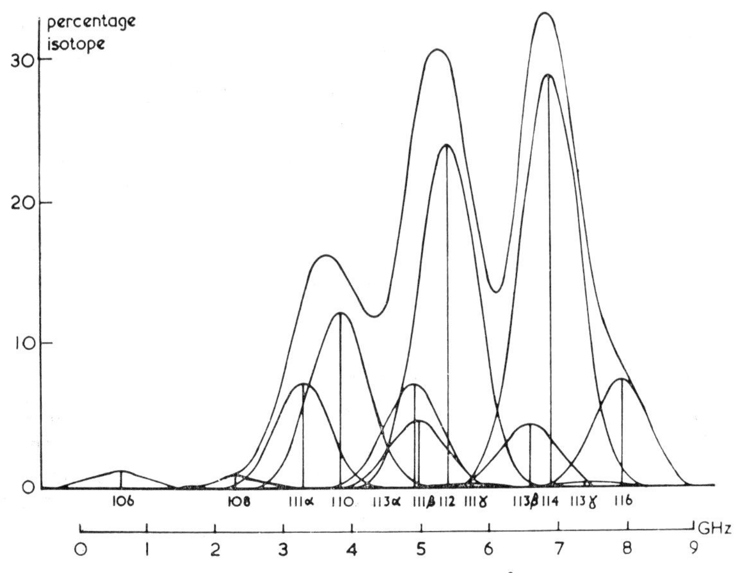

FIG. 28. Theoretical gain profile for the He–Cd 4416 Å laser transition for natural isotopic abundance Cd.

1970). Alternatively the gain of the laser can be increased by using single isotope cadium (Silfvast, 1968).

G. LIFETIMES IN NOBLE GAS ION LASERS

The small signal gain g (further discussed in Section XI, A) of any laser, whether it is homogeneously or inhomogeneously broadened, is proportional to

$$\left[\frac{1}{4\pi}\left(\frac{\ln 2}{\pi}\right)^{1/2}\right]\frac{\lambda^2 A_{21}}{\Delta\nu}\left(\frac{N_2}{g_2}-\frac{N_1}{g_1}\right)g_2$$

where N_2, N_1, g_2 and g_1 are the populations and degeneracy factors of the upper (2) and lower (1) laser levels, A_{21} is the Einstein coefficient for spontaneous emission for the transition at wavelength λ and $\Delta\nu$ is the linewidth. In a homogeneously broadened laser $\Delta\nu$ is the Lorentzian linewidth, $\Delta\nu_N$, whilst in an inhomogeneously broadened laser it is $\Delta\nu_D$, the Doppler broadened linewidth.

If the total pumping rates of the upper and lower levels of the laser transition are R_2 and R_1, and their effective lifetimes are τ_{2_e} and τ_{1_e} (allowing for both radiative and non-radiative deactivation). Then $N_2 = R_2\tau_{2e} N_1 = R_1\tau_{1e}$ that is the gain

$$g \propto \frac{\lambda^2 A_{21}}{\Delta\nu}\left(\frac{R_2}{g_2}-\frac{\tau_{ie}R_1}{\tau_{2e}g_1}\right)g_2\tau_{2e}$$

If the decay of the laser levels is dominated by radiative processes, $\tau_{2e} = \tau_2$ and $\tau_{1e} = \tau_1$ where τ_2 and τ_1 are the radiative lifetimes of the upper and lower laser levels. The gain of the transition increases as the ratios $R_2/g_2 : R_1/g_1$ and $\tau_{2e} : \tau_{1e}$ increase.

In the noble gas ion lasers, not only is there generally a favourable value of the pumping ratios of the upper and lower laser levels but there is also a very favourable radiative lifetime ratio $\tau_2 : \tau_1$. This is best illustrated by listing some strong laser transitions for which the upper and lower level lifetimes have been measured or calculated (see Table XIII). The table also lists the natural linewidth calculated from the upper and lower level lifetimes. These very favourable lifetime ratios which do exist in noble gas ion lasers are of great importance in determining their desirable operating characteristics.

Although there is a wealth of data available in the literature on the lifetimes of levels in other ion lasers [see for examples references listed in Miles and Wiese (1970) and Fuhr and Wiese (1971) and data in Wiese *et al.* (1966, 1969)] surprisingly both the upper and lower level lifetimes are known for few transitions in ion lasers. The main exceptions are the noble gases where an abundance of lifetime and transition probability data are available, both experimental and theoretical. However, by analogy with the noble gases where very favourable lifetime ratios exist it can probably be concluded that similar favourable lifetime ratios exist in the other ion lasers (with the exception of those few ion laser transitions which are self-terminating).

1. *Neon*

In Table XIII the lifetimes of the upper levels are taken from the experimental measurements of Hesser (1968), Brand *et al.* (1970) and Denis (1969). The value of 8·3 ns taken for the lifetime of the upper level of 3713 Å is a mean of the two values given by Brand *et al.* (1970) and Denis (1969). The lower level lifetimes are calculated (Hodges *et al.*, 1970).

2. *Argon*

Although very many measurements are available for the lifetimes of the upper laser level of the strong visible transitions in argon II, the measurements of Bennett *et al.* (1964) are probably the most reliable and these are the ones given in Table XIII.

For the lifetimes of the upper levels of the two strong argon III transitions mean values obtained from the data of Frick *et al.* (1970), Ceyzeriat *et al.* (1971), Kernahan *et al.* (1970) and Denis and Dufay (1969) are given. The lower levels of the argon II transitions are calculated values from Luyken (1972) and Statz *et al.* (1965), with corrections pointed out by Kitaeva *et al.* (1969) and Marantz (1968) (see also Koster *et al.*, 1968).

3. *Krypton*

The upper levels of the strong singly- and doubly-ionized krypton transitions have been measured by Fink *et al.* (1970). The lower levels of the singly-ionized transitions have been calculated by Marantz (1968) and Marantz *et al.* (1969).

4. *Xenon*

The upper level lifetimes of the lines listed have been measured by Allen *et al.* (1969). No information appears to be available on the lower level lifetimes of the transitions, but they are almost certainly short (< 1 ns).

X. Ion Laser Technology

The most important ion laser at present, of those being discussed in this review, is the argon ion laser. This is largely due to its capability of producing CW output power greater than 1 W; it is the highest powered CW gas laser operating in the visible and it emits strongly in the blue–green region (where sensitive photodetectors are available). In view of this, more attention will be paid here to the engineering of this laser, although consideration will be given to other ion lasers of interest, usefulness or possible future development. These include the Kr II laser giving strong red and weaker blue and yellow lines, the UV emission from Ne II, Ar III and Kr III, the UV and blue–green emission from Xe IV, pulsed ion lasers and metal vapour lasers. Despite the sensitivity of its excitation process and problems of plasma confinement, it has been possible to develop the argon laser technically into a relatively small, reliable and inexpensive device. Operating lifetimes in excess of several thousand hours can be obtained from these devices.

The technology required for the CW and pulsed ion lasers and noble gas and metal vapour lasers differs considerably and so the techniques for each will be discussed separately. More attention will be paid to CW lasers, this emphasis arising from their greater importance in applications. It should be added that some of the material described here has been selected to illustrate the variety and ingenuity of the techniques used in the construction of ion lasers.

A. TUBE DESIGN FOR CW NOBLE GAS ION LASERS

The technology pertinent to these CW lasers is mainly determined by their excitation mechanism, involving electron impact as described earlier. This form of excitation is characterized by the need to have high current densities, around 200–1000 A cm^{-2}, in the gas at a filling pressure of about 0·1 to 1 torr. The basic feature of a continuous noble gas ion laser is a high current density discharge obtained either by confining the discharge to a capillary of a few mms diameter and passing a current of a few tens of amps, or in a wider bore tube

(> 1 cm) but using currents in excess of 100 A. In narrow bore tubes the discharge is often further confined by applying a longitudinal magnetic field. With input powers in the region of 3 to 15 kW the gas temperature, as shown in Section VIII, is high. Almost all the input power of several kW is dissipated thermally, since the laser efficiency is $< 1\%$, and hence water cooling is required, except in the case of low power air-cooled ion lasers. The large thermal gradients across the confining tube walls means that a high degree of thermal stress is induced in the tube material. Also gas pumping effects, which are discussed in Section X, B 3, lead to a pressure gradient along the discharge which can become undesirably large without the use of a gas return path.

Of prime importance in the construction of CW noble gas ion lasers of around 1 W output power is the discharge confining tube. A large amount of development effort has been expended in evolving a tube design which has a tolerable working life under these conditions. The development of the techniques used in ion lasers over the last eight years is an interesting story in its own right. The construction of conventional CW argon lasers of about 1 W output power will be described first before a discussion of other types of argon laser.

Because of its characteristic operating conditions, special techniques for containment of the ion plasma are necessary. Typical requirements for the plasma containment tube are: length 30–50 cm; bore 2–4 mm (i.e. almost capillary tube dimensions); able to withstand thermal stress and dissipate a few 100 W cm^{-2}; withstand ion bombardment with minimal tube erosion (which can lead to failure, as shown in Section VIII, the walls being bombarded by ions having energies of tens of eV); have a low gas permeation and a low gas clean-up rate; emit little gas contaminant; exert no major influence on laser performance by affecting the electrical characteristic detrimentally, e.g. by modified wall sheaths or radial field, and have a tolerable working life, e.g. > 1000 hours. The necessary properties for the material of the plasma containment tube may be summarized as: (1) low thermal expansion; (2) vacuum tight; (3) maintain thermal conductivity to high temperatures; (4) available in a useful form and at an acceptable price; (5) resistant to ionic sputtering and (6) good thermal shock resistance. The designs can be characterized into two general types, (a) continuous tubes or (b) segmented tubes. In (a) materials for tube construction have been ceramic, of which beryllia is now mainly used, or fused silica. For segmented tubes graphite or beryllia segments and refractory metal segments are most usual. The material of the confining tube should have a good thermal conductivity and a high ratio of thermal conductivity to expansion coefficient. As we will see these requirements are satisfied by beryllia, graphite, fused silica and some metals.

In all tubes, except the anodic-bore graphite tube, the walls acquire a negative sheath which attracts positive ions which impact onto the wall with quite high energy. This causes tube damage by sputtering material from the tube wall which can lead to local tube failure directly by wall decomposition at the point of most damage or by creating "hot spots". The bombardment of the walls by ions, in addition to wall erosion and the creation of weak spots, can cause release of impurity gases from the walls which may act as quenchers, cause poisoning of the cathode or be deposited on the windows at the tube ends. Other forms of degradation in CW ion lasers are electrode sputtering, cathode poisoning, gas clean-up and the optical degradation effects considered in Section X, E 5.

Main consideration will be given to the continuous beryllia, the segmented graphite and segmented metal types of tubes. The first two are common in commercial types of argon lasers and will each be considered here in detail. Several other materials, e.g. in our laboratory silicon carbide, have been investigated.

(1) *Segmented Graphite*

The advantages of graphite as a material for a plasma containment tube are that it can withstand high temperatures (in excess of 1000°C), has reasonable thermal conductivity and thermal surface emissivity, can withstand the discharge conditions of an ion laser without rapid erosion since it is exceptionally sputter resistant, is easily machined and is of low cost. The design of a typical tube is shown in Fig. 29; this is a commercial design (Coherent Radiation Inc. Model 52G) based on graphite segments separated by ceramic spacers and contained in a fused silica tube which is surrounded by a water jacket. The segment structure is kept at low temperature by radiation cooling to the water jacket.

The ion laser illustrated is also surrounded by a water-cooled solenoid which provides the axial magnetic field and acts as the jacket for water cooling of the tube. This design uses a directly-heated tungsten dispenser cathode and a radiatively-cooled molybdenum anode. The graphite discs have off-axis gas return channels to maintain uniform gas pressure, which is also helped by gas ballast tubes. A gas reservoir and refill mechanism, either manual or automatic, to allow long-term replenishment of gas cleaned up by the discharge is added. The solenoid is compact and light weight and necessarily of high power dissipation, which is easily accommodated by water cooling. Heat transfer to the resonator and bulk of the laser is minimized by an outer water-cooled heat shield. The graphite is segmented since the laser arc discharge has a low axial field (Henqvist and Fendley, 1967) and the graphite segments are close enough such that the voltage drop between discs is $<$ that needed to produce cold cathode emission from them. As will be discussed later, the laser is usually run with an axial magnetic field along the tube axis to increase the excitation electron density, which helps to confine the discharge.

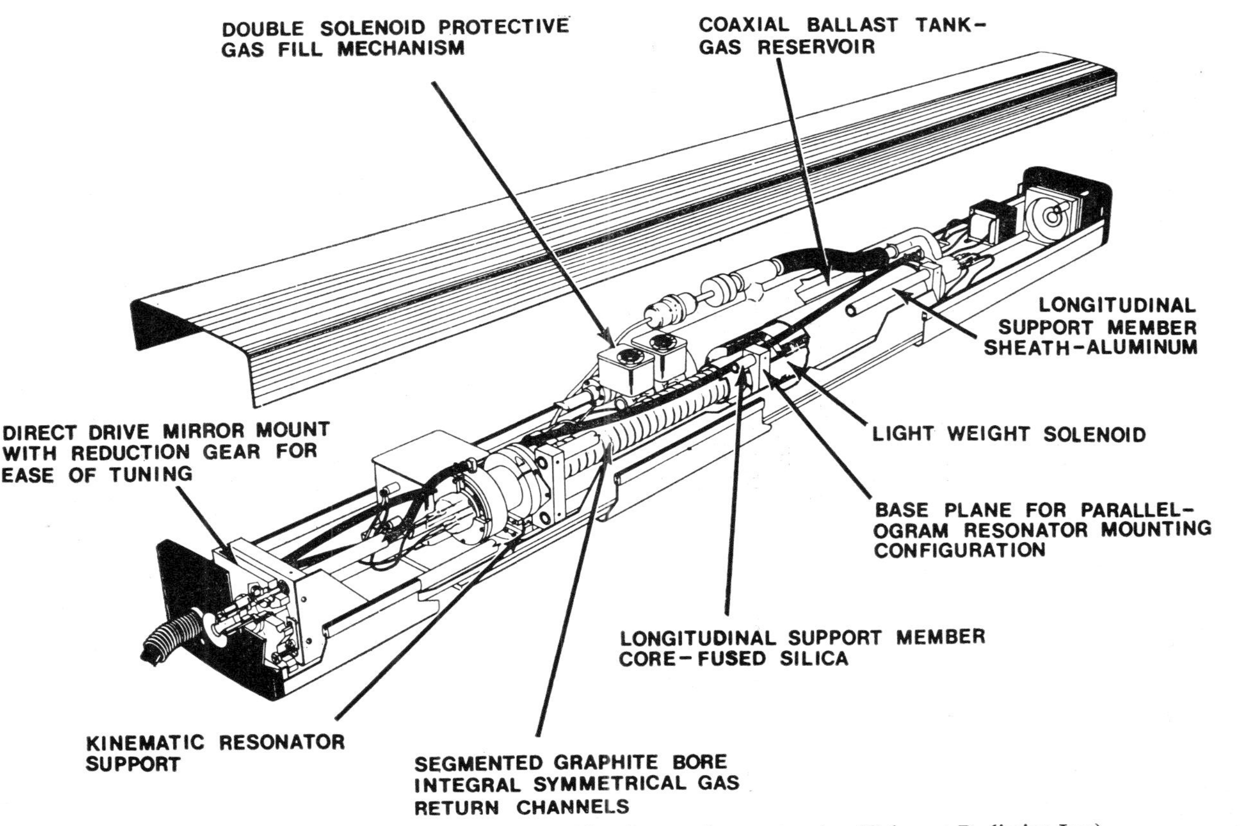

FIG. 29. CW argon ion laser of a segmented graphite plasma tube construction (Coherent Radiation Inc.).

There are some disadvantages of the graphite structure, the main one being that the graphite cleans up the gas and hence a replenishment system is necessary. The rate of gas clean up is lower than for fused silica but not as low as for BeO. Also the graphite needs careful processing, including high temperature vacuum outgassing and cleaning. This particularly helps in minimizing release of gas, dust or fine particles which would contaminate the argon gas or settle on the output windows. The life expectancy of the graphite tube is several years of normal use.

As illustrated in Fig. 29 the graphite segments are cooled by radiation to a surrounding water jacket rather than by conduction. The high emissivity of graphite enables this process to be efficient; and the equilibrium temperature of the graphite is maintained at a reasonable value at about 1000°C and a power dissipation capability is possible of about 50 W cm^{-1} (Hernqvist and Fendley, 1967). Careful cooling of the graphite tube leads to a low noise content and better frequency stability of the output laser beam.

To minimize the damage by sputtering at the ends of the confining structure as the discharge is reduced from several centimeters diameter to a few millimeters, a taper is put on the ends of the tube aperture at the cathode and anode ends and at the ends of the individual segments. The lengths of the graphite segments should also be short to reduce sputtering. A radiation-cooled segmented graphite system is also shown in Fig. 30 (Bridges *et al.*, 1971) where the tapered segments and the temperature distribution along the segments are illustrated.

(2) *Beryllia*

Beryllia (beryllium oxide, BeO) has come into use for ion laser tubes in the last few years, later than several other materials because it was considered technically difficult for use in this application, although it had been used in electronic applications previously. Beryllia is treated here separately from other ceramic materials which have been used in ion lasers because of its especially useful properties and because it is now a commonly used material in commercial ion laser tubes. Its main advantages are its high thermal conductivity, good thermal shock resistance, high electrical resistivity and low expansion coefficient. Also it has a low gas consumption (gas clean-up) and together with low release of contaminating gas or particles, means that it can operate over a long period of time.

The choice of the beryllia starting material itself is important and its preparation, e.g. possibly involving ceramic metallizing and high temperature soldering, needs well developed techniques. The material and its thorough processing are more costly than graphite. The thermal properties of the material have several consequences: a greater wall thickness can be used giving added strength and the high heat dissipation rate produces a lower inner wall temperature (~ 100°C) and allows thermal equilibrium to be reached more quickly. Some other properties are influenced by the change in operating conditions since higher operating currents are possible for a given output power—it operates with a smaller length than, say, a fused silica tube and at lower discharge operating voltage, so that a smaller size laser can be made. It

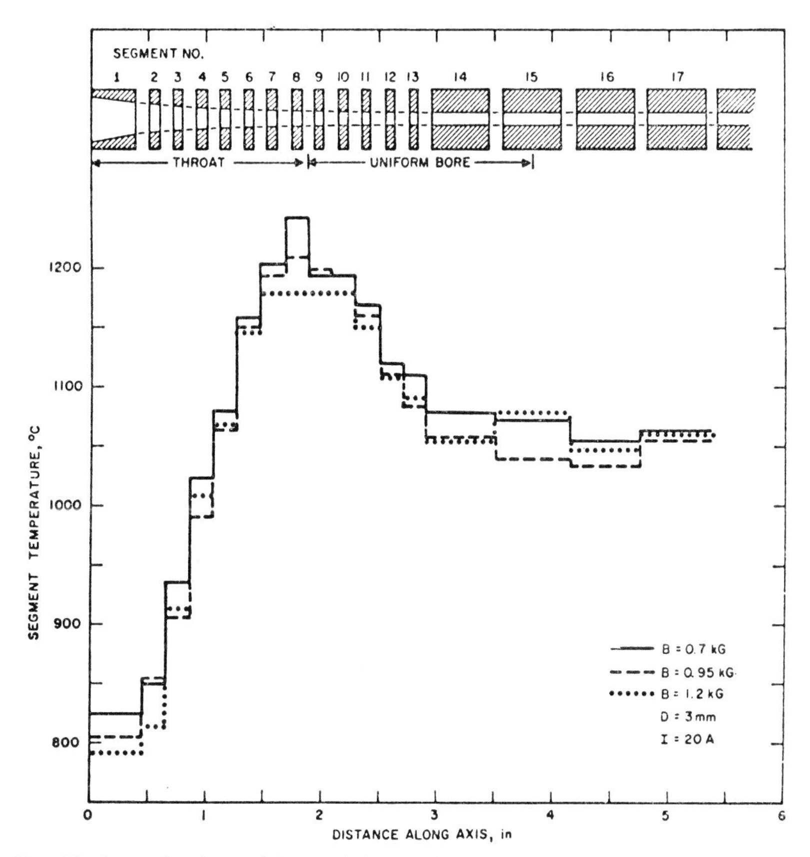

FIG. 30. Argon ion laser with a radiation-cooled segmented graphite tube (bore = 3 mm, current = 20 A) showing the distribution of temperature for different axial magnetic fields (B) (Bridges *et al.*, 1971).

seems to be confirmed that the possibility of toxic hazards arising from the use of BeO in the normal form and operation of noble gas ion lasers is small.

A modern commercial version of the continuous argon laser based on a BeO tube is shown in Fig. 31 (this is a schematic diagram of the Spectra–Physics model 165 argon laser). The detail in this diagram of optical and mechanical features will be described later; here attention is drawn to the use of BeO in segmented form which is surrounded by a water cooling jacket. The

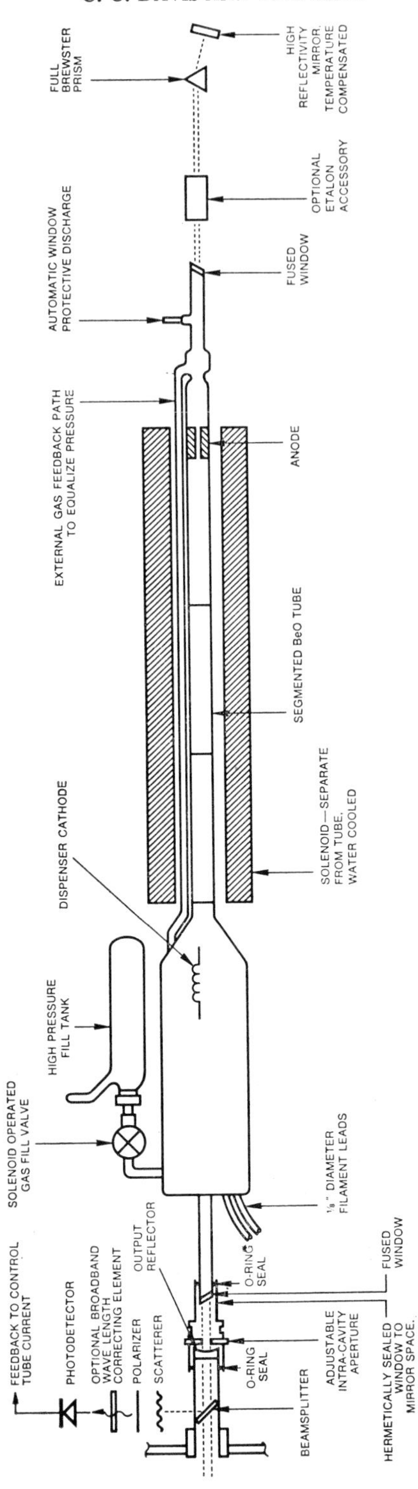

FIG. 31. CW argon ion laser having a segmented BeO plasma tube (Spectra-Physics, Inc.).

segments of BeO are fitted together with a non-metallic bond, this form of bonding may have some advantages over metallic bonding where electrolysis may take place and which requires closed-loop water cooling with a heat exchanger. As well as its use in segmented structures, beryllia also may be used as a continuous tube. A design has been based on a continuous tube of length 30 cm and bore 3 mm by Conder and Foster (1970). Metallized connections at each end have been made which can reasonably withstand high and low temperature recycling. The beryllia is metallized and nickel plated and materials selected to minimize stress from unequal thermal expansion coefficients.

Beryllia can be used at high input powers > 300 W cm^{-1}, in tubes of 2–3 mm bore. Under these conditions tube erosion or disintegration and gas clean-up are much less than for fused silica which was the common low electrical conductivity material used for continuous tubes before BeO. A more recent design of a segmented BeO tube has been reported by Ferrario and Sona (1972) which uses simplified construction techniques giving it a high degree of mechanical ruggedness.

(3) *Segmented Metal*

The use of metal tubes would seem to provide a neat and inexpensive solution to the plasma containment problem. Metals can be easily fabricated, are robust and they have high thermal conductivity (e.g. $100 \times$ fused silica). In order to be directly useable with radiative cooling refractory metals, tungsten or molybdenum must be used. To prevent the input current simply travelling through the tube, the metal must be segmented with the combined equivalent anode and cathode potential falls of a segment being significantly greater than the potential difference in the discharge between segments. Discharges in metal containing tubes have been considered by Sheidlin and Asinovskii (1963) and Maecker (1956) has operated a segmented metal structure at 10 kW cm^{-1}. Other non-laser applications of metal segments in confining discharges have been reported by Shumaker (1961) and Emmons (1963). The application of metal segments to argon laser tubes was first described by Labuda *et al.* (1965).

The disadvantage of metals, assuming the difficulty of maintaining the gas discharge in the electrical conducting tube is overcome by using short insulated segments, is their low threshold to sputtering (Stuart and Wehner, 1962; Wehner, 1954, 1955, 1959) where atoms and ions striking the metal knock off metal atoms. This process can lead to tube wall erosion and gas clean-up. Most refractory metals have sputtering energy thresholds in the range 25–30 eV when bombarded with Ar^+. The sputtering threshold for low energy ions decreases from normal incidence to grazing incidence by a factor of around 10; for aluminium the sputter threshold at grazing incidence is 3·2 eV (rising to 82 eV at normal incidence) (Wehner, 1955) which accounts for the large degree of sputtering observed in aluminium. Because of the radial fields inside the segmented metal tube (see Section VIII) ion collisions near normal incidence are likely to be the case but at the segment ends the field distribution may become non-radial and cause erosion from sputtering to be greater. Even for relatively low mean ion energies ($\sim 0{\cdot}1$ eV) and normal incidence sputtering threshold of 15 to 20 eV (A1), the sputtering rate can still be significant because of the high ion density ($> 10^{13}$ ions per cm^3). Hernqvist and Fendley (1967) have compared the sputtering effects of Mo, Ta and high purity graphite for a 5 A current discharge in argon, the greater resistance of high purity graphite was evident. The use of metal segments coated with sputter resistant material has been mentioned by Seelig and Banse (1970).

Metal segments can be radiation cooled by containing them in a water-cooled fused silica envelope or by direct water-cooling. In radiation cooling the segments reach 1000–1500°K and emit largely in the near IR. Only refractory materials such as Mo or W can then be used. Water cooled segments can be made from Cu, Al, etc. Insulation between segments can be by ceramic spacers, plastic O-rings or anodized films, and vacuum seals can be by epoxy resin, Viton O-rings or ceramic-metal brazed seals. A composite structure has been proposed by Paananen (1966) of a molybdenum sleeve pressed into an aluminium holder. A metal segment laser using Al discs insulated by 0·5 mm thick Delrin gaskets was developed by Rigden (1965). A central aperture in each disc defines the laser beam axis and this is surrounded by twelve holes for water cooling. In a similar manner, holes can be provided for equilization of gas pressure. The potential difference between the tube segments was 7 V. Some sputtering damage was observed in this structure after 20 hours operation. Aluminium discs have also been used by Maitland and Cornish (1972) of length 2·5 cm and insulated by neoprene O-rings and nylon spacers.

Bridges *et al.* (1969) report the development of a metal disc plasma confining device with the metal discs separated by silica spacers and placed together in a silica tube. This device was reported as capable of sustaining discharge currents of 1600 A cm^{-2} and dissipating 600 W cm^{-1}. These authors also investigated gas pumping effects and the radiative transfer of heat from the structure.

A high power radiation-cooled segmented metal laser similar to the design of Labuda *et al.* (1965) has been developed by Latimer (1968). This was constructed with tungsten discs 0·05-in. thickness and 3-in. diameter spaced with BeO rings which were mounted on three $\frac{1}{4}$-in. diameter BeO rods; this structure was mounted inside a 3-mm i.d. fused silica tube. It was capable of dissipating an input power of over 700 W cm^{-1} by thermal radiation from the discs, calculated to have a temperature of 1400°C, which is absorbed in a surrounding water jacket. A segmented metal tube design using wide-bore 4 cm long segments in high current operation has been used by Wang and Lin (1972) giving laser output powers greater than 50 W (see also Section X, G). Donin (1970) has investigated Ar laser operation in a metal segment tube. He found that the characteristics of the discharge were largely similar to a dielectric walled tube although the different heat removal from the walls affected the spectral characteristics of the laser such as the Doppler width. A possible non-equilibrium ionic velocity distribution is set up which causes distortion in the inhomogeneous broadening of the spontaneous radiation.

The electrical characteristics of a partially conducting structure are interesting. For a metal segment electrically floating in equilibrium the rate of arrival of electrons and positive ions is equal. The segment assumes a negative charge which attracts positive ions and a positive ion space charge sheath forms at the surface of the segment. With respect to the undisturbed plasma the segment adopts a negative potential with a value such that the electron and positive ion currents are equal. There is a net flow of negative charge from the plasma to the metal wall at the cathode end of the segment and an equal net flow of positive charge from the plasma to the metal wall at the anode end. However, the average state of the plasma in the volume of the discharge is not much disturbed by the field distortions near the wall. Short segment lengths reduce sputtering damage. Positive bias on the segments removes the electron sheath and reduces ion bombardment. This then will reduce the rate of sputtering and tube erosion.

A study of the operation of segmented metal laser tubes has been described by Maitland (1971) and Maitland and Cornish (1972). The maximum length of the segment which allows an axial discharge to be maintained after it has been established, d_{max}, is given as

$$d_{max} = \frac{kT_e}{eX}\left(\frac{2M_i}{\pi m_e}\right)^{1/2}$$

in which T_e is the electron temperature, M_i and m_e are the ion and electron masses and X is the mean potential gradient along the axis of the tube. For $T_e = 2 \times 10^4$ °K and X = 300 V m^{-1} a large value of d_{max} is predicted of about 1 m. In the presence of a longitudinal magnetic field, to provide enhancement of current density (see Section X, F), the potential difference between the plasma and the segment is decreased and the segment may even reach positive potential. The maximum length with magnetic field, B, is then

$$d_{max} = \frac{kT_e}{eX}\left(\frac{2T_i}{\pi T_e}\right)^{1/2} F$$

where T_i is the ion temperature and $F = B_m/B$ = ratio of (field for which positive ion speed = mean speed of electron diffusion across the magnetic field) to (magnetic field).

For $X = 300$ Vm^{-1}, $T_i = 3000$°K, $T_e = 20{,}000$ K and $F = B_m/B = 10$ a value of d_{max} (with magnetic field) = 1·8 cm is obtained. The starting and operation of arc discharges through metal segments with $L/d > 100$ is not possible. With

$L/d \sim$ (5 to 10) cold cathode emission cannot be sustained and the gas has a greater conductance.

Some emphasis has been placed here on metal containment tubes because of their more recent development and also because of their potential for development in the future.

(4) *Other Materials*

(a) *Fused silica.* This has largely been replaced in commercial lasers for which high power or long life are required by tubes of beryllia, graphite or metal segments as described earlier. But for quick and inexpensive fabrication of an argon laser, particularly for laboratory purposes, fused silica has proved a suitable material. A thorough study of the properties of argon lasers with fused silica tubes has been carried out by Labuda *et al.* (1965), Webb (1968) and Gorog and Spong (1967).

Its disadvantages are erosion by ion bombardment and sputtering leading to failure and poor thermal conductivity such that even with an outside water cooling jacket the inner wall temperature is high. The gas clean-up rate of fused silica is fairly high necessitating a gas replenishment system for long life. The tube may need replenishment after about 50 hours running, depending on the power input and gas ballast volume. Also gases or material released into the laser can deposit on the beam exit windows or the tube walls causing a "hot spot" to develop which can create a local weak point in the tube. Erosion is more severe at the cathode end of the tube due to a greater density of higher energy positive ions and a correspondingly high rate of sputter damage. In our laboratory we have found that severe damage occurred to a tube which had been operating at relatively low current density ($\sim$ 300 A cm^{-2} for about 250 hours. Fused silica is inferior to BeO, graphite or metals for sputter damage when used at current levels in excess of 500 A cm^{-2} by about 3 times. For example, the sputter threshold of fused silica is about half that of tungsten and in a high current gas discharge will have a much higher sputter yield. The fused silica tube is cooled by conduction to a concentric water jacket. Water cooled silica tubes can dissipate around 140 W cm^{-2} of wall surface (Boersch *et al.*, 1967; Herziger and Seelig, 1969). This is less than the dissipation possible with segmented metal.

(b) *Ceramic.* Ceramics other than BeO have been used, the most common being alumina (Al_2O_3, aluminium oxide). This material is intermediate between fused silica and BeO for thermal conductivity and melting point but inferior to these materials for thermal shock behaviour. This means that careful temperature control is necessary. Also the ceramic tube is subject to discharge damage. Segmented ceramic tubes have been used by Paananen (1966) who has reported severe damage at the cathode end in a 3·5-mm bore tube run at discharge currents up to 90 A. Techniques of using alumina for ion lasers have been developed by Nasini *et al.* (1969) and further by Redaelli (1970). The design has aimed at being simple and low cost and based on a commercial form of alumina, Sintox. The operating characteristics of this tube are claimed to be comparable to those of beryllia tubes. An added advantage over fused silica tubes is a low gas clean-up rate (Nasini *et al.*, 1969). A high temperature laser tube for metal vapour lasers was constructed by Piltch and Gould (1966) which used alumina as a liner whose ends fitted inside another alumina tube connected to electrode assemblies by alumina-kovar brazed joints.

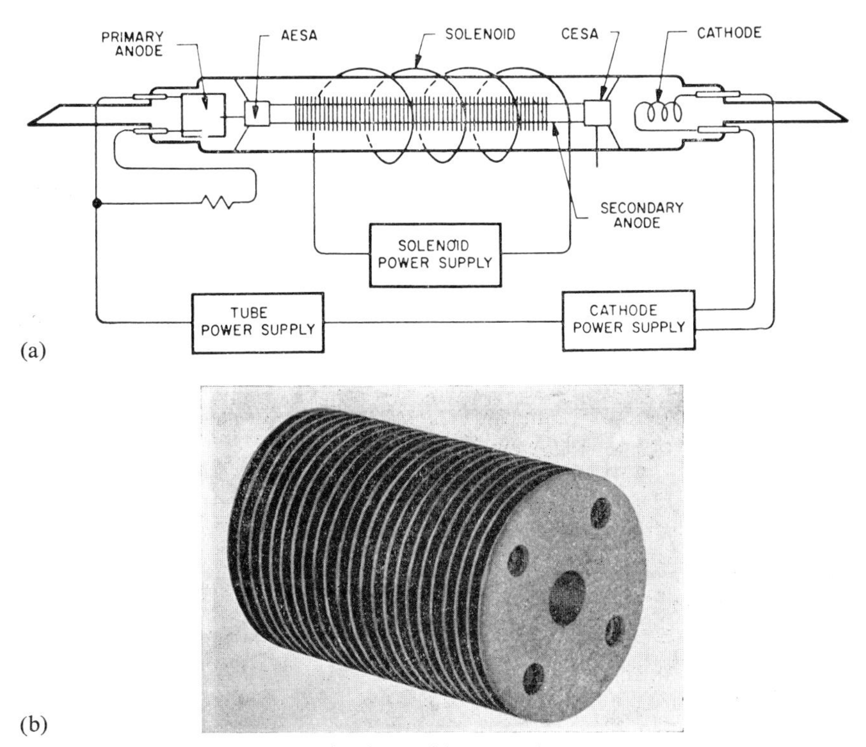

FIG. 32. CW argon ion laser with an anodic bore graphite tube.

(c) *Anodic bore graphite.* A form of this tube is shown in Fig. 32a (McMahon, 1968). The tube itself is shown in more detail in Fig. 32b as a continuous cylinder of pyrolytic graphite made up such that the less conducting c-axis of the graphite is oriented parallel to the axis of the tube. The wall of the graphite is arranged to be at a positive potential, this has the effect of reducing ion bombardment. The cross-sectional area of the tube is gradually decreased such that the resistance of the tube is increased towards the cathode to ensure that all points of the tube are at a positive potential relative to the plasma.

This tube has a small electron current into the wall and no negative sheath as is the case with tubes with non-conducting walls. Without the negative sheath, which induces positive ion radial flow in non-conducting tubes (as described in Section VIII), some other properties of the laser are affected. For example, the degree of resonance trapping within the ion is increased since the range of ion velocities is less and the low conductivity near the wall may alter the stability of the arc discharge.

(d) *Anodized aluminium.* A continuous tube of aluminium with the inside of the tube coated with an anodized layer of mainly Al_2O_3 has proved effective as an argon ion laser tube (Paanenen, 1966; Medicus and Friar, 1967; Donin *et al.*, 1969; Schafer and Seelig, 1970). Paananen (1966) has reported work on an anodized film prepared on the laser tube using sulphuric-oxalic acid and water electrolyte at 15°C and optimization of durability, pore size and crack resistance was attempted. Cracks in the anodic film at curved surfaces can cause arc breakdown. The films tend to be attacked at the cathode and anode ends. Paananen found that the film can have reasonable life if the voltage stress at corners is overcome and the tube length kept below about 50 cm. Medicus and Friar (1967) who used a layer thickness of 0·05 mm found that soft layers were more durable than hard impervious layers because they allowed absorbed gases to be removed by vacuum processing. Donin *et al.* (1969) have used an anodized Al tube up to current densities of 3500 A cm^{-2} with a confining capillary of Al segments 5·5-mm thick. He reports that the tube life was short for current densities > 1300 A cm^{-2}.

B. D.C. GAS DISCHARGES IN NOBLE GAS ION LASERS

(1) *Electrical characteristics*

The properties of the plasmas of ion lasers have been discussed previously in Section VIII. Here we will briefly describe the electrical techniques of exciting the discharge. Electric discharge excitation is almost always used for D.C. ion lasers. The discharge is of high current for conventional noble gas ion lasers and is of the low pressure type with the voltage/current characteristic showing a typical negative resistance form at lower currents as shown in Fig. 33. This indicates operation at a low and approximately constant discharge voltage. The typical arc discharge impedance is $\sim 1\ \Omega\ cm^{-1}$ at around 250 V, 30 A with a mean field gradient 10 V cm^{-1}. Most of the voltage drop occurs in the long positive column of the narrow bore and this region determines almost entirely the V/I characteristic, with the anode and cathode regions only playing minor roles.

The discharge is normally driven from a constant current source which can be provided by a variety of rectifier circuits. Because of the high current and smooth output requirements, 3-phase input and inductive, or for low currents, capacitive smoothing is most often used. Current changes due to supply line or plasma tube voltage can be overcome by the use of active current regulation using series transistor pass banks, saturable reactors or thyristors. The discharge is usually driven in series with a ballast resistance to make the net characteristic positive at low currents and generally to prevent oscillation caused by the negative resistance of the discharge.

Compact and well regulated power supplies have been developed commercially for the inert gas ion lasers. The laser shown in Fig. 29 is designed to run at a voltage about 25 V below that obtained by direct rectification of

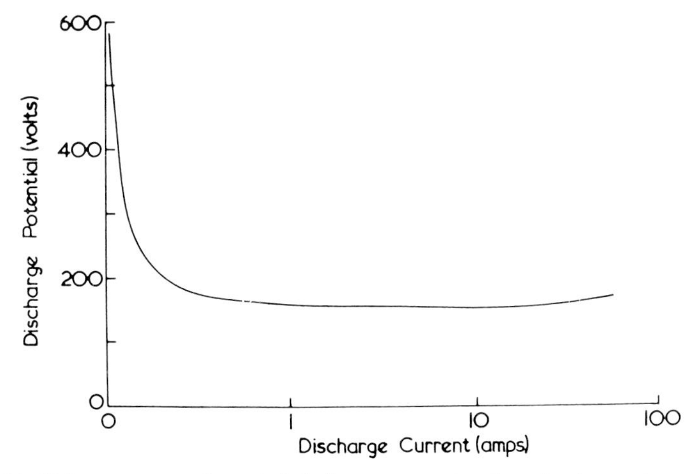

FIG. 33. Voltage–current characteristic for an argon ion laser having a plasma tube of 3 mm bore, 30 cm length at a filling pressure of 0·7 torr (Conder and Foster, 1970).

230 V A.C. The 25 V is dropped across a multiple transistor passbank which can be feedback driven to provide either constant tube current or constant light output. To compensate for variations in line voltages a small buck-boost autotransformer is used. This type of supply will provide typical running characteristics of about 30 A current and at a potential of about 260 V. To allow regulation of the output intensity a light sampling device can be used to provide feedback to control the tube current.

Threshold current densities for noble gas ion lasers Ar II, Kr II, Xe II and Ne II are ~ 20 A cm^{-2}. In Ar II, the transition 4880 Å has the lowest threshold current. The data of various workers is broadly in agreement when allowance for the different operating conditions of their lasers is made. For example Paananen (1966) finds for the Ar II 4880 Å line a current threshold in a 2·2-mm bore tube of 65 cm length of 17·6 A cm^{-2} at a tube potential of 480 V for high reflectance mirrors ($>99{\cdot}6\%$). Threshold currents observed by Gorog and Spong (1967) for Ar 4880 Å were about four times greater for a laser with 5%

output coupling. These authors find only a small dependence of threshold current on magnetic field but for the Ar 4880 Å line in a variety of laser tubes a minimum is apparent in the region of 400–800 gauss.

(2) *Gas Filling Pressures*

The optimum operating pressure depends on (i) discharge current, increasing with increase in current; (ii) bore diameter, decreasing with increase in bore diameter and (iii) tube length, increasing with tube length. All these observations are influenced by the gas drive-out and gas pumping effects discussed in Sections VIII and X, B (3).The pressure dependence of the output power for an Ar 4880 Å laser is shown in Fig. 34. In common with most gas lasers the output power is strongly dependent on pressure—the laser only operating over a narrow pressure range.

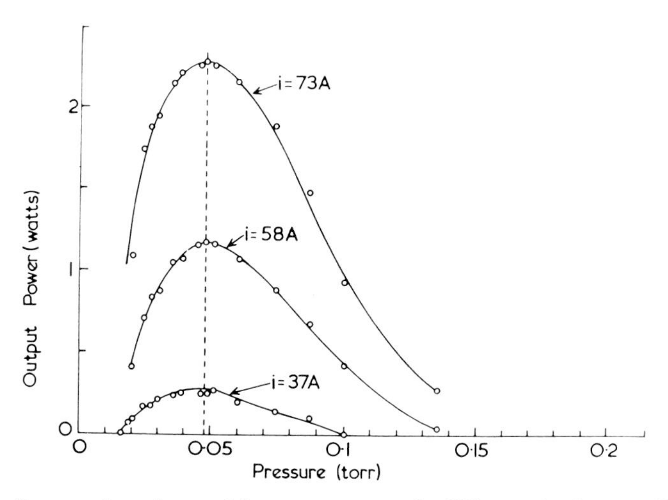

FIG. 34. Pressure dependence of the output power of a CW argon ion laser at 4880 Å for various discharge currents *i* in a 1-cm bore tube (Boersch *et al.*, 1967).

The optimum pressures of segmented tubes are significantly lower than for continuous bore tubes (Gorog and Spong, 1967)—0·1 torr rather than 0·4 torr—due to more effective elimination of gas pumping effects by efficient gas by-pass systems. Gas pumping leading to large pressure differences in the continuous bore tubes occurs. The active gas density is much lower than the filling pressure due to thermal drive-out effects. The high pressure region in the presence of high axial magnetic fields has been studied by Gorog and Spong (1966). For $p \gtrsim 0{\cdot}2$ torr and $B \gtrsim 400$ gauss the optimum field was more critical than at lower pressures. Pulsed inert gas ion lasers have about an order of magnitude lower cold filling pressure than CW lasers. The optimum filling pressure for CW operation of an Ar laser is $\simeq 0{\cdot}4$ torr (for a 3-mm bore, 45 cm length, 5% output coupling, 15 A current and 1000 gauss axial magnetic field).

One might expect some effects due to additional mechanisms of ion production (such as Penning or charge exchange collisions) in a mixture of Ar and He, or changes due to modification of the electron temperature. Goto *et al.* (1970) have reported that a decrease in relative peak intensity of the Ar 4880 Å line, with current (> 200 A) observed in pure Ar, was not observed in a mixture of Ar and He under pulsed excitation; of several possible effects the results indicated that radiation trapping was not very prominent in the Ar–He mixture and the electron temperature is only slightly altered. In the authors' laboratory some effects of added He have been observed in D.C. argon lasers, such as operation over a wider pressure range and a slight apparent increase in output intensity.

(3) *Gas Pumping*

In an ion laser the condition of quasineutrality in the discharge plasma leads to both electrons and ions taking up equal amounts of momentum from the axial electric field. However, ions within a mean free path of the wall deliver a considerable proportion of their momentum to the walls because of their radial motion whilst the electrons transfer essentially all their momentum to the gas. This results in a net anode directed force being exerted on neutral atoms within an ion mean free path of the wall. This flow of neutral atoms is offset slightly by the cathode directed drift of the ions, so a net pressure differential is set up between anode and cathode to balance the net anode directed force on the gas produced by the electrons. This is known as gas pumping, as distinct from gas drive-out which is caused by temperature differences between the hot capillary region and the cooler regions of the laser. In small bore discharge tubes at high currents gas pumping is a large effect, as was originally noted by Gordon and Labuda (1964) who found that it caused their early model CW argon ion laser to extinguish itself after a few seconds of current flow. Anode–cathode pressure ratios of 10:1 or more could build up in less than a minute. Subsequently all high current CW ion lasers have incorporated some form of gas bypass to allow gas which would otherwise build up at the anode to flow back to the cathode end of the laser. However this bypass must be arranged so that flash-over between anode and cathode does not occur through it. In early CW ion lasers the bypass usually satisfied these conditions by taking the form of a wide bore tube wound into a spiral to provide sufficient length to prevent flash-over. In some recent segmented lasers multiple bypass holes of smaller diameter than the main capillary are provided in each segment. There is generally insufficient time for either gas pumping or drive-out to occur in pulsed discharges, unless the laser is operated repetitively (Neusel, 1966) or with long pulses (> several ms) (Redaelli and Sona, 1967; Ahmed *et al.*, 1966, 1967).

A technique of starting an ion laser which has a BeO discharge tube with a central aperture for the laser medium and surrounding smaller gas return paths has been described by Wexler (1972). In this method the laser can be struck and the discharge maintained in the central region without flash-over to one or more of the gas return tubes. In this way the gas return path can be made to have a larger conductance than the laser bore.

Although the observed pumping at normal pressure is from the cathode to the anode with the high pressure developing at the anode end, at low pressures pumping in the opposite direction to the cathode can occur. A CW ion laser has a higher filling pressure than a pulsed ion laser, but in comparing the operating pressures of CW and pulsed ion lasers when allowance is made for the different filling pressures, thermal gas drive-out and gas pumping effects, the operating pressures (which are quite different from the filling pressures) are about the same. Higher current operation requires higher filling pressures because of increased gas pump-out effects.

Gas pumping is well known in non-laser discharges (Langmuir, 1923; Francis, 1956; Cairns and Emeleus, 1958; Leiby and Oskram, 1967). Cairns and Emeleus described the origin of the pumping effect in terms of (i) transport of the ions, (ii) action of the longitudinal field on the positive space-charge in the plasma and sheath and (iii) the difference in momentum given to the wall by positive ions and electrons which give a compensatory reaction on the gas. Effects (i) and (ii) increase the pressure at the cathode and occur at low pressures. Effect (iii) increases pressure at the anode and occur at higher pressures. Langmuir (1923) and Druyvesteyn (1935) comment on effect (iii) which is the more common effect in ion lasers since the change over pressure from effects (i) and (ii) to (iii) was at about 0·1 torr (Francis, 1956).

Theories developed for gas pumping usually ignore thermal and mechanical effects at the ends of the positive column and also assume a uniform positive column. Moving striations may affect the pumping by, for instance, producing a periodic pumping action. The manifestation and importance of gas pumping effects in noble gas ion lasers has led to renewed consideration of its origin and investigations of its properties.

Chester (1968a, b) has made a thorough theoretical and experimental investigation of the gas pumping effect and has introduced a modification to the Druyvesteyn approach to include a gas pumping force from the radial motion of the ions. He has checked this treatment by measuring cathode and anode pressures in a CW Ar II laser of 2·5-mm bore, 30 cm length (Chester, 1968b). At currents of several amps, the anode–cathode pressure difference (Δp) $\simeq$ 0·25 torr for $p < 1{\cdot}2$ torr and $\Delta p \propto 1/p$ for $p > 1{\cdot}2$ torr. He found that his theory was able to predict both the magnitude and pressure and current dependence of Δp.

In the high current region Kitaeva *et al.* (1970) report gas pumping from anode to cathode. To counteract gas pumping the use of continuous gas flow has been investigated by See *et al.* (1967). This was intended to remove impurities and increase the life of the tube. At currents < 50 A in a 6 mm bore tube gas flows from anode to cathode or cathode to anode gave equal power output and presumably maintained near uniform pressure. But at currents > 50 A the gas flow cathode to anode gave significantly higher powers, which is countering the anode to cathode gas pumping in this current range in agreement with Kitaeva *et al.* (1970).

Tombers and Chanin (1970) investigated the effect of the conductance of the gas return path on He, Ne and He–Ne gases. In order to reduce the pressure gradient by 100 times they find the bypass radius should be at least three times the discharge tube radius.

C. ELECTRODES

(1) *Cathodes*

The CW ion lasers, being excited by high current discharges require cathodes capable of delivering currents between 30 and 100 A. These cathodes should be: (i) reliable; (ii) able to withstand the environment of the ion plasma; (iii) be relatively immune to poisoning from gases evolved over time; (iv) not themselves producing quenching gases and (v), for convenience, be compact. Thermionic cathodes are almost always used and several types have been adopted in ion lasers, the most common being cathodes of the dispenser and oxide types. Other, less common, types that have been used are arc cathodes, hollow cathodes and mercury pool cathodes. General difficulties with thermionic cathodes are that they are susceptible to contamination and the formation of arc spots on the cathode can damage the surface. Cold cathodes are usually adequate for pulsed ion lasers.

Since the operating gas pressure must be selected to optimize the laser action characteristics almost no facility is available to adjust the pressure to suit the cathode running properties. A good description of cathodes and cathode materials is given by Kohl (1960), and MacNair (1969) has studied cathodes for use in gas lasers.

(a) *Dispenser cathodes.* This type of cathode has a porous sintered matrix structure, usually of tungsten, which is impregnated with barium calcium aluminates and is used in most commercially available argon lasers. Direct heating of the cathode maintains a supply of low work function electron emitting material at the active cathode surface. The cathode provides high currents, up to 10 A cm^{-2}, at medium temperatures around 1100°C. It is resistive to sputtering and is robust and able to be let up to atmospheric air when cold. The use of sintered cathodes under gas discharge conditions has been investigated by Kucherenko and Yavorskiy (1965). Also Hernqvist and

Fendley (1966) have obtained very good results using these cathodes in ion lasers. The directly heated LaB_6 cathode in an improved form has been used by Redaelli (1970) in an argon laser. This cathode can give currents up to 3 A cm^{-2} at a temperature of 1450°C and with an emission efficiency of ~ 120 mA/W. The standard L-cathode type of dispenser cathode has been used in a long study of the argon laser (Gorog and Spong, 1967).

(b) *Oxide cathodes.* These have been widely used in ion lasers because of their reasonable emission currents for relatively low temperature working (~ 1000°C), acceptable sizes, availability and low cost. However, it needs careful activation and is easily poisoned so that, for instance, its emission current reduces after being let-up to air when cold—although this may be done several times without making it useless. Compared with the dispenser cathode it has lower current emission for equal area, is therefore larger and also less robust and comparatively easily damaged by arc spots.

(c) *Arc cathodes.* A cathode based on the use of the cathode spot of an independent arc struck between shaped copper or Al electrodes has been described by Donin *et al.* (1967). This produced currents up to 90 A and was capable of higher values; and it was suggested that the cathode could operate with D.C. currents up to 10^3 A. One design problem was the sputtering of the cathode material, which may contaminate the laser medium. It was found that Cu was better than Al in this respect and its effect was reduced further by using a surrounding cavity shield.

A robust cathode is formed from a plasma jet which has been adopted as a cathode for ion lasers by Maitland (1969). The construction of this demountable cathode is shown in Fig. 35a and b. The plasma formed between a cathode and nearby anode issues as a jet from a hole in the anode. One property of the plasma jet is to introduce excited states from the jet into the main discharge unlike other cathodes whose function is only to produce electrons. Gas is driven past the cathode through the hole in the anode and is pumped away from the anode region. A gas flow of 400 cm^3 (STP) min^{-1}, is used in starting the plasma jet and a lower rate for normal operation. Jet currents up to 50 A are used with an arc voltage of 8–20 V. This cathode is able to deliver currents up to 50 A, is robust, not easily poisoned or contaminated, can be let-up to air and is relatively cheap but its use necessitates a continuous gas flow through the laser.

(d) *Hollow cathodes.* A hollow cathode discharge (HCD) was developed for use in highly ionized plasmas by Lidsky *et al.* (1962). In a cathode with a hollow cylindrical structure, the emission current is greater than from a thermionic emitter of the same surface area. These cathodes have been used in high output power ion laser devices producing powers in excess of 1 W. The electron energy distribution in a hollow cathode discharge has a higher electron temperature (Kretschmer *et al.*, 1968) than in a conventional positive column.

These cathodes are simple; no heater supply is required for them and in addition they are not easily posioned, not subject to sudden failure, easily demountable, more robust than oxide cathodes and inexpensive. Their performance in an argon laser is comparable to that of an oxide cathode for total

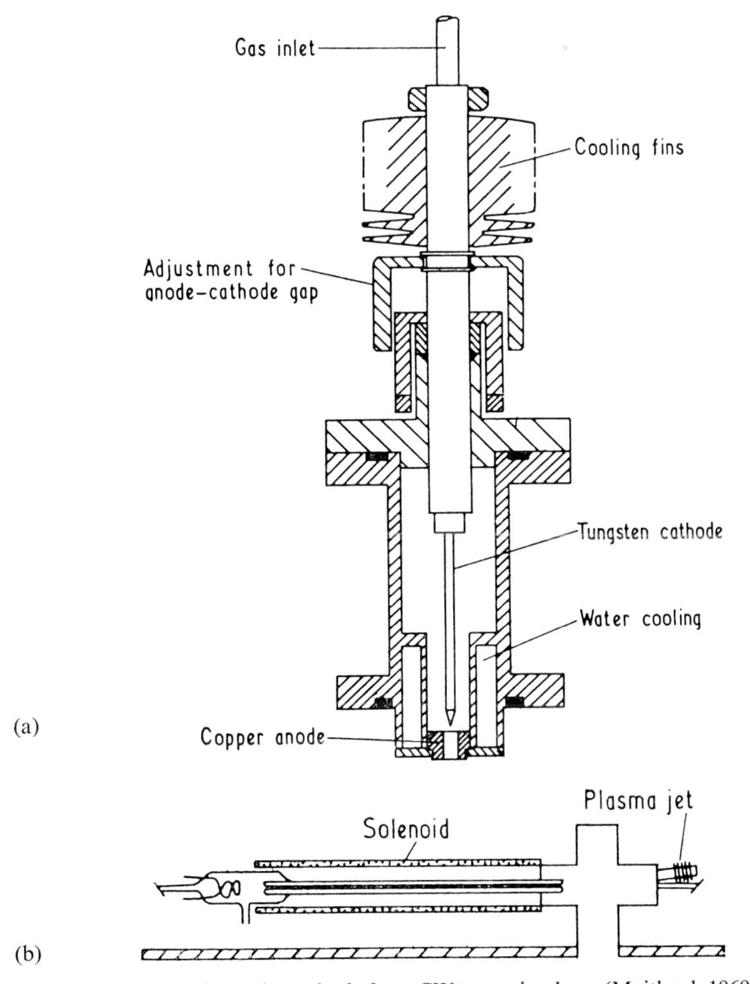

FIG. 35. Plasma jet cathode for a CW argon ion laser (Maitland, 1969).

output power, operating gas pressure, magnetic field and threshold current. Jennings *et al.* (1970) make the point that the cascade pumping contribution from the $4d$ and $5s$ levels into the upper $4p$ laser levels is about the same in the hollow cathode discharge as with the oxide cathode discharge. They comment that, since in a hollow cathode discharge the electron energy distribution is more pronounced at higher energies than for an oxide cathode discharge, it might be expected that the cascade contribution would be increased. This effect may be masked by the structure of the laser where the higher electron energy distribution does not hold over the entire active discharge region. In the hollow cathode ion bombardment liberates electrons from the cathode surface. The primary electrons begin an electron avalanche from further ionization and the increased ion bombardment rapidly heats the cathode to temperatures greater than 2000°K and the arc current rises. Jennings *et al.* (1970) have shown that the hollow cathode discharge, when based on cylinders of boriated tungsten, can be operated at about the same electrical efficiency as the heated oxide cathode. Also they show that a flowing gas system is not necessary. Their form of hollow cathode assembly is shown in Fig. 36a. For this laser the cathode and anode are similar and are made from coppper which is water cooled. The cathode includes a boriated tungsten dispenser cathode in its assembly and the electrodes are sealed to the fused silica tube simply by shielded O-rings. Notice that in the laser of Jennings *et al.* the discharge is constricted by molybdenum discs (see Section X, A) with central 6·35 mm apertures held in place by fused silica spacers at an optimum spacing of 3 cm.

Huchital and Rigden (1967) used a tantalum tube 8 cm long by 5 mm diameter in a metal flange as part of a flowing gas argon laser (Fig. 36b). A glow discharge is initiated externally, e.g. by R.F. ionization. A potential of 300 V across the laser causes ion flow to the cathode. They found that the best operating pressure was about 1 torr at the cathode surface. Sputtering effects in this design are reduced by being confined within the cathode and by the high operating gas pressure. However, this cathode had an operating lifetime of only tens of hours. Carbonne and Witteman (1970) used a tantalum hollow cathode and an auxiliary anode to act as starting electrode and evaporation shield. They found this structure was particularly suitable for high current operation and would provide currents up to 150 A.

(e) *Mercury pool.* The use of a mercury pool as a cathode surface which is capable of emission currents > 100 A has been described by Ferrario and Sona (1969), the current limit is set by the pool temperature. The use of mercury pools has also been described by Allen (1954, 1962). Their advantages are that they need no activation, are not readily poisoned, are easily constructed, have a long life and can be let-up to air in demountable structures. Some precautions are necessary to prevent the mercury vapour from contaminating the discharge. This can be done by water cooling but this makes the system bulky.

(f) *Cold cathodes.* Cold cathodes are commonly used in pulsed gas lasers. They have the disadvantage of needing high voltage supplies and may produce sputtered elements or gas clean-up but they are simple and less prone to attack by reactive gases. Some specially developed forms of cold cathode have been

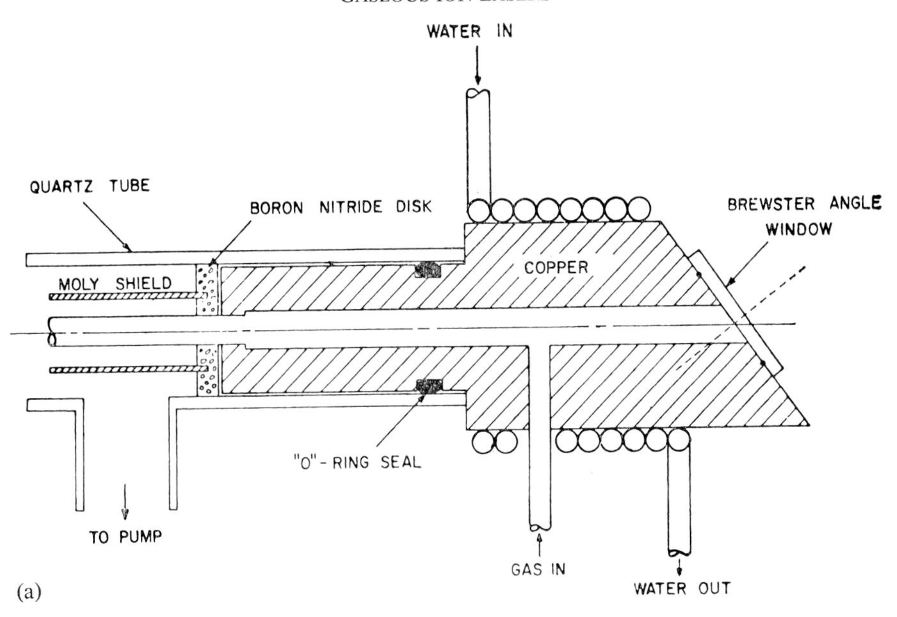

(a)

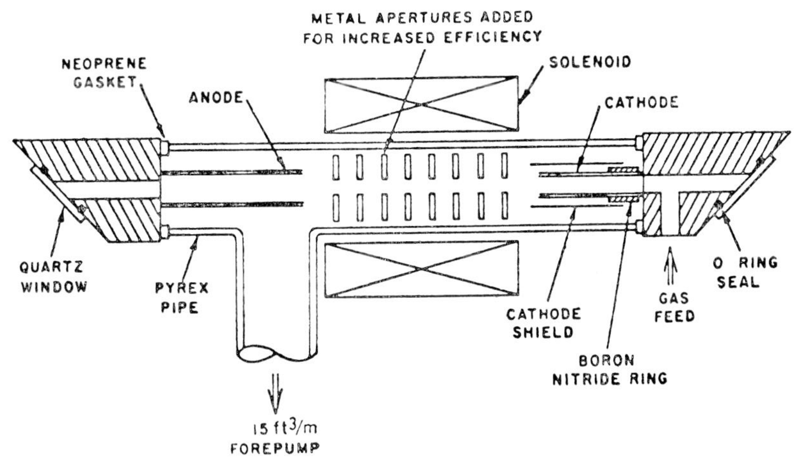

(b)

FIG. 36. Hollow cathodes for noble gas ion lasers: (a) boriated tungsten (Jennings *et al.* 1970); (b) tantalum (Hutchital and Rigden, 1967).

used. Hernqvist (1969) has described a cold cathode for ion lasers—termed an adconductor—based on a thick alumina coating deposited on a glass envelope which is made conductive and emissive by potassium metal.

A cold cathode made up of a pellet of indium does not need high anode–cathode voltages for reliable triggering (Barker and Jarrett, 1969; Simmons and Witte, 1970). Emission currents up to 2000 A and long pulse lengths ~7 ms have been obtained. Also the cathode is insensitive to contamination. It has been used by Simmons and Witte (1970) in the high power pulsed Xe IV laser (see Section XI, A (3)).

(2) *Anodes*

Because of the high currents in ion lasers the thermal dissipation in the anode can be high. The anode, which is usually metallic (e.g. stainless steel or copper) can be cooled by radiation from its surface which may be made large and can also, if necessary, easily be water cooled.

The position of the anode and cathode in relation to the active constricted discharge region is not unimportant. Damage to the electrodes or their surrounds can be minimized and instabilities, particularly at the anode end, can be reduced (Gorog and Spong, 1967). This dependence of the noise and frequency components in the laser output on the plasma has been confirmed (see Section XI, C). The various types of cathode can be expected to behave differently and an investigation of noise components produced in the plasma by different types of cathodes may provide design information for a reduction of plasma noise.

D. R.F. EXCITED NOBLE GAS ION LASER DISCHARGES

A number of reports have been made of the use of R.F. excitation in ion lasers with external electrodes (Bell, 1965; Renton and Zelby, 1965; Goldsborough, 1966; Zarowin and Williams, 1967; Paik and Creedon, 1968; Ahmed *et al.*, 1969 and Ling *et al.*, 1970). This method of excitation removes some of the detrimental effects of D.C. excitation such as cataphoresis, some striations and instabilities, and gas pumping; with no internal metal electrodes reactive gases such as chlorine or phosphorus can be excited. Capillary bore erosion, which is a major problem in D.C. discharge ion lasers, is much reduced and of course electrode sputtering is also eliminated. The R.F. energy is coupled into the gas by accelerating electrons at a frequency, 5–50 MHz, such that the electron collision frequency is $\sim$ the frequency of the applied field. The frequency may need to be chosen such that R.F. modulation of the laser output does not restrict the application of the laser. A typical R.F. power source using a class C power oscillator can produce R.F. power with up to 80% efficiency and deliver this to the discharge with a coupling efficiency of about 50%. The R.F. field can be coupled to the discharge either capacitively (E-field) or inductively (H-field). Some methods of pumping ionized argon lasers with R.F. excitation are shown in Fig. 37.

The first report of R.F. excitation of the argon laser was by Bell (1965) and his laser is shown in Fig. 37a. A ring discharge was used and pulsed power coupled in by a ferrite toroid when circulation currents of hundreds of amps was obtained. The total voltage drop around the arc ring was of the order of

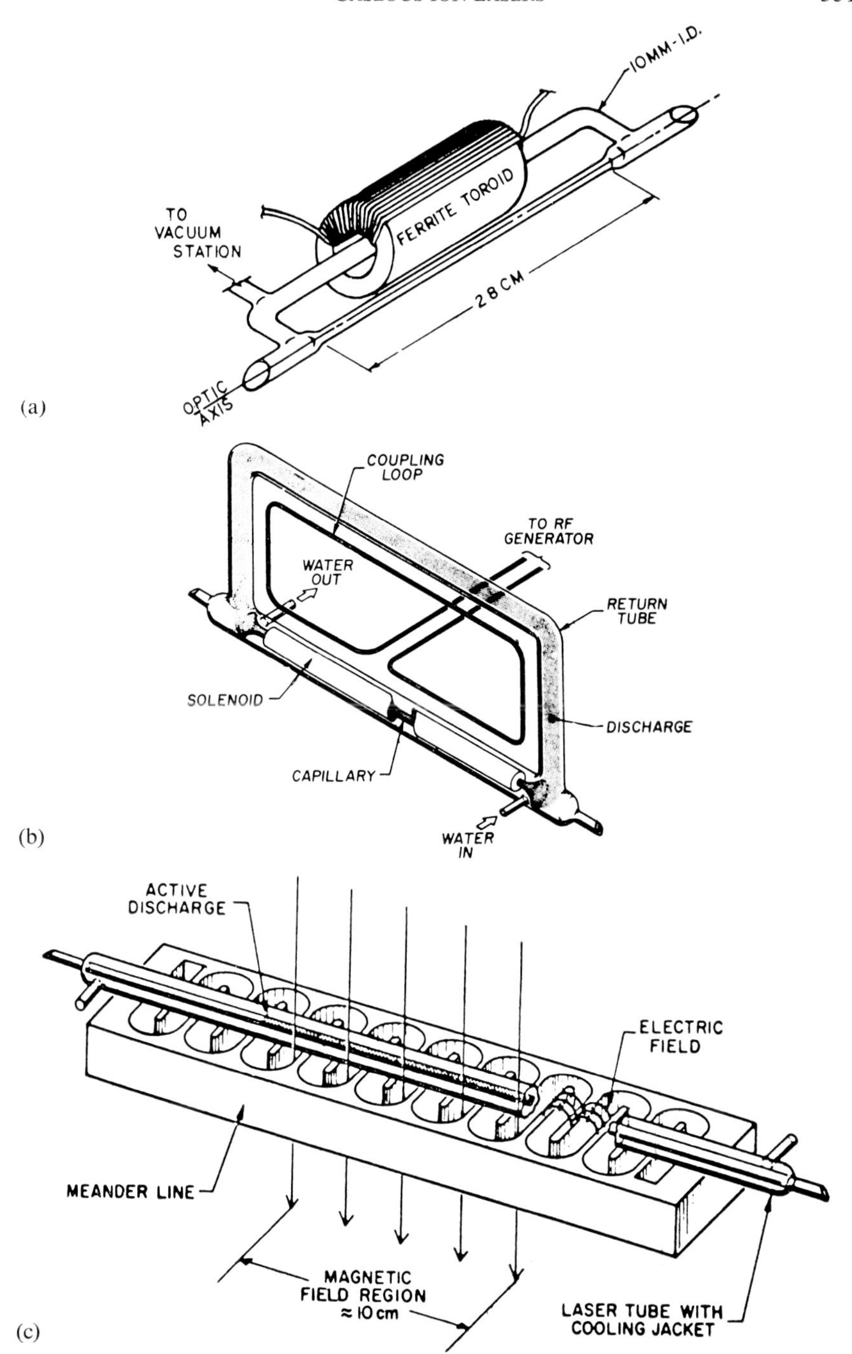

FIG. 37—*continued on page 352*

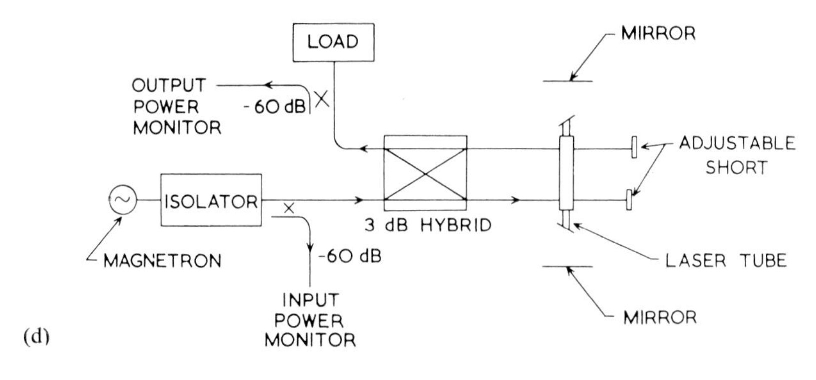

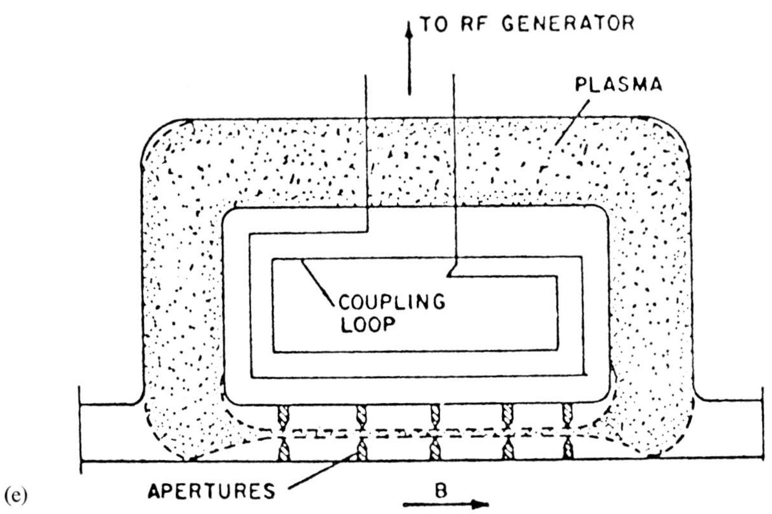

FIG. 37. RF excited ion lasers using (a) ferrite toroid (Bell, 1965); (b) coupling loop (Goldsborough *et al.*, 1966); (c) cyclotron resonance (Goldsborough, 1966); (d) microwave pump (Paik and Creedon, 1968); (e) coupling loop with metal apertures and magnetic confinement (Ahmed *et al.*, 1969).

the ionization potential of the gas. Laser lines in Ar II, Kr II, Hg II and S II were obtained and later also in Cl II and Br II. The operating conditions of gas pressure, voltage drop and the effect of axial magnetic fields were similar to D.C. discharge lasers. A similar laser used in the pulsed mode was described by Taylor *et al.* (1966) producing 5–10 μs pulses at a duty cycle of 10^{-4}. About 10 μW output on Ar II lines for 40–100 W input was obtained.

Goldsborough *et al.* (1966) used an inductively excited ring discharge to produce CW laser action in Ar II. Power was coupled into the low resistance plasma loop by a parallel air core coil from a low impedance R.F. source. This is shown in Fig. 37b this system was the basis of a commercial laser which is now no longer marketed. This method of excitation is more efficient than that used by Bell (1965) as it eliminated the losses in the ferrite core.

Microwave excited discharges have been used generally in gaseous excited state and radical formation work with high coupling efficiency to the gas (Fehsenfeld *et al.*, 1965). A novel but less efficient laser using microwave cyclotron resonance has been described by Goldsborough (1966). This discharge provided a high E/p ratio, and since the energy gained by electrons in the discharge is $\propto E/p$, the average electron energy is higher than in D.C. excitation. The cyclotron resonance frequence is $f_c = eB/2\pi m$ and is typically ~ 3 MHz/gauss. Goldsborough used a 1-kW CW magnetron feeding a non-radiating meander line which produced an E-vector parallel to the axis of the cavity as shown in Fig. 37c. The active region is 10 cm long and 3 mm bore. The threshold input power was 50 mW and 5 mW output was obtained for 500 W in giving a very low efficiency $\sim 10^{-3}\%$. An electron in orthogonal electric and magnetic fields undergoes spiral motion and absorbs energy from the microwave field over many excitation cycles. The cyclotron radius, $r = (2mE/eB^2)^{1/2}$, for typical fields is $\sim 0{\cdot}2$ mm which is $<$ the dimensions of the laser tube bore.

Excitation by pulsed cyclotron resonance was also used by Paik and Creedon (1968). As shown in Fig. 37d a microwave C-band, 1260 MHz, source was used with a long active region and a wide bore to accommodate the tube. The peak power into the tube was 200–400 kW and average power 200–400 W. With an absorption coefficient of about 10% the average power absorbed into the discharge was about 50 W. The threshold for the Ar 4880 Å line was a peak power ~ 100 kW at the 10^{-3} duty cycle and an average power ~ 10–20 W. The R.F. penetration into the gas was small and discharge and laser action occurred near the walls. It was suggested that resonant ring or matched cavity coupling would improve excitation uniformity.

Zarowin and Williams (1967) inductively excited a wide bore (1 cm) by 15 cm long argon laser using quasi-CW 90% duty cycle excitation from low frequency oscillation at 2·5 kHz. This device had a large plasma current (~ 300 A) at 15 kW since only low field strengths are needed to maintain the wide bore plasma (~ 1 V cm^{-1}). This large plasma current is in phase with the applied voltage when the leakage reactance $(\omega(1 - K^2)L_s$ (K = coupling coefficient, L_s = inductance of plasma tube) is small compared with the plasma resistance R_L. An output > 250 mW was obtained with a 2% transmitting mirror which was probably not providing optimum transmission. Another laser based on R.F. excitation in a wide bore tube is that of Ling *et al.* (1970) who used helical inductive pumping. R.F. power was at 13·5 MHz with a characteristic impedance of 50 ohms. Output powers up to 0·8 W were obtained for an input power of 8 kW and mirror transmission of 0·35%.

The inductively-coupled R.F. excited toroidal laser has been combined with plasma confinement by a magnetic field and apertures (Ahmed *et al.*, 1969)—this is illustrated in Fig. 37e. The active laser region is a set of 8–10 graphite discs with a central 0·25 cm hole spaced by 6 cm and placed in a fused silica tube of 4 cm diameter. The efficiency of this laser was high, 0·17%, and output powers ~ 1·5 W for a 5 kW R.F. input power were obtained. An optimum magnetic field > than for D.C. discharges was found. The efficiency is comparable or even better than conventional D.C. argon discharges.

R.F. excitation has not been used as extensively as D.C. excitation for the noble gas ion lasers and even its use with reactive gases has been scant. This is largely due to the marginal improvements and some disadvantages in general of this method over D.C. excitation but the results of Ahmed *et al.* indicate that it is still worthy of development.

E. OPTICAL RESONATORS FOR ION LASERS

(1) *Mechanical Design*

The stability of the cavity length and the angular alignment of the cavity reflectors are very important. Length stability is related to the stabilization of axial frequency modes which will be discussed further in Section XI, F. Stable operation requires that fluctuations in this length are less than about $\lambda/100$. The angular mirror stability defining the laser axis will affect the laser output power fluctuations and this alignment should be kept within 10–100 microradians (Freiberg and Halsted, 1969). In general the extent of stabilization required for the cavity depends on the time range over which disturbances are to be minimized. For short times, external vibrations are important while for long times stabilization against internally generated thermal gradients is required. The resonator, which is normally kinematically mounted, is subject to transverse thermal gradients and ambient temperature changes which lead to angular distortion and dimensional changes. The thermal gradients arise from the need for power dissipation ~ few kW in the laser.

Commercial argon ion lasers normally use in their resonator structure a carefully chosen set of materials to preserve cavity length and mirror alignment using, for example, fused silica rods to define the length and angles with the cavity reflectors directly connected to the fused silica components to reduce vibrational distortion. Metals may be used to provide components having thermal conductivity and mechanical rigidity.

(2) *Mirrors*

These are normally long radius plane and concave spherical mirrors made with multilayer dielectric coatings deposited on fused silica substrates. The mirror radii need to be selected to compromise between mode volume (or laser power in the mode) and resonator stability.

Because in general inert gas ion lasers can emit many wavelengths, there has been some interest in making an optical resonator to act simultaneously over the entire visible wavelength region. Perry (1965) made an early report of multi-layer dielectric coated mirrors with high broad band reflection and low transmission over the range 4300 to 7400 Å. This was achieved by careful refinement of the evaporation process in the film deposition.

(3) *Transverse Modes*

For most applications, the uniphase TEM_{00} mode is required. This can be selected by an in-cavity aperture or iris which is arranged to pass the TEM_{00} mode with little attenuation and to attenuate higher order transverse modes. In small bore lasers the plasma confinement tube bore itself may act as the limiting aperture. Highest output power is obtained if full use is made of the volume of the laser medium. This may be achieved by oscillation in TEM_{00} mode with large radius mirrors or, if uniphase output or minimum beam divergence is not required, oscillation in a higher order transverse mode, since this will have a larger diameter. These considerations are particularly relevant to the wide bore, high current type of ion laser discussed in Section X, G.

The low order transverse modes of the medium power 1–3 W Ar II laser have been studied by Freiberg and Halsted (1969) to determine the relationships between output power, beam divergence and mirror alignment tolerances. They found that the measured intensity distribution of higher order modes differed considerably from the theory of Fox and Li (1966). The main effect is the enhancement of the off-axis peaks relative to the centre peak. This is attributed to the high value of the small signal gain and the saturable gain characteristics of the medium. They found that the free-space beam divergence agreed closely with that for passive cavity modes. The mirror alignment tolerances for different transverse modes are less severe for double concave than for plano-concave cavities as expected. For higher order transverse modes the alignment tolerances become less strict. High order transverse modes have a larger beam diameter and use a greater volume of the laser medium and hence produce the higher output power. The fractional power in transverse modes as a function of mirror radius for plano and double concave cavities has been derived by Freiberg and Halsted and is shown in Fig. 38a. For a double concave cavity the power in the transverse modes as a function of mirror radii and mirror alignment is shown in Fig. 38b.

The Ar II laser has been found to operate fairly readily in the "doughnut mode" (Goldsborough, 1964); this can be considered as a hybrid mode made up of TEM_{01} and TEM_{10} modes, for convenience it is designated TEM^*_{10}. The change of the pulsed ion laser output beam cross-section from a disc

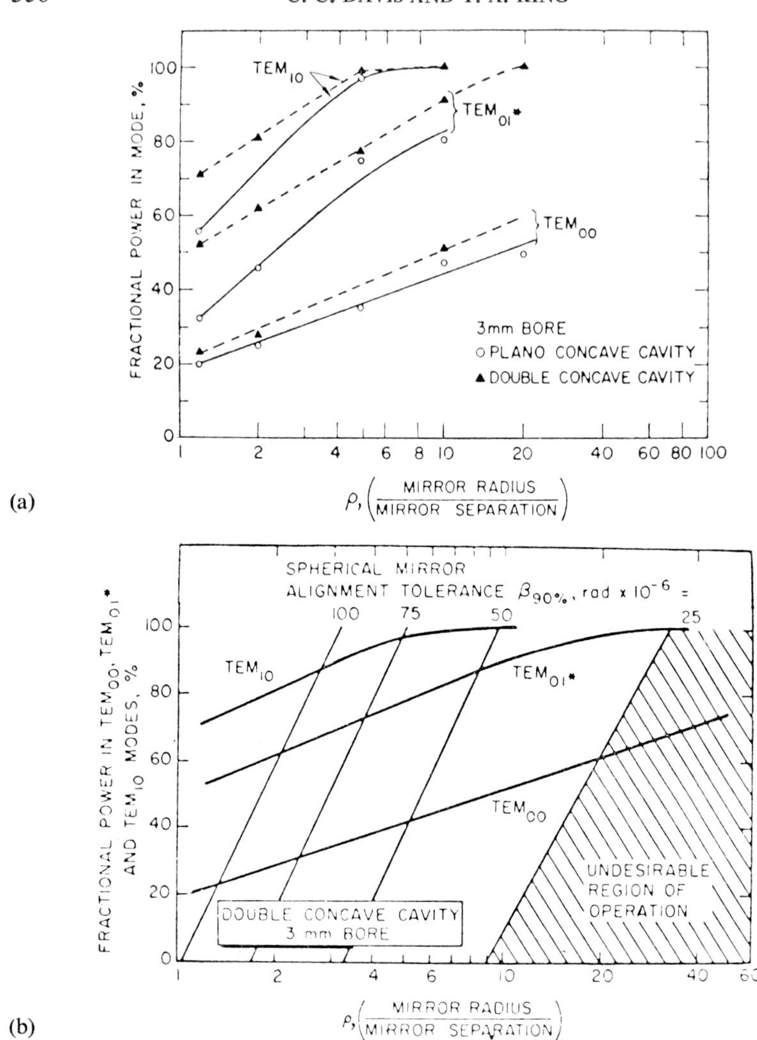

FIG. 38. (a) Dependence of percentage power in transverse modes on mirror radii for plano–concave and double-concave cavities. (b) Relative power in transverse modes with mirror radii and mirror alignment (Freiberg and Halsted, 1969).

pattern to a ring pattern with reduced laser intensity in the centre was observed by Bloom *et al.* (1964) in Hg II, and in Ar II by Cheo and Cooper (1965). Some ouput radiation patterns observed by Gorog and Spong (1966) for an argon laser operating CW at 30 A in a 28-cm length, 4-mm bore laser with an axially applied magnetic field are shown in Fig. 39. Bloom attributed the doughnut mode to selective pumping of the laser levels by less energetic electrons near the tube walls. In Hg the effect occurred at quite low currents ~ 10–50 A but in Ar the currents required were much higher ~ 1000 A in 4–10 mm bore tubes. Cheo and Cooper interpreted the effect as arising from radiation trapping between the lower laser level and the ion ground state since it occurred at the

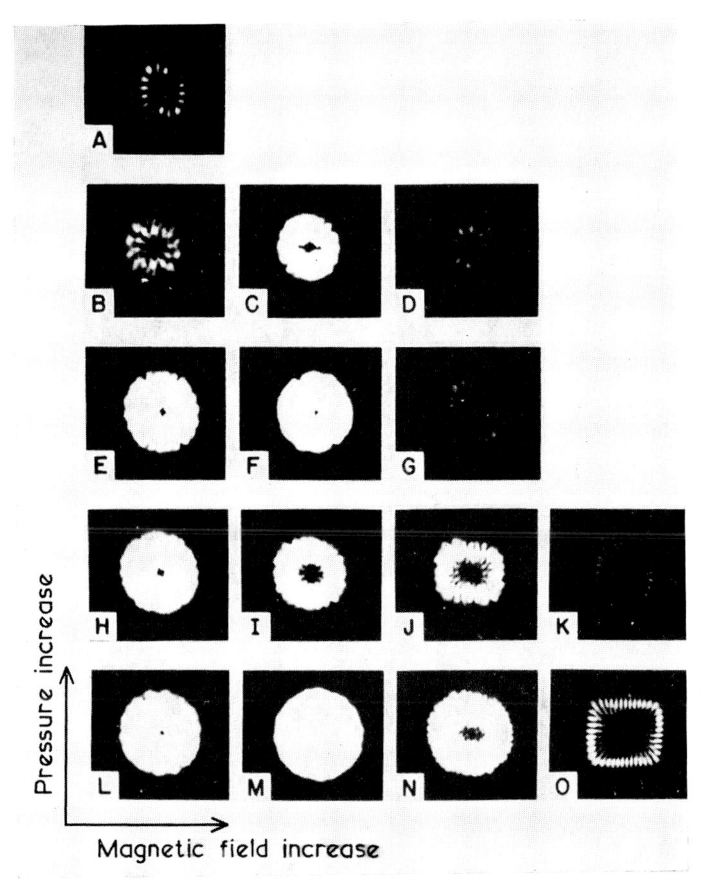

FIG. 39. Illustrative radiation patterns for a CW argon ion laser operating at 250 A cm^{-2} (Gorog and Spong, 1966).

onset of quenching in the laser. Ring-like modes have been previously discussed in Section IV, C(2), in connection with the role of radiation trapping in the argon laser. The presence of the doughnut TEM_{10}^* mode is often difficult to detect.

The use in an argon laser of a central hole in the output mirror as a means of coupling out power and selecting TEM_{00} mode has been investigated by Gordeyev *et al.* (1969). This method is more commonly used in I.R. gas lasers.

They claim more output power for a hole of 0·4 mm diameter than for transmission coupling but a thorough comparison with the case of no output coupling aperture was not made. The ease of selecting TEM_{00} rather than TEM^*_{10} depended on the optical resonator parameters.

The stabilization of the transverse mode spectrum in the Ar II laser by removing the competition on the 4880 and 5145 Å lines when they are operating in high order transverse modes has been achieved by Forsyth (1969). Stable transverse modes would be useful in a scanning output laser application (Auston, 1968). This is discussed further in Section XIII.

(4) *Wavelength Selection*

Since most ion lasers will oscillate at several wavelengths either separately or simultaneously a means of selecting a single wavelength is often necessary. This is usually achieved opto-mechanically with a rotatable internal Brewster-angle prism. It has been reported (Tandler, 1969) that Littrow prisms are not satisfactory for wavelength selection since they reduce the output power by a factor between 0·7 and 0·9 compared to Brewster angle prisms and contribute power dependent mode distortion. These effects are attributed to strain or absorption in the fused silica-dielectric film interface. Gratings, although having much higher dispersion in general than prisms in the long wavelength or visible region, are not used since they have too great a loss. They can be used in high gain infrared and dye lasers.

Non-mechanical methods of selecting wavelengths have been described by Habegger *et al.* (1966), Wen *et al.* (1966) and Habegger (1967). These authors used an electro-optic method based on the use of a crystalline quartz rod, a $\lambda/4$ plate, two KD*P crystals and the laser tube Brewster windows. Rotatory dispersion of polarization in the quartz rod is acted on by the electro-optic crystals, which rotate the polarization, and the Brewster windows act as analyser in which the plane of polarization in the plane of incidence has very low loss while other wavelengths are attenuated. This dynamic wavelength selector can switch between wavelengths in a time $\sim 1\ \mu s$ and has been used to switch between the Ar II lines at 5145, 4965, 4765 and 4979 (Habegger *et al.*, 1966). Also the primary colours in the Kr II laser at 6764 (red) 5208 (green) and 4762 (blue) have been switched (Wen *et al.*, 1966).

A novel method to increase or decrease intensity in Ar II or Kr II lasers using an intra-cavity electron beam tube has been described by Hammer and Wen (1965). The laser tube is brought near to threshold by a conventional discharge and its laser character modified by driving the electron beam tube. A He–Ne laser based on a similar electron excitation system was described by Tien *et al.* (1964). Strong absorbing effects in Kr II were observed by Hammer and Wen for the lines 4619, 4762 and 5208 Å while the 4765 Ar II line showed strong excitation in this system in agreement with the discussion in Section IV regarding the greater contribution to the excitation of its upper level by single-step processes.

(5) *Optical Degradation Effects*

Generally argon lasers can operate for several thousand hours at output power levels in the watt region without degradation. If particularly high powers in excess of 10 W or UV emission are required, or if care is not taken to remove some sources of contamination, severe optical degradation of Brewster end-windows and laser mirrors can occur since the internal cavity radiation power may be greater than 1 kW cm^{-2}. The onset of degradation in medium power visible argon lasers has been thoroughly investigated by de Mars *et al.* (1968). Their general finding is that contaminants from the discharge tube and electrode erosion are transported to the Brewster windows of the tube where they act as absorption centres. The Brewster windows become non-uniformly heated and distort. For an argon laser operating at 70 A and producing an intra-cavity power of 1 kW cm^{-2} a reduction to half power occurred in about 15 minutes. The output power can also become self-limited; de Mars *et al.* found their 100 W laser limited at 20 W under some conditions. Optical degradation power levels in the same range were found by Gorog (1971) who also mentions an optical power limitation for damage in laser mirrors of 1 kW cm^{-2}. Paananen (1968) finds mirror degradation at laser output power levels of 50 W cm^{-2}. Damage in dielectric-coated surfaces has been discussed by Bass and Barrett (1971).

Contaminants can be liberated by sputtering, particularly from the cathode. Bass *et al.* (1968) have detected six elements emitted from the cathode structure in running. Two obvious sources of rapid degradation which can easily be avoided are the use of low vapour pressure greases or from poorly cleaned Brewster windows.

Latimer (1968) has found that fast degradation of an Ar II laser operating on UV lines 4067, 4131, 3638, 3507 and 3511 Å occurs. The degradation was attributed to the processes described by de Mars *et al.*, or to the decomposition of the mirror coatings. With internal mirrors and a 100 A discharge current, the laser output on the Ar II 4067 Å line decreased by 50% in 2 minutes.

A damage threshold to standard commercial mirrors, of hard dielectric coatings on fused silica substrates, in an argon laser has been given by Wang and Lin (1972). In a cavity made up of a plane and a concave mirror the threshold power to burn at the mode spot on the plane mirror was 7300 W cm^{-2} over the calculated spot diameter or a power density of 700 W cm^{-2} in the output multimode beam. The power density to cause optical degradation was lower than this.

Little effort has yet been expended on trying to overcome these degradation effects on dielectric mirrors and undoubtedly improvements in damage threshold can be made. For example the immersion of the dielectric mirror in a liquid has been used by Gregg and Thomas (1966), in this case for a Q-switched ruby laser.

F. MAGNETIC FIELD PLASMA CONFINEMENT AND POWER ENHANCEMENT

Longitudinal magnetic fields along the laser axis up to about 1000 gauss increase the output power of CW gas ion lasers. This occurs because of an increase in the axial charged particle density as discussed in Section VII, C, which increases the laser pumping rate and gain. The effect of large axial magnetic fields in increasing the charged particle density in discharges was described and investigated by Bickerton and von Engel (1956) and in increasing

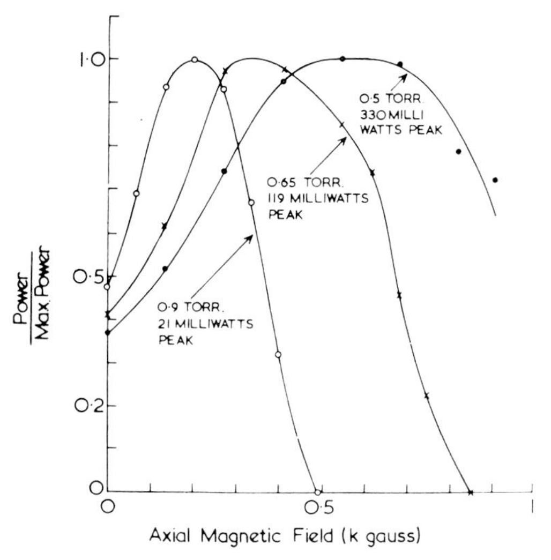

FIG. 40. Output power dependence on axial magnetic field for a CW argon ion laser with discharge current of 30 A (Gorog and Spong, 1966).

the spontaneous emission intensity from ions by Francis (1956). Enhancement of laser output power by axial magnetic fields was first described by Labuda *et al.* (1965) and Bridges *et al.* (1966).

The dependence of CW output power on magnetic field for the Ar II laser is shown in Fig. 40. The optimum magnetic field is ~ 1000 G for a 3-mm bore tube. For large bore tubes the optimum field is lower (Ahmed *et al.*, 1967) such that for wide bore tubes (> 1 cm diameter) there is no power enhancement (Boersch *et al.*, 1970). The magnetic field can be considered to have the effect of giving the laser a smaller effective diameter. In argon all the laser lines appear to optimize at about the same value of magnetic field but in Kr and Xe different transitions optimize at quite different values. This can be used as a means of tuning the laser to select operating wavelengths. In general the heavier noble gases Kr and Xe are more critically dependent on the magnetic field setting and have optimum magnetic fields lower than Ar—presumably because of their lower ion mobility.

The effect of the magnetic field on the polarization of the laser reported by Gorog and Spong (1966), Sinclair (1966) and Fotiadi and Fridrikhov (1970) are discussed in Section IX, E. There are different optimum field values for lasers with Brewster windows and with internal mirrors. The dependence of the output power on magnetic field is influenced by the gas pressure and discharge current. Higher pressures reduce the rate of fall-off of the enhancement factor after the peak. The optimum field decreases with increase in discharge current (Toyoda and Yamanaka, 1967) and the enhancement is greater if the laser is operating near threshold. This decrease with current increase may be because the self-generated magnetic field of the discharge is becoming important, particularly for example in the wide bore high current lasers.

In pulsed operation (Section X, H) little enhancement with magnetic field has been reported for a laser at optimum current and pressure (Jarrett and Barker, 1968) but Birnbaum (1968) obtained increased output power and higher efficiency and Toyoda and Yamanaka (1967) observed power enhancement and a lower threshold current. In order to investigate ion drift gas pumping Ahmed *et al.* (1967) have compared CW and pulsed measurements in magnetic fields for tubes having diameters > 5 mm. In the pulsed mode the optimum field was 400 gauss compared with optimum fields of 2800 gauss for CW operation at the same low pressure. It is to be expected that optimum magnetic fields in pulsed operation, where gas pumping effects are small (see Section X, B), will be lower than in CW operation.

Alternating axial magnetic fields were compared with D.C. fields by Konkov *et al.* (1968). The field frequency was 50 Hz with no D.C. level applied to a 40-cm length, 2·5-mm bore tube with Brewster windows. For the Ar 4880 Å line it was found that the A.C. field gave higher mean laser intensity than the D.C. field by a factor of about 1–2, depending on the field, and also a lower threshold for laser action.

The increase in charged particle density on the axis holds under conditions in which classical collisional diffusion theory holds (see Section VII). It has been pointed out by Powers (1966) that at sufficiently high magnetic fields an "anomalous diffusion" may dominate in the discharge which may have the effect of reducing the charged particle density and laser output power. The quenching of laser action at the higher fields, greater than the peak output in pulsed Ar II lasers, has been attributed by Toyoda and Yamanaka (1967) to the development of plasma instabilities and alteration of the spatial distribution of the inversion density leading to quenching. Quenching processes have also been discussed by Vladimirova *et al.* (1970). Plasma instabilities may arise in the laser, e.g. plasma turbulence described by Hoyaux (1968) which may be undesirable in laser applications. Another of these is the helical instability which gives non-uniform gain (Kadomtsev and Nedospasov, 1960). Fotiadi and Fridrikhov (1970) report a subsidiary peak in power output at fields about one-ninth of that for the main peak for the Ar 4880 Å line, i.e. 80–90 G compared with the main peak occurring at 700 G for their laser. This was in a narrow bore tube (1·2 mm) with a short active length (12 cm) and in a resonator of 90 cm length with Brewster windows. They were unable to account for the extra peak in terms of Zeeman splitting of the σ_+ and σ_- components. A secondary peak at much higher fields, in a laser with internal mirrors but under non-optimum pressure conditions, found by Borisova *et al.* (1967) was attributed to broadening of the Doppler line profile by Zeeman effects.

For lasers with external mirrors and Brewster windows the plane of the linearly polarized standing wave set up is rotated by the Faraday effect when an axial magnetic field is applied. This can lead to additional losses at the Brewster windows. Faraday rotation in resonant cavities has been discussed by Rosenberg *et al.* (1964). Fotiadi and Fridrikhov (1970) have observed the dependence of output power with variation of axial magnetic field and have calculated the rotation of the plane of polarization per unit optical length from $\theta = VH$ where $V =$ Verdet constant and $H =$ magnetic field. Using the analytic expression for V of Tobias and Wallace (1964) ($V = 0{\cdot}3\mathrm{g}\,(\nu_0)$ deg. (gauss cm)$^{-1}$ where $\mathrm{g}(\nu_0)$ is the line centre gain coefficient) they find the gain coefficient peaks at about 1100 gauss roughly in agreement with experiment. The Faraday rotation resulting from the induced CW or pulsed self-generated axial magnetic fields in high current lasers may contribute to magnetic field or high current quenching. Wang and Lin (1972) have suggested the use of stable magnetic confinement to reduce the loss rate of ions in an attempt to improve the laser efficiency.

G. HIGH POWER WIDE BORE NOBLE GAS ION LASERS

The use of wider bore tubes (> 1 cm) filled to lower gas pressure in the argon ion laser was first tested by Paananen (1966) and Boersch *et al.* (1966 and 1967). The use of wider bore tubes was discussed by Paananen (1966) with proposed limiting values and by de Mars *et al.* (1968) who, when investigating optical degradation in ion lasers, obtained ~ 100 W from an intermediate sized bore (8 mm) 2·7 m long argon laser. This form of argon ion laser has been further developed to produce power outputs up to 150 W (Boersch *et al.*, 1970; Seelig and Banse, 1970). The main differences between these tubes and conventional Ar lasers is that high CW currents are used, up to 500 A cm^{-2} in the wide bore tube, and no axial magnetic field is necessary. The need is still, as in conventional noble gas ion lasers, to increase the electron density whilst maintaining an optimized electron energy distribution. A further useful feature is that heat flow to the tube walls is smaller than in higher pressure operation, tube wall damage is less and the laser has a simple construction.

Radiation trapping has been proposed as a saturation mechanism in the argon ion laser. However, there is now strong evidence that radiation trapping effects are smaller than was anticipated from early work. The successful use of these wide bore structures suggested that radiation trapping effects were not so important (this is also discussed in Section IV, C).

Scaling relationships for the Ar ion laser indicate that excitation–rate-limited power output was scaled as $(pR)^2$ and the peak generation efficiency as (pR)

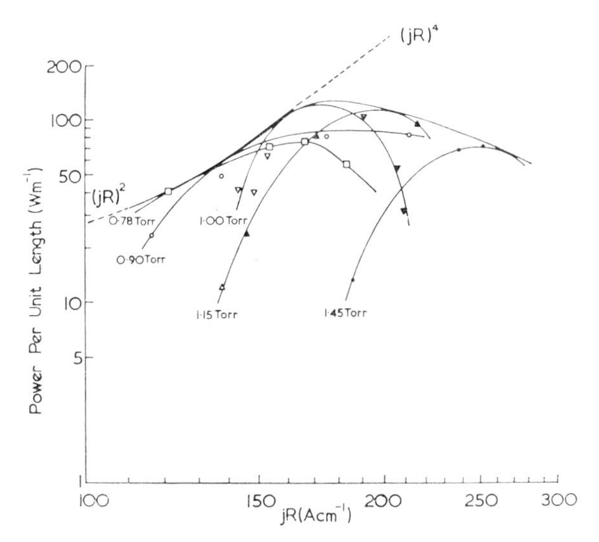

FIG. 41. Total power per unit length for the seven dominant Ar II lines with product current density (j) × tube radius (R).

which means that large bore tubes at higher pressure are favoured for high power and efficiency up to the possible radiation trapped limit. The pR values used in these lasers are $\sim 2{\cdot}5 \times 10^3$ torr cm rather than 0·05–0·5 torr cm in conventional lasers. Calculations by Herziger and Seelig (1968) of gain on a rate equation model replacing atomic parameters by experimental values indicated that the wide bore tube would be of advantage for high power operation.

Further development was reported by Boersch *et al.* (1970) using current densities up to 500 A cm^{-2}. An output power of 90 W was obtained from a tube of length 85 cm. Significantly this work reported saturation of the Ar II or Kr II emission at these high current levels although no saturation was observed for Ar III. Figure 41 shows the dependence of laser output per unit length as a function of [(discharge current density) × (tube radius)]. Here saturation of the laser output is evident with an optimum pressure near 1 torr and $(jR_{opt})) \simeq 170$ A cm^{-1}. This current dependence is discussed also in Section XI, A.

The wide-bore laser operates under wall-stabilized arc discharge conditions and provides a similar electron density to discharges in small bore tubes. High power outputs have been obtained from this type of laser (Banse *et al.*, 1969; Herziger and Seelig, 1969), 120 W total power from Ar II transitions and 20 W from Kr II for an input of 160 kW with a maximum current density of 265 A cm^{-2}. Boersch *et al.* (1967) suggested that it might be possible to increase the output power up to 500 W.

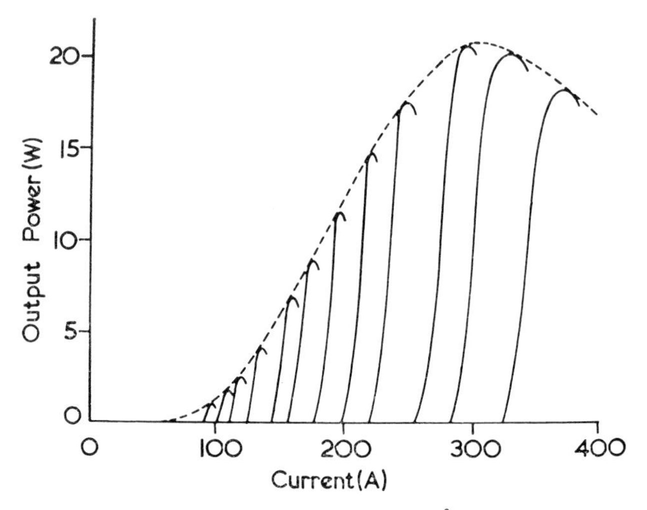

FIG. 42. Current dependence of output power at 4880 Å for a CW argon ion laser; the optimum power at any current is indicated by the dashed line and the maximum power for this tube is 21 W at 295 A (Boscher *et al.*, 1971).

Saturation of CW laser power in large bore lasers has been reported (Boscher *et al.*, 1971). Saturation was seen in argon, krypton and xenon in the high current region. This is shown in Fig. 42 for the Ar 4880 Å line together with the current dependence data of Boersch *et al.* (1967) for a 1-cm bore tube. Saturation has been discussed by Boersch *et al.* (1970) with possible mechanisms of (i) resonance trapping (ii) limitation of Ar II species and increase in Ar^{n+} $(n > 1)$ and (iii) presence of impurities such as O_2 and metal atoms. See *et al.* (1967) have used continuous argon gas flow to remove impurities. Seelig and Banse (1970) have obtained 150 W at an efficiency of 0·13% on the Ar II transitions around 5000 Å. For this output the conditions were tube bore 1·2 cm, active length 2·2 m, mirror reflectivities 99·9 and 90%; the tube operated with a potential of 400 V and current 300 A. Thus from these wide bore wall-stabilized arc discharges overall powers $\geqslant 1$ W cm^{-1} of active length are obtainable. The attainment of output powers in excess of 150 W is limited by current optical technology. These high power tubes have used symmetrically placed axial cathodes and anodes which could handle 300 A current in the tube bores in the range 0·7 to 1·5 cm.

Using a wide bore tube (i.d. = 1·2 cm) of this form, Banse *et al.* (1968) obtained continuous UV output powers in the watt range in argon (3511 and 3638 Å, 1·5 W) and krypton (3507 Å, 1·3 W). Herziger and Seelig (1969) also obtained 1·5 W CW on the argon UV lines at 3638 and 3511 Å.

Herziger and Seelig (1968, 1969) have shown that output power $P \propto$ [(volume) $\times$ (number of excited states)] $\propto R^2 N$, but $N \propto j^2$ for 2-step excitation and then $P \propto (jR)^2$. For $jR \leqslant 50$ A cm^{-1}, i.e. conventional small-bore ion lasers, the gas discharge is determined by electron–atom collisions. For $jR > 150$ A cm^{-1} the discharge is more influenced by electron–ion collisions (particularly as the degree of ionization of the plasma increases); this is the region of the high power wide bore laser. No saturation is observed for jR up to 200 A cm^{-1}. Experiments on fused silica tubes have been conducted up to $jR \sim 250$ A cm^{-1} and as yet no experiments appear to have been done in the region above this.

The scaling laws of Herziger and Seelig (1968, 1969) are useful up to current densities of a few hundred A cm^{-2}. They can be interpreted to mean that since $j \propto p$ then output power $\propto (pR)^2$. They show that the electron density is greater in wide bore rather than narrow bore tubes. Skurnick and Schacter (1972) have reasoned that since the electron density increases in wide bore tubes, this could counter the radiation trapping effects. The (pR) scaling relationship has also been obtained by Wang and Lin (1972).

The radial potential in a laser tube leads to large radial ion velocities (see Section VII, A) which produce a radial Doppler profile dependence. The ion velocity will increase in value with distance from the tube centre. This is discussed in Section VII, A, and has been mentioned by Hernqvist and Fendley (1967). This broadening of the Doppler profile will lead to reduced reabsorption and radiation trapping effects.

In a large bore segmented metal plasma tube Wang and Lin (1972) report output powers up to 50 W cm^{-1}. They have carried out a thorough parametric investigation and observe no radiation trapping limitation for (pressure $\times$ radius) values below 0·5 torr cm and additionally conclude that the power output of this laser is not limited by gain saturation effects. They claim the limitation is from inefficient use of the available inversion and by optical degradation. This suggests that higher efficiency and output power may be obtained by optimizing gain matching and the extraction of power from the gain profile by optimizing the axial mode frequency spacing.

Banse *et al.* (1968) report optimum pressures around 0·3 torr and also find that a good by-pass is useful in removing pressure gradients to give a more stable output and reduce plasma oscillations. A relationship between optimum pressure p_{opt} and bore diameter d was found by Boersch *et al.* (1967) of the form

$$p_{opt} . d = 5 \times 10^{-2} \text{ torr cm.}$$

and this was confirmed by Herziger and Seelig (1969). However Wang and Lin (1972) find that at a fixed current density the output power depended strongly on the filling pressures with an optimum value at each current density. The optimum pressure increased with current rather than remained constant as found by Boersch *et al.* (1967) and Herziger and Seelig (1969).

It is interesting to note that argon lasers operating at high inversion density of about 10^{10} cm^{-3} (for Ar II at 4880 Å) are expected to induce amplified spontaneous emission (ASE, super-radiance or non-resonant laser oscillations) (Banse *et al.*, 1968). Amplified spontaneous emission was obtained at 4880 and 5145 Å in a tube with tilted plane mirrors with a single pass gain of 20 dB and a power output of up to 20 W with a beam divergence of 10^{-4} rad (Banse *et al.*, 1969). Thus the wide bore laser is capable of high population inversion and high gain and, without the resonant optical cavity, high powers are attainable with no problem of optical degradation at resonator mirrors.

As discussed by Seelig and Banse (1970) wide bore, high current wall-stabilized discharges can provide high power continuous lines in the visible and UV, not only in argon but other noble gases. Very little work seems to have been carried out on this type of laser for gases other than Ar and Kr. Their application to the other noble gases and other gaseous ion laser systems might be an interesting future development.

H. PULSED NOBLE GAS ION LASERS

Pulsed operation enables high peak currents to be easily reached and has been used extensively to observe new laser lines in gas lasers. At higher currents transitions in higher states of ionization can be excited and laser action can be obtained in the pulsed mode on transitions which do not lase in the CW mode because of self-terminating effects. However, much less development effort has been expended on pulsed ion lasers than on CW ion lasers.

The average input power is much lower than for conventional D.C. ion lasers and consequently simple glass discharge tubes can be used. The pulsed mode enables higher input peak powers to be used ~ 100 times greater than in the D.C. laser (Goto *et al.*, 1971; Kitaeva *et al.*, 1969), and the output radiation peak powers are also higher but surprisingly not usually by a large factor. The gas pressures for optimum performance are generally lower around 20 mtorr or less compared with conventional D.C. lasers, partly reflecting the reduced effect of gas pumping. High E/p values are used, typically in the range $E/p \sim (10^2 \text{ to } 10^4)$ V cm^{-1} torr^{-1} compared with D.C. lasers in the range $E/p \sim (10 - 10^2)$ V cm^{-1} torr^{-1}.

In the pulsed argon laser there are three distinctive regions of lasing the first two of which are illustrated in Fig. 43 (Davis and King, 1971):

(1) Very near the beginning of the current pulse is a very transient spike, pulse A. This is the only pulse observed at low pressure (Kobayashi *et al.*, 1966). This transient laser action is most prominent for the Ar II 4754 Å line and is also observed for the other prominent Ar II lines. This pulse favours lower pressures which is consistent with the single-step high T_e excitation model discussed in Section IV, B.

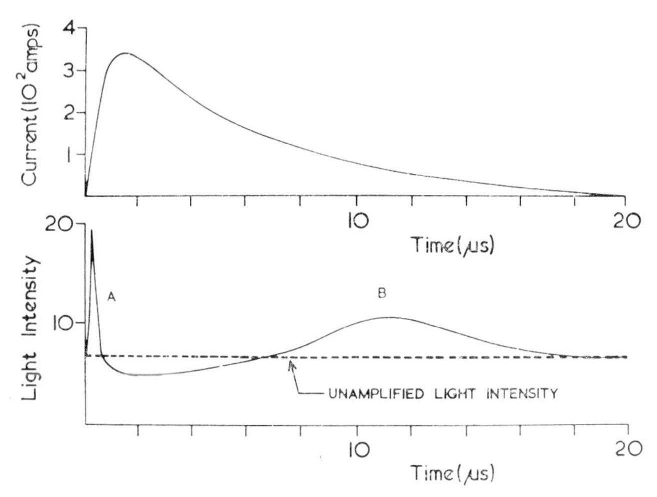

FIG. 43. Gain and exciting current time dependence in a pulsed argon ion amplifier at 4880 Å and pressure of 15 μtorr. (Davis and King, 1971).

(2) Later in the current pulse near the peak there is a longer duration pulse, pulse B, sometimes termed quasi-CW since its duration can be increased by increasing the excitation pulse duration and its shape follows that of the current pulse. This longer duration output pulse may sometimes break up into two. Peak output increases, with pressure (Kobayashi *et al.*, 1966), and at higher pressures this pulse becomes much more intense than the first pulse.

(3) A pulse can occur in the afterglow after the current pulse has ceased; this was observed by Neusel (1966) in a 400–12,000 A cm^{-2} discharge with E/p from 500 to 10^4 V cm^{-1} $torr^{-1}$ at 10 mtorr pressure and by Smith and Dunn (1968). The afterglow pulse begins after a delay of 2 to 10 μs after the first pulse and lasting up to 60 μs after a 2-μs current pulse.

Double pulse behaviour has been observed by several workers (Bridges and Chester, 1965; Demtroder, 1966; Klein, 1970; Davis and King, 1971). The second pulse has been interpreted as an effect due to plasma heating which occurs during the pulse, increasing the Doppler width and removal of resonance trapping (Gordon *et al.*, 1966; Bridges and Chester, 1965; Husain, 1968; Skurnick and Schacter, 1972). Gordon *et al.* showed that the quenching of laser action could be reduced by heating the discharge tube. They attributed this reduction to the removal of resonance trapping by increasing the Doppler width. It is not clear from this work whether all parts of the system were heated or if the reduction in quenching could not be attributed to a reduction in pressure by gas drive-out. In this second pulse Klein (1970) has found that the ion temperature can increase by up to two times while for shorter current pulses ($\sim 10\ \mu s$) Hattori and Goto (1969) found T_e to be constant. They suggested that the double pulse behaviour might be explained by electron destruction of the upper laser level but Klein observed that the spontaneous emission remained constant during the current pulse.

Cottrell (1968) and Davis and King (1971) have interpreted this pulse as arising from the electron excitation mechanism itself on the upper and lower laser levels (see also Demtroder, 1966; Demtroder and Elendt, 1966; Kobayashi *et al.*, 1966; Leonov *et al.*, 1966; Glazunov *et al.*, 1967; Cheo and Cooper, 1965). The current dependence of the peak output power for the first pulse has been measured by Cottrell (1968) and fits a function of the form: power $= W = aI - bI^2 - I_t$ where a and b are constant and I_t is a threshold current. Davis and King (1971) found a similar dependence for the gain of the second pulse for an Ar II amplifier. They found that the output power was proportional to gain in the second quasi-CW pulse over a wide range of currents. This shows that the Ar II 4880 Å line in pulsed operation is homogeneously broadened. The form of the output power and gain current dependence are in agreement with an apparent single-step excitation of the upper level and 2-step excitation of the lower level.

The first pulse shows linear dependence of laser output power on peak pulsed current density. A saturation of output power is predicted by the intermediate broadening theory with increasing saturation parameter depending on the parameter

$$\frac{2\gamma}{\Delta\nu_D} = \frac{\text{collision broadened linewidth}}{\text{Doppler width}}$$

However, Cottrell found a contradiction in the small signal gain showing no sign of saturation up to an excitation parameter of $X = 96$.

The dependence of the peak power of pulsed argon ion lasers on current has been discussed by Hattori and Goto (1968a, b, 1969). In their inductively excited argon ion laser, various dependences were observed at higher currents: (i) the output power at 4765 Å did not decrease while it did from other Ar II transitions; (ii) at pressures above 30 mtorr the output at 4880 Å excited by a double pulse of 10 and 40 μs only appeared in the second 40 μs pulse; (3) at 26 m-torr the laser output first increased, then decreased and then increased again. In the case of this laser the parameters T_e, cascade rate, neutral atom density and $\Delta\nu_D$ are all changing during the pulse in the high current region.

1. *Designs of Pulsed Ion Lasers*

There is interest in pulsed ion lasers for applications where their wavelength or power level is required but not continuous emission. These lasers have the advantage of not requiring high power dissipation and simple constructions can be used.

Hernqvist (1970) has described a pulsed Ar laser operating at 5145 Å producing pulse emission several ms duration at a repetition rate of 2 pps at 1 W peak power levels. The emission is of the second pulse type in this time range. A neat and simplified pulsed gas laser for general use has been described by Lomnes and Taylor (1971) using a cold cathode and demountable electrodes. The laser has operated at repetition rates of a few 100 pps producing a few watts in Ar II 4880 and 5145 Å and 10 W in OIII at 5592 Å. In order to generate short and intense pulses with a high and constant repetition rate a coaxial gas laser was constructed by Herziger and Theiss (1970). In this laser an output pulse was generated with the characteristics: risetime = 12 ns, peak power $\sim$ 0·5 kW and repetition rate = 10 Hz in Ar II. Pulsed argon ion lasers excited at low voltage ($\sim$ 120 V) and low current (< 100 A) have been made (Kitaeva and Osipov, 1968). Up to seven lines were observed with a maximum output $\sim$ 1 W. Relatively long pulse operation was observed up to 70 μs and the pulse under certain conditions broke up into two.

A pulsed ion laser which can operate at very high repetition rates for short periods of time has been developed [the "burst" ion laser: (Smith, 1969)]. In a sense it is intermediate between the pulsed and CW lasers, based on the simple, small, lightweight and inexpensive structures of the pulsed laser, but being able to deliver time-average powers equal to that of the CW lasers with a duty cycle up to 0·25. This laser should be of use in applications requiring high powers for short times without the need to use a CW ion laser. In normal pulsed mode a repetition rate of 360 pps can be increased to 1200 pps in the "burst" mode and typical power performance figures are for Ar 5145 Å in normal pulse mode: peak power 2·6 W, average power 65 mW; in burst mode: peak power 2·6 W, average power 200 mW (for the Britt Electronic Products Corporation pulsed burst laser). This laser is available commercially operating on the Ar 5145 Å, Xe 5353 Å and the UV lines Ar 3638 and 3511 Å.

Gas pumping in pulsed Ar lasers has been discussed by Neusel (1966). Improvement in laser power can be obtained by using gas return paths in pulsed lasers as in D.C. lasers. But gas pumping is less of an effort in pulsed lasers because of the short duration of the excitation pulse compared with the gas pumping drive-out time ~ 1 ms.

Most experimental observations have been conducted over a wide range, up to about 5×10^4 A cm^{-2} in current density, but mainly in lasers with narrow bore tubes (Bridges and Chester, 1965; Cheo and Cooper, 1965; Leonov *et al.*, 1966; Kulagin *et al.*, 1966; Demtroder, 1966; Glazumov *et al.*, 1967; Cottrell, 1968; Smith and Dunn, 1968; Hattori and Goto, 1968 and 1969; Davis and King, 1971). Wider bore tubes (6–10 mm) have been investigated by Hattori and Goto (1968) in the higher current region > 500 A and higher current behaviour up to 3×10^4 A cm^{-2} has also been investigated by Davis and King (1971) and Kulagin *et al.* (1966). The laser of Hattori and Goto (1967, 1968) had a ring discharge tube with peak discharge currents between 500 and 1000 A in a double 10 and 40 μs pulse. For Ar 4880 Å the optimum pressure was 15 mtorr. The saturation behaviour of the Ar 4880 and 4765 Å lines was quite different in a 1-cm bore tube at 15 mtorr. The 4765 Å line was not saturated at 900 A while the 4880 Å line saturated at 500 A. Saturation was found for the 4765 Å line at 880 A for a higher gas pressure of 22 mtorr.

The behaviour of the pulsed ion laser in axial magnetic fields has been observed by Labuda *et al.* (1965), Gorog and Spong (1966) and Jarrett and Barker (1968). Jarrett and Barker have reported that under optimum pressure and current conditions for the Ar 4880 and 5145 Å lines there is no magnetic field enhancement, and Ames (1972) has found similar behaviour for the xenon ion laser. However Birnbaum (1968) and Toyoda and Yamanaka (1967) have reported power enhancement, higher efficiency and lower threshold (see Section X, F). We can conclude that the magnetic field enhancement effect is smaller in pulsed lasers and its appearance depends on the gas and excitation conditions of the laser. The effect of an axial magnetic field on the pulsed xenon ion laser operating on the six strong lines at 4954, 5008, 5159, 5260, 5353 and 5395 Å showed that power enhancement was found for non-optimum or low power operation but not for operation under optimum conditions (Ames, 1972).

Several special forms of pulsed lasers have been proposed. The use of dynamic pinch discharges by a number of authors is discussed more fully in Section IV, E. Pinching effects in pulsed argon lasers were first reported by Bennett *et al.* (1964, 1965). The use of the high frequency fields accompanying the development of these instabilities as a means for laser pumping has been attempted (Tsytovich and Shapiro, 1964). At high currents and low pressure Klein (1970) observed a discharge instability which might have been an unstable plasma pinch. The use of collective interactions of electron beams with the plasma has been used to produce laser action in Ar II, Kr II, Xe II, Ne II and Cl II, (Tkach *et al.*, 1967, 1969 and 1972). The interaction produces a plasma-beam discharge with a high electron temperature (even up to 10–10^2 KeV) in a large column highly ionized plasma. Electron pulses of 40 keV energy at 35 A current and 90 μs duration have been used (Tkach *et al.*, 1972). Plasma electrons gain energy in the electric field of the oscillations excited by the electron beams in the plasma. The production of the plasma-beam discharge has been studied by Fainberg and Shapiro (1966, 1968), and Shapiro (1965). High pumping rates are available in this type of discharge.

The central dark region in the output beam (ring-mode) reported by Cheo and Cooper (1965) has been interpreted by Skurnick and Schacter (1972) by considering radiative and collisional de-excitation. The output beam is made up of high order modes which do not derive intensity from the central region of the laser tube which is quenched by these effects. Non-uniform gain across the bore and an optimum gain varying as r^{-1} for Ar 4880 and 5145 Å has been reported by Jarrett and Barker (1968). Cottrell (1968) has given an explanation of this effect in terms of the radial dependence of the current density. The earlier explanation of Cheo and Cooper based on radiation trapping may only be partly contributing to these effects (Goto *et al.*, 1971). High order modes in a circular dot pattern have been reported by Cottrell (1968) in pulsed krypton. Ring modes observed by Gorog and Spong (1966) and some higher order transverse modes are shown earlier in Fig. 39.

A cavity dumping method for an argon laser in which the stored energy held in a resonator with high reflectance mirrors is extracted with an in-cavity acousto-optic modulator has been reported by Maydan (1970). Pulses of up to 60 W peak power with pulse widths from CW to 15 ns and repetition rates from DC to 20 MHz have been obtained. A peak power of about 30 W is obtainable at repetition rates up to 1 MHz for a 1 W Ar 5145 Å laser operating in 15 ns pulses.

I. UV NOBLE GAS ION LASERS

Ion lasers provide most of the useful short wavelength coherent transitions available, the shortest being CIV at 1548·2 Å. In other gas lasers this is exceeded only by H_2 at 1161 and 1236 Å (Hodgson and Dreyfus, 1972; Waynant, 1972). Some selected values of the prominent wavelengths and powers of UV ion laser lines are given later in Table XIV. UV emission was investigated by several early workers (Cheo and Cooper, 1965a, b; Heard and Peterson, 1964; Paananen, 1966; Bridges *et al.*, 1966). However the He–Cd metal vapour line at 3250 Å is the shortest CW wavelength directly available.

Three basic methods of obtaining UV emission ($\lambda < 4000$ Å) have been developed.

(1) Extending the techniques used in common narrow bore ion lasers in the visible to produce UV output.
(2) The use of wide bore high current lasers.
(3) SHG of visible lines.

UV emission from neon has been investigated by Paananen (1966) and Hernqvist and Fendley (1967). Neon appears to be the only singly-ionized noble gas that lases CW in the UV. Optimum power output is obtained at lower current densities than for doubly-ionized systems and will contribute less wall erosion because of their lower impact energy at the wall. In singly-ionized neon Paananen (1966) reported that the two characteristic lines at 3324 and 3378 Å lased, 3324 Å CW with 30 mW and 3378 quasi-CW; Hernqvist

TABLE XIV. *CW UV ion laser lines. Some representative reported values of output power are given for typical noble gas ion lasers as described in Section I, A*

Wavelength (Å)	Ion	Typical output power (W)
2572·66	Ar II (5145 Å) + SHG	0.5
3250·29	Cd II	0·02
3323·73	Ne II	0·05
3336·73	Ar III	
3344·72	Ar III	
3358·49	Ar III	
3378·22	Ne II	
3454·24	Xe III	
3507·42	Kr III	1·5
3511·12	Ar III	0·5
3564·23	Kr III	0·5
3637·89	Ar III	2
3713·08	Ne II	
3780·97	Xe III	
4060·41	Xe III	
4067·37	Kr III	0·5
4131·33	Kr III	0·2

and Fendley (1967) reported similar behaviour and observed optimum UV power at relatively lower pressure than visible transitions.

Other UV lines were recorded in CW operation by Fendley (1968) who has conducted a systematic search using narrow bore segmented graphite tubes and Bridges *et al.* (1967) with low output powers ($\leqslant$ 70 mW). Fendley did not carry out his work at high enough currents to observe the onset of saturation except for Xe III and possibly Kr III. A general feature is that high current densities are required with the strongest lines being in the doubly-charged ions Ar III and Kr III. Bridges *et al.* (1969) reported 2·3 W CW UV output—presumably over several lines in argon—at an efficiency of about 10^{-3}%; 31 violet and UV lines were observed in Ne II, Ar III, Kr III and Xe III. For this work segmented metal ion laser tubes capable of running at 1600 A cm^{-2} with a power input of 600 W cm^{-1} were constructed (the construction of these lasers is described in Section X, A (3)).

The use of wide bore high current lasers has produced a continuous UV output of 3 W from Ar III (Boersch *et al.*, 1970) at 3511 and 3638 Å. The properties of this type of ion laser are described in Section X, G. Other reports of UV emission from this type of ion laser have been given by Banse *et al.* (1968), Seelig and Banse (1970), Bridges *et al.* (1969). Banse *et al.* (1968) obtained maximum powers of 1·5 W from Ar III (3511 and 3638 Å) and 1·3 W from Kr III (3507 Å) at current densities of 250 A cm^{-2} in a 1·2-cm i.d. tube. Similar values were also obtained by Herziger and Seelig (1968). The current density thresholds were about 170 A cm^{-2} for both Ar and Kr for a resonator made up of two 99% reflection and 0·5% transmission mirrors. They suggest that the efficiency may be enhanced by increasing the value of (current density × radius).

The Ar III lines at 3638 (~ 1·35 W) and 3511 (0·35 W) have been obtained in a very long pulse (quasi-CW) mode; outputs have been measured in CW operation from Kr III transitions for times ~ sec at 3507 (0·44 W), 4067 (0·37 W) and 4131 (0·15 W) Å (Latimer, 1968). Latimer attributes 80% of the Ar III UV power to the 3638 transition and in Kr III the power is in the ratio 3507 Å : 3664 Å of 7 : 3. The line at 3638 Å in Ar III was found to be the strongest line in the range 3500–4500 Å (Latimer, 1968). This was confirmed by Bridges *et al.* (1969) and it is generally the strongest CW UV line in ion lasers. A CW argon laser of 15 W total output power produces about 0·7 W CW on the combined 3511 and 3638 Å lines.

(1) *Second Harmonic Generation from Ar* 5145 Å

Efficient second harmonic generation (SHG) of the CW Ar 5145 Å line been achieved producing reasonable CW UV power at 2573 Å. We have seen earlier that the ion lasers can produce directly useful emission in the UV, particularly from multiply-ionized species. These direct methods, however, are inefficient although some recent advances have been made with the Xe ion laser as described in Section XI, A. The use of SHG with ion lasers appears to have been confined up to now mainly to conversion of the Ar 5145 Å line to 2573 Å.

Dowley (1968) has reported SHG with conversion efficiencies from 5145 Å of up to 50% using frequency doubling crystals of ADP and KDP inside the laser cavity in the form of resonant SHG to increase the incident power. The system is shown in Fig. 44 with the in-cavity crystal having surfaces cut at Brewster's angle and highly polished. The laser optical resonator was made up of a 30 cm radius output mirror M_1 coated to be > 99% reflecting at 5145 Å and 75% transmissive at 2573 Å and a Littrow prism M_2 highly reflecting at 5145 Å. The cavity contained the argon plasma tube P and a 30 cm focal length converging lens with anti-reflection coatings for 5145 Å. Single axial mode operation was obtained by placing a mode selecting etalon between P and M_2 (see Section XI, D).

The fundamental optical electric field and the second harmonic propagate in a direction normal to the crystal optic axis, the fundamental as an ordinary ray inducing a second harmonic polarization along the optic axis which propagates as the extraordinary ray. The angle between the wave vector of the incident field and the x, y crystallographic axis in a plane perpendicular to the optic axis is near to 45° (45° Z-cut). For phase matching the doubling crystal is temperature tuned by cooling for 90° phase matching, with an optimum temperature near −10°C for both ADP and KDP. To obtain maximum phase-

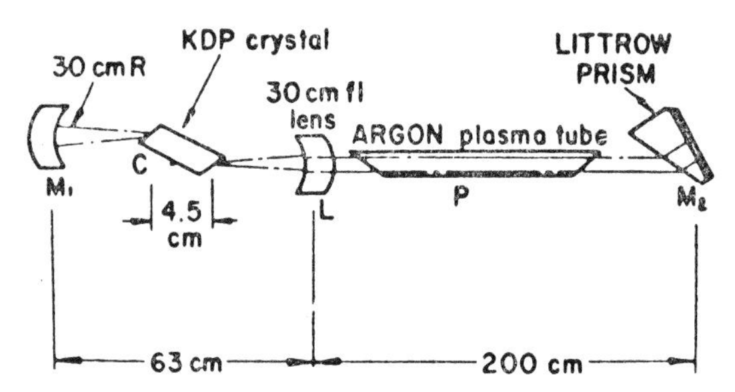

FIG. 44. Second harmonic generation from Ar 5145 to 2573 Å using an in-cavity KDP crystal (Dowley, 1968).

matched second harmonic power the incident field is focused. For a weakly focused beam the SH power is given by

$$P(2\omega) = 2\left(\frac{2N-1}{N}\right)\left(\frac{\mu_0}{\varepsilon\varepsilon_0}\right)^{3/2} \frac{\omega^2 d_{36}^2 l^2}{\pi\omega_0^2} \frac{\sin^2[\Delta k(l/2)]\, P^2(\omega)}{[\Delta k(l/2)]^2}$$
$$= KP^2(\omega)$$

in which N is the number of longitudinal modes of the fundamental, l is the length, d_{36} the non-linear polarizability, ε the dielectric constant of the crystal at frequency ω and Δk = phase mismatch between ω and $2\omega = (4\pi/\lambda)(n_\omega^0 - n_{2\omega}^e)$ and ω_0 is the beam waist radius. An optimum value is obtained for K depending on the laser saturation parameter and the cavity losses, and for the laser of Dowley had a value $K = 2 \times 10^{-3}\ \mathrm{W}^{-1}$. With a crystal of length $l = 5$ cm maximum SH power at 2573 Å was obtained of 415 mW.

The in-cavity method is very sensitive to stray losses from the doubling crystal and the focusing elements being in the cavity which reduces the laser fundamental power and therefore the SH power. Expected conversion efficiencies near 100% were not realized. The total insertion loss from scatter, reflection, astigmatism and absorption at frequency ω was 1·5% and an additional smaller loss from absorption at frequency 2ω. These losses were dominated by a further loss in the presence of the SH which was attributed to absorption of the fundamental and SH in the crystal leading to thermal refractive index distortions. Dowley and Hodges (1968) used an AR coated lens for focusing giving a loss $\simeq 0{\cdot}6\%$ and AR coated quartz windows index matched to the crystals or directly AR coated crystals which gave much higher CW second harmonic power. In this way they have reduced losses to $< 1\%$ with up to 30 W in-cavity power. With $\sim$ 10 W in-cavity power a SH CW power up to 50 mW was obtained. This was less than the calculated 360 mW and the reduction was attributed in this case to UV damage causing the laser to oscillate in high order transverse modes which are not as efficient as the zero order transverse mode in SH production.

Frequency doubling of the Ar 5145 Å line to 2573 Å using a doubling crystal outside the laser cavity was described by Labuda *et al.* (1966a) and Labuda and Johnson (1967). They calculated a possible 20% conversion efficiency for the SH using a KPD crystal and a maximum second harmonic power for doubling outside the cavity using a focused beam under optimum signal frequency conditions as

$$P(2\omega) = 0{\cdot}6\, l P^2(\omega) \quad \text{mW}$$

where l (cm) = crystal length and P (watts) = laser input power. The conversion efficiency for an average primary input power level of 50 mW is 0·8% rising to between 1·1 and 1·6% for laser peak powers of 1 W. The maximum crystal length is determined by absorption at the SH and the difficulty in producing long crystals with limiting lengths around 5 cm. Enhancement of the SH power by up to 30 times was obtained by Labuda and Johnson using a mode-locked input. Under these conditions an average power of 40 μW was obtained at 2573 Å for a 100 mW fundamental frequency TEM_{00} beam at 5145 Å with a KDP crystal of 3 cm length. For a fundamental frequency power of 1 W at 5145 Å it is calculated that 100 mW would be obtained at 2573 Å under mode-locked conditions.

Crystals of ADP and KDP for SHG with the argon ion laser have been compared by Dowley and Hodges (1968) in terms of transparency at ω and 2ω, 90° phase matching conditions, power at the fundamental frequency and focusing. Both ADP and KDP showed wide variation of absorption at 2ω for different samples, selected crystals of ADP and KDP appeared equally suitable for low absorption at ω (with absorption coefficients of 0·002 cm^{-1} for ADP and 0·01–0·02 cm^{-1} for KDP). Dowley and Hodges obtained 90° phase matching for Ar 5145 Å at −11°C for KDP and −9·2°C for ADP and found ADP more able to match over a wide wavelength range but KDP less sensitive to crystal temperature fluctuations. They estimate that the doubling crystals can sustain 40 W total CW incident power at 5145 Å and sustain at 2573 Å 100 mW for KDP and 500 mW for ADP. They found optical elements in the cavity other than the doubling crystal was limiting the attainable SH power.

This SHG source of CW power levels near 1 W at 2573 Å has several potential applications. These include the study of UV induced processes in analysis and spectroscopy, time resolved spectroscopy with mode-locking producing nanosecond pulses, Rayleigh, Brillouin and Raman scattering with a gain in scattering efficiency, as a pump for a visible parametric oscillator and for use with photoresist and photochromic materials.

XI. Output Characteristics of Noble Gas Ion Lasers

A. OPERATING LINES AND POWER OUTPUTS

The most useful and important laser lines from ion lasers are those from the argon ion laser which occur around 5000 Å. For a typical 2 W argon ion laser the lines, powers and distribution of powers obtainable are shown in Table XV together with those from a 1 W Kr laser. The argon laser provides the highest CW power of the ion lasers of up to 150 W at about 0·1 % efficiency. Commercial CW argon lasers are available which provide total powers over several wavelengths of up to 15 W. A commercial 15 W argon laser could deliver from tens of milliwatts on a number of weaker lines to 5 and 6 W respectively on each of the stronger lines at 4880 and 5145 Å. Although krypton gas produces lower power in a comparable laser than argon it has the advantage of providing wavelengths covering a wider range in the visible region. The highly ionized states of atoms may prove a useful source of UV and vacuum UV lines, probably of the fast pulse type as in H_2 or D_2. The shortest laser wavelength at present of any medium is the 1161 Å transition in the Werner bands of gaseous H_2 (Hodgson and Dreyfus, 1972; Waynant, 1972). A report of laser action in triply-ionized C IV at 1548 and 1550 Å in a travelling wave low pressure discharge, has been given recently (Waynant, 1973).

1. *CW Noble Gas Lasers*

A comparison of the power in the stronger lines at 5145 and 4880 Å in the CW argon ion laser is of interest. The powers are roughly comparable in a typical 2W argon laser but although the 4880 Å line has a lower threshold greater power is obtainable at 5145 Å, as indicated by the greater saturation intensity for the 5145 Å line compared with the 4880 Å line [about 50 W cm^{-2} (5145 Å) as against 5 W cm^{-2} (4880 Å) (Zory, 1967)]. The two lines have about the same value of the product (saturation parameter × gain). The output power on the 5145 Å line in a CW laser well above threshold, say in the 1 W region, is about 10 % greater as indicated in Table XV; in the pulsed laser it is about 50 % greater. The broadening mechanisms of the argon laser are discussed in Section IX, the conclusion being that the lines are effectively homogeneously

TABLE XV. *Prominant operating lines and typical powers of CW* TEM_{00} *laser lines in argon and krypton ion lasers. Lines are selected by an in-cavity prism; (a) UV lines need special optics, (b) balanced blue (4880 Å), green (5745 Å) and red (6471 Å) laser.*

Ar: 2W nominal power			Kr: 1W nominal power		Mixed Ar/Kr, 1W nominal power[b]	
Wavelength (Å)	Typical power (mw)	% of power	Wavelength (Å)	Typical power (mw)	Wavelength (Å)	Typical power (mw)
3511 + 3638[a]	10	0·5	3507 + 3564[a]	30		
4545	40	2	4619	5	4579	20
4579	110	5·5	4680	5	4765	60
4658	50	2·5	4762	40	4825	10
4727	65	3	4825	25	4880	200
4765	225	11	5208	55	4965	50
4880	580	29	5309	150	5017	20
4965	180	9	5682	110	5145	200
5017	110	5·5	6471	380	5208	20
5145	630	30	6764	90	5309	80
			7525	75	5682	80
			7931	8	6471	200
			7993	25	6764	20

broadened. The dependences of the gain of the laser $g(\nu)$ is described in Section XI, A (4), and will saturate for homogeneous broadening at a value given by

$$g(\nu) = \frac{g_0(\nu)}{1 + (I/I_s)}$$

where $g_0(\nu)$ is the unsaturated gain at frequency ν and I_s is the saturation intensity. The optimum power depends on the three parameters I_s, g_0 and total losses L. If g is low, e.g. a low power Ar laser, or L high the 4880 Å line has the greater power. At high g and low L the saturation intensity I_s, being higher for 5145 Å, enables higher output power at 5145 Å to be obtained; this is particularly so in pulsed operation.

Argon and krypton can be excited in the same laser to provide a balanced red (Kr 6471 Å), green (Ar 5145 Å) and blue (Ar 4880 Å) output (see Table XV). Commercial lasers are available providing 0·25 W simultaneously at each of these wavelengths. A white light output based on argon and krypton ion lasers has been investigated by Leonard *et al.* (1970). The laser was made to have equal intensities of red (Kr 6471 and 6764 Å), yellow (Kr 5682 Å), green (Kr 5208 Å) and blue (Ar 4880 Å) achieved by adjusting partial and total pressure and the axial magnetic field. Broad band dielectric mirrors suitable for a laser of this type are available, an early report on such mirrors for the range 4300 to 7400 Å was given by Perry (1965). It has been shown that in the mixed gas laser the argon and krypton levels do not derive excitation independently. For example, the argon 4880 Å line was found to increase and the Kr 5208 Å line to decrease with the other gas added (Sasaki *et al.*, 1973).

The current dependence of the output power of an ion laser is complex, not only from the excitation and laser processes involved, but also from other processes such as gas pumping and removal of neutral particles. We can arbitrarily divide the behaviour into three regions

(i) At low current density, say $j < 100$ A cm^{-2}, the dependence is $P \propto I^m - I_t$ where $m > 2$. At low power the power broadening is small and the laser is inhomogeneously broadened.

(ii) At intermediate current densities, say $j = 100$ A cm^{-2} to 400 A cm^{-2}, the dependence is $P \propto I^2 - I_t$. This behaviour, found by several workers in conventional small bore argon lasers over a range of current, gas pressure and tube dimensions including observation of spontaneous emission, is consistent with a 2-step excitation mechanism and a homogeneously broadened line.

(iii) A region of saturation and power decrease occurs at high current densities $j > 700$ A cm^{-2}.

Evidence for saturation with current in the CW output from argon has been found by Paananen (1966), with saturation occurring at about 800 A cm^{-2}, and Donin (1969) who investigated current densities up to 3500 A cm^{-2}. Other evidence for current saturation is from Fendley (1968) and Boersch *et al.* (1970). The pulsed laser study of Cheo and Cooper (1965) yielded a saturation current in the same region as that of Paananen. Jarrett and Barker (1968) report pulsed gain saturation current densities of 630 A cm^{-2} (0·5 cm bore, 47 mtorr), 380 A cm^{-2} (0·8 cm, 32 mtorr) and 220 A cm^{-2} (1·5-cm bore, 17 mtorr). Current saturation in pulsed argon ion lasers has been found also by Hattori and Goto (1968, 1969), Cottrell (1868), Klein (1970) and Davis and King (1971). The quenching of laser action with increased pressure and current has been attributed to a number of phenomena and has been discussed in detail in Section IV, C.

Most of the input power is dissipated in the arc column. The efficiency found by Labuda *et al.* (1965) was ~0·1% in large diameter (8 mm) tubes. With small bore tubes the efficiency may be as low as 0·01%. Wang and Lin (1972) find that the efficiency scales with pressure and tube radius as the product (pR) and large bore tubes are better for higher efficiency up to a possible radiation-trapped limit which has not yet been attained. Comparison of the relative merits of an oscillator with a combination oscillator–amplifier system for high power CW outputs of 1 kW has been carried out by Gorog (1971). For an oscillator the saturated gain equals the losses, while in an amplifier the gain saturation is determined by the light flux. Gorog concludes that a simple oscillator is better than a combined oscillator–amplifier at this power level with an efficiency factor better by a factor of 2. He further concludes that a 1 kW argon ion laser can be constructed using a large-bore tube with internal mirrors (i.e. no Brewster windows) and his estimates indicate that the intra-cavity power could be handled by available laser optics. At the other extreme low power argon lasers ~10 mW can be constructed which are small, relatively simple and air-cooled.

2. *Pulsed Argon Lasers*

The peak power of pulsed ion lasers is greater than CW operation, but not as substantially so as in several other lasers. Some representative performance values are that maximum peak power over all wavelengths is 10 W for 10 μs pulses, 50 pps multiphase output or 3 W uniphase. The average power in 50 μs pulses, 50 pps is 15 mW multiphase or 5 mW uniphase. With a single set of mirrors the maximum peak power in 10 μs pulses at 50 pps is 700 mW multiphase or 200 mW uniphase.

3. *High Power Pulsed Xe IV Laser*

The six strong laser transitions in Xe at 4954, 5008, 5159, 5260, 5353 and 5395 Å have produced 0·4 J/pulse with a 20 kW peak power (Bridges *et al.*, 1971). This high power mode is unlike pulsed laser action in the other noble gases. The high power operation was also reported by Simmons and Witte (1970a) and Papayoanou and Gumeiner (1970). These powers, and also that reported by Jarrett and Barker (1969) for Xe 5353 Å of 110 W peak for 50 pulses, were substantially higher than those earlier reported by Dahlquist (1965) and Demtroder and Elendt (1966). Papayoanou and Gumeiner found the optimum pressures were low (~ 10 mtorr) and optimum current densities high. They observed pinch effects with laser action beginning and continuing through the dynamic part of the pinch reaching maximum intensity at the minimum radius. Jarrett and Barker (1969) found pulsed saturation currents about 3 times higher for Xe IV 5353 Å compared with the strong Ar II 5145 Å line. Ames (1972) has found that an axial magnetic field only enhances the xenon laser power when it is operating at low power levels and does not improve the maximum output power.

The coherence properties of a pulsed xenon laser were investigated by Moskalenko *et al.* (1971). The lines investigated were at 4310, 4954, 5008, 5260, 5353, 5397 and 5955 Å. The output pulse duration was 0·4 μs and the pulse repetition frequency 500 Hz. Spatial and temporal coherences were investigated with Young's double slits and Michelson interferometer experimental arrangements.

The distribution of energy over the output cross-section was a slightly distorted Gaussian. The degree of spatial coherence varied between lines. For 5955 Å the output was spatially coherent with γ near 1, over about the central 0·5 mm of the beam. The coherence length of the laser was found to be 20 to 30 cm but dependent on excitation conditions. Beam divergences between 3′ and 10′ were found depending on the mirrors defining the optical cavity. They concluded that the output was substantially single mode.

4. *Gain in Ion Lasers*

Descriptions of gain and gain saturation in gaseous lasers has been given by Gordon *et al.* (1963), White *et al.* (1963), Rigrod (1963, 1965, 1968), Laures (1968), Smith (1966a) and several other workers. They consider the interaction of a Doppler broadened transition at a centre frequency ν_0 with an incident plane wave at frequency ν. The stimulated emission cross-section has a Lorentzian lineshape corresponding to a natural linewidth which is pressure or power broadened. The intensity in the laser cavity stabilizes when the gain coefficient equals the loss coefficient.

For inhomogeneous broadening we can regard the hole width γ_h' as $\gamma_h' \ll \Delta\nu_D$, γ_h' is a measure of the region over which the gain saturates and can be related to the hole width of the homogeneous broadening γ' in

$$\gamma_h' = \gamma'\left(1 + \frac{I}{I_s}\right)^{1/2}$$

Here I is the in-cavity intensity and I_s is termed the saturation intensity. For a uniform plane wave the gain $g(\nu)$ at frequency ν for inhomogeneous broadening may be obtained as (Smith 1966a):

$$g(\nu) = g_0\left(1 + \frac{I}{I_s}\right)^{-1/2} \exp -\left(\frac{\nu - \nu_0}{\Delta\nu_D}\right)^2 \qquad \ldots(11.1)$$

in which g_0 = gain at line centre

$$\nu_0 = \frac{1}{4\pi}\left(\frac{\ln 2}{\pi}\right)^{1/2} \frac{\lambda^2 A_{21}}{\Delta\nu_D}\left(N_2 - \frac{g_2}{g_1}N_1\right).$$

For homogeneous broadening:

$$g(\nu) = g_0\left(1 + \frac{I}{I_s}\right)^{-1/2} \mathscr{R}W\left[\left(\frac{\nu - \nu_0}{\Delta\nu_D}\right) + i\left(\frac{\gamma_h'}{\Delta\nu_D}\right)\right] \qquad \ldots(11.2)$$

in which $\mathscr{R}W(y)$ = real part of the error function for a complex argument defined as $W(y) = [1 + \mathrm{erf}(iy)]\mathrm{e}^{-y^2}$. The plasma dispersion function, defined as

$$Z(y) = \frac{2y}{\sqrt{\pi}} \int_0^\infty \frac{\mathrm{e}^{-x_2}\,\mathrm{d}x}{x^2 - y^2}$$

is related to the error function as $Z(y) = i\sqrt{\pi}W(y)$.

The saturated gain per pass G at line centre $\nu = \nu_0$ is related to the unsaturated gain G_0 as (Gordon *et al.*, 1963; Rigrod, 1963):

$$G = G_0\left(1 + \frac{I}{I_s}\right)^n \qquad \ldots(11.3)$$

with $n = -\frac{1}{2}$ for inhomogeneous broadening and $n = -1$ for homogeneous broadening. The ratio of the maximum unsaturated gain at $\nu = \nu_0$ to the total losses a (which include diffraction, absorption and scattering losses), termed the excitation parameter X, can be given as

$$X = \frac{G_0}{a + t} = \left(1 + \frac{I}{I_s}\right)^{1/2} \frac{Z_i[i(\gamma'/\Delta\nu_D)]}{Z_i\{i(\gamma'/\Delta\nu_D)\,[1 + I/I_s]^{1/2}\}} \qquad \ldots(11.4)$$

Here $(a + t)$ represents the total losses per pass. For inhomogeneous broadening $\gamma' \ll \Delta\nu_D$ equation (11.4) reduces to the equation of Rigrod (1963) and gives agreement with the oscillation threshold derived from Lamb theory.

In single mode operation the output power at $\nu = \nu_0$ can be obtained from equation 11.4 and is equal to tI in which t = output coupling per pass and for an inhomogeneously broadened transition is then

$$I_{\mathrm{out}} = \frac{tI_s}{2}\left[\left(\frac{G}{a + t}\right)^2 - 1\right] \qquad \ldots(11.5)$$

The optimum transmittance t_{opt} is obtained from

$$t_{\text{opt}} = (aG)^{1/2} - a \quad \text{...(11.6)}$$

The in-cavity intensity in a homogeneously broadened laser operating well above threshold is

$$I = I_s \frac{G}{a+t} = I_s(X-1) \quad \text{...(11.7)}$$

Then the output intensity becomes

$$I_{\text{out}} = tI_s\left[\frac{G}{a+t} - 1\right]. \quad \text{...(11.8)}$$

The optimum transmittance is given by

$$\frac{G}{a}\left(1 - \frac{t_{\text{opt}}}{a}\right) = \left(1 + \frac{t_{\text{opt}}}{a}\right)^3 \quad \text{...(11.9)}$$

The common case in gas lasers of intermediate broadening has not been treated analytically.

Multimode operation has been described using various methods. If the axial modes are well separated, e.g. in a short laser, each mode can be considered to saturate the gain medium independently (Rigrod, 1963). Also uniform gain saturation has been discussed by White *et al.* (1963) in which the frequency dependence of the in-cavity intensity is approximated by a smoothed average distribution and Smith (1966a) adopted a model in which the many axial modes are assumed equivalent to a set of modes of width $2\gamma_h'$. For an inhomogeneously broadened laser the output intensity in a symmetrical resonator with losses per pass $= a$, mirror reflectances r_1 and r_2 giving $r^2 = r_1 r_2$ and transmission t can be written as (Rigrod, 1963)

$$I_{\text{out}} = I_s\left(\frac{4L}{c}\right)\Delta\nu_D\left\{\left(\frac{\pi}{8}\right)^{1/2}\left[\left(\frac{G}{a+t}\right)^2 \operatorname{erf}\left[2\ln\left(\frac{G}{a+t}\right)\right]^{1/2} - \left[\ln\left(\frac{G}{a+t}\right)\right]^{1/2}\right\} \quad \text{...(11.10)}$$

and for homogeneous broadening

$$I_{\text{out}} = \frac{I_s}{2}\left(\frac{t}{a+t}\right)[G + \ln(1-a-t)] \quad \text{...(11.11)}$$

On the model of Smith an approximate equation for optimum transmission is obtained as for the single mode inhomogeneous broadened case of the form of equation (11.6). This suggests that as far as output power is concerned the multimode laser would behave as an inhomogeneously broadened laser. The assumption of the model that $2\gamma' > c/2L$ usually applies in the case of the argon ion laser; the results of Byer and Costich (1967), described later, which indicate an apparent inhomogeneous nature of the power output with mirror transmittance, are in agreement. For ion lasers near threshold the power output from equation (11.10) will be proportional to $(\text{gain})^2$, further above threshold the power will be proportional to gain from equation (11.11).

Using Lamb theory Allen (1968) obtained for single mode or multi-mode operation an expression of the form

$$I = I_0 t \left(\frac{2t_{\text{opt}} - t}{(t_{\text{opt}})^2} \right) \qquad \ldots(11.12)$$

in which t = transmission of output mirror with an optimum value t_{opt} and I_0 = peak output intensity. By measuring the loss (say t_m) that must be inserted into the laser cavity to extinguish the laser when using high reflectivity mirrors the optimum transmittance is obtained as: $t_{\text{opt}} = \frac{1}{2}t_m$.

The output intensities gain and optimum output coupling, at 4880 and 5145 Å of a typical medium power argon ion laser have been studied by Byer and Costich (1967) using a variable output coupler based on a MacNeille beam splitter and a retardation plate. For the 4880 Å line operating with dielectric mirrors at 800 mW a single pass gain of 45·5% was measured and an optimum output mirror transmittance of 5·5% when other additional cavity losses of 6·2% were present. For the 5145 Å output the figures were: output power = 735 mW; single pass gain = 16·3%, with single pass loss = 5·2%; optimum mirror transmittance = 2% and additional in-cavity losses = 1%. We note that the optimum output couplings for the strong CW argon ion lines are relatively low. The asymmetry in the output power curves with mirror transmission in the results of Byer and Costich only approximately fit the prediction of equation (11.12) of Allen (1968). The dependence of the output power for variation of output coupling indicated (in the results of Byer and Costich) inhomogeneous rather than homogeneous broadening when compared with the analysis of Smith. This may result from axial mode competition preventing some adjacent modes from oscillating (Bridges and Rigrod, 1965; Borisova and Pyndyk, 1968) and may also result in an effective inhomogeneous form of behaviour as far as output power is concerned. However, from the whole gain profile contributing to the power output, an effective homogeneous broadening is active.

For lasers with high gain and requiring high output coupling, e.g. the CO_2–N_2–He laser, optimum coupling has been investigated by Rigrod (1965) and Meneely (1967). The analysis of Meneely is for homogeneously broadened lasers and may be considered relevant to the argon ion laser at high input power levels. The values of the optimum mirror transmission obtained by these methods approximately fits the experimental data of Byer and Costich. Other treatments of optimum output coupling have been made by Arecchi *et al.* (1963), Yariv (1963, 1967), Cabezas and Treat (1966), Miles and Lotus (1968),

Bogatkin *et al.* (1968), Rosenberger (1969), Sinclair and Bell (1969) and Siegman (1971).

Various measurements of the gain of the argon laser have been made. Some typical results are those of Freiberg and Halsted (1969), for a CW 1–3 W argon laser at 4880 Å with a gain of 3 dB m^{-1}, and those of Gorog (1971), of 1·7 dB m^{-1} for an argon amplifier at 5145 Å. Byer and Costich (1967) report the following values: 4880 Å line operating at 800 mW 2·7 dB m^{-1}; 5145 Å line operating at 735 mW, 1·1 dB m^{-1}. Sayers (1969) has measured the single pass gain directly for the CW Ar 4880 Å line using an inserted near-Brewster angle flat for attenuation. For a 60-cm long 2-mm bore laser at 100 mtorr filling pressure operating without an axial magnetic field he obtaines a maximum gain of 0·8 dB m^{-1} at a pressure of 0·1 torr and 7 A current. Close comparison of the gain measurements of different workers is not worthwhile because of the strong dependence of gain on the exact laser operating conditions. Also caution is necessary if small-signal gain is measured near threshold in a laser with Brewster angle end windows since Gorog (1971) has shown that the losses at the Brewster windows are not negligible in this case.

TABLE XVI. *Gain measurements using a pulsed oscillator-amplifier combination for pulsed laser lines in an argon ion laser. Laser tube of* 3 *mm bore,* 50 *cm active length with no applied magnetic field.* (Davis and King, 1971).

Wavelength (Å)	Gain (10^{-2} cm^{-1})	N_2 (10^{10} cm^{-3})	Conditions
Initial transient, pulse A			
4765	7·8	2	28·5 mtorr, 25 kW
4880	5·4	0·96	15 mtorr, 12·5 kV
5145	1·7	2·05	17 mtorr, 20 kV
Quasi-CW, pulse B			
4880	1·4	0·25	20 mtorr, 20 kV

Direct gain measurements on a pulsed argon laser have been made by Jarrett and Barker (1968) and Davis and King (1971) using an oscillator and amplifier. The results of Davis and King are shown in Table XVI for the 4765, 4880 and 5145 Å lines in the first transient pulse together with values of the upper laser level population N_2. In the quasi-CW region a maximum value of 6 dB m^{-1} for the Ar 4880 Å line is found with no magnetic field. A value for Ar 4880 Å has also been obtained by Klein (1970) of 20 dB m^{-1} in a 4-mm bore. For pulsed Ar 4880 and 5145 Å lines the optimum gain is proportional to $(r)^{-1}$ (Jarrett and Barker, 1968). No magnetic field enhancement is found under optimum conditions but enhancement under non-optimum conditions is found. The gain is non-uniform over the tube bore as would follow from the type of excitation acting.

Observation of CW amplified spontaneous emission (superradiance) on the Ar 4880 and 5145 Å lines in a wide bore high current tube has been reported (Banse *et al.*, 1968). In a non-resonant but multipass structure made up of two tilted mirrors a single pass gain of 20 dB over 220 cm in a 1·2-cm bore tube was obtained giving an output of 20 W cm^{-2}.

5. *Saturation Parameter*

Little comparative data on this parameter for ion lasers is available in the literature. Zory (1967) has made the most extensive study and has determined this parameter indirectly for the argon laser by two methods based on line-width and maximum single frequency power measurements. He investigated a narrow range of pressure and current and found only a small dependence of the saturation intensity on these parameters. Defining the saturation intensity as $I_s = (\sigma_0 T_1)^{-1}$ (method a) with σ_0 = resonance cross-section = $\lambda_0^2 A/4\pi r$ (A = total transition rate), $\gamma = (\gamma_a + \gamma_b)/2 = \frac{1}{2}$ [sum of spontaneous decay rates of upper (a) and lower (b) levels] and T_1 = relaxation time = $[\delta(\gamma_a - A) + \gamma_b]/\gamma_a\gamma_b$ where δ = degeneracy a/degeneracy b, γ_a and γ_b are total decay rates for a and b. For Ar II $\gamma_b = 2{\cdot}78 \times 10^9$ s^{-1} (corrected value of Statz *et al.*, 1965) and γ_a values (Bennett *et al.*, 1965, 1966) give $T_1 = 0{\cdot}97 \times 10^{-8}$ sec (4880 Å) and $T_1 = 1{\cdot}01 \times 10^{-8}$ sec (5145 Å). Alternatively (method b) I_s is obtained from the relation

$$I_s = \mathscr{P}_{tm}(t\hbar\omega A P_m)^{-1}$$

where $\mathscr{P}_{tm}$ = maximum single frequency output power, t = transmission coefficient of output mirrors, A = average cross-section area occupied by the laser mode in the discharge bore and P_m = maximum normalized internal laser intensity = $\bar{I}_m/I_s$.

The method based on single frequency output power measurements is more indirect and uses some parameters which are not obtained easily. For the linewidth method Zory has used values of γ_b from Statz *et al.* (1965) which were later corrected (see Section IX, G, e.g. γ_b for the 4880 and 5745 Å transitions which have the same lower level = $5{\cdot}52 \times 10^8$ sec^{-1} rather than the corrected value of $27{\cdot}60 \times 10^8$ sec^{-1}). An approximate result is that $T_1 \simeq (\gamma_a)^{-1}$ and a 20% correction must be made in the value of I_s from Zory's data. At a gas pressure of 0·3 torr and 10 A current Zory finds I_s (4880 Å) = 6·8 [method (a) corrected] and 3·7 (method b) W cm^{-2} and I_s (5145 Å) = 63 [method (a) corrected] and 16·0 (method b) W cm^{-2}. For a higher current density region (42 A in a 2·5-mm more tube) and a single frequency mode Donin *et al.* (1967) find a higher value for Ar 4880 Å of 50 W cm^{-2}, although a lower value more in agreement with Zory is found by Odintsov *et al.* (1968) for multimode operation.

Odintsov *et al.* have also shown that uniform saturation of the gain occurs in the argon laser when operating multimode and they have determined saturation parameters for Ar II 4880, 4765 and 5145 Å lines (Odintsov *et al.*, 1968). The laser was 23 cm long with a 2-mm bore, had a 170 G external magnetic field and an Ar pressure of 0·3 torr with discharge currents up to 10 A. The saturation power was 120 mW (3·8 Wcm^{-2}) at single frequency and 220 mW (7 W cm^{-2}) multifrequency. For a cross-sectional area of the TEM_{00} mode of 0·57 mm^2 they obtained a saturation power density of $W_0 = 21$ W cm^{-2} (single frequency) and 3·9 W cm^{-2} (multifrequency). As expected, the value of the saturation parameter depended significantly on the ratio $(\Delta\nu_L(\ln 2)^{1/2}/\Delta\nu_D)$ characteristic of the widths of the Lorentzian and Doppler contributions to the lineshape.

6. *Hole Widths and Lamb Dip*

From the following observations we can conclude that the argon laser exhibits a hybrid form of broadening and does not show true homogeneous or inhomogeneous broadening:

(1) the large Lorentzian widths $\gtrsim 500$ MHz;
(2) that the gain in a multimode argon laser saturates uniformly (Odintsov *et al.*, 1968);
(3) the high conversion efficiency of multi-frequency power into a single frequency with efficiency $\gtrsim 0{\cdot}5$, probably from a high degree of cross-relaxation, e.g. as seen in the uniform depletion of the gain profile;
(4) for the pulsed argon laser: output power $\propto$ gain (Davis and King, 1971);
(5) several axial modes oscillate with spacings $> c/2L$ due to mode competition—giving a measure of inhomogeneous broadening.

In the discharge the ions are perturbed by long-range Coulomb forces from other ions and electrons in the high density plasma in addition to power broadening caused by the in-cavity laser field. Bennett *et al.* (1966) have reasoned that the net effect of these perturbations is to increase the hole linewidth without greatly changing its Lorentzian character. In addition to possible large linewidths from Stark broadening the short lower state lifetime will contribute a large natural broadening (see Section IX, B and G).

Using a scanning Fabry–Perot interferometer with measurements reduced according to a Voigt profile [using a method of Ballik (1966)] Bennett *et al.* (1966a, b) have measured the homogeneous linewidth. These are shown in Table XVII. The strong lines in Ar II are found to have similar values. There is agreement between these measurements and those of Zory (1966), who has measured large Lorentzian widths for the Ar II 4880 Å line within 10% of those of Bennett but which differ by up to 50% for Ar II 5145 Å. Both Bennett *et al.* and Zory find that the linewidth can be increased by increasing the

TABLE XVII. *Homogeneous (Lorentzian) linewidths in the argon ion laser.*

Pressure (torr)	Current (A)	Lorentzian widths (MHz)			
		4880 Å		5145 Å	
		Ballik *et al.*	Zory (1967)	Ballik *et al.*	Zory
0·3	5	500[a]	540	680[a]	
0·3	7	510[b]	585	800[c]	600
0·3	10		700		660
0·5	5	360[c]	405	600[c]	
0·5	7	430[b]	450	900[c]	560
0·5	10				540

[a] Bennett *et al.* (1966).
[b] Ballik *et al.* (1966).
[c] Ballik's measurements quoted by Zory (1967).

current and decreasing the pressure due to increased ion–electron density. The Lorentzian width measurements of Donin *et al.* (1967) for the Ar 4880 Å line are even greater than those measured by Bennett *et al.* and Zory. In a 2·5-mm bore, 8 cm length tube, run at 42 A and producing an ion temperature $\sim 17{\cdot}5 \times 10^3$ K Donin has measured widths of 900–1100 MHz. These values are for currents much larger than those used by Bennett *et al.* and Zory and the trend with current increase is confirmed. Large Lorentzian linewidths have also been reported from Fabry–Perot studies on argon hollow cathode discharges at relatively low currents (Kreye, 1971).

The width of the Lamb dip in an argon ion laser has been measured by selecting a single axial frequency by the Fox-Smith interferometric method (see Section XI, E) and scanning this mode. The width obtained is between 400–500 MHz at low excitation and is pressure and current dependent—decreasing with pressure increase at high pressures and increasing with current. Lamb dip measurements have been made by Ballik *et al.* (1966), Zory (1966; 1967), Odintsov *et al.* (1968) and Donin *et al.* (1969) (who did not report a linewidth).

The Lamb dip was measured by Odintsov *et al.* (1968) for the Ar II 4880 Å line in a 2-mm bore, 23 cm long tube operating at 10A with no axial magnetic field. His results were

Ar II 4880 Å Current (A)	Lamb dip FWHM (MHz)
7	570
8·5	670
10	760

The ratio of $\varepsilon = \dfrac{\Delta\nu_L\sqrt{\ln 2}}{\Delta\nu_D}$ was found to be $\varepsilon \simeq 0{\cdot}1$.

B. LINE AND MODE INTERACTIONS

1. *Line Interactions*

The most direct form of this interaction is when separate laser transitions share a common upper or lower level—termed normal competition. Some examples of common levels are the level $4s\,^2P_{3/2}$ which is a common lower level for the Ar 4880 and 5145 Å transitions and the common lower level of the Kr II 5682 and of the 6471 Å lines. The Ar II 4880 and 5145 Å lines do not compete strongly compared with the Kr II 5682 and 6471 Å lines. Three other argon levels which share a common level are the 4579, 4765 and 4965 Å lines with the common lower level $4s\,^2P_{1/2}$. Statz *et al.* (1969a, b) observed that for an argon laser oscillating simultaneously at 4880 and 5145 Å each of these lines contained beats characteristic of the mode spacing of the other line. This interaction between the lines occurred because these two laser lines have a common lower level. Here the modes of the 5145 Å line interfere and modulate the induced transition rate between the upper and lower laser states at the difference frequency. Statz *et al.* mention the possible role that non-linear polarization side-bands may have in the occurrence of self-mode-locking discussed in Section XI, F. Other discussions of non-linear polarization side-bands are to be found in Stetser and de Maria (1966), Boersch *et al.* (1967) and Statz *et al.* (1967 and 1969).

A form of anomalous interaction has been reported by Ferrario and Sona (1969) in Ar II and Kr II where an increase in intensity of one line results in an increase in intensity of another line, e.g. in argon the lines 4880 and 4965 Å show this anomalous interaction. Normal and anomalous competition was observed by Merkelo *et al.* (1968) and an explanation for normal competition was provided. The anomalous competition occurs mainly near threshold where the homogeneous contribution to the line broadening may be reduced.

2. *Mode Interactions.*

Because of the large homogeneous linewidth (large hole width) in the argon laser, neighbouring axial modes compete for the same excited states. The gain on the weaker of the competing modes is reduced and becomes insufficient for them to oscillate. The presence of axial mode competition can lead to the suppression of noise in the free running laser. This mode competition can arise from (1) free running modes competing for the same atomic population as described above, which can lead to random mode pulling and amplitude fluctuations, and (2) combination tones as described by Lamb (1964). Bridges and Rigrod (1965) in a study of the Ar 4880 and 5145 Å lines observed stable operation only near threshold. Above a critical current or axial magnetic field unstable amplitude oscillations occurred. This stable region [also discussed by Borisova *et al.* (1968) and Lebedeva *et al.* (1970)] may be a manifestation of mode-locking near threshold or at low levels of excitation. For an ion laser operating in an axial magnetic field when the axial field is below a critical value, the stable spectrum is from a single axial mode; above the critical field it is from two axial modes whose minimum frequency separation is of the order of the homogeneous linewidth (Gorog and Spong, 1967). Near the critical field the spectrum is very sensitive to perturbations such as current changes and the laser switches randomly between single frequency and two frequency operation. This critical field may be that value at which the σ_+ and σ_- Zeeman components are separated by a finite fraction of their linewidths. For fields below this the components compete particularly if thermalization between Zeeman sub-levels by collisions is rapid.

The removal of mode competition between the Ar 4880 and 5145 Å lines, in a CW laser has been achieved by Forsyth (1967, 1969) when the axial mode frequency spacing was less than the homogeneous hole width for both TEM_{00} and higher order transverse mode operation, and his method can be used to stabilize other pairs of transitions in Ar and Kr ion lasers which share a common lower level. Both transitions are operated simultaneously and the mode frequency spacings are arranged to be sufficiently different for each of the transitions.

The low power single frequency observations of Bridges and Rigrod (1965) for Ar 4880 Å, and for 5145 Å by Forsyth (1967) and the two-frequency operation at 4880 Å observed by Bass *et al.* (1968) has been extended to a high power stable mode of operation by Yarborough and Hobart (1968). They observed high power stable modes of operation in a graphite tube having a 2·25 mm bore with a magnetic field, and obtained 90% of the total output power of 2 W in a single frequency with the other 10% being in side-bands which were phase-locked to the centre frequency. The spectrum was similar to an FM output having a modulation depth of 0·7 (see Section XI, E). They observed the effect of axial mode competition with the frequency interval between axial modes increasing in steps of mode spacing as the input power was increased At high magnetic fields the mode frequency spacing was equivalent to $5 \times c/2L$ ($\equiv 500$ MHz), the behaviour being similar to that reported by Gorog and Spong (1967). Similar behaviour was observed at 5145 Å and on other argon laser lines. This behaviour seems closely related to FM and self-mode-locked operation. Bridges and Rigrod (1965) and Sedelnikov *et al.* (1971) report beat signals at frequencies of $nc/2L$ with $n = 2$, 4 or 6 in the argon laser.

C. NOISE AND AMPLITUDE INSTABILITIES

There are several contributions to the total output noise of noble gas ion lasers, which may be summarized in three categories: (i) mechanical, thermal or power supply noise, (ii) plasma and discharge noise and (iii) noise from axial and transverse mode interactions. For argon and krypton ion lasers the laser output power instability in the region, 1 Hz to 2 MHz is < 1%. Some typical short and long term average noise figures are given in Table XVIII and a noise curve in Fig. 45 for an Ar laser at 30 A current. For the He–Cd metal vapour

TABLE XVIII. *Noise instability figures for an argon ion laser after* 30 *mins. warm-up* (Tandler, 1969)

Frequency Range	Approx. % RMS Noise: With current regulation	Approx. % RMS Noise: With additional feedback from light regulation
1 Hz–1 kHz	0·6	0·2
10 Hz–2 MHz	0·6	0·5
1 sec–1 hr	1·0	0·2
1 sec–10 hr	3·0	0·5

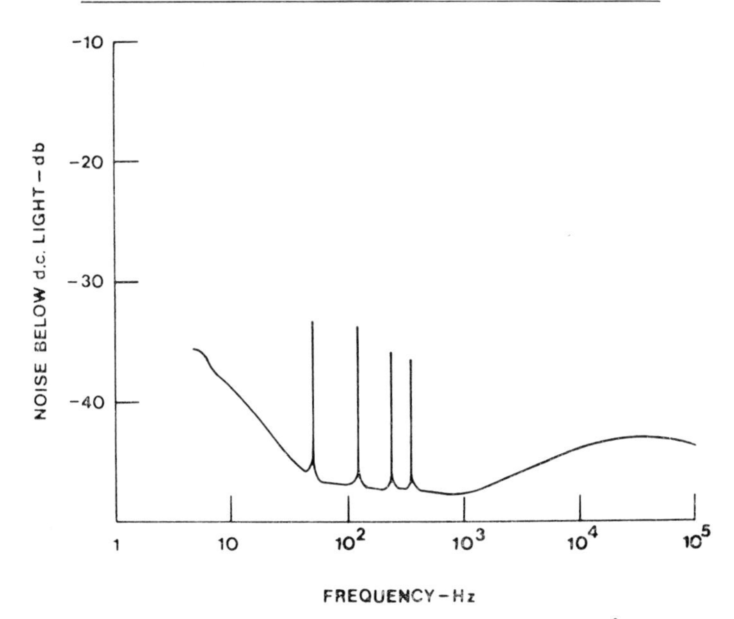

FIG. 45. Low frequency noise in an argon ion laser output at 4880 Å operating at 30 A regulated current from 60 Hz A.C. (Tandler, 1969).

laser the power instability is ~ 10% in the region 1 Hz to 2 MHz. Instabilities in noble gas ion lasers mainly occur in the frequency range 10 kHz to 1 GHz. Investigations of noise in argon ion lasers have been carried out by Gorog and Spong (1967), Fendley (1967) and Jackson and Paul (1969). Noise in segmented graphite tubes for ion lasers has been discussed by Targ *et al.* (1968) and Jackson and Paul (1969) report that the noise content of such a laser is less than that of one using a fused silica or beryllia tube. They attributed the noise to mechanical oscillation instabilities in the laser tube. Under some circumstances power supply generated noise can be dominant; however, this source of noise can be largely eliminated by the use of well-stabilized current sources. Mechanically or acoustically generated oscillations can be generated in the laser cavity structure. Cavity fluctuations, such as changes in the resonator length, which are usually of thermal origin, occur on a longer time scale. These noise effects are short term (times < 1 sec) with characteristic frequency > 1 Hz. Other longer term drifts can be superimposed on these short term charges by slow thermal changes in the material of the resonator structure or slow changes in the gas properties.

Low frequency noise in ion lasers is of somewhat small amplitude and is derived mainly from harmonics of the power line frequency. These fluctuations are shown in Fig. 45. The plasma also contributes discharge noise with a $1/f$ frequency spectrum whose magnitude is dependent on the discharge conditions of current, pressure and the tube dimensions. Discharge noise typically contributes about 5% rms fluctuation. Suzuki (1971) has discussed how this can be reduced by the use of short sections in the laser. He finds a noise component due to anode oscillations proportional to (ballast resistance)$^{-1}$ in the range 1–10 kHz and proposes the use of a cylindrical anode close to the capillary and coaxially on the tube axis to reduce this noise component. These plasma oscillations produce fluctuations in voltage at the anode and cathode which are impressed on the laser output. The amplitude of these oscillations can also be influenced by the axial magnetic field. Gorog and Spong (1967) and Bensimon and Divonne (1970) report oscillations in discharge current in the range 10–25 kHz which were influenced by the position of the anode with respect to the laser bore. These instabilities occurred when the anode was more than a few cms from the bore and could be quenched by adjusting the magnetic field or discharge current. Hernqvist and Fendley (1967) also report a dependence of noise on the position of the cathode.

Oscillations in the range of several kHz to 100's of kHz attributed to anode oscillations have been investigated by Galehouse *et al.* (1971) and suppressed by the use of an auxiliary cathode. Bensimon and Divonne (1970) have also obtained a reduction in low frequency noise by the use of an auxiliary cathode. These noise components give coherent fluctuations in the output laser light. Galehouse *et al.* also identify other noise sources arising from axial magnetic field effects and striations. These can be eliminated by suitable selection of the tube operating conditions. Thus to obtain quiet operation of the laser, the choice of pressure, current, magnetic field and electrode position must be made with noise suppression in mind. The role of these fluctuations, including striations, is not well investigated in ion lasers; their role is clearer in neutral gas lasers such as the He–Ne laser. There is a further noise effect due to the presence of an external magnetic field known as the Kadomtsev instability which may have been observed by Galehouse *et al.*

In wide bore tubes Wang and Lin (1972) have found very strong low frequency oscillations for filling pressure below a critical value. Similar oscillations have been seen in other low pressure discharges (Crawford and Freeston, 1963) and are attributed to ion waves. These cause a reduction in the output intensity and fluctuations.

Targ and Yarborough (1968) obtain 20–30 dB suppression of low frequency noise by internal phase modulation. The low frequency noise in multimode free running lasers results from the time varying amplitude and random phases of the modes. Some methods of suppressing this are internal phase modulation (Massey, 1966), running the laser to encourage self-mode-locking or internal loss modulation (Hodara and George, 1966; Uchida and Ueki, 1967). In pulsed lasers at high currents and low pressure a discharge instability has been observed which might have been from an unstable plasma pinch (Klein, 1970; Bennett *et al.*, 1965).

In general, the high gain and large oscillation bandwidth of ion lasers allow many axial modes to oscillate simultaneously. This multimode laser is unstable in operation due to mode fluctuations and large amplitude fluctuations can develop, with the mode interactions occurring over a short time scale (Bridges and Rigrod, 1965). Near threshold the argon laser exhibits stable frequency behaviour where, in the absence of an axial magnetic field, discrete frequencies separated by several axial mode spacings are observed. Well above threshold these frequencies fluctuate greatly and appear as noise amplitude fluctuations. When an axial magnetic field is applied the near-threshold sprectrum shows only two axial modes whose frequency separation depends on the applied field and g-values of the laser levels. Transverse mode competition may also lead to amplitude fluctuations. Gorog and Spong (1969) found strong frequency fluctuations in an argon laser developed when it was converted from single to multi-transverse mode operation.

D. SINGLE FREQUENCY OPERATION

A very useful characteristic of noble gas ion lasers is that high power can be obtained in a single axial mode. Ideally this should be possible with a minimal reduction from the multimode output power. Of concern here are methods used for single frequency operation of noble gas ion lasers (particularly argon, but almost all the methods are generally applicable). The high gas temperatures in ion laser plasmas means that their Doppler linewidths are large $\sim$ (5–10) GHz. Although the hole widths in the lasers are large and suppression of neighbouring axial modes can occur, several axial modes still operate simultaneously. For many applications single frequency operation is necessary.

The most common methods of obtaining a single frequency are to use an etalon in the cavity or an additional reflector. Other methods that have been used include an absorbing thin film or absorption cell, a deliberately small resonator length, a highly dispersive cavity and frequency modulation techniques. The use of a short resonator length (Gordon and White, 1964) is not practical for ion lasers because of their large Doppler linewidth and the consequent large reduction in their output power and so this method will not be discussed further here. Each of the other methods will be briefly discussed. Since the ion lasers are effectively homogeneously broadened well above threshold, high power in a single axial mode can be obtained. In any event for normal operation of an argon ion laser mode competition ensures that the mode spacing is > 500 MHz (see Bass *et al.*, 1968). Generally between 0·5 to 0·8 of the multimode power can be obtained in a single axial frequency. This degree of conversion has been obtained in the noble gas ion lasers and the He–Cd metal vapour laser. This figure also implies that the mode interaction width or cross-relaxation within the gain profile are large.

1. *Interferometric Methods*

(a) *Etalon*. Operates by suppressing all axial modes expect one. Single frequency operation from an in-cavity etalon was first reported by Manger and Rothe (1963), Collins and White (1963) and, using solid etalons, by Peterson and Yariv (1966). This technique was later developed into a commercial device and incorporated directly into CW noble gas ion lasers.

The technique is illustrated in Fig. 46a. For the etalon the maximum transmission occurs at wavelength λ or etalon frequency ν_E when

$$m\lambda = 2\mu t \cos\theta = m\frac{c}{\nu_E}$$

where m is an integer, θ is the etalon tilt angle, μ is the refractive index between the plates and t is the plate spacing. The laser cavity axial mode frequencies are given by

$$\nu_n = n\frac{c}{2L}$$

where n is an integer and L is the resonator length. The laser is induced to oscillate at ν_n near ν_E i.e. near the etalon transmission peak. If either ν_n or ν_E is not stable then the laser maintains oscillation such that $\nu_n - \nu_E$ is a minimum. This may mean that jumping (mode-hopping) between frequencies can occur.

Tuning of the etalon is accomplished by changing the etalon angle or temperature. An air-spaced etalon needs to use a low thermal expansion coefficient spacer while a solid etalon requires good temperature stabilization to minimize thermal refractive index variations. Frequency stability ~20

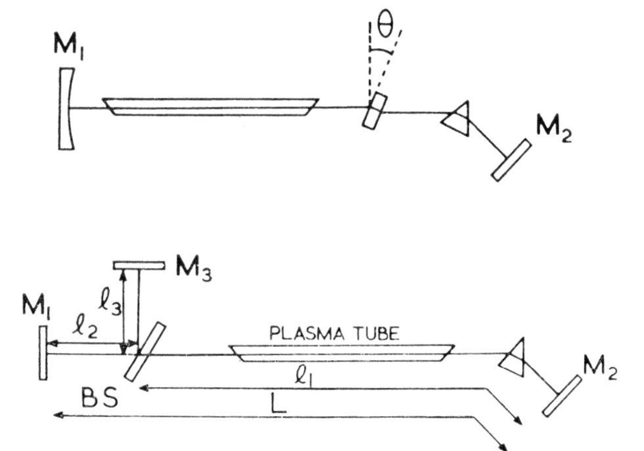

FIG. 46. Single axial mode, single frequency selector (a) in-cavity etalon; (b) Fox-Smith interferometer.

MHz (°C)$^{-1}$ is attainable with the air-spaced etalon. For angular tuning, in order to tune between modes, the change in angle of the etalon is required to be

$$\Delta\theta_E = \Delta\nu_E \frac{\mu}{\nu\theta}$$

For $\nu = 6 \times 10^{14}$, $\mu = 1{\cdot}46$ for fused silica and $\Delta\nu_E = 10$ GHz then the angular change required is $\Delta\theta = 8 \times 10^{-3}$ radian $\simeq 0{\cdot}5°$; for an air-spaced etalon $\Delta\theta \simeq 0{\cdot}25°$.

This method is simple and efficient. A conversion efficiency of about 50% multimode power into a single frequency can be obtained. This depends on the insertion loss of the etalon in the cavity and on other factors such as scatter, absorption and walk-off as well as the degree of homogeneous broadening in the laser. The free spectral range of the etalon is chosen to be greater than the effective gain bandwidth. For example, for the argon ion laser $\Delta\nu_E > 10$ GHz. The etalon can be used to achieve discontinuous tuning over the gain bandwidth.

(b) *Additional resonator* (*Fox–Smith*) *interferometer*. The Fox–Smith interferometric method (Smith, 1965, 1966) is shown in Fig. 46b and has been well applied particularly to single frequency laser operation of the He–Ne laser. This method has been applied to argon by Zory (1966, 1967), Gorog and Spong (1969), Sinclair (1968), Odintsov *et al.* (1968), Belyaiev *et al.* (1969) and Smith *et al.* (1969).

An additional beam splitting mirror BS is introduced into the laser cavity and a fully reflecting mirror M_3 added. The beam splitter in the interferometer made up of M_1 M_3 and BS is rotated by 90° from the usual arrangement in a Michelson interferometer. If $l_2 + l_3$ contains an integral number of half wavelengths, a standing wave builds up between M_1 and M_3 and this tends to cancel light reflected from BS away from M_3 and all the light incident on BS is reflected back into the laser. The beam splitter reflectance affects the Q of the secondary resonator $M_1 M_3$ and BS but not the cancelling of interference. To minimize reflection loss at the front surface of BS this can be set at the Brewster angle. The interferometer insertion loss is between 5 and 10%. The wave reflected from the interferometer is analogous to the transmitted wave in the Fabry–Perot interferometer. Belyaiev *et al.* (1969) have found that for effective frequency selection in long argon lasers, in their case 235 cm active length, the reflectivity of BS should be > 0·5. The reflection coefficient of the beam splitting mirror BS is chosen to provide sufficient selection in the interferometer. The additional resonator and the two resonators with respect to each other need to be stabilized against length variation.

If the additional resonator does not introduce any loss, then 0·6–0·7 of the power is obtained in a single frequency. For example Gorog and Spong (1969) obtained 0·52W at 4880 Å using a Fox–Smith interferometer (from a 3-mm bore, 46 cm long laser operating at 35 A in a magnetic field of 1100 G). Smith *et al.* (1969) obtained conversion ratios of 0·7 at 4880 Å and 0·5 at 5145 Å in the argon laser.

Rigrod and Johnson (1967) have proposed the use of an in-cavity resonant prism mode selector to overcome some of the problems of construction, alignment and stabilization of the 3 mirror cavity of the Fox–Smith interferometer. For slow tuning they suggest thermal tuning with temperature changes of only 1·6°C for tuning over one order. They have maintained a single near centre mode of the Ar 4880 Å line for up to 6 hours with a temperature stabilized prism resonator and for this line have obtained 0·6 to 0·7 of the multi-frequency power in a single frequency for a laser operating at the 200 mW level, a value comparable to etalon or 3-mirror operation.

Zory (1966) has reported some interesting comparative figures for multi-frequency to single frequency conversion. These are shown below for a 2-mm bore, 50 cm long, 15 A, 0·5 torr argon laser.

	Power (mW)	
	4880 Å	5145 Å
Basic laser, multi-frequency	490	330
Single frequency by Fox–Smith interferometer	130	120
Multi-frequency with interferometer replaced by flat and 6% loss	250	140

Other forms of the Fox–Smith interferometer have been proposed, based on multiple parallel mirrors (Kogelnik and Patel, 1962; Collins and White, 1963) and of the Michelson interferometer type (DiDomenico, 1966a, b). An on-line confocal reflection system was used by Sinclair (1968) with an extra on-line mirror which forms an additional confocal cavity. In the Fox–Smith interferometer $d = l_2 + l_3$ is chosen so that the free spectral range is > Doppler width. The transverse modes of the interferometer must match the transverse modes of the laser otherwise spurious resonances can occur. The difference between the Fox–Smith interferometer and the interferometer of Sinclair is that mode matching is not completely necessary. The transverse modes for a confocal resonator are degenerate in frequency and the interferometer of Sinclair is easier to use than that of Fox and Smith. Sinclair obtained single frequency from the 5145 Å line in argon with an extra on-line mirror having 30 % reflectivity forming a 7·5 mm confocal cavity with a mode match lens in cavity and obtained 150 mW single frequency.

An application of the Fox–Smith interferometric method to single frequency selection in the ring laser has been given (Smith, 1968) in which stabilization of an ion laser is proposed (see Section XI, E (7)).

Other frequency selection systems are now briefly reviewed.

2. *Absorbing Film*

The absorbing film is positioned to coincide with a nodal plane of the electric field and the losses for the selected mode can be very low ~ 0·25 %. This method is helped in the argon laser by mode competition effects arising from the large hole widths such that neighbouring modes are extinguished or have very low gain. Then an axial mode separation of say $c/2L \sim 150$ MHz in a 1-m long laser does not have to be discriminated against. This method was first proposed by Troitskii and Goldina (1968). In their system they used a thin nickel film on a quartz substrate which had an absorption coefficient of 0·14 at the laser wavelength.

A full study has been made by Smith *et al.* (1969) who obtained 0·7 of the multimode power in a single frequency, e.g. 360 mW from a line oscillating at 500 mW. The film was pure Ni or nichrome on fused silica plates and best results were obtained with a nichrome film of 150 Å thickness. The front and back surfaces of the plate were antireflection coated and contacted with oil to the back surface of the metal coated plate. This method was applied to the argon laser by Donin *et al.* (1969) using a thin absorbing silver film on a silica substrate; the film transmission was 52 % and absorption 30 %.

The performance of the absorbing film system is similar to that for the Fox–Smith interferometer and may be more used in the future.

3. *Competition Between Modes*

This method was devised by Forsyth (1967, 1969) to obtain single frequency operation at 5145 Å in an argon laser. The laser must be working in a magnetic field on the 4880 and 5145 Å lines simultaneously (the method requires that the two transitions should share a common level as is the case here) and the cavity length for the two lines must be different. A complete explanation for the effect has not been found. But since the 4880 and 5145 Å lines share a common lower level it is probably to be found in terms of mode competition between the lines. The population of the common level will vary at a frequency equal to the difference in the axial mode frequencies of the two transitions. For a common level having a short lifetime the difference frequency fluctuations are absent and the population variations are not characteristic of either transition. Then single mode operation is encouraged. The short lifetimes of the lower levels of the noble gas ion lasers will enhance this behaviour. Forsyth obtained 70 mW single mode at 5145 Å with this method. It was found that the 4880 Å line need not be oscillating single transverse mode or even stably. Gorog and Spong (1969) find that under these conditions the 4880 Å line oscillates in only 2 axial modes separated by 1·5 GHz. When one of the lines is blocked the other returns to unstable multimode operation.

It has been reported that the Ar 4880 Å line can be made to operate on a single axial mode frequency by running it in the TEM_{00} transverse mode and carefully controlling the axial magnetic field (Tandler, 1971). Yarborough and Hobart (1968) have observed a high power stable mode of operation in a segmented DC excited tube. With the laser operating simultaneously at 4880 and 5145 Å stable multifrequency operation occurred. When the laser was oscillating well above threshold at 4880 Å only in an axial field of 1 kG, 3 stable modes were found with 90% of the power in the central one. Similar performance was found for 5145 Å at low power levels with a change to multifrequency operation at higher powers. The influence of the segmented bore appears to be important as this effect has not been observed with continuous bore tubes.

4. *FM laser*

FM operation of the He–Ne laser was first reported by Harris and Targ (1964) and an early review has been given by Harris (1966). The mode amplitudes and mode phases take on the parameters appropriate to the side-bands of an FM signal when the laser is internally phase-modulated at a frequency almost equal to the intermode frequency. In the time domain the output is an FM signal with a carrier frequency near line centre and a modulation frequency equal to the mode spacing. The effect was explained by Harris and McDuff (1964, 1965) in both a linear and non-linear theory. A periodic pulsed output discussed by Harris and McDuff (1965) was observed by Amman *et al.* (1965), as illustrated in Fig. 47.

The coupling between the axial modes of a laser which is internally phase modulated can be described by a set of equations which include the mode amplitudes and phases, a mode coupling coefficient and take into account mode pulling and pushing. The form of the equations has been set out by Gordon and Rigden (1963), Harris and McDuff (1964, 1965), Yariv (1965), DiDomenico (1964) and Crowell (1965) and in a non-linear theory by Harris and McDuff (1965) for FM locking and McDuff and Harris (1967) for AM locking.

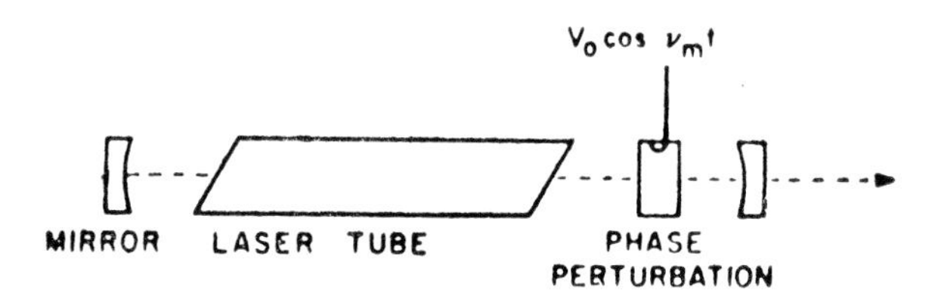

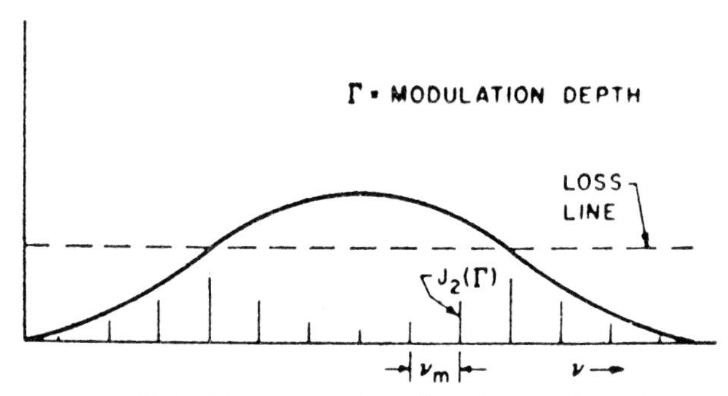

FIG. 47. FM laser system (Harris and McDuff, 1965).

Following the description of Harris and McDuff (1964) the phase perturbing element allows the path length of the cavity to be rapidly varied. With the frequency almost equal to the axial mode frequency spacing the output laser oscillation is a set of modes with nearly Bessel function amplitudes and FM phases. These are scanned over the spontaneous line profile at the drive frequency. In order to get FM oscillation the strength of the phase perturbation must be sufficient for the parametric gain to be large compared with the net saturated atomic gain.

A physical description is that the phase perturbation associates with each of the free-running modes a set of FM side bands. The free-running mode becomes the centre frequency (or carrier) of an FM signal. The FM oscillations compete for the inverted atomic population since they see the same population. The free-running modes may see different populations. The competing FM oscillations are strongly coupled and the FM oscillation at the centre of the atomic line will quench the weaker ones. The output of the FM laser is made up of a large number of optical frequencies. The FM laser has been used to achieve frequency stabilized, single frequency operation by Osterink and Targ (1967).

The FM single frequency can be stabilized relative to the centre of the Doppler line (Harris *et al.*, 1965; Targ *et al.*, 1967). By the use of an error signal derived from the beat frequency generated if the centre of the FM signal deviates from the line centre, automatic frequency control and long term stability is possible. Using an argon FM laser Targ *et al.* (1968) stabilized the 5145 Å line to 1

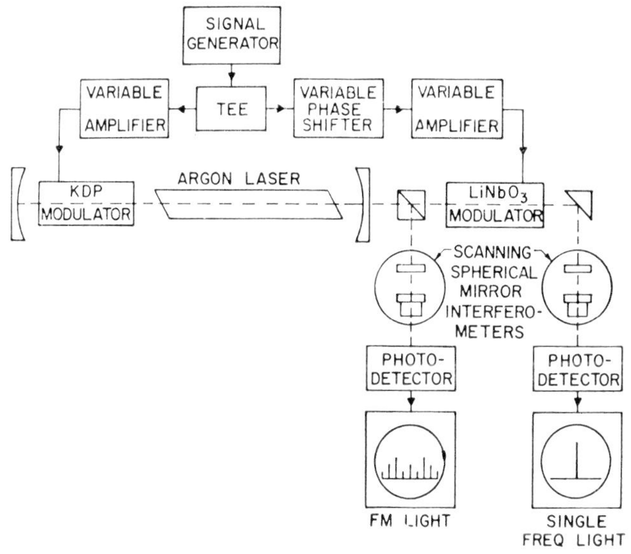

FIG. 48. Single frequency operation by the super-mode technique (Osterink and Targ, 1967).

part in 10^8 obtaining a relatively high power (400 mW) single frequency output. In this laser they report a 20–30 dB reduction in low frequency mode competition noise.

The output from an FM laser can be converted into a high power single frequency by passing the output through an external phase modulator out of phase but at the same frequency as the internal modulator. This single frequency selection method has been described by Massey *et al.* (1965) for the He–Ne laser. It is sometimes known as the "super-mode" technique, and is shown schematically in Fig. 48. The external modulator operates at a depth of modulation equal to the internal modulator and the phase is adjusted to bring that modulation depth $\Gamma = 0$. The energy in the FM output side bands is obtained in a single frequency output. Single frequency operation of the argon ion laser was obtained in this manner by Osterink and Targ (1967). These authors used a R.F. excited laser having an intermode spacing of 128 MHz. Since alternate modes were suppressed by mode competition this was internally modulated by a KDP crystal at 256 MHz. The FM sideband spacing was the same 256 MHz. 350 mW of FM signal power at 5145 Å was obtained from an original multifrequency power of 400 mW with the internal modulator in cavity. The method was not found suitable for the Ar 4880 Å line. The external modulator was a $LiNbO_3$ crystal 2 cm long by 5 mm^2 through which two passes were made providing a depth of modulation of 4·5. An efficiency ~ 0·45 of the free running multimode power to single frequency was achieved. A theoretical efficiency of 1 is not obtained because of losses in the modulators and distortion in the FM output. Gilson and Stoll (1969) have shown that fine FM detection resolution is possible from the free-running multimode, uniphase argon laser with modes distributed over about 5 GHz.

5. *Single Frequency Conversion Efficiency*

Zory (1967) and Gorog and Spong (1969) have calculated the expected efficiency of conversion from multi axial mode (multifrequency) to single frequency operation in the argon laser. The conclusion from this work is that over a large range of excitation more than half the multifrequency power can be obtained in a single frequency. This emerges from the large homogeneous linewidth and also from the probable high degree of cross-relaxation which should exist in this laser. Similar conversion efficiencies are found in the Cd metal vapour laser and also in the He–Ne laser where cross-relaxation is now known to be quite large (Smith, 1972). The ratio of the single frequency to multifrequency saturation powers for the Ar 4880 Å line has been calculated and measured by Odintsov *et al.* (1968). The measured value was 1/0·55 for a 2-mm bore, 23 cm long, 0·3 torr argon, 10 A laser.

6. *Tuning*

A proposal for tuning ion lasers over a range greater than the Doppler profile by accelerating the ions has been made by Singer (1968). With accelerating potentials of 2×10^4 V cm^{-1}, argon ion velocities of 10^8 cm sec^{-1} are achieved in 2×10^{-7} sec over a distance of 10 cm. Frequency shifts of $\Delta\nu \simeq \pm 3 \times 10^{-3}\ \nu_0$ (~ 15 Å) might be obtainable.

The tuning of ion lasers in axial magnetic fields has been studied by Gorog and Spong (1967) for the case of the stable single mode output observed below certain critical fields. A tuning characteristic for the Ar 5145 Å line of 3·73 MHz/G was obtained.

7. *Ring Laser*

Rigrod and Bridges (1965) found that a ring laser based on argon oscillates only in one circulatory direction for all except a narrow range of gain near threshold. Above threshold the travelling wave oscillation occurred in the direction opposite to the ionic drift velocity. At higher excitation levels travelling waves could occur independently in either of the two directions. Their explanation was that in the argon laser the hole width is greater than the axial inter-mode spacing. If spatial cross-relaxation depended only on the ion motion then only running waves can oscillate in the ring laser which oscillate in the same direction. This unidirectional behaviour has been seen also in the He–Ne laser (Moss *et al.*, 1964) but in the He–Ne laser it is uncommon whereas in the Ar laser it is the rule.

In a ring laser the longitudinal modes are separated by c/R where R is the total path length in the ring cavity. A technique of single frequency selection and stabilization of an ion ring laser has been proposed by Smith (1968). Two additional partially transmitting mirrors are added to the ring whose function is to form an auxiliary cavity to select out a particular resonant frequency. With two $\lambda/4$ plates and a Faraday rotator the path length and resonant frequency of the second cavity is different for clockwise and counter-clockwise waves depending on the magnetic field on the Faraday rotator. If the frequency difference is a multiple of c/R the laser will oscillate at 2 frequencies, clockwise (C) and counter-clockwise (CC). If these frequencies are symmetric with respect to the centre of the Doppler profile they will compete through interaction with the same group of atoms. This competition can be used to stabilize the laser. In an ion laser ions will have different drift velocities in the clockwise and counter-clockwise directions along the tube and the gain profile will depend on the direction of the wave along the tube. The difference I_C–I_{CC} near line centre will be a rapid varying function of ν and a frequency discriminant is obtained without the need for the Faraday rotator.

E. FREQUENCY STABILIZATION

There are several general methods of frequency stabilization of the output of a laser. These may be listed as:

(1) stabilization to the atomic gain characteristic
(2) mode competition or Lamb-dip frequency stabilization
(3) stabilization to a frequency defined by a reference cell
(4) a combination of (1) or (2) and (3)
(5) stabilization using FM techniques.

Despite the real or potential interest in the application of ion lasers, particularly the argon ion laser, to spectroscopy, metrology and communications, until recently little effort has been expended on frequency stabilization of these lasers. However, the more recent experiments on the argon laser using an iodine absorption cell (Ryan *et al.*, 1972; Hohimer *et al.*, 1972), described later in more detail, are most interesting developments. Most effort in this area has been given to stabilization of the He–Ne laser; however, the techniques used there are in general applicable also to the ion lasers. Various review papers on frequency stabilization in lasers have been produced by White (1965), Birnbaum (1967) and Hall (1968). Some of the frequency stabilization methods used with the He–Ne laser, such as the use of the Lamb-dip and a saturable absorber, are not as suitable for ion lasers because of the large linewidth and small depth of the Lamb dip.

The instabilities in oscillating frequency which arise when a laser is oscillating at a single axial frequency arise from a variety of sources which include:

(1) Mechanical resonances, air convection currents and sound disturbances.
(2) Changes in the ambient temperature and pressure which contribute mainly low frequency long term fluctuation (White, 1965).
(3) Variation in the position of the Brewster windows (Mielenz *et al.*, 1966).
(4) Spontaneous emission noise contributes the ultimate limit (Schawlow and Townes, 1958; Yariv, 1967; Birnbaum, 1967).

Commercial noble gas ion lasers have long term frequency drift rates < 100 MHz over tens of minutes if the laser tube is operating in a thermally stabilized cavity. Low level plasma oscillations contribute short term ($\sim$ ms) frequency jitter $\sim$ (5–15) MHz.

Two reports (Ryan *et al.*, 1972, Hohimer *et al.*, 1972) have appeared recently on stabilizing the frequency of an argon laser line to an absorption line following the work of Ezekiel and Weiss (1968). A commercial high power single transverse and single axial mode argon laser operating at 5145 Å has been stabilized by deriving a multiple feedback loop from an optical cavity discriminator stabilized against an iodine vapour absorption line (Hohimer *et al.*, 1972). The Ar laser (Coherent Radiation 52B) operating at 1·6 W at 5145 Å has a temperature stabilized solid intracavity etalon as a single frequency selector. With optimized temperature control and orientation of the etalon a single mode power > 800 mW was obtained. The principle of the frequency stabilization is described by Smith (1965) and White (1967). The laser light is incident on a reference cavity made up of a Fabry–Perot optical spectrum analyser which is temperature stabilized to have a long term frequency stability better than 20 MHz/hr. The resonance frequency of this reference cavity is made to coincide with that of the laser by piezoelectric tuning and its resonance frequency is modulated at 2 kHz. When the laser output differs from the centre frequency of the reference cavity an amplitude modulated signal is detected which is used to correct the laser frequency. There is an additional feedback loop to improve the long term frequency response which uses the absorption of I_2 as a frequency reference cell. Here the R(15)–P(13) rotational line of the $B(^3\Pi^+_{ou})$–$X(^1\Sigma^+_g)$ transition of $^{127}I_2$ is used. The short term frequency stability is ± 1 MHz or $\Delta\nu/\nu = \pm 2 \times 10^{-9}$. The long term frequency variations including drift are less than 6 MHz over 2 hours. Hohimer *et al.* (1972) suggested the use of I_2 hyperfine absorption to improve long term stability and this was subsequently done by Ryan *et al.* (1972).

The argon ion transition at 5145 Å was locked to an absorption line in an external reference of a molecular beam of I_2. The drift of the laser frequency was < 20 Hz giving $\Delta\nu/\nu < 3 \times 10^{-11}$ for a time period up to 20 minutes. The experimental system is shown in Fig. 49. The 5145 Å transition was operated using a Fox–Smith interferometer and is tuneable across the gain profile by a piezoelectric mirror movement. This excites a I_2 molecular beam and the laser-induced fluorescence is detected with a lock-in amplifier synchronously with modulation of the laser. The laser frequency is locked to the centre of one

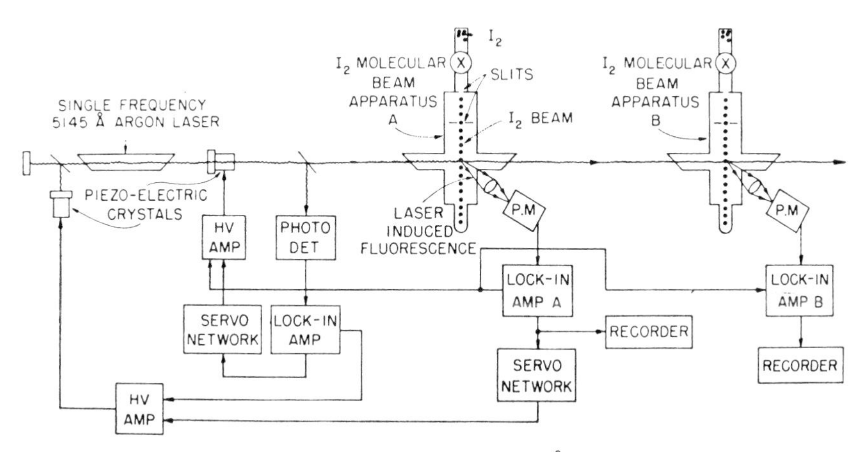

FIG. 49. Frequency stabilization system for the Ar 5145 Å laser locked to an I_2 molecular beam absorption line (Ryan *et al.*, 1972).

of the I_2 transitions by maintaining the laser frequency at the zero-crossing of the intensity derivative. The output of lock-in amplifier B gives a measure of the drift of the laser frequency. Argon laser plasma oscillations were suppressed using the techniques of Galehouse *et al.* (1971) described in Section XI, D, although Ryan *et al.* find adequate short term laser frequency stability is still a problem.

The advantage of the molecular beam external reference is that the I_2 molecule is under isolated conditions in the beam and does not undergo any significant frequency shifts from collisions and has minimal collisional broadening. By using the excitation at right angles as shown in Fig. 49 Doppler broadening is also almost eliminated. The $^{127}I_2$ transitions excited were the hyperfine components of the P(13) R(15) (43-0) lines of the transition

$$^1\Sigma_g^+\,(X) - {}^3\Pi_{ou}^+(B).$$

As Ryan *et al.* point out the molecular beam reference promises a high degree of long-term stability and resetability of the laser frequency. The main difficulty is the finite Doppler shift, the effects of which may be capable of elimination by locking the laser to the saturated absorption dip at the centre of the I_2 transition in the beam.

Stabilization of the 4880 Å line of a commercial argon ion laser over long periods of time but with lower stability than that reported by Hohimer *et al.* (1972) or Ryan *et al.* (1972) has been achieved by Maischberger (1971). Stabilization is obtained here by controlling the mirror separation of a 3-mirror single mode selector cavity using a Lamb-dip stabilized He–Ne laser as a reference and a stability of $\Delta\nu/\nu = 2 \times 10^{-8}$ over 30 minutes was obtained.

Stabilization by the use of FM laser techniques (the FM laser is described in Section XI, D) has been carried out. Frequency stabilization of the argon laser at 5145 Å by Targ *et al.* (1968) used a 400 mW FM laser with the line centre as a reference. He achieved a stabilization of 1 in 10^8.

The large spontaneous linewidth of the argon laser means that there are many free running modes and modulation has to be carried out at a multiple of the intermode frequency and a high modulation depth is necessary. Targ *et al.* modulated at $3/2\ (c/2L)$ and mode competition suppressed neighbouring modes, this led to low distortion FM oscillation; the apparatus is shown in Fig. 50. This method may not be so easily applicable to the Ar II 4880 Å transition.

Use of the rotation of the plane of polarization in a magnetic field as a means of stabilizing an argon ion laser has been proposed by Burmakin *et al.* (1971). In an axial magnetic field the real and imaginary parts of the polarizability of the gas medium contain resonance terms depending on the size of the Zeeman splitting and on the detuning of the lasing frequency with respect to the centre of the Doppler line $(\nu - \nu_0)$. These resonance terms lead to (1) dips in the power against detuning curve and (2) a frequency dependent rotation of the plane of polarization. The magnitude of this rotation of the plane of polarization on field can be quite large. For a Brewster window laser operating at 4880 Å producing plane polarized light at a frequency near line centre and for low gain saturation of the medium taking the Landé g-factors of the levels to be identical, Burmakin *et al.* give expressions for the rotation angle with and without gain saturation. This rotation effect could be used to produce an error signal in an automatic frequency control system for the 4880 Å transition with a Fox–Smith interferometer single mode selection system. An absolute frequency stabilization within $\pm$ 10 MHz or $(\Delta\nu/\nu) \sim 10^{-8}$ is expected.

A stabilization method for an ion ring laser has been proposed by Smith (1968) and is discussed in Section XI, D (7). The wide bore high current wall-stabilized argon ion laser (as described in Section X, G) exhibits superradiance.

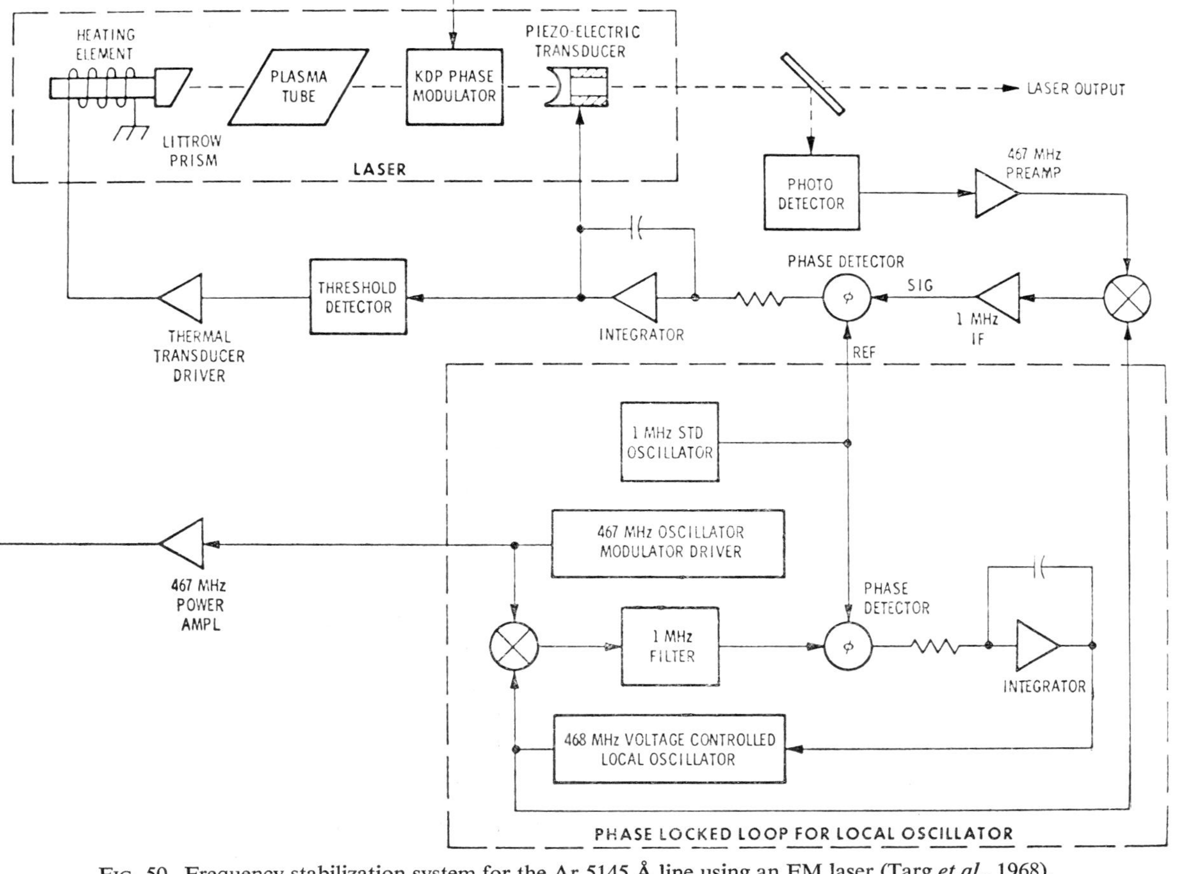

FIG. 50. Frequency stabilization system for the Ar 5145 Å line using an FM laser (Targ *et al.*, 1968).

A gain of about 10 dB m^{-1} and a power output of 20 W cm^{-2} has been measured by Banse *et al.* (1968). This emission should be quite stable since it is non-resonant and independent of the cavity. The attainment of stable frequency sources by the use of superradiant emission has been proposed for several other lasers including the Hg^+ laser. The frequency stability is almost independent of vibrational and thermal fluctuations as found in the normal laser output. The output frequency is self-regulating to the centre of the gain profile, although its linewidth is related to the operating conditions.

F. MODE-LOCKING

Several thorough review articles on mode-locking in gas lasers have been written (Harris, 1966; Smith, 1970; Allen and Jones, 1971) which describe the general features of pulse width, pulse repetition frequency and pulse spacing. In the argon ion laser, having an oscillating linewidth up to 10 GHz, a minimum pulse width of about 10^{-10} sec is possible which is comparable to the lifetime of the lower laser level. Possible application to time dependent spectroscopy is evident, particularly because of the wide range of wavelengths available. The main characteristic of the mode-locked train of pulses are:

Pulse-width $= \dfrac{1}{N\Delta\nu}$ N = number of coupled axial modes, $\Delta\nu$ = intermode frequency spacing

Peak pulse power $= N$ (CW unmode-locked power)

Pulse repetition frequency $= c/2L$ L = resonator length

For an argon ion laser having a 1 m resonator length typical values are: pulse width 2×10^{-10} sec, $N \sim 30$; and p.r.f. $\sim$ 150 MHz.

Forced mode-locking in an argon ion laser was described by Crowell (1965) who at that time also reported the absence of self-mode-locking. Now however, both active and self-mode-locking are well known in ion lasers. De Maria and Stetser (1965) were able to obtain shaped pulses from an argon laser (square, stepped or saw-tooth) by combining travelling and stationary waves. This was achieved by modulation using a fused silica acousto-optical cell. Gaddy and Schaefer (1966) observed self-mode-locking of the Ar 4880 Å line producing $\sim$ 0·5 ns pulses separated by 4·2 ns in a low power laser. Self-mode-locking was easily obtained but the output showed large fluctuations with irregular pulses and indicated incomplete mode-locking. In applications this may be a problem if the power level needs to be kept fairly constant, particularly when losses are put in the cavity. The above authors also reported self-mode-locking with repetition frequencies of $c/2L$ and $2c/2L$. This mode-locking at multiples of the basic $c/2L$ repetition frequency has been observed in several gas laser systems (McClure, 1965; Uchida and Ueki, 1967; Baker and King, 1973).

The Ar II 4880 Å has also been reported to show complete self-mode-locking by Borisova and Yasinskii (1971) who used a short cavity to obtain a large intermode spacing (530 MHz). This eliminated competition between modes, which can lead to instability (Section XI, B). Very stable oscillation involving about 8 axial modes was obtained which was free of low frequency noise. The beat frequency between different pairs of overlapping oppositely travelling modes coincided—this is characteristic of self-mode-locking.

Both forced and self-mode-locking have also been observed in the krypton ion laser (Heising *et al.*, 1971). With forced mode-locking by an acousto-optic fused silica modulator pulses of width 0·2 ns were obtained at a p.r.f. of 88 MHz. These authors found that krypton operated in the self-mode-locked mode more readily than argon. The self-mode-locked pulse width was about 0·4 ns in krypton at 6471 Å and other lines over a wide range of power levels. The conditions for self-mode-locking in argon have been discussed by Statz *et al.* (1965). They have calculated a degree of mode pulling of 100 kHz which needs to be balanced by mode repulsion to achieve mode regularity.

The UV lines at 3511 and 3638 Å in Ar III have been mode-locked by Heising *et al.* (1971) using an acousto-optic modulator. For these lines pulse widths of 0·2 ns and average powers of 10 mW were obtained, under similar conditions 0·17 ns pulses at an average power of 0·75 W were obtained from the Ar II 5145 Å line. Forced mode-locking on the Ar II 5145 Å line has also been studied by Scavennec (1971). Using the same non-linear crystal Gurskii (1969) achieved simultaneous forced mode-locking and second harmonic generation on the Ar 4880 Å line.

There is a connection between the regions of stable or unstable mode operation and mode-locking effects. The stable region of operation discussed by Rigrod and Bridges (1965), Borisova *et al.* (1968) and Lebedeva *et al.* (1970) may be due to mode-locked quieting of the laser. These effects have been investigated by Bass *et al.* (1968) who found stable pulses $\sim$ 0·2 ns wide at in-cavity power levels of 11 W cm^{-2}; on increasing the in-cavity power the laser changed into unstable operation. At in-cavity power levels of 15 W cm^{-2} these mode-locked pulses became broader and frequencies developed comparable to the frequencies of amplitude fluctuations characteristic of the unstable region.

As discussed in Section XI, B, mode interactions and fluctuations in multimode lasers contribute fluctuations in the frequency spectrum and lead to a high degree of noise. In the mode-locked laser there is no noise due to the combination tones since the modes are equally spaced. The mode-locking leads to quiet operation of the laser with no mode competition, with fixed phases between modes, no mode pulling and no large amplitude fluctuations of the individual modes. The residual noise after mode-locking is due to random fluctuations in the laser elements and spontaneous emission into the modes. This suppression of noise by mode-coupling has been reported by Targ and Yarborough (1968) who used internal phase modulation and achieved 20 to

30 dB suppression of low frequency AM noise in both phase-locked and FM operation with no lower loss. With in-cavity phase modulation of the laser, depending on the detuning $\Delta\nu$ from the fundamental cavity reference, the output will be either an FM output ($\Delta\nu \neq 0$) or a phase-locked pulse output if $\Delta\nu = 0$. The phase-locked pulsed output will have the properties of the mode-locked pulses described above for the case of loss-modulated cavities. The relationship of mode-locking and the FM laser is discussed in Section XI, E. A thorough description of FM (internal phase modulation) and AM (amplitude modulation) for a homogeneously broadened laser has been given by Kuizenga and Siegman (1970a, b).

A higher output power is obtained if the modes are not coupled in phase and are not operating in the self-mode-locked condition. This is because better use of the inversion distribution is made; Smith (1972) observed this effect for the He–Ne laser.

A description of self-mode-locking has been given in terms of π-pulses having the condition $\pi = Ep_{12}\tau/\hbar$ where E is the electric field amplitude, p_{12} is the dipole moment of the transition and τ is the pulse duration (Smith, 1967; Fox and Smith, 1967; Frova *et al.*, 1969) and by which the inversion in the medium can be removed. Bass *et al.* (1968) have reported that the pulse is not completely a π-pulse but a $0{\cdot}3\ \pi$ pulse following the definition of Fox and Smith (1967). It is unlikely that it is as much as a π-pulse because of the relaxation of the lower level (having a life-time $\sim 0{\cdot}3$ ns) and to a lesser extent the upper level. These effects were observed by Bass *et al.* (1968) when the amplitude was too low for π-pulse operation.

XII. Techniques and Operating Characteristics of Metal Vapour Lasers

A. Techniques

Metal vapour lasers which operate on transitions in the metal ions in low current density discharges in helium or neon buffer gases are of much simpler construction than the noble gas ion lasers. A discussion of the non-technological aspects of these lasers is given in Sections III and V.

The first reports of ion laser action in metal vapours were pulsed emission from cadmium, observed by Fowles and Silfvast (1965) and Silfvast *et al.* (1966), and in quasi-CW operation by Fowles and Hopkins (1967). The emphasis has shifted from pulsed to D.C. operation in the subsequent development of metal vapour lasers and tubes for D.C. operation will be described here. These D.C. lasers operate with low input powers similar to He–Ne lasers and simple glass structures can be used. In these lasers the dominant requirement is to maintain a uniform distribution of the vapour in the discharge at a pressure of about 10^{-2} torr. The output is quite strongly pressure dependent as with most gas lasers and therefore temperature control of the metal source to better than 1°K is necessary. He is added at a pressure of a few torr as a source of excitation, to sustain the discharge and to act as a buffer gas.

(1) *Multisource Cadmium Lasers*

Studies of the He–Cd laser have been made using positive column discharges inside heated ovens (Fowles and Silfvast, 1965; Fowles and Hopkins, 1967; (Silfvast *et al.*, 1966, Csillag *et al.*, 1970a, b). Silfvast (1968) used a silica discharge tube inside a tubular furnace running the length of the laser tube with electrodes and Brewster windows external to the furnace. The Cd

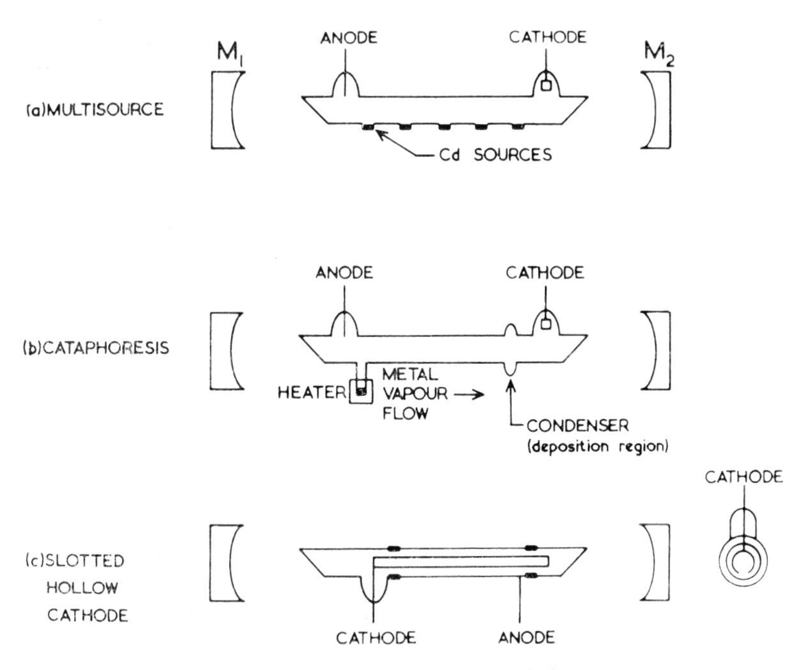

FIG. 51. Metal vapour laser tube designs.

was placed directly in the discharge tube or in small side cups at about 15 cm intervals. The use of side tubes allowed better control of the Cd pressure. A schematic diagram of a multisource metal vapour laser is shown in Fig. 51a. The Cd vapour pressure is difficult to control because of the discharge heating in these arrangements. With several sources of Cd in the tube, a low noise and more stable output is obtained. Also a long time is required for a uniform Cd density to be set-up by diffusion.

A small and simplified Cd laser has been produced by Silfvast and Szeto (1971) based on a segmented bore tube. The Cd is introduced into the discharge from heated annular Cd metal segments equally spaced through the bore of the tube. Good uniformity of Cd distribution is obtained with this laser by diffusion and cataphoresis. The Cd segments containing a few gm of metal were 4 mm width and spaced 6–8 cm apart in a 2-mm bore tube of length 26 cm. The discharge has the dual purpose of heating the Cd segments and as a means of excitation, producing He metastables. The geometry of the tube can be chosen to match the Cd temperature and excitation requirements. The optimum current found by Silfvast was 65 mA and the tube had a switch-on time of 3–4 minutes and optimum power was reached after about 5 minutes. The laser is simplified since it only requires one power supply and no additional heaters. The several Cd segments produce a more uniform distribution of vapour which leads to a more stable output with low noise content (< 1 % at a power slightly below maximum). From a 38-cm laser Silfvast (1968) obtained about 10 mW at 4416 Å and 1 mW at 3250 Å using a natural isotropic Cd mixture. Dyatlov *et al.* (1970) have excited laser action in an extended transverse diode structure with a long planar oxide cathode.

Some of the characteristics of the Cd laser are similar to those of the He–Ne laser. The optimum pressure, axial voltage gradient and gain are inversely proportional to the bore diameter and the optimum current is proportional to diameter. The saturation current density for the He–Cd laser is rather lower than for the He–Se laser in most laser arrangements, about 2 A cm^{-2} for the He–Cd compared with about 7 A cm^{-2} for He–Se.

Low frequency (50 Hz) A.C. driven discharges have been used by Csillag *et al.* (1970) in which cataphoresis effects are small and the optimum operating conditions are similar to D.C. excitation conditions; they found an optimum source temperature of 200°C corresponding to 5×10^{-3} torr Cd pressure which gave an optimum gain of 5·4 % per metre. In this work an intermediate saturation region with current was observed and then a fast increase in laser power at higher currents, such that saturation was only seen for low Cd pressures. The possible role of direct electron excitation at low currents and dominant Penning ionization at higher currents was considered. With RF excitation, matching into the discharge is difficult as the discharge impedance changes with Cd density.

(2) *Cataphoresis Metal Vapour Lasers*

In gas mixtures in narrow tubes component atoms of low ionization potential and of lower pressure can be driven to the cathode by the cataphoresis effect (see Shair and Remer, 1968). The resulting ion current is $C_i = N_i \mu_i E$ where E is the mean electric field along the tube, μ_i is the metal ion mobility and N_i is the metal vapour ion density, and C_i is approximately proportional to discharge current as $I(\mu_i/\mu_e)$ where μ_i is the electron mobility and I is the discharge current. The cataphoresis flow increases until balanced by anode back diffusion which is smaller in small bore tubes. When the lower ionization potential atoms are continually fed into the discharge at the anode end only a small concentration gradient is set up in the tube, depending on the discharge current, being less above a critical current (about 50 mA in a 4-mm tube). Sosnowski (1969) has reported some departure from a uniform distribution by observations of spontaneous sidelight but for a current > 50 mA an almost uniform distribution is obtained.

A schematic diagram of a cataphoresis laser is shown in Fig. 51b. The He pressure is sufficiently high to prevent Cd diffusing to the Brewster windows. The Cd density is controlled by the temperature of the small oven surrounding the Cd reservoir at the anode end with temperatures around 280°C as optimum. The Cd flow rate is $\simeq 1{\cdot}5$ mg/hr at 100 mA current, and other typical parameters are voltage 5 kV, current 100 mA, bore 2 mm, length 100 cm, pressure about 3×10^{-3} torr and He pressure about 3 torr. The optimum pressure, electric field and gain are proportional to $(r)^{-1}$ while the optimum current is proportional to r.

A cataphoresis laser of 143 cm length, 2·4 mm bore, 3·4 torr He, 110 mA current and 3–6% transmission mirrors was constructed by Goldsborough (1969) who obtained an output power of 200 mW at 4416 Å with natural isotope Cd. About 20 mW was obtained at 3250 Å with non-optimized mirrors. This design proved relatively stable in operation with a long tube lifetime. Cataphoresis lasers have also been described by Silfvast (1969), Sosnowski (1969) and Fendley *et al.* (1969), Hernqvist (1970), Hodges (1970), Stefanov and Petrova (1971), Giallorenzi and Ahmed (1971), Hernqvist and Pultorak (1972) and Baker and King (1973). The laser of Sosnowski used two anodes and a central cathode to prevent Cd deposition on the windows. Sosnowski showed that a good uniformity of Cd density could be obtained in the cataphoresis laser. This laser operated for over 500 hours with little deterioration. Short discharges at each end of a tube could be used to protect the windows. The laser of Fendley *et al.* had dimensions 1·6 mm bore, 41 cm length and running parameters: reservoir temperature ~270°C, current 50 mA, voltage 2200 volts and He pressure 7 torr. The output powers were 12 mW (TEM_{00}) 4416 Å and 6 mW (3250 Å). A cataphoresis laser with a heated and baffled wide bore diffusion return tube has been reported by Hernqvist (1970). This laser had a low noise output when the temperature of the condenser was optimized. Hernqvist and Pultorak have also used a cataphoresis laser for He–Se (1972). A cataphoresis laser which included a furnace over the whole of the active bore has been described by Silfvast and Szeto (1970). The heating from the discharge current keeps the bore at a higher temperature than the Cd source and hence the furnace temperature controls the Cd vapour pressure.

A disadvantage of the cataphoresis laser is that an instability occurring near the Cd source at the anode will propagate over the length of the tube and be impressed on the laser output. Also the Cd supply is limited; Fendley *et al.* (1969) reported 50 mg of Cd lasting for 50 hours in a cataphoresis tube but it was possible to use a tube in which the roles of condenser and oven are reversible. No change in output was found for 11 reversals corresponding to 550 hours of use but some loss of He occurred.

The theory of cataphoresis of Sosnowski has been extended to include the dependence of the electric field on the cadmium density (Baker, 1973) and a very uniform cadmium distribution is still predicted in the steady state. This theory also shows that a disturbance at the anode, e.g. an oven instability giving a brief increase in Cd density, propagates to the cathode with the slope of the trailing edge being amplified until it has a sharply rising "shock wave" distribution. This form of instability can be expected to be more likely to occur at low frequency, as is found in the He–Cd laser.

(3) *Hollow Cathode Metal Vapour Lasers*

The hollow cathode discharge (HCD) is a transverse discharge between an outer anode tube or pin and a cylindrically symmetric inner cathode tube. The cathode is slotted along its length; an illustration of the HCD in a metal vapour laser is shown in Fig. 51c. Metals can be evaporated and can diffuse in a uniform distribution through the slot into the central hollow cathode negative glow region. This slotted form of HCD is different from that described in Section X, C (1), for use as the argon ion laser cathode. The HCD has a non-Maxwellian electron energy distribution which, with a more pronounced high energy tail, is particularly suitable for exciting laser oscillation (see Hofmeister *et al.*, 1969). Piper *et al.* (1972) found that on introducing the metal vapour into the HCD the overall discharge voltage at a given current increases, unlike the behaviour of positive column metal vapour lasers. The HCD provides a uniform distribution of the vapour which is efficiently excited and the discharge needs only low voltage power supplies. Because of its unique excitation properties Schuebel (1970a, b, c) and Sugawara and Tokiwa (1970) have been able to find new laser transitions in the metal vapours Cd and Zn, aided also by the higher gain of HCD lasers. Laser transitions pumped by charge transfer collisional reactions are more efficiently pumped in the HCD than the positive column discharge as can be expected from the higher energy of the charged pumping state and enhanced high energy electron distribution. Many transitions are only observed in HCD lasers, for example in the He–Cd laser, 5337, 5378, 6355, 6360, 7237, 7284, 8067 and 8530 Å (Piper and Webb, 1973).

Schuebel (1970a) describes a HCD He–Cd laser with a cylindrical slotted cathode and cylindrical anode made of kovar (cathode 5·6 mm i.d. and anode 9·4 mm i.d.) 46 cm long. Cd wire was placed in the anode tube and evaporated by heating tape to pressures between 10^{-2}–1 torr, relatively higher than for the positive column laser. The design features of the transverse discharge slotted

HCD laser have been discussed by Schuebel (1970). Piper and Webb (1973) report discharge current instabilities and contraction of the discharge in the laser for values of currents greater than 1·5 A.

The use of HCD in metal vapour lasers in which sputtering of the cathode itself acts as the source of metal vapour has been described by Karabut *et al.* (1970). The cathode surface in this laser was at about 50°C and direct vapourization was small. Its ability to be controlled and optimized has not yet been reported on.

The HCD of Sugawara and Tokiwa (1970a, b) was made from pin anodes spaced about 8 cm apart with holes in the cathode structure directly opposite the pins. The inside of the cathode surface was coated with the metal vapour, Zn or Cd in this case. Laser oscillation was observed on 15 lines in Zn II and Cd II. A simple, inexpensive multiple anode–hollow cathode structure has been used by Piper *et al.* (1972) and Piper and Webb (1973) in observing metal vapour laser transitions in the He–Cd, He–Zn, He–Se, He–I_2 and He–As. This laser, designed for currents greater than 1 A, used a cathode of heavy-wall stainless steel tubing allowing easy external cooling and flowing gas to overcome impurity effects.

An advantage of the HCD is that moving striation instabilities may be much less than for positive column discharges and noise content in the kHz–MHz region [of the kind observed by Garscadden *et al.* (1966) in He–Ne lasers] should be less.

B. OPERATING CHARACTERISTICS OF METAL VAPOUR LASERS

These lasers have been developed more recently than the noble gas ion lasers and have some properties which are complementary. They can provide relatively high output power, for example the He–Cd laser at 4416 Å has the power output capability of a similarly sized He–Ne laser at 6328 Å, regrettably with the efficiency of typical non-molecular gas lasers. These lasers provide numerous wavelengths spaced throughout the visible region, including the significant transition in the He–Cd laser at 3250 Å—the shortest CW wavelength laser line.

1. *He–Cd Laser*

The gain of the Cd II 4416 Å line in natural cadmium has been measured by Silfvast (1968, 1969) and Fendley *et al.* (1969), values of 4–5 % for a 2-mm bore tube of 26 cm length were obtained and 2·1 % for the 3250 Å line; for single isotope cadmium at a current of about 60 mA it is about 20 % m^{-1}. The use of single isotope cadmium increases the gain by a factor of about three since the isotope shift is greater than the Doppler width (see Fig. 28, p. 326). Power outputs on the cadmium II line at 4416 Å of up to 200 mW have been obtained from a 1 m tube (Goldsborough, 1969a) and 20 mW at 3250 Å from a similar size of laser although the mirrors used were not optimized for this wavelength. Efficiencies at 4416 Å of up to 0·05% have been reported and for the 3250 Å line about four times less.

The typical optimum helium pressure in a He–Cd laser is about 2 torr and laser action up to about 8 torr is possible. Laser action in Ne–Cd mixtures has also been reported (Csillag *et al.*, 1971) although the excitation process is much less efficient. The Ne metastable levels are lower in energy than the cadmium ion upper laser levels and hence a different excitation process from Penning ionization is involved. Surprisingly Stefanov and Petrova (1971) find a 25% improvement in output power by using a 10:1 He–Ne mixture over pure He. In a mixed Cd–He–Ne mixture simultaneous laser oscillation at 4416 Å (Cd^+) and 6328 Å (Ne) has been demonstrated (Ahmed and Campillo, 1969). The maximum available powers per unit volume for the He–Cd lines at 4416 and 3250 Å in natural cadmium have been calculated by Hodges (1970) to be 4416 Å: 12 mW cm^{-3} and 3250 Å: 2 mW cm^{-3}. For the same discharge conditions Goldsborough (1969) has found 4416 Å: 31 mW cm^{-3} and 3250 Å: 0·5 mW cm^{-3} (for non-optimum output coupling).

At a given He pressure the output power peaks at higher current as the cadmium concentration is increased and the optimum current decreases with increasing helium pressure; there is a direct connection between E/p and Cd^+ density. Threshold current for the He–Cd 4416 Å line using natural cadmium is about 40 mA in a 4 mm bore tube. The output power is proportional to current, saturation of laser power with current increase arises from saturation of the helium metastables by electron collisional de-excitation. Saturation currents are about 150 mA in a 3 mm bore tube and 200 mA in a 4 mm bore tube, saturation current densities being about 2 A cm^{-2}. The saturation process has been considered by Giallorenzi and Ahmed (1971) using a rate equation approach. The model can predict the experimental saturation curves and is based on Penning excitation and saturation predominantly from collisional de-excitation of helium metastables by electrons. Penning excited lines do saturate with current increase but no saturation of power with discharge current up to 5 A has yet been observed in the hollow cathode discharge He–Cd laser for charge transfer excited lines (Piper and Webb, 1973).

2. *He–Se Laser*

The strong (> 5 mW) CW lines in a 1 m tube are at 4976, 4993, 5069, 5176, 5228, 5306, 10409 and 12588 Å. Up to now 46 lines have been found spanning the spectrum from 4468 A to 12588 Å. Saturation with current in this laser occurs at about 4 times higher current density than with the He–Cd laser. For example, in a 3-mm bore tube saturation occurs at 0·5 A with a saturation current density of 7 A cm^{-2}. Output powers from 3 to 50 mW per line have been measured over the entire spectral range of operation in a 2-m discharge tube. This tube, which had a centrally mounted cathode and anodes at each end, has produced a combined output power of 250 mW on the six strongest blue–green transitions (Klein and Silfvast, 1971).

3. *He–Zn Laser*

The strong lines from this laser are at 4911, 4924, 6021, 6102, 7588 and 7757 Å. The very short lifetime of the lower laser level contributes a natural Lorentzian width $\sim$ Doppler width. This high degree of homogeneous broadening allows most of the output power to be converted into a single axial mode frequency.

Hernqvist and Pultorak (1972) have made a comparison of the merits of the He–Cd, He–Se, He–Zn and He–Ne lasers. They conclude that the efficiency of the He–Se laser estimated from its output on the six strongest blue–green transitions is about half that of the He–Cd laser and comparable with the He–Ne laser, although its output power/volume is more than twice that of the He–Ne laser. However the He–Se and He–Cd lasers have high output noise ($\sim 10\%$) although this can be reduced in He–Cd lasers using a recirculating laser structure (Hernqvist, 1970a) or segmented bores with distributed cadmium (Silfvast and Szeto, 1971). The He–Cd laser has the outstanding advantage of having all its output in two short wavelength lines. Table XIX

TABLE XIX. *Approximate outputs for helium excited lasers (having* 50 *cm length by* 1·5 *mm active bore)* (Hernqvist and Pultorak, 1972)

Type	Wavelength (Å)	Output (mW)	Current (mA)
He–Ne	6328	12	25
He–Cd	4416	25	55
	3250	4	55
He–Zn	7479	15	70
	5894	4	70
He–Se	Six lines 4976–5305	28	110

compares the power outputs and required discharge currents for four different types of helium excited laser operating with approximately the same voltage drop. The prominent He–Se laser lines lie very close to those of argon ion lasers and in the power range below a few tens of milliwatts the efficiency of the He–Se laser is some three to four times better.

4. *Noise in Metal Vapour Lasers*

Light emitted from a gas discharge positive column has large intensity fluctuations in both spontaneous (Donahue and Dieke, 1951; Pekarek, 1968) and stimulated (Prescott and der Ziel, 1964; Garscadden *et al.*, 1964) emission. In the metal vapour lasers noise is introduced by fluctuations in the helium metastable or neutral cadmium density. These fluctuations may become stable over the length of the tube which then results in low noise. Also fluctuations in the form of striations moving from the anode to the cathode are common in noble gas discharges where striations may be introduced by, for example, the helium metastables. It has been shown (Baker, 1973) that the neutral cadmium pressure gradient is amplified in passing along the laser tube. Silfvast (1971) suggests that the electrons emitted by the Penning process are coupled to the helium metables which leads to large fluctuations in the metastable density, electron density and possibly electron temperature. Energy may be transferred between the metastables and the electrons which could provide an explanation of the large discharge noise seen in sidelight emission from helium and cadmium ion levels and in the laser output on the cadmium ion lines when Penning ionization is optimized. Hernqvist (1972) has found evidence that noise originates at the cathode and probably from discharge striations as found also in He–Ne lasers (Suzuki, 1971).

Long term stability (times > 1 ms) of the He–Cd laser is good; however, in the frequency range 10–100 kHz there are output power fluctuations of 10–15% peak-to-peak (Fendley *et al.*, 1969) which are attributed to neutral cadmium vapour density fluctuations. These are only partially correlated with the noise in the discharge current. In the segmented bore He–Cd laser of Silfvast and Szeto (1971) operating at helium pressures < 3 torr, the laser noise was very small, < 1%, while the spontaneous side-light noise remained near 10%. The side-light fluctuations were caused by regular stable plasma oscillations and overall gain fluctuations were small. Silfvast has also seen stable plasma oscillations in other He–Cd lasers when the cadmium pressure was higher than optimum. These stable oscillations are attributed to fluctuations in helium metastable density resulting from two-way transfer between metastables and Penning electrons which become regular under the right conditions of tube size, position and temperature of the cadmium source and helium pressure. The recirculation cataphoresis laser of Hernqvist (1970a) reduced the noise from neutral cadmium density fluctuations, which had a 10% peak-to-peak amplitude and were strong at 200 kHz and 10 MHz by improving the uniformity of the cadmium density to less than 1%. This noise level is then comparable to the He–Ne and argon ion lasers. In a cataphoresis laser Baker (1973) has found peaks in the frequency spectrum at about 100 and 200 kHz under optimum conditions.

Noise figures have also been discussed for other metal vapour lasers. In the He–Se laser Herqvist and Pultorak (1972) found noise amplitudes slightly less than for the He–Cd laser at 10% peak-to-peak. In the He–Zn laser Jensen *et al.* (1971) find a noise amplitude < 10% in a hollow cathode discharge, a value which was about 25 dB lower than from the conventional cataphoresis laser over the same bandwidth. The high noise level in the He–Zn cataphoresis laser is more likely to be due to zinc density fluctuations than to the nature of the excitation process or properties of the Zn II laser transitions.

5. *Effect of Transverse Magnetic Fields*

Hernqvist (1969) and Fendley *et al.* (1969) found that a transverse magnetic field can have a large effect on the He–Cd laser. With magnetic field B parallel to the electric vector E—determined by the Brewster windows—only the central plane polarized component oscillated. The power output increases monotonically with B probably due to plasma effects. With B perpendicular to E the right and left handed circular polarized components oscillate. At 450 G the splitting is comparable to the Doppler width; however, at 800 G the components are separated and the laser output is lower than in zero field. The width of the zero magnetic field power dip for the He-Cd 4416 Å transition has been measured (Dienes and Sosnowski, 1970); this dip may be useful in frequency stabilization by placing the gas laser in an axial magnetic field.

6. *Mode-locking*

Forced mode-locking at 4416 and 3250 Å in the He–Cd laser using intracavity quartz acoustic loss modulation was reported by Silfvast and Smith (1970). Using a special isotope mixture of the six cadmium isotopes, which gave an oscillation linewidth of about 8·5 GHz, mode-locked pulses of 150 ps width were obtained with greater than 1·6 W peak power at 4416 Å; this pulse width corresponded to full mode-locking over the whole gain bandwidth. At 3250 Å pulses of width 400 ps and peak power 0·1 W were obtained using natural Cd. The pulses were separated by $2L/c$ for both the 4416 and 3250 Å lines and stable operation over several hours was obtained. Laser action at 3250 Å could not be obtained with the special wideband isotope mixture through insufficient gain but it should be possible to use special isotope mixtures providing slightly narrower bandwidths to obtain 200 ps pulses at this wavelength. The mode-locked pulsewidths obtained by Silfvast and Smith are the narrowest so far reported for visible or UV CW gas lasers.

Self-mode-locking of the 4416 Å transition in a He–Cd laser using natural cadmium has been reported by Faxvog *et al.* (1970) and was observed under all conditions of operation. The mode amplitude were quite stable and the characteristic laser noise reduced. Observed pulse widths were $\sim$ 2 ns limited by detector time response. Stable self-mode-locking has been found also in the single isotope He–Cd^{114} laser (Willenberg and Carruthers, 1970); this mode-locking did not occur when the amplifying medium was centrally placed in the optical cavity but stable operation was found when it was displaced by 15–20 cm from the centre towards one of the mirrors.

Self-locked pulses at sub-multiples of the cavity axial mode frequency separation $c/2L$ has been observed in the He–Cd laser (Baker and King, 1973 and in press) using both natural and single isotope Cd^{114}. This has been analysed in terms of mode-locking of two dominant adjacent order transverse modes TEM_{mn} and $TEM_{m'n'}$ with $m + n + 1 = m' + n'$. Stable self-locking in this system can be maintained almost continuously apart from vibrational disturbances. Self-mode-locking of low order transverse modes has also been observed in the He–Se laser on the 4976, 4993, 5069, 5176, 5228 and 5305 Å transitions by Baker and King (1973). The repetition frequency of the self-locked pulses was a multiple n of the cavity axial mode frequency separation with $n = 2$, 3 or 5 depending on the transition. This effect was attributed to axial mode competition arising from the large natural linewidth (the lower laser level has a strong vacuum UV transition draining the level to the ion ground state and is very short lived) and pressure broadening from the relatively high helium pressure normally used of about 8 torr.

Phase modulation of the He–Se laser using an ADP transverse field modulator has been carried out by Ferrario and Sona (1972) who investigated the AM and FM regions of operation. Their main results were (i) on the 5227 Å line a strong third order beat component in the RF spectrum was observed and the laser had a double FM structure due possibly to an isotope shift of about 350 M Hz between Se^{80} (50% abundance) and Se^{78} (25%), (ii) on the 5227 Å line pulsed emission with a second harmonic repetition rate (c/L) which was stable over a large range of detuning $d = (\nu_m - \nu_0/\nu_0($ (ν_m = modulation frequency, ν_0 = modulation frequency giving max amplitude of pulses in the AM region) (iii) the 5305 Å transition showed strong mode competition giving very stable single mode operation (iv) the line at 5176 Å exhibited similar behaviour to the 5227 Å line.

XIII. Applications

The CW and pulsed noble gas ion lasers, particularly the argon ion laser with its blue–green output, may be considered superficially to complement the CW, low power, red wavelength, high coherence of the He–Ne laser output and the pulsed (or CW), high power, low coherence of the solid state and dye lasers. Ion lasers at present provide as their main radiation characteristics:

(i) Several spectral lines with watts of CW power in the visible region with reasonable coherence properties.
(ii) Additionally, many CW and pulsed laser lines from the UV to the near IR in powers ranging from mW to watts.

(iii) Some useful CW UV lines from transitions in Ar III, Kr III, Ne II and He–Cd.
(iv) Some fairly high energy visible pulsed lines in Xe IV.

These laser transitions are listed in Table I (pp. 177–212) and further described in Section III. The properties of the more useful laser lines have been discussed in Sections XI, A, and XII.

This brief review of applications of these lasers necessarily will be highly selective because of the large body of work over many diverse fields which has been carried out. We will be more concerned here with the relatively recent literature, mainly from 1969 to the present, and attempt simply to highlight a few areas of application of ion lasers. To achieve an effective review of all the applications of ion lasers one or more specific application reviews would be justified.

The main applications of ion lasers, principally the argon ion laser, are in:

(i) Inelastic light scattering—Rayleigh linewidth, Brillouin and particularly Raman scattering.
(ii) Pumping of CW dye lasers.
(iii) Holography.
(iv) Optical memories and optical data processing.
(v) Biological and medical applications.

In addition there have been a myriad of other applications some of which have been mentioned in the introduction. We will for convenience divide this discussion of applications into three sections: (1) physical and chemical research, (2) engineering and (3) biological and medical. Some of the experiments we describe could have been carried out with lasers other than ion lasers, e.g. Nd:YAG or dye lasers, but generally the applications draw on the particular properties of the ion lasers which are described in Sections XI and XII.

A. PHYSICAL AND CHEMICAL RESEARCH

1. *Light Scattering*

Rayleigh (linewidth), Brillouin and Raman inelastic light scattering, using mainly the argon ion laser as an excitation source, is a flourishing field. No attempt will be made here to be complete in this review in covering the large amount of work reported in the literature. Several reviews specifically devoted to light scattering have appeared. Collected papers and reviews on light scattering have been given by, for example, Wright (1969), Chu (1970), Balkanski (1971) and Peticolas (1972). A collection of papers on Rayleigh linewidth scattering has appeared in the Journal de Physique (Vol. 33, Supplement Cl, 1972) and reviews by Cummins and Swinney (1970) and Pike (1970).

In Brillouin scattering reviews have appeared by Fleury (1970), Mountain (1970) and Burstein *et al.* (1971). For Raman scattering two recent collected works are by Gilson and Hendra (1970) and Szmansky (1970). There are reviews by Mooradian (1970b), on semiconductors, Koenig (1971), on synthetic polymers, and (1972), on biological polymers, and Peticolas (1972). At the second International Conference on Light Scattering in Solids (Balkanski, 1971) light scattering using argon ion lasers was reported in systems involving

(i) resonant Raman scattering in perfect crystals and crystals with impurities,
(ii) electronic excitations,
(iii) magnetic excitations,
(iv) polaritons,
(v) phonons and
(vi) phase transitions.

The ability to tune to discrete wavelengths in the ion lasers is useful in distinguishing weak Raman and Brillouin scattering spectra from fluorescence. We can expect that more use will be made of frequency stabilized argon ion lasers (stabilization methods are desctibed in Section XI, E) to improve the laser single mode stability from about 100 MHz to about 50 kHz. The intense Rayleigh wings affect the direct measurement of Raman and Brillouin spectra when $\Delta\bar{\nu} \leqslant \pm 50$ cm^{-1}. An iodine cell filter is often used to filter the scattered light when the argon laser is operating single mode on the 5145 Å line. The argon mode can be tuned, usually by an in-cavity etalon, to match an iodine absorption line of the vibrational–rotational transition 0–43 P(12), R(14) $\mathrm{X}\,^1\Sigma_g^+ - \mathrm{B}^3\Pi_{\mathrm{ou}}^+$ at 19,429·27 cm^{-1} (5145·36 Å). The iodine cell is typically 10 cm in length and held at about 70°C which provides an attenuation of about 10^3 to 10^4 (Harting *et al.*, 1970; Lippert *et al.*, 1971; Peticolas *et al.*, 1971; Hibler *et al.*, 1971; Schoen and Jackson, 1972). Raman spectra down to within 2 cm^{-1} of the exciting line have been recorded (Lippert *et al.*, 1971) with an argon laser and an iodine filter system. The complication of the frequency of the laser mode being time dependent and drifting out of resonance with the iodine line can be overcome by feedback control of the laser. Some of the problems associated with the use of the iodine filter have been described by Hibler *et al.* (1971) and Lippert *et al.* (1971). Structure in the iodine absorption spectrum near 5145 Å has been studied by Ezekiel and Weiss (1968) and Schoen and Jackson (1972) have shown that absorption lines at −3 and +6 GHz limit the useful range of the filter in Brillouin scattering and other absorption lines can slightly distort the Raman spectrum. Other methods of reducing stray light in addition to the iodine cell technique have been compared by Landon and Reed (1972).

Rayleigh linewidth spectroscopy (intensity fluctuation or light beating spectroscopy) is still in a developing stage. Application is being found in sensing fluctuation phenomena in pure liquids, double or triple component solutions and possibly in solids. For a particle or macromolecule in solution the basic forms of motion detectable are translational, rotational and intramolecular fluctuations. The collection of papers on Rayleigh linewidth scattering given at the Colloquium held in Paris in 1971 (*Journal de Physique*, **33**, 1972) has already been cited. The Rayleigh depolarized spectrum from liquids in the range 0 to 150 cm^{-1} has been observed. In molecular liquids the linewidth is made up of a sharp central Lorentzian from reorientation of permanently anisotropic molecules and a broad background due to either intermolecular polarization or interrupted molecular vibration. Fine structure has been seen and attributed to pair correlations between molecules in a velocity gradient produced by a propagating mode (Ben-Reuven and Gershon, 1971) and Rytov theory considers coupling between shear waves and orientational fluctuations. The broad spectrum has been attributed to:

(i) in the Rytov theory, orientation fluctuations and coupling between shear waves,
(ii) rotational motion of permanently anisotropic molecules and intermolecular elastic collisions,
(iii) collisions producing an induced polarizability depending on the intermolecular separation.

The measurement of the temperature, density and dynamics of the lower troposphere has been studied (Benedetti-Michelangeli *et al.*, 1972). The Doppler shifts associated with echoes from atmospheric aerosols when illuminated at night time with a single frequency argon ion laser have been measured. The detection system is made up of a small telescope and a scanning Fabry–Perot interferometer with photon counting and atmospheric Doppler shifts of 1 ms are resolved. The application of the measurement of Doppler frequency shifts impressed on a laser beam to many moving systems is developing to include laser Doppler anemometry, wind tunnel velocities, vibrational analysis and machine tool control.

Brillouin scattering experiments using Fabry–Perot interferometers with plane or spherical reflectors and scanned by pressure changes or piezoelectric movement have been done in many laboratories. Some commercially available Fabry–Perot interferometers use typically a 5-cm gap with spherical reflectors which are piezoelectrically scanned providing a free spectral range of 0·05 cm^{-1} and allow the measurement of Brillouin shifts down to 10^{-2} cm^{-1} (300 MHz). Sandercock (1970) has described the use of stabilized multipass Fabry–Perot interferometers having a contrast ratio several orders of magnitude greater than a single pass interferometer of the same finesse. Also Fabry–Perot interferometers in series have been used (Cannell and Benedek, 1970; Cummins, 1971; Ford *et al.*, 1968). The double-passed stabilized scanning interferometer has been applied by Sandercock to study elastic constants in SbSI (1970), acoustic mode anomalies in the Jahn–Teller phase transition in $DyVo_4$ and $TbVo_4$ (1972b) and to Si and Ge (1972a). Brillouin scattering experiments were described at the latest Paris Conference on Light Scattering in Solids (Balkanski, 1971) on acoustic phonons in metals; rare gas single crystals; CdS and NH_4Br; polymethyl acrylate and succinonitrile MnF_2, CoF_2, $KMnF_3$ and $RbMnF_3$ crystals and rotational contributions to Brillouin spectra.

The argon laser has been used to scatter from excitations in materials where the beam only penetrates a few hundred angstroms (Mooradian, 1970a). When the laser frequency lies close to some interband transitions in the material strong resonance enhancement of the scattering intensity occurs. Application of resonant microwave power at 25·6 GHz to the Ni^{2+} spin system at 2°K heats to 60°K a narrow band of transverse acoustic mode phonons travelling in a given direction (Brya *et al.*, 1968) and observation by Brillouin scattering of a microwave phonon bottleneck in 1 % Ni^{2+} : MgO has been made. Steady state heating of the phonons by the relaxing spins over 20,000 °K in regions of the crystal occurs attributed to a form of stimulated emission of the phonon. The phonon bottleneck was observed in the Brillouin spectrum to persist even at temperatures as high as 37°K.

Several commercial versions of laser Raman spectrometers with double or triple monochromators are available to which the laboratory can add their own laser. This is most often an argon ion or krypton ion laser. The commercial ion lasers have some space for in-cavity insertion of additional elements, e.g. gas cells for Raman spectroscopy. We will consider a few examples from the large amount of Raman scattering research data. Low frequency Raman spectroscopy has been described by Scott (1972). Some of this work includes observation of structural phase transitions in solids and soft modes or temperature dependent phonons in $SrTiO_3$ (Fleury *et al.*, 1968) down to 10 cm^{-1} (2·4 Å), and soft modes in CsH_2As_4 and KH_2As_4 (Katiyar *et al.*, 1971; Cowley *et al.*, 1971). The interaction between optic and acoustic modes have been observed in $BaTiO_3$ by Lazay and Fleury (1971). Other low frequency spectra within 3 Å of the exciting line have been investigated such as in crystals with structural phase transitions, polaritons, electrons in wide-gap semiconductors, phonons and spin-flip scattering.

The high resolution Raman scattering of gases has been considered in a series of papers by Weber *et al.* (1965, 1967), Barrett and Weber (1970) and Walker and Weber (1971) and Weber and Schlupf (1972). The later techniques involve excitation by a single mode argon laser, a multiple-pass Raman cell inside the laser cavity and the use of baked Kodak IIIa-J photographic plates. Changes in the dispersion of the spectrograph by variation in barometric pressure are compensated by changing the wavelength of the exciting single mode of the laser by tuning an in-cavity etalon. The pure rotational and rotational–vibrational spectrum of small molecules (e.g., O_2, N_2, CO_2 and NO) have been recorded. For large molecules rotational spectra have been obtained with a resolution of 7×10^5. The potentiality of the laser linewidth resolution of 10^{-3} cm^{-1} has not yet been reached. The high resolution Raman spectroscopy of gases has been reviewed by Clements and Stoicheff (1970). Gases in combustion engines have been measured to determine spatially resolved distribution of concentration and temperature in a flame.

Lippert and Peticolas (1972) have discussed Raman scattering in polymer systems. They have measured vibrational frequencies in a polymer backbone and calculated local conformational changes. They observed the temperature dependence of the unwinding of the double-helix molecule in double-stranded polynucleotides of DNA and RNA. Also they have studied membranes made up of half protein and half lipid with added cholesterol. The crystalline lattice modes of a polymer were reported by Berio and Cornell (1972) studying orthorbombic polyoxymethylene and attributed the spectrum to out-of-phase translational lattice modes along the crystal *a* and *c* axes. Polymer scattering applications have been reviewed by Koenig (1971, 1972), Schaufele (1970) and Peticolas (1972).

The modulation of electron beams at optical frequencies using a CW argon ion laser has been reported (Schwarz and Hora, 1969) where the light and electron beams cross orthogonally in a dielectric film. The visible light colour was recorded by impinging the electron beam on a nonfluorescent screen. Subsequent attempts to reproduce these experimental results by other workers (for example Hadley *et al.*, 1971) have not been fruitful. However, a body of theoretical work has appeared, stimulated by the reported observation, such that there is good theoretical evidence for various photon-electron interactions under certain conditions.

2. *Spectroscopy*

A thorough review by Moore (1971) has considered the chemical applications of lasers generally. These uses are likely to continue to expand as CW lasers which are tuneable become available over a greater range of wavelength. Several interesting experiments have been made such as the laser-induced high resolution absorption spectroscopy using molecular beams. For example a high resolution spectrum of the hyperfine structure of I_2 has been obtained in a beam excited by the Ar 5145 Å line (Ryan *et al.*, 1972). Also coherent interaction phenomena has been studied using the $^{202}Hg^+$ laser on a ^{87}Rb vapour cell (Gibbs and Slusher, 1970) in which self-induced transparency, pulse break-up and peak intensity amplification by the absorber was reported. We will here mention two more areas in which other work has been carried out using ion lasers: fluorescence spectroscopy and molecular lifetime studies.

Laser-induced fluorescence is useful in allowing molecules to be prepared in known vibrational–rotational states, and their subsequent energy transfer processes followed, and to provide information on molecular structure. The CW tuneable dye lasers now becoming more available will enable these techniques to be extended. Resonance fluorescence of the I_2 molecule has been investigated by Ezekiel and Weiss (1968). Using the Ar 5145 Å line, excitation of the $X\,^1\Sigma_g^+ - B^3\Pi_{ou}^+$ was made and the resonance fluorescence observed from a molecular beam of I_2. The extension of this work to stabilization of the argon laser has been described in Section XI, E. The fluorescence in I_2 has also been studied by Sakurai and Broida (1969a, 1970) and information of the assignment of absorption transitions obtained. Resonance fluorescence of I_2, Br_2, Cl_2, BrCL, ICL and IBr has been obtained by Holzer *et al.* (1970). The high resolution spectroscopy of I_2 has been studied by Ryan *et al.* (1972) and energy transfer in I_2 by Kurzel and Steinfeld (1970).

Demtroder *et al.* (1969) and McClintock *et al.* (1969) have studied resonance fluorescence in Na_2 in the $X\,^1\Sigma_g^+ - B\,^1\Pi_u$ blue–green band system. They have determined spectroscopic constants, potential curves and Franck–Condon factors for the transition and using level-crossing techniques determined the lifetime of the $B^1\Pi_u$ state ($6{\cdot}41 \times 10^{-9}$ sec) and studied molecular hyperfine structure effects. Also radiative lifetimes in K_2, Rb_2 and Cs_2 using an argon laser modulated at 18 MHz and the phase shift technique have been measured (Baumgartner *et al.*, 1970). Fluorescence in Cs_2 and Rb_2 has been studied by McClintock and Balling (1969) using focused argon radiation; the intensity dependence on the incident power was studied and this is related to collisional effects. Saturation of absorption and fluorescence in Na_2 using polarized Ar 4880 Å excitation has been observed by Drullinger and Zare (1969) with saturation occurring at 10 mW cm^{-2}. From these polarized excitation and fluorescence measurements collisional data can be obtained. Laser induced fluorescence with argon ion laser excitation has been used by Sakurai *et al.* (1970) to study spectra of the $A\,^1\Sigma - X\,^1\Sigma$ system of BaO. Rotational and vibrational assignments were made and rotational and vibrational constants for the lower electronic state obtained. Also fluorescence bands in NO_2 have been studied using krypton ion laser radiation (Sakurai and Broida, 1969b). The fluorescence of CuO held in noble gas matrices at 15°K has been studied (Shirk and Bass, 1970). They observed $v' = 1$ fluorescence and compared relaxation times of vibrational internal conversion and radiative decay.

The mode-locked noble gas ion lasers emit a train of pulses typically with a repetition rate $\sim 10^8$ pulses per second, and each pulse having a width as low as 0·2 ns and a peak power $\sim$ 50 W (equivalent to about 10^{-7} J or about 10^{11} quanta per pulse at 5000 Å). The use of this type of source for excitation in fluorescent lifetime studies, either using the pulse train directly in a phase-modulation experiment, or by repetitively selecting single mode-locked pulses, should become common techniques. This is particularly so now that the argon laser can be simultaneously mode-locked and the second harmonic of 5145 at 2573 Å generated, or the 3250 Å line of the He–Cd laser used, to provide near UV source wavelengths. The use of cavity dumping techniques (Maydan, 1970) to produce pulses of width 10 ns, peak power ~ 60 W and at repetition rates up to 20 MHz are also notable. These methods are superior to the direct modulation of the CW laser (Haug *et al.*, 1969). The use of mode-locked lasers in fluorescent lifetime studies has been briefly described by Mack (1968) and Backrach (1972) for lifetimes in the range 1 ns to 80 μs. Backrach has used an acousto-optic deflection system for modulation giving mode-locked pulse widths of 0·25 ns and a time resolution of 0·5 ns. The time dependence and quenching of fluorescence of NO_2 showed that the sharp filament structure was quenched ten times faster than the overall fluorescence (Sakurai and Broida, 1969). Lifetimes of diatomic species have been measured by McClintock *et al.* (1969), Shirk and Bass (1970) and Drullinger and Zare (1969).

3. *Pumping CW Dye Lasers*

The CW dye laser is an important laser system because of its ability to provide reasonable CW laser power ~ watts which is tunable over most of the visible spectrum. Much of the work achieved to date in fluorescence spectroscopy can be extended to many more molecules and interest in extension of the tuning range is great. The attainment of CW dye laser action was for a long time hampered by the problem of build-up of molecules in the first triplet state which absorbed at the laser wavelength and also absorption inducing temperature dependent refractive index changes which caused scatter losses. The first CW dye lasers were reported by Peterson *et al.* (1970a) and Banse *et al.* (1970) and later by Peterson *et al.* (1970b), Hercher and Pike (1971), Kohn *et al.* (1971) and Tuccio and Strome (1972). An argon laser is used as a longitudinal pump for a very small dye cell. The output characteristics are power levels up to a few watts and an efficiency for the dye cell ~ 10%. For example, a rhodamine 6G dye in water excited by a 4 W Ar 5145 Å laser produces about 400 mW output. For longitudinal excitation the threshold pumping rate depends on the optical cavities and the dye, and a value for rhodamine 6 G in methanol is about 30 kW cm^{-2} for 2% loss per pass in the dye optical cavity. For an argon laser operating at 4W, which can be focused to a diffraction limited spot of a few microns, pumping at greater than 1MW cm^{-2} is obtained.

The dye cell resonator is usually of spherical or hemispherical geometry to allow TEM_{00} operation for low loss and a small focal spot size. The absorption length for the radiation is about 0·1 cm for the optimum concentration. The resonator needs to be of sufficient length if tuning is to be possible. Tuccio and Strome (1972) and Hercher and Pike (1971) have designed a resonator of hemispherical form and with an internal lens to ease the position and alignment tolerances on the mirrors. An excitation system in which the dye cell is placed inside the laser cavity has been studied by Kohn *et al.* (1971).

The rhodamine 6 G laser has been tuned over the range 5250 to 7010 Å. Prisms rather than gratings are used as tuning elements because of the need for low loss in the cell resonator. Tuned output linewidths ~ 5 Å (Tuccio and Strome, 1972) for TEM_{00} mode can be further narrowed by an in-cavity etalon. Operation in a single longitudinal mode has been obtained by Hercher and Pike (1971) giving a linewidth of about 35 MHz at 5900 Å. Forced mode-locked pulse trains from an argon pumped rhodamine 6 G CW laser have been measured by Kuizenga (1971) and Dienes *et al.* (1971). Pulse widths of 55 ps were obtained by Dienes *et al.*, for a system with the dye cell inside the argon laser and acousto-modulated by a fused silica crystal. Narrower pulse widths down to a few ps can occur if the whole bandwidth is locked. Kuizenga used a $LiNbO_3$ crystal as a modulator and pulse widths of 100 ps duration were obtained. Passive mode-locking of the CW dye laser has been reported by Shank *et al.* (1972) with pulse widths as low as 1·5 ps in a rhodamine 6G laser mode-locked by a DODC dye.

CW laser pumping beams providing ~ 0·7 W at the UV wavelengths of 3511 and 3638 Å are now available which should allow the extension of the tuneable CW dye systems to shorter wavelengths. The CW dye laser will undoubtedly develop into a powerful new tool for studying high resolution spectroscopy. The use of purpose built lasers, additional to the Ar 5145 Å line, providing high pumping rates at shorter wavelengths should extend the CW dye tuning range to shorter wavelengths.

Other applications of the argon laser as a pump for other lasers have been proposed. Birnbaum *et al.* (1968a, b, 1970) has pumped ruby, Nd^{3+} glass and Ca WO_4 lasers with a pulsed argon laser. The threshold for the ruby laser was 13 mJ and Q-switched operation was obtained with 15 ns output pulses at 300°K and ~ 6 ns at 77°K; the threshold for the Nd^{3+} glass laser was 4 mJ. CW single mode operation of a ruby laser cooled to 77°K and also at 4·2°K has been obtained by end-pumping the ruby laser with Ar 5145 Å CW output. The ruby output was in the TEM_{00} mode with a full spectral width of 35 kHz with no amplitude modulation (spiking) being present compared with that previously observed in CW ruby lasers.

4. *Miscellaneous*

It has been suggested that the mercury–helium laser could be used as a secondary frequency standard both in the visible (Byer *et al.*, 1965) and in the microwave region (Beterov *et al.*, 1969). Because of its large atomic mass, this laser exhibits a very small Doppler width with clearly resolved isotope structure, the Doppler width observed in a pulsed hollow cathode laser is $\sim$ 500 MHz for each isotope component. This is about one-third the Doppler width in a helium–neon laser and is the smallest Doppler width for any gas laser operating in the visible. Beterov *et al.* (1969) obtained a microwave frequency standard by operating a mercury laser on two of the mercury isotope frequencies simultaneously and utilizing the resulting microwave beat frequency. They conclude that the laser frequencies themselves should be capable of stabilization to 1 part in 10^{12} and that a microwave standard better than 1 part in 10^8 should be obtainable.

Up-conversion of IR photons to blue in a low noise, narrow bandwidth detector system practically approaching the theoretical performance limit has been carried out by Smith and Mahr (1970). The mixing crystal was $LiNbO_3$ and the up-converted IR could be tuned between 2·5 and 4 μm by temperature tuning of the crystal. In a focusing system with a matched telescope optimization was achieved with light being up-converted in the diffraction limited overlapping spot. In their experiment a 0·5 W, 5145 Å argon ion laser and a 3·39-μm beam were used to generate light at 4470 Å. The measured efficiency $\sim 10^{-4}$ agreed with theory and a low noise background was found such that possibly 10^{-14} W could be detected in the IR in 1 sec in a bandwidth of 1 cm^{-1}.

Thermal steady-state self-trapping and transient self-focusing are important considerations in parametric oscillators and harmonic generators. Thermal self-focusing of CW argon laser beams has been studied by Dabby and Whinnery (1969), Carman *et al.* (1969), Akhmanov *et al.* (1970) and Dabby *et al.* (1970). Akhmanov *et al.* found that the critical power is only a few hundred mW in glasses and crystals and a minimum diameter developed $\sim$ 50 μm in optical glasses. The investigation of absorption by the focusing or defocusing of the laser beam (Akhmanov *et al.*, 1970; Dabby *et al.*, 1969; Smith, 1969) is sensitive and useful, particularly if the dependence of the refractive index on the electric field of the light is small. Mooradian (1970a) has discussed thermal trapping in argon laser beams in glass rods. The CW beam propagating in the slightly absorptive medium collapses to a small diameter as a radial termpeature gradient develops across the beam and a refractive index which increases with temperature. Dabby *et al.* (1970) have shown that self-phase modulation of CW 1 W argon ion laser beams in a medium which possesses a temperature-dependent refractive index produces far-field aberration rings. This self-induced frequency modulation was also observed with a pulsed 1·5 W argon laser and chirping was detected.

The argon laser has been used in radiation pressure (optical levitation) experiments by Ashkin (1972). Small particles have been moved in liquid media and accelerations up to 10^6 g obtained. For the Ar 5145 Å line having a

1W output focused to a diameter of 2μm a force on a particle would be exerted of 10^{-3} dyne if the particle were perfectly reflecting. If the particle diameter was also 2 μm than the acceleration of 10^6 g would follow. The separation of 0·5 μm plastic spheres from 2·5 μm spheres has been observed. A possible extension is to the separation of viruses, macromolecules or cells in liquids. The separation effect would also depend on refractive index, shape and orientation and hence may be more specific than the ultracentrifuge. The optical levitation may be applicable to holding small particles for experiments, such as D-T pellets for laser produced plasmas and for particle size analysis by scattering. For absorbing particles Ashkin has mentioned application in gas separation and atomic beam velocity analysis. Other possible applications for these forces include the acceleration to high velocity of neutral particles and the separation of isotopes.

Cross-linking of polymethylmethacrylate has been induced by 3250 Å radiation from the He–Cd^+ laser (Tomlinson *et al.*, 1970). An earlier report was made of polymerization using an Ar 4880 Å laser acrylamide with an added sensitizer; the authors suggest oxidation products in the polymer UV, producing free radicals which form cross-links between adjacent chains.

The use of an argon laser in the study of gas flow by a Schlieren technique has been described by Webster and Graham (1968). High speed (10^7 pps) Schlierin photography can be carried out on gas flows with exposures down to 50 ns.

B. ENGINEERING APPLICATIONS

Some industrial applications of lasers have been dealt with in previous books and review articles (Gaglinano *et al.*, 1969; Charschan, 1973; Beesley, 1971; Shulman, 1970; Preston, 1972). We will briefly discuss here holographic memories and a laser TV display system as examples of industrial applications.

1. *Holography*

Holography with argon lasers has been carried out by Moran (1971a, b) in which he considers the aberrations involved in changing from a recording wavelength, say Ar 4880 Å, to a reconstruction wavelength, say ruby 6943 Å. A thorough review of current and proposed technology for large storage systems has been given by Matick (1972) in which optical and other memory systems are compared. The design principles of two holographic memories have been considered by Mikaeliane *et al.* (1970) based on: (i) consecutive storage of individual numbers on the same hologram and (ii) parallel storage of many binary numbers on individual holograms. The argon laser is usual for optical memory use because of its fairly high CW power and useful wavelengths. Optical memories have been described by La Macchia (1969)—a random access holographic storage optical memory with storage of > 10^8 bits which is better than the limit of other non-optical systems; by Meyerhofer (1970)—a pulsed argon laser and holographic storage; and by MacDonald and Beck (1969). Some commercial systems, e.g. the UNICORN and FM 390, are available. The advantages of holography over simple microimaging techniques is mainly of spatial redundancy and simplified optics but there are difficulties in the uniformity of the holographic reconstruction.

A read-only holographic optical memory providing fairly fast (μs) random access to large quantities of data has been constructed (Anderson *et al.*, 1967). A high speed light deflection system directs an argon laser beam to one hologram in a large two-dimensional array. The contents of the addressed hologram in the form of a dot pattern are projected onto a photodetector array for read-out. The system of Anderson *et al.* used cascaded lead molybdate acousto-optic deflection with the holograms constructed in dichromated gelatin and a phototransistor array or photosensitive flip-flop for read-out. A holographic memory system for reading, storage, writing and erasing based on an argon laser, a liquid crystal page composer, a thermoplastic-photoconductor array and electro-acousto deflection provided random access at high speed (Rajchman, 1973). Reading is by detecting real images from the storage hologram with an array of PIN diodes and erasure is by heating the necessary part of the holographic array. Recording and reproducing digital data holographically at 300 Mbits/sec using mode-locked cavity dumped argon lasers has recently been carried out by Roberts *et al.* (1973) with a laser producing 2 ns pulses at $2{\cdot}3 \times 10^5$ pulses per second.

The first multi-colour image to be reconstructed with laser light from a transmission hologram was by Pennington and Lin (1965) using the He–Ne 6328 Å and the Ar 4880 Å lines to reconstruct the image. Also the Lippman process was used by Lin *et al.* (1966) to achieve a multi-colour image from a reflection hologram. Ion lasers offer several, medium power, CW wavelengths for multi-colour holography. The availability of quite high CW powers up to $\sim$ 6 W at 5145 Å in the argon laser allows shorter exposure times or larger objects to be used in holographic applications.

2. *Display*

A large screen laser colour TV projector has been used to obtain projection of bright images with high resolution onto a large screen (Yamada *et al.*, 1970). Three lasers were used to generate the three primary colours, red (Kr 6471 Å), green (Ar 5145 Å) and blue (Ar 4880 Å) with a total CW power of 12 W. Each beam was intensity modulated separately by a birefringent electro-optic modulator made up of DKDP crystals in a five-stage cascade structure. The modulator had an effective diameter of 4 mm and a half-wave voltage of 800 V at 6471 Å. The three beams were combined into one by dichroic mirrors.

This multi-colour beam was deflected by a rotating mirror scanner to form a raster on a 3 × 4 metre screen. The horizontal scanner is a 16-face polygon rotating at 5900 rpm and the vertical scanner a 24-face polygon driver at 150 rpm. The projector had a resolving power of 500 lines vertically and 400 lines horizontally with a resolution depending on the laser spot size and video-amplifier bandwidth. The illumination on the screen was 50 lux and a contrast of 30 to 1 was obtained.

The scan laser has been discussed earlier in Section XI as a possible scanning system based on the use of wide bore ion laser tubes and has been considered by Pole and Myers (1968) and Forsyth (1969).

Other possible engineering applications of ion lasers are in optical data processing and computer links; machining and cutting; production of micro-circuits and resistor trimming; high speed photography; underwater viewing; surveying, radar and ranging; communications (particularly deep-space) and guidance systems.

C. BIOLOGICAL AND MEDICAL APPLICATIONS

The argon ion laser has several useful properties which make it suitable for investigation of biological systems. These include the pulsed and CW capability, the availability of many wavelengths which fall in the sensitivity region of biological materials and a wide and suitable range of output powers. Also there are available several specific organelle binding dyes which absorb in the blue-green region and which are non-toxic and which allow selectivity in the activated site.

1. *Photocoagulation*

Although photocoagulation by lasers is still in an investigatory stage photo-coagulators using the argon ion laser have been developed (for example, the Coherent Radiation Inc. model 800) to perform the functions performed previously by xenon arc lamps. The blue-green light of the argon ion laser is absorbed by the pigment epithelium and choroid and by haemoglobin in the blood. The advantages of the argon ion laser photocoagulator over the xenon arc have been discussed by L'Esperance (1968) and Little *et al.* (1970) and may be listed as:

(i) High CW powers.
(ii) No IR or long wavelength background radiation so that the total incident power is lower by a factor of about 6.
(iii) High spatial coherence provides light collimation and low divergence.
(iv) Ability to focus a variable spot size from μm to mm with uniformity of exposed areas.

(v) Able to use electro-mechanical control of the exposure; since the laser can be operated CW, longer exposures than with, for example, the pulsed ruby laser can be made with reduced shock wave effects.

The blue-green wavelengths are at the peak of absorption of the pigment epithelium and haemoglobin such that nerve fibres and nuclear layers are largely unaffected. Coagulation is made with lower energy than with the xenon arc and the lower total powers into the eye means that the danger of zone effects is less. Vessels in the vitreous and on the papilla can be treated with little risk of destroying the underlying retina or optic nerve. High transmission in the ocular media and retinal pigment epithelium mean that the retina can be treated with less power. The argon ion laser has been applied to such disorders as holes and retinal tears, macular disorders, ophthalmic vascular diseases (mainly diabetic retinopathy, sickle-cell retinopathy and Eale's disease). Complications such as secondary macular edema and vitreous retraction are absent and no retrobulbar anesthesia is necessary. Commercial argon ion laser photocoagulators are available complete with an aiming beam, an articulating arm and a delivery system.

2. *Cells*

The selective irradiation of cells can be accomplished fairly readily with laser beams because of their coherence and low divergence, allowing easy focusing and directionality, with high enough intensity to interact with a low concentration of natural or artificial chromophores and monochromatic for matching to absorption bands or staining dyes. A tuneable laser would also be useful for selective excitation. The argon ion laser has main wavelengths which match absorption peaks in natural chromophores such as haemoglobin, photosynthetic pigments, carrotenoids and cytochromes. Since the laser output is visible alignment can be made directly.

A pulsed argon ion laser producing about 35 W peak power at 4880 and 5145 Å has been used to study the formation of lesions *in vitro* on predetermined sites of selected chromosomes (Berns *et al.*, 1969a, b, c; Berns and Rounds, 1973). A culture system was used of salamander lung tissue explants and acridine orange as a photosensitizing agent which binds to the lysosomes and chromosomes. Lesions of varying severity were produced on selected sites *in vitro* of up to 1 μm size from the focused beam. The system is also being used to selectively place lesions in large sarcosomes (mitochondria) of beating rat myocardial cells in tissue culture without prior treatment of the photosensitizing agent. Sacrosomes containing the natural chromophores cytochrome c and c_1 absorbing at 5200 and 5500 Å from the heart muscle of rats have been studied (Berns and Rounds, 1973). Possible areas of application are in studying genes in chromosomes, mapping genes and induction of changes in beat frequency and contractibility in mitochondria in studying cell function and pathological response. The potential of the argon laser to perform selective morphological and functional alteration of nucleoli of tissue culture cells has been investigated by Berns *et al.* (1969b). The action of the laser can be made highly selective to the photosensitized structure and entire nucleoli can be "turned off" functionally or individual parts of a single nucleolus altered. The quantitative determination of the fluorescence from cells containing a fluorochrome and excited by an argon ion laser in a high speed flow system has been described by Vandilla *et al.* (1969).

3. *Miscellaneous*

Holograms have been made of the fundus of the eye of an anaesthetized cat using a 2 W CW argon laser (Wiggins *et al.*, 1972). As used in this work the 2 W power level is within the damage threshold and hence holograms of human fundi may possibly be made safely without anaesthesia.

Health hazards associated with the argon ion laser have been discussed by L'Esperance (1968, 1969), Campbell *et al.* (1969) and Clarke (1970). Collected works on biological and medical applications have appeared by McGuff (1966), Goldman (1967), Goldman and Rockwell (1972) and Wolbarsht (1971).

In this brief review of applications we have omitted to describe several fields where ion lasers have been applied. Some of these are interferometry and optical testing; local heating; applications in solid state physics and gaseous electronics; communications (particularly deep-space), optical ranging, alignment and surveying; underwater TV; production of microcircuits and resistor trimming; guidance systems and rotation sensing ring lasers; cutting and machining and the entertainment industry. Several of these are discussed in quite recent publications such as Charschan (1973), Beesley (1971), Bloom (1971), Shulman (1970) and Gagliano *et al.* (1969).

References to Tables I.1–I.33

1. Cheo, P. K. and Cooper, H. G. (1965a). *J. Appl. Phys.* **36**, 1862–1865.
2. Goldsmiths and Kaufman, A. S. (1963). *Proc. Phys. Soc.* **81**, 544–552.
3. de Bruin, T. L. (1932). *Z. Phys.* **77**, 505–514.
4. Bridges, W. B. and Chester, A. N. (1965a). *Appl. Opt.* **4**, 573–580.
5. Persson, W. (1971). *Physica Scripta* **3**, 133–155.
6. Dana, L., Laures, P. and Rocherolles, R. (1965). *C.R. Acad. Sci. Paris*, **260**, 481–484.
7. Bridges, W. B., Freiberg, R. J. and Halsted, A. S. (1967). *I.E.E.E. J.Q.E.* (Notes and Lines) **QE-3**, 339.
8. Bridges, W. B. and Chester, A. N. (1965b). *I.E.E.E. J.Q.E.* **QE-1**, 66–84.
9. Rosenthal, A. H. (1930). *Ann. Phys.* **4**, 49–81.
10. Minnhagen, L. and Stigmark, L. (1957). *Ark. Fys.* **13**, 27–36.
11. McFarlane, R. A. (1964). *Appl. Optics* 3, 1196.
12. Minnhagen, L. (1963). *Ark. Fys.* **25**, 203–284.
13. Harrison, G. R. (1969). "M.I.T. Wavelength Tables." M.I.T. Press, Cambridge, Mass.

14. Denis, A. and Dufay, M. (1969). *Phys. Lett.* **29A**, 170.
15. Davis, C. C. and King, T. A. (1972). *I.E.E.E. J.Q.E.* (Corr.) **QE-8**, 755–757.
16. Striganov, A. R. and Sventitski, N. S. (1968). Tables of Spectral Lines of Neutral and Ionised Atoms. Trans. from Russian IFI/Plenum, New York.
17. Bell, W. E. and Bloom, A. L. Unpublished data supplied to Bridges and Chester, 1965. A. L. Bloom, private communication to Bridges and Chester, 1969.
18. Bridges, W. B. (1964). *Appl. Phys. Lett.* **4**, 128–130 and (1964) *Appl. Phys. Lett.*, **5**, 39 (Erratum).
19. Convert, G., Armand, M. and Martinot-Lagarde, P. (1964). *C.R. Acad. Sci., Paris* **258**, 4467–4469.
20. Gordon, E. I., Labuda, E. F. and Bridges, W. B. (1964). *Appl. Phys. Lett* **4**, 178–180.
21. Bennett, W. R., Jr., Knutson, J. W., Jr., Mercer, G. N. and Detch, J. L. (1964). *Appl. Phys. Lett.* **4**, 180–182.
22. Heard, H. G. and Peterson, J. (1964a). *Proc. I.E.E.E.* (Corr.) **52**, 1049–1050.
23. Convert, G., Armand, M. and Martinot-Lagarde, P. (1964). *C.R. Acad. Sci. Paris* **258**, 3259–3260.
24. Birnbaum, M. and Stocker, T. L. Unpublished.
25. Kayser, H. and Konen, H. (1932). "Handbuck der Spectroscopie." S. Hirzel Valaj, Leipzig, Germany, Vol. 8, Pt. 1, p. 97.
25a. Ibid., Vol. 6. p. 819.
26. Sinclair, D. C. (1965). *J. Opt. Soc. Am.* **55**, 571–572.
27. Horrigan, F. A., Koozekanani, S. H. and Paananen, R. A. (1965). *Appl. Phys. Lett.* **6**, 41–43.
28. Paananen, R. A. (1966). *Appl. Phys. Lett.* **9**, 34–35.
29. Hernqvist, K. G. and Fendley, J. R., Jr. (1967). Proc. Symp. Modern Optics (Ed. J. Fox), Vol. 17, pp. 383–387. Polytechnic Press, Brooklyn, New York.
30. Fendley, J. R., Jr. (1968). *I.E.E.E. J.Q.E.* **QE-4**, 627–631.
31. Bridges, W. B. and Halsted, A. S. (1966). *I.E.E.E. J.Q.E.* (Notes and Lines) **QE-2**, 84.
32. Neusel, R. H. (1967). *I.E.E.E. J.Q.E.* (Notes and Lines) **QE-3**, 207–208.
33. Ericsson, K. G. and Lidholt, L. R. (1967). *I.E.E.E. J.Q.E.* (Corr.) **QE-3**, 94.
34. Hodges, D. T. and Tang, C. L. (1970). *I.E.E.E. J.Q.E.* (Corr.) **QE-6**, 757–758.
35. Minnhagen, L., Strihed H. and Petersson, B. (1969). *Ark. Fys.*, **39**, 471–493.
36. Humphreys, C. J. Unpublished data supplied to Bridges and Chester (1965).
37. Laures, P., Dana, L. and Frapard, C. (1964). *C.R. Acad. Sci. Paris* **258**, 6363–6365.
38. Dana, L. and Laures, P. (1965). *Proc. I.E.E.E.* (Corr.) **53**, 78–79.
39. Bridges, W. B. (1964). *Proc. I.E.E.E.* (Corr.) **52**, 843–844.
40. Moore, C. E., Natl. Bur. Std. (U.S.), Art. No. 467, Vol. 1 (1949); Vol. II (1952); Vol. III (1958).
41. Coleman, C. D., Bozman, W. R. and Meggers, W. F., Table of Wavenumbers *Natl. Bur. Std.* (*U.S.*), Monograph 3 (1960).
42. Dahlquist, J. A. (1965). *Appl. Phys. Lett.* **6**, 193–194.
43. de Bruin, T. L., Humphreys, C. J. and Meggers, W. F. (1933). *J. Res. Natl. Bur. Std.* (*U.S.*) **11**, 409–440.
44. Labuda, E. F. and Johnson, A. M. (1966). *I.E.E.E. J.Q.E.* (Corr.) **QE-2**, 700–701.
45. Johnson, A. M. and Webb, C. E. (1967). *I.E.E.E. J.Q.E.* (Notes and Lines) **QE-3**, 369.
46. Neusel, R. H. (1966e). *I.E.E.E. J.Q.E.* (Notes and Lines) **QE-2**, 758.
47. Bridges, W. B. (unpublished); Humphreys, C. J. Unpublished.
48. Cottrell, T. H. E., Sinclair, D. C. and Forsyth, J. M. (1966). *I.E.E.E. J.Q.E.* (Notes and lines) **QE-2**, 703.
49. Neusel, R. H. (1966d). *I.E.E.E. J.Q.E.* (Notes and Lines) **QE-2**, 334.
50. Neusel, R. H. (1966b). *I.E.E.E. J.Q.E.* (Notes and Lines) **QE-2**, 106.
51. Bell, W. E. (1965). *Appl. Phys. Lett.* **7**, 190–191.
52. der Agobian, R., Otto, J.-L., Cagnard, R., Barthelemy, J. and Echard, R. (1965). *C.R. Acad. Sci.* (*Paris*), **260**, 6327–6329.
53. Bridges, W. B. and Chester, A. N. (unpublished).
54. Laures, P., Dana, L. and Frapard, C. (1964). *C.R. Acad. Sci. Paris*, **259**, 745–747.

55. Gallardo, M., Garavaglia, M., Tagliaferri, A. A. and Gallego Lluesma, E. (1970). *I.E.E.E. J.Q.E.*, **QE-6**, 745–747.
56. Humphreys, C. J. (1936). *J. Res. Natl. Bur. Std.* (*U.S.*) **16**, 639–648.
57. McFarlane, R. A. (1964). *Appl. Phys. Lett.* **5**, 91–93.
58. Heard, H. G., and Peterson, J. (1964). *Proc. I.E.E.E.* (Corr.) **52**, 1050.
59. Bridges, W. B. and Mercer, G. N. (1969). *I.E.E.E. J.Q.E.* (Corr.) **QE-5**, 476–477.
60. Bridges, W. B. and Chester, A. N. (1971b). *I.E.E.E. J.Q.E.* (Notes and Lines) **QE-7**, 471–472.
61. Jarrett, S. M. and Barker, G. C. (1969). *I.E.E.E. J.Q.E.* (Notes and Lines) **QE-5**, 166.
62. Humphreys, C. J. (1939). *J. Res. Natl. Bur. Std.* (*U.S.*) **22**, 19–53.
63. Humphreys, C. J., Meggers, W. F. and de Bruin, T. L. (1939). *J. Res. Natl. Bur. Std.* (*U.S.*) **23**, 683–699.
64. Tell, B., Martin, R. J. and MacNair, D. (1967). *I.E.E.E. J.Q.E.* (Notes and Lines) **QE-3**, 96.
65. Goldsborough, J. P. and Bloom, A. L. (1967). *I.E.E.E. J.Q.E.* (Notes and Lines) **QE-3**, 96.
66. Wheeler, J. P. (1971). *I.E.E.E. J.Q.E.* (Notes and Lines) **QE-7**, 429.
67. Neusel, R. H. (1966a). *I.E.E.E. J.Q.E.* (Notes and Lines) **QE-2**, 70.
68. Hoffmann, V. and Toschek, P. (1970). *I.E.E.E. J.Q.E.* (Notes and Lines) **QE-6**, 757.
69. Andrade, O., Gallardo, M. and Bockasten, K. (1967). *Appl. Phys. Lett.* **11**, 99–100.
70. McFarlane, R. A. (1964). *Appl. Optics.* **3**, 1196.
71. McFarlane, R. A. Unpublished data supplied to Bridges and Chester (1965b).
72. Kiess, C. C. and de Bruin, T. L. (1939). *J. Res. Natl. Bur. Std.* (*U.S.*) **23**, 443–470.
73. Bell, W. E., Bloom, A. L. and Goldsborough, J. P. (1965). *I.E.E.E. J.Q.E.* (Corr.) **QE-1**, 400.
74. Cheo, P. K. and Cooper, H. G. (1965b). *Appl. Phys. Lett.* **7**, 202–204.
75. Zarowin, C. B. (1966). *Appl. Phys. Lett.* **9**, 241–242.
76. Goldsborough, J. P., Hodges, E. B. and Bell, W. E. (1966). *Appl. Phys. Lett.* **8**, 137–139.
77. Carr, W. C. and Grow, R. W. (1967a). *Proc. I.E.E.E.* (Lett.) **55**, 726.
78. Martin, W. C. and Corliss, C. H. (1960). *J. Res. Natl. Bur. Std.* (*U.S.A.*) **64A**, 443 477.
79. Jensen, R. C. and Fowles, G. R. (1964). *Proc. I.E.E.E.* (Corr.) **52**, 1350.
80. Fowles, G. R. and Jensen, R. C. (1964). *Proc. I.E.E.E.* (Corr.) **52**, 851–852.
81. Fowles, G. R. and Jensen, R. C. (1964). *Appl. Optics* **3**, 1191–1192.
82. Koval'chuk, V. M. and Petrash, G. G. (1966). *J.E.T.P. Lett.* **4**, 144–146.
83. Fowles, G. R. and Jensen, R. C. (1965). *Phys. Rev. Lett.* **14**, 347–348.
84. Collins, G. J., Kuno, H., Hattori, S., Tokutome, K., Ishikawa, M. and Kamiide, N. (1972). *I.E.E.E. J.Q.E.* (Notes and Lines) **QE-8**, 679–680.
85. Piper, J. A., Collins, G. J. and Webb, C. E. (1972). *Appl. Phys. Lett.* **21**, 203–205.
86. Willett, C. S. and Heavens, O. S. (1966). *Optica Acta* **13**, 271–274.
87. Willett, C. S. (1967). *I.E.E.E. J.Q.E.* (Notes and Lines) **QE-3**, 33.
88. Willett, C. S. and Heavens, O. S. (1967). *Optica Acta* **14**, 195–197.
89. Berezin, I. A. (1969). *Opt. Spect.* **26**, 466–467.
90. Cooper, H. G. and Cheo, P. K. (1966). *I.E.E.E. J.Q.E.* (Notes and Lines) **QE-2**, 785.
91. Rao, A. B. (1938). *Ind. J. Phys.* **35**, 399–405.
92. Martin, W. C. and Tech, J. L. (1961). *J. Opt. Soc. Am.* **51**, 591–594.
93. Keeffe, W. M. and Graham, W. J. (1965). *Appl. Phys. Lett.* **7**, 263–264.
94. Keeffe, W. M. and Graham, W. J. (1966). *Phys. Lett.* **20**, 643.
95. Ölme, A. (1970). *Phys. Scripta.* **1**, 256–260.
96. Carr, W. C. and Grow, R. W. (1967). *Proc. I.E.E.E.* (Lett.) **55**, 1198.
97. Silfvast, W. T., Fowles, G. R. and Hopkins, B. D. (1966). *Appl. Phys. Lett.* **8**, 318–319.
98. Silfvast, W. T. (1969). *Appl. Phys. Lett.* **15**, 23–25.
99. Hodges, D. T. (1971). *Appl. Phys. Lett.* **18**, 454–456.
100. Cahuzac, P. (1972). *I.E.E.E. J.Q.E.* (Notes and Lines) **QE-8**, 500.
101. Risberg, P. (1955). *Ark. Fys.* **9**, 483–494.
102. Li, H. and Andrew, K. L. (1971). *J. Opt. Soc. Am.* **61**, 96–109.

103. Palenius, H. P. (1966). *Appl. Phys. Lett.* **8**, 82–83.
104. Bowen, I. S. (1934). *Phys. Rev.* **45**, 401.
105. Toresson, Y. G. (1960). *Ark. Fys.* **18**, 389–416.
106. Shenstone, A. G. (1961). *Proc. Roy. Soc.*, **A261**, 153–174.
107. Toresson, Y. G. (1959). *Ark. Fys.* **17**, 179–192.
108. Palenius, H. P. (1970). *Phys. Scripta.* **1**, 113–135.
109. Palenius, H. P. (1969). *Ark. Fys.* **39**, 15–64.
110. Martin, W. C. (1959). *J. Opt. Soc. Am.* **49**, 1071–1085.
111. Shenstone, A. G. (1963). *Proc. Roy. Soc.* **A276**, 293–307.
112. Bell, W. E., Bloom, A. L. and Goldsborough, J. P. (1966). *I.E.E.E. J.Q.E.* (Notes and Lines) **QE-2**, 154.
113. Deech, J. S. and Sauders, J. H. (1968). *I.E.E.E. J.Q.E.* (Notes and Lines) **QE-4**, 474.
114. Cooper, H. G. and Cheo, P. K. (1966). Physics of Quantum Electronics Conference (San Juan, Puerto Rico 1965), pp. 690–697. McGraw-Hill.
115. Fowles, G. R., Silfvast, W. T. and Jensen, R. C. (1965). *I.E.E.E. J.Q.E.* (Corr.) **QE-1**, 183–184.
116. Gilles, M. (1931). *Ann. Phys.* **(10)**, **15**, 267–410.
117. Bloch, L. and Bloch, E. (1929). *Ann. Phys.* **(10)**, **12**, 5–22.
118. Hunter, A. (1934). *Phil. Trans. Roy. Soc.* **A233**, 303–326.
119. Bowen, I. S. and Millikan, R. A. (1925). *Phys. Rev.* **25**, 591.
120. Berry, H. G., Schectman, R. M., Martinson, I., Bickel, W. S. and Bashkin, S. (1970). *J. Opt. Soc. Am.* **60**, 335–344.
121. Bromander, J. (1969). *Ark. Fys.* **40**, 257–274.
122. Heard, H. G. and Peterson, J. (1964). *Proc. I.E.E.E.* (Corr.) **52**, 1258.
123. Birnbaum, M., Tucker, A. W., Gelbwachs, J. A. and Fincher, C. L. (1971). *I.E.E.E. J.Q.E.* (Notes and Lines) **QE-7**, 208.
124. Gadetskii, N. P., Tkach, Yu. V., Slezov, V. V., Bessarab, Ya. Ya. and Magda, I. I. (1971). *J.E.T.P. Letts.* **14**, 101–103.
125. Bockasten, K. (1955). *Ark. Fys.* **9**, 457–481.
126. Eriksson, K. B. S. (1958). *Ark. Fys.* **13**, 303–329.
127. Hallin, R. (1966). *Ark. Fys.* **32**, 201–210.
128. Hashino, Y., Katsuyama, Y. and Fukuda, K. (1972). *Jap. J. Appl. Phys.* **11**, 907.
129. Allen, R. B., Starnes, R. B. and Dougal, A. A. (1966). *I.E.E.E. J.Q.E.* (Notes and Lines) **QE-2**, 334.
130. Gerritsen, H. J. and Goedertier, P. V. (1964). *J. Appl. Phys.* **35**, 3060–3061.
131. Gerritsen, H. J. Unpublished data supplied to Bridges and Chester (1965b).
132. Bell, W. E. (1964). *Appl. Phys. Lett.* **4**, 34–35.
133. Bloom, A. L., Bell, W. E. and Lopez, F. O. (1964). *Phys. Rev.* **135**, A578–A579.
134. Heard, H. G., Makhov, G. and Peterson, J. (1964). *Proc. I.E.E.E.* (Corr.) **52**, 414.
135. Byer, R. L., Bell, W. E., Hodges, E. and Bloom, A. L. (1965). *J. Opt. Soc. Am.* **55**, 1598–1602.
136. Bockasten, K., Garavaglia, M., Lengyel, B. A., and Lundhohn, T. (1965). *J. Opt. Soc. Am.* **55**, 1051–1053.
137. Goldsborough, J. P. and Bloom, A. L. (1969). *I.E.E.E. J.Q.E.* **QE-5**, 459–460.
138. Schuebel, W. K. (1971). *I.E.E.E. J.Q.E.* (Corr.) **QE-7**, 39–40.
139. Cahuzac, Ph. (1971). *J. Phys.* **32**, 499–505.
140. Meggers, W. F. (1967). *J. Res. Natl. Bur. Std.* (*U.S.*) **71A**, 396–544.
141. Cahuzac, Ph. (1968). *Phys. Lett.* **27A**, 473–474.
142. Cahuzac, Ph. and Brochard, J. (1969). *J. Phys.* **30**, Cl-81–Cl-82.
143. Cahuzac, Ph. (1970). *Phys. Lett.* **31A**, 541–542.
144. Crooker, A. M. and Dick, K. A. (1968). *Can. J. Phys.* **46**, 1241–1251.
145. Martin, W. C. and Kaufman, V. (1970). *J. Res. Nat. Bur. Std.* (*U.S.*) **74A**, 11–22.
146. Fowles, W. R. and Silfvast, W. T. (1965). *I.E.E.E. J.Q.E.* (Corr.) **QE-1**, 131.
147. Jensen, R. C., Collins, G. J. and Bennett, W. R., Jr. (1969). *Phys. Rev. Lett.* **23**, 363–367.
148. Bloom, A. L. and Goldsborough, J. P. (1970). *I.E.E.E. J.Q.E.* (Notes and Lines) **QE-6**, 164.

149. Sugawara, Y. and Tokiwa, Y. (1970). *Jap. J. Appl. Phys.* **9**, 588–589.
150. Sugawara, Y., Tokiwa, Y. and Iijima, T. (1970). *Jap. J. Appl. Phys.* **9**, 1537.
151. Schuebel, W. K. (1970b). *I.E.E.E. J.Q.E.* (Notes and Lines) **QE-6**, 654–655.
152. Jensen, R. C., Collins, G. J. and Bennett, W. R., Jr. (1971). *Appl. Phys. Lett.* **18**, 50–51.
153. Handrup, M. B. and Mack J. E. (1964). *Physica* **30**, 1245–1275.
154. Webb. C. E. (1968). *I.E.E.E. J.Q.E.* (Corr.) **QE-4**, 426–427.
155. Silfvast, W. T. and Klein, M. B. (1972). *Appl. Phys. Lett.* **20**, 501–504.
156. Watanabe, S., Chihara, M. and Ogura, I. (1972). *Jap. J. Appl. Phys.* **11**, 600.
157. Bockasten, K. (1956). *Ark. Fys.* **10**, 567–582.
158. Walter, W. T., Solimene, N., Piltch, M. and Gould, G. (1966). *I.E.E.E. J.Q.E.* **QE-2**, 474–479.
159. Edlen, B. and Risberg, P. (1956). *Ark. Fys.* **10**, 553–566.
160. Fowles, G. R. and Hopkins, B. D. (1967). *I.E.E.E. J.Q.E.* (Notes and Lines) **QE-3**, 419.
161. Goldsborough, J. P. (1969). *I.E.E.E. J.Q.E.* (Notes and Lines) **QE-5**, 133.
162. Silfvast, W. T. (1968). *Appl. Phys. Lett.* **13**, 169–171.
163. Csillag, L., Janossy, M. and Salmon, T. (1970). *Phys. Lett.* **31A**, 532–533.
164. Schuebel, W. K. (1970). *Appl. Phys. Lett.* **16**, 470–472.
165. Tibiloz, A. S. (1965). *Opt. Spec.* **19**, 463–464.
166. Schuebel, W. K. (1970). *I.E.E.E. J.Q.E.* (Corr.) **QE-6**, 574–575.
167. Csillag, L., Itagi, V. V., Janossy, M. and Rozsa, K. (1971). *Phys. Lett* **34A**, 110–111.
168. Silfvast, W. T. and Klein, M. B. (1970). *Appl. Phys. Lett.* **17**, 400–403.
169. Klein, M. B. and Silfvast, W. T. (1971). *Appl. Phys. Lett.* **18**, 482–485.
170. Martin, D. C. (1935). *Phys. Rev.* **48**, 938–944.
171. Krishnamurty, S. G. and Rao, K. R. (1935). *Proc. Roy. Soc.* **A149**, 56–70.
172. Bartelt, O. (1934). *Z. Phys.* **91**, 444–470.
173. Hernqvist, K. G. and Pultorak, D. C. (1972). *Rev. Sci. Inst.* **43**, 290–292.
174. Ferrario, A. and Querzola, B. (1971). *Opt. Comm.* **3**, 161–164.
175. Karabut, E. K., Mikhalevskii, V. S., Papakin, V. F. and Sem, M. F. (1970). *Sov. Phys. Tech. Phys.* **14**, 1447–1448.
176. Keidan, V. F. and Mikhalesvkii, V. S. (1968). *J. Appl. Spec.* **9**, 1154.
177. Bridges, W. B. and Mercer, G. N. (1969). Tech. Rept. ECOM-0229-F Hughes Research Laboratory, Malibu, California (DDC No. AD-861927).
178. Jensen, R. C. and Bennett, W. R., Jr. (1968). *I.E.E.E. J.Q.E.* **QE-4**, 356.
179. Aleinikov, V. S. (1970). *Opt. Spec.* **28**, 15–17.
180. *Electronics*, page 17, Jan. 24, 1964.
181. Silfvast, W. T. Private communications to Willett (1971) and Bridges and Chester (1971a).
182. Bloom, A. L. (1968). Private communication to Bridges and Chester (1971a) (9/5/1968).
183. Levinson, G. R., Papulovskiy, V. F. and Tychinskiy, V. P. (1968). *Rad. Eng. Elec. Phys.* **13**, 578–582.
184. Tolkachev, V. A. (1968). *J. Appl. Spect.* **8**, 449–451.
185. Papayoanou, A. and Gumeiner, I. M. (1970). *Appl. Phys. Lett.* **16**, 5–8.
186. Simmons, W. W. and Witte, R. S. (1970). *I.E.E.E. J.Q.E.* **QE-6**, 466–469.
187. Leonov, R. K., Protsenko, E. D. and Sapunov, Yu. M. (1966). *Opt. Spect.* **21**, 141–142.
188. Piper, J. A. and Webb, C. E. (1973). *J. Phys.* **B. 6**, L116–L120.
189. Hashino, Y., Katsuyama, Y. and Fukuda, K. (1973). *Jap. J. Appl. Phys.* **12**, 470.
190. Waynant, R. W. (1973). *Appl. Phys. Lett.* **8**, 419–420.
191. Gallego Lluesma, Tagliaferri, A. A., Massone, C. A., Garavaglia, M. and Gallardo, M. (1973). *J. Opt. Soc. Anm.* **63**, 362–364.
192. Papayoanou, A., Buser, R. G. and Gumeiner, I. M. (1973). *I.E.E.E. J.Q.E.* **QE-9**, 580–585.

References

Ahmed, S. A. and Faith, T. J., Jr. (1966). *Proc. I.E.E.E.* (Lett.) **54**, 1470–1471.
Ahmed, S. A. and Campillo, A. J. (1969). *Proc. I.E.E.E.* (Lett.) **57**, 2084–2085.
Ahmed, S. A., Campillo, A. J. and Cody, R. J. (1969). *I.E.E.E. J.Q.E. QE*-5, 267–271.
Ahmed, S. A., Faith, T. J. Jr. and Hoffman, G. W. (1967). *Proc. I.E.E.E.* (Lett.) **55**, 691–692.
Akhmanov, A. A., Gorokhov, Yu. A., Krindach, D. P., Sukhorukov, A. P. and Khokhlov, R. V. (1970). *Soc. Phys. J.E.T.P.* **30**, 9–12.
Allen, J. E. (1954). *Proc. Phys. Soc.* **67B**, 768–774.
Allen, J. E. and Magistrelli, F. (1962). *Nature* (*Lond.*) **194**, 1167.
Allen, L. (1968). *J. Phys. E.* **1**, 794–795.
Allen, L. and Jones, D. G. C. (1971). In "Progress in Optics" (Ed. E. Wolf), Vol. 9, pp. 181–234. North-Holland.
Ames, H. S. (1972). *I.E.E.E. J.Q.E.* **QE-8**. 808–809.
Ammann, E. O., McMurtry, B. J. and Oshman, M. K. (1965). *I.E.E.E. J.Q.E.* **QE-1**, 263–272.
Anderson, O. A., Baker, W. R., Colegate, S. A., Firth, H. P., Ise, J., Pyle, R. V. and Wright, R. E. (1958). *Phys. Rev.* **110**, 1375–1387.
Anderson, L. K., Brojdo, S., La Macchia, J. T. and Lin, L. H. (1967). Proc. I.E.E.E. Conf. on Laser Eng. and Appl., Washington, D.C.
Arecchi, F. T., Sacchi, C. A. and Sona, A. (1963). *Alta Freq.* **32**, 183.
Ashkin, A. (1972). *Sci. Am.* **226**, No. 2, 63–71.
Auston, D. H. (1968a). *I.E.E.E. J.Q.E.* **QE-4**, 420–422.
Auston, D. H. (1968b). *I.E.E.E. J.Q.E.* **QE-4**, 471–473.
Backrach, R. Z. (1972). *Rev. Sci. Inst.* **43**, 734–737.
Baker, H. J. (1973). Ph.D. Thesis, Uhiversity of Manchester, unpublished.
Baker, H. J. and King, T. A. (1973). *J. Phys. D*, **6**, 395–399.
Balkanski, M. (1971). (Ed.) Proc. 2nd Conf. "Light Scattering in Solids". Paris, 1971, Flammarion Sciences, Paris.
Ballik, E. A. (1966). *Appl. Opt.* **5**, 170–172.
Ballik, E. A., Bennett, W. R., Jr. and Mercer, G. N. (1966). *Appl. Phys. Lett.* **8**, 214–216.
Banse, K., Borsch, H., Herziger, G., Schafer, G. and Seelig, W. (1969). *Z. Angew. Phys.* **26**, 195–200.
Banse, K., Bret, G., Fruer, J., Gassmann, H. and Seelig, W. (1970). Proc. 6th Quan. Elec. Conf., Kyoto, Sept. 1970.
Banse, K., Herziger, G., Schafer, G. and Seelig, W. (1968). *Phys. Lett.* **27A**, 682–683.
Barker, G. C. and Jarrett, S. M. (1969). *I.E.E.E. J.Q.E.* **QE-5**, 363.
Barrat, M. and Barrat, J. P. (1963). *C.R. Acad. Sci.* (*Paris*) **257**, 1463–1465.
Barrett, J. J. and Weber, A. (1970). *J. Opt. Soc. Am.* **60**, 70–77.
Bartlett, O. (1934). *Z. Phys.* **91**, 444–470.
Bass, M. and Barrett, H. H. (1971). *In* "Damage in Laser Materials" (Ed. A. J. Glass and A. H. Guenther). N.B.S. Special Publ. No. 356.
Bass, M., de Mars, G. and Statz, H. (1968). *Appl. Phys. Lett.* **12**, 17-20.
Baumann, S. R. and Smith, W. H. (1970). *J. Opt. Soc. Am.* **60**, 345–347.
Baumgartner, G., Demtröder, W. and Stock, M. (1970). *Z. Phys.* **232**, 462–472.
Beesley, M. J. (1971). "Lasers and Their Applications." Taylor and Francis, London.
Beigman, E. L., Vainshtein, L. A., Rubin, P. L. and Sobolev, N. N. (1967). *J.E.T.P. Lett.* **6**, 343–345.
Bell, W. E. (1964). *Appl. Phys. Lett.* **4**, 34–35.

Bell, W. E. (1965). *Appl. Phys. Lett.* **7**, 190–191.
Bell, W. E. and Bloom, A. L. (1965). Private communication to W. B. Bridges and A. N. Chester.
Bell, W. E., Bloom, A. L. and Goldsborough, J. P. (1965). *I.E.E.E. J.Q.E.* (Corr.) **QE-1**, 400.
Bell, W. E., Bloom, A. L. and Goldsborough, J. P. (1966). *I.E.E.E. J.Q.E.* (Notes and Lines) **QE-2**, 154.
Belyaiev, V. P., Burmakin, U. A., Evtyunin, A. N., Kovolyov, F. A., Lebedeva, V. V. and Odintsov, A. I. (1969). *I.E.E.E. J.Q.E.*, **QE-5**, 589–591.
Benedetti-Michelangeli, G., Congeduti, F. and Fiocco, G. (1972). *J. Atmos. Sci.*, **29**, 906–910.
Bennett, W. R., Jr. (1965). *Appl. Opt. Suppl. on Chemical Lasers*, 3–33.
Bennett, W. R., Jr., Ballik, E. A. and Mercer, G. N. (1966a). *Phys. Rev. Lett.* **16**, 513–605.
Bennett, W. R., Jr., Knutson, J. W., Jr., Mercer, G. N. and Detch, J. L. (1964). *Appl. Phys. Lett.* **4**, 180–182.
Bennett, W. R., Jr. and Lichten, W. (1965). Unpublished.
Bennett, W. R., Jr., Mercer, G. N., Kindlmann, P. J., Wexler, B. and Hyman, H. (1966b). *Phys. Rev. Lett.* **17**, 987–991.
Ben-Reuven, A. and Gershon, N. D. (1971). *J. Chem. Phys.* **54**, 1049–1053.
Bensimon, J., Breton, J., Cotte, M. and Karar, A. (1971). *CR. Acad. Sci.(Paris)* **272B**, 615–618.
Bensimon, J. and Divonne, A. de al Forest (1970). *J. Phys. D.* **3**, L58–L61.
Berezin, I. A. (1969). *Opt. Spec.* **26**, 466–467.
Berio, F. J. and Cornell, D. D. (1972). *Am. Chem. Soc. Meeting*, Boston, April 1972.
Berns, M. W., Olson, R. S. and Rounds, D. E. (1969a). *Nature, Lond.* **221**, 74–75.
Berns, M. W., Olson, R. S., and Rounds, D. E. (1969b). *J. Cell. Biol.* **43**, 621–626.
Berns, M. W., Olson, R. S. and Rounds, D. E. (1969c). *Expl Cell. Res.* **56**, 292.
Berns, M. W. and Rounds, D. E. (1973). *Sci. Am.* **222**, 99–110.
Beterov, I. M., Klement'ev, V. M. and Chebotayev, V. P. (1969). *Rad. Eng. Elec. Phys.* **14**, 1790–1793.
Bhaumik, M. L., Hughes, W., Jensen, R., Kolb, A. and Shannon, J. (1973). Opt. Soc. Am. Meeting, Denver, Mar. 14.
Bickerton, R. J. and von Engel, A. (1956). *Proc. Phys. Soc.* **69B**, 468–481.
Birnbaum, G. (1967). *Proc. I.E.E.E.* **55**, 1015–1026.
Birnbaum, M. (1968). *Appl. Phys. Lett.* **12**, 86–89.
Birnbaum, M. and Fincher, C. L. (1968a). *Appl. Phys. Lett.* **12**, 225–227.
Birnbaum, M. and Fincher, C. L. (1968b). *Proc. I.E.E.E.* **56**, 1096–1097.
Birnbaum, M., Wendzikowski, P. H. and Fincher, C. L. (1970). *Appl. Phys. Lett.* **16**, 436–438.
Bloch, L. and Bloch, E. (1929). *Ann. der. Phys.* (10) **12**, 5–22.
Bloom, A. L. (1963). *Appl. Phys. Lett.* **2**, 101–102.
Bloom, A. L. (1968). Private communication to W. B. Bridges and A. N. Chester 1968, see, Bridges, W. B. and Chester, A. N. (1971).
Bloom, A. L. (1971). *In* "Progress in Optics" (Ed. E. Wolf), Vol. 9. North-Holland.
Bloom, A. L., Bell, W. E. and Lopez, F. O. (1964). *Phys. Rev.* **135**, A578–A579.
Bloom, A. L., Byer, R. L., and Bell, W. E. (1965). Proc. Phys. Quan. Elec. Conf., Puerto Rico. (Eds Kelley, P. L., Lax, B. and Tannerwald, P. E.), McGraw-Hill, 1966.
Bockasten, K. (1955). *Ark. Fys.* **13**, 303–329.
Bockasten, K. (1956). *Ark. Fys.* **10**, 567–582.

Boersch, H., Boscher, J., Hoder, D. and Schafer, G. (1970). *Phys. Lett.* **31A**, 188–189.
Boersch, H., Herziger, G., Seelig, W. and Volland, I. (1967). *Phys. Lett.* **24A**, 695–696.
Bogaslovskii, A. A., Gunger, T. T., Didrickill, L. N., Novikova, V. A., Kune, V. V. and Stepanov, A. F. (1967). *Electronnaya Tekhn.* ser. **3**, No. 1, p. 8.
Bogatkin, V. I. *et al.* (1968). *Rad. Eng. Elec. Phys.* **13**, 134–136.
Borisova, M. S., Ischenko, Ye. F., Ladygin, M. V., Molchashkin, M. A., Nasedkin, Ye. F. and Ramazanova, G. S. (1967a). *Rad. Eng. Elec. Phys.* **4**, 526–528.
Borisova, M. S., Ischenko, Ye. F., Ladygin, M. V., Molchashkin, M. A., Nasedkin, Ye. F. and Ramazanova, G. S. (1967b). *Rad. Eng. Elec. Phys.* **4**, 529–533.
Borisova, M. S. and Pyndyk, A. M. (1968). *Rad. Eng. Elec. Phys.* **13**, 658–660.
Borisova, M. S. and Yasinskii, V. M. (1971). *Opt. Spec.* **27**, 231–232.
Boscher, J., Kindt, T. and Schafer, G. (1971). *Z. Phys.* **241**, 280–290.
Bowen, I. S. and Millikan, R. A. (1925). *Phys. Rev.* **25**, 591–599.
Brand, J. H., Cocke, C. L., Carnutte, B. and Swenson, C. (1970). *Nucl. Inst. Meth.* **90**, 63–70.
Brandi, H. S. (1972). *Phys. Rev. Lett.* **29**, 1539–1541.
Bridges, W. B. (1964a). *Appl. Phys. Lett.* **4**, 128–130.
Bridges, W. B. (1964b). *Proc. I.E.E.E.* (Corr.) **52**, 843–844.
Bridges, W. B. and Chester, A. N. (1965a). *Appl. Opt.* **4**, 573–580.
Bridges, W. B. and Chester, A. N. (1965b). *I.E.E.E. J.Q.E.* **QE-1**, 66–84.
Bridges, W. B. and Chester, A. N. (1971). Ionised Gas Lasers. *In* "Handbook of Lasers," CRC Press, Cleveland, Ohio.
Bridges, W. B., Chester, A. N., Halsted, A. S. and Parker, J. V. (1971). *Proc. I.E.E.E.* **59**, 724–737.
Bridges, W. B., Clark, P. O. and Halsted, A. S. (1966). *I.E.E.E. J.Q.E.*, **QE-2**, 9.
Bridges, W. B. and Halsted, A. S. (1967). Hughes Res. Lab. Malibu. Col. Tech. Rep. A.F.A.C.-TR-67-89 (DDC Accession No. AD-814-897).
Bridges, W. B., Halsted, A. S. and Mercer, G. N. (1969). *I.E.E.E. J.Q.E.* **QE-5**, 365.
Bridges, T. J. and Rigrod, W. W. (1965). *I.E.E.E. J.Q.E.* **QE-1**, 303–308.
Brown, S. C. (1966). "Basic Data of Plasma Physics." M.I.T. Press, Cambridge, Mass.
Brya, W. J., Geschwind, S. and Devlin, G. E. *Phys. Rev. Lett.* **21**, 1800–1802 (1968).
Burmakin, V. A. and Evtyunin, A. N. (1969). Proc. 9th Int. Conf. on Phen. in Ionised Gases, Bucharest, p. 257 (Editura Academici Republici Socialiste Romania).
Burmakin, V. A., Korolev, F. A., Lebedeva, V. V., Odintsov, A. I., Salimov, V. M. and Sinitsa, L. N. (1971). *Rad. Eng. Elec. Phys.* **16**, 1228–1231.
Burstein, E., Itou, R. Pinczuk, A. and Shand, M. (1971). *J. Acoust. Soc. Am.* **49**, 1013–1025.
Byer, R. L., Bell, W. E., Hodges, E. and Bloom, A. L. (1965). *J. Opt. Soc. Am.* **55**, 1598–1602.
Byer, R. L. and Costich, V. R. (1967). *Appl. Opt.* **6**, 578–579.
Cabezas, A. Y. and Treat, R. P. (1966). *J. Appl. Phys.* **37**, 3556–3563.
Cahuzac, Ph. (1968). *Phys. Lett.*, **27A**, 473–474.
Cahuzac, Ph. and Brochard, J. (1969). *J. Phys.* **30**, C6-67.
Cahuzac, Ph. (1970a). *Phys. Lett.* **31A**, 541–542.
Cahuzac, Ph. (1970b). *Phys. Lett.* **32A**, 150–151.
Cahuzac, Ph. (1971). *J. Phys.* **32**, 499–505.
Cairns, R. and Emeleus, K. G. (1958). *Proc. Phys. Soc.* **71**, 694–698.
Campbell, C. J., Rittler, M. C. and Swope, C. H. (1969). *Am. J. Ophthal.* **67**, 671.

Cannell, D. S. and Benedek, G. B. (1970). *Phys. Rev. Lett.* **25**, 1157–1161.
Carbonne, R. T. and Witteman, W. J. (1970). *Rev. Sci. Inst.* **41**, 689–690.
Carman, R. L., Mooradian, A., Kelley, P. L. and Tufts, A. (1969). *Appl. Phys. Lett.* **14**, 136–139.
Carr, W. C. and Grow, R. W. (1967a). *Proc. I.E.E.E.* (*Lett.*) **55**, 726.
Carr, W. C. and Grow, R. W. (1967b). *Proc. I.E.E.E.* (*Lett.*) **55**, 1198.
Ceyzeriat, P., Denis, A. and Dufay, M. (1971). *J. Opt. Soc. Am.* **61**, 641–642.
Charschan, S. S. (Ed.) (1973). "Lasers in Industry," Van Nostrand-Reinhold Co.
Cheo, P. K. and Cooper, H. G. (1965a). *J. Appl. Phys.* **36**, 1862–1865.
Cheo, P. K. and Cooper, H. G. (1965b). *Appl. Phys. Lett.* **6**, 177–178.
Cheo, P. K. and Cooper, H. G. (1965c). *Appl. Phys. Lett.* **7**, 202–204.
Chester, A. N. (1968a). *Phys. Rev.* **169**, 172–183.
Chester, A. N. (1968b). *Phys. Rev.* **169**, 184–193.
Chu, B. (1970). *Ann. Rev. Phys. Chem.* **21**, 145–174.
Clarke, A. M. (1970). C.R.C. Crit. Rev. in Environmental Control, **1**, 307.
Clements, W. R. L. and Stoicheff, B. P. (1970). *J. Mol. Spec.* **33**, 183–186.
Clout, P. N. and Heddle, D. W. O. (1969). Proc. VIth. Int. Conf. Phys. El. Atomic Coll. pp. 290–293. M.I.T. Press, Cambridge, Mass.
Clout, P. N. and Heddle, D. W. O. (1971). *J. Phys.* B, **4**, 483–493.
Cobine, J. D. (1958). "Gaseous Conductors," Dover Publications Inc. N.Y.
Cohen, I. M. and Kruskal, M. D. (1965). *Phys. Fluids*, **8**, 920–934.
Coleman, C. D., Bozman, W. R. and Meggers, W. F. (1960). Table of Wavenumbers, Natl. Bur. Stad. (U.S.). Monograph 3.
Colli, L. and Facchini, U. (1952). *Rev. Sci. Inst.* **23**, 39–42.
Collins, G. J., Jensen, R. C. and Bennett, W. R., Jr. (1971a). *Appl. Phys. Lett.* **18**, 282–284.
Collins, G. J., Jensen, R. C. and Bennett, W. R., Jr. (1971b). *Appl. Phys. Lett.* **19**, 125–128.
Collins, G. J., Kanu, H., Hattori, S., Tokutome, K., Ishikawa, M. and Kamiide, N. (1972). *I.E.E.E. J.Q.E.* (Notes and Lines) **QE-8**, 679–680.
Collins, S. A. and White, G. R. (1963). *Appl. Opt.* **2**, 448–449.
Conder, P. C. and Foster, H. (1970). *Radio Electronic Eng.* (G.B.) **39**, 97–103.
Convert, G., Armand, M. and Martinot-Lagarde, P. (1964a). *C.R. Acad. Sci. Paris* **258**, 3259–3260.
Convert, G., Armand, M. and Martinot-Lagarde, P. (1964b). *C.R. Acad. Sci. Paris* **258**, 4467–4469.
Copper, H. G. and Cheo, P. K. (1966a). Phys. Quan. Elec. Conf. (Puerto Rico, 1965), pp. 690–697. McGraw-Hill.
Cooper, H. G. and Cheo, P. K. (1966b). *I.E.E.E. J.Q.E.* (Notes and Lines) **QE-2**, 785.
Corliss, C. H. and Bozman, W. R. (1962). "Experimental Transitions Probabilities for Spectral Lines, of Seventy Elements." N.B.S. Mono. 53.
Cotman, R. N. and Johnson, W. B. (1967). Case Inst. of Tech., Tech. Report A–54.
Cottrell, T. H. E. (1968). *I.E.E.E. J.Q.E.* **QE-4**, 435–441.
Cowley, R A., Coombs, G. J., Katiyar, R. S., Ryan, J. F. and Scott, J. F. (1971). *J. Phys. C.* **4**, L203–207.
Crawford, F. W. and Freeston, I. L. (1963). Int. Conf. Ion. Phen. in Gases, Paris, 1, pp. 461–464.
Crowell, M. H. (1965). *I.E.E.E. J.Q.E.* **QE-1**, 12–20.
Csillag, L., Janossy, M., Kantor, K., Rozsa, K. and Salamon, T. (1970a). *J. Phys. D.* **3**, 64–68.
Csillag, L., Janossy, M. and Salamon, T. (1970b). *Phys. Lett.* **31A**, 532–533.

Csillag, L., Itagi, V. V., Janossy, M. and Rozsa, K. (1971). *Phys. Lett.* **34A**, 110–111.
Cummins, H. Z. (1971). *In* "Light Scattering in Solids" (Ed. M. Balkanski), pp. 2–8.
Cummins, H. Z. and Swinney, H. L. (1970). *In* "Progress in Optics" (Ed. E. Wolf), Vol. 8, pp. 134–200. North-Holland.
Dabby, I. W., Boyko, R. W., Shank, C. V. and Whinnery, J. R. (1969). *I.E.E.E. J.Q.E.* **QE-5**, 516–520.
Dabby, F. W., Gustafson, T. K., Whinnery, J. R., Kohanzadeh, Y. and Kelley, P. L. (1970). *Appl. Phys. Lett.* **16**, 362–365.
Dabby, F. W. and Whinnery, J. R. (1969). *Appl. Phys. Lett.* **13**, 286.
Dahlquist, J. A. (1965). *Appl. Phys. Lett.* **6**, 193–194.
Dana, L. and Laures, P. (1965). *Proc. I.E.E.E.* (Corr.) **53**, 78–79.
Dana, L., Laures, P. and Rocherolles, R. (1965). *C.R. Acad. Sci. Paris* **260**, 481–484.
Davis, C. C. (1970). Ph.D. Thesis. Manchester University.
Davis, C. C. and King, T. A. (1971). *Phys. Lett.*, **36A**, 169–170.
Davis, C. C. and King, T. A. (1972a). *I.E.E.E. J.Q.E.* **QE-8**, 755–757.
Davis, C. C. and King, T. A. (1972b). I.E.E. 2nd Int. Conf. on "Gas Discharges". London.
Deech, J. S. and Sanders, J. H. (1968). *I.E.E.E. J.Q.E.* (Notes and Lines). **QE-4**, 474.
de Lang, H. (1967). *Physica* **33**, 163–173.
De Maria, A. J. and Stetser, D. A. (1965). *Appl. Phys. Lett.* **7**, 71–73.
de Mars, G., Seiden, M. and Horrigan, F. A. (1968). *I.E.E.E. J.Q.E.* **QE-4**, 631–637.
Demtröder, W. (1966). *Phys. Lett.* **22**, 436–438.
Demtröder, W. and Elendt, E. (1966). *Z. Naturforsch.* **21a**, 2047–2054.
Demtröder, W., McClintock, M. and Zare, R. N. (1969). *J. Chem. Phys.* **51**, 5495–5508.
Denis, A. (1969). *C.R. Acad. Sci. Paris* **B268**, 383–385.
Denis, A. and Dufay, M. (1969). *Phys. Lett.* **29A**, 170.
Devlin, G. E., Davis, J. L., Chase, L. and Geschwind, S. (1971). *Appl. Phys. Lett.* **19**, 138–141.
DiDomenico, M., Jr. (1964). *J. Appl. Phys.* **35**, 2870–2876.
DiDomenico, M., Jr. (1966a). *Appl. Phys. Lett.* **8**, 20–22.
DiDomenico, M., Jr. (1966b). *I.E.E.E. J.Q.E.* **QE-3**, 11.
Dienes, A. and Sosnowski, T. P. (1970). *Appl. Phys. Lett.* **16**, 512–514.
Dienes, A., Ippen, E. P. and Shank, C. I. (1971). *Appl. Phys. Lett.* **19**, 258–260.
Donahue, T. and Dieke, G. H. (1951). *Phys. Rev.* **81**, 248–261.
Donin, V. I. (1970a). *Opt. Spec.* **26**, 160–161.
Donin, V. I. (1970b). *Opt. Spec.* **26**, 128–131.
Donin, V. I., Klement'ev, V. M. and Chebotayev, V. P. (1967). *Inst. Exp. Tech.* **No. 4**, 932–933.
Donin, V. I., Troitskii, Yu. V. and Goldina, N. D. (1969). *Opt. Spec.* **25**, 64–65.
Dowley, M. W. (1968). *Appl. Phys. Lett.* **13**, 395–397.
Dowley, M. W. and Hodges, E. B. (1968). *I.E.E.E. J.Q.F.* **QE-4**, 552–558.
Dreyfus, R. W. and Hodgson, R. T. (1972). *Appl. Phys. Lett.* **20**, 195–197.
Drullinger, R. E. and Zare, R. N. (1969). *J. Chem. Phys.* **51**, 5532–5542.
Druyvesteyn, M. J. (1935). *Physica* **2**, 255–266.
Dunn, M. H. (1968). *I.E.E.E. J.Q.E.* **QE-4**, 357–358.
Dunn, M. H. (1972). *J. Phys. B* **5**, 665–672.
Dyatlov, M. K., Ostapchenko, E. P. and Stepanov, V. A. (1970). *Opt. Spec.* **29**, 1014–1015.
Dyson, D. J. (1965). *Nature, Lond.* **207**, 361–363.
Emmons, H. W. (1963). "Modern Developments in Heat Transfer" (Ed. W. E. Ibele), pp. 401–478. Academic Press, London and New York.

Ericsson, K. G. and Lidholt, L. R. (1967). *I.E.E.E. J.Q.E.* (Corr.) **QE-3**, 94.
Ezekiel, S. and Weiss, R. (1968). *Phys. Rev. Lett.* **20**, 91–93.
Fainberg, Ya. B. and Shapiro, V. D. (1966). *Sov. Phys. J.E.T.P. Lett.* **4**, 20–23.
Fainberg, Ya. B. (1968). *Sov. Phys. Uspekhi* **10**, 750–758.
Faxvog, F. R., Willenbring, G. R. and Carruthers, J. A. (1970). *Appl. Phys. Lett.* **16**, 8–10.
Faxvog, F. R., Chow, C. N. Y., Bieker, T. and Carruthers, J. A. (1970). *Appl. Phys. Lett.* **5**, 48.
Fehsenfeld, F. C., Evenson, K. M. and Broida, H. P. (1965). *Rev. Sci. Inst.* **36**, 294–298.
Felt'san, P. V. and Povch, M. M. (1970). *Opt. Spect.* **28**, 119–121.
Fendley, J. R., Jr. (1967). *I.E.E.E. J.Q.E.* **QE-3**, 66.
Fendley, J. R., Jr. (1968). *I.E.E.E. J.Q.E.* **QE-4**, 627–631.
Fendley, J. R., Jr., Gorog, I., Hernqvist, K. G. and Sun, C. (1969). *R.C.A. Rev.* **30**, 422–428.
Ferrario, A. and Sona, A. (1969). *I.E.E.E. J.Q.E.* **QE-5**, 124–125.
Ferrario, A. and Sona, A. (1972). *Rev. Sci. Inst.* **43**, 1216–1218.
Fink, U., Bashkin, S. and Bickel, W. S. (1970). *J. Quart. Spec. Rad. Trans.* **10**, 1241–1256.
Fleury, P. A. (1970). *Phys. Acoustics* (Ed. W. P. Mason), Vol. 6, pp. 2–64.
Fleury, P. A., Scott, J. F. and Worlock, J. M. (1968). *Phys. Rev. Lett.* **21**, 16–19.
Ford, N. C., Jr., Langley, K. H. and Pugielli, V. G. (1968). *Phys. Rev. Lett.* **21**, 9–12.
Forrest, J. R. and Franklin, R. N. (1966a). *Br. J. Appl. Phys.* **17**, 1061–1067.
Forrest, J. R. and Franklin, R. N. (1966b). *Br. J. Appl. Phys.* **17**, 1569–1574.
Forrest, J. R. and Franklin, R. N. (1968). *J. Phys. D.* **1**, 1357–1368.
Forsyth, J. M. (1967). *Appl. Phys. Lett.* **11**, 391–394.
Forsyth, J. M. (1969). *J. Appl. Phys.* **40**, 3049–3051.
Fotiadi, A. E. and Fridrikhov, S. A. (1970). *Sov. Phys. Tech. Phys.* **14**, 1292–1293.
Fowles, G. R. and Jensen, R. C. (1964a). *Proc. I.E.E.E.* (Corr.) **52**, 851–852.
Fowles, G. R. and Jensen, R. C. (1964b). *Appl. Opt.* **3**, 1191–1192.
Fowles, G. R. and Hopkins, B. D. (1967). *I.E.E.E. J.Q.E.* (Notes and Lines) **QE-3**, 419.
Fowles, G. R. and Silfvast, W. T. (1965). *I.E.E.E. J.Q.E.* (Corr.) **QE-1**, 131.
Fowles, G. R., Silfvast, W. T. and Jensen, R. C. (1965). *I.E.E.E. J.Q.E.* (Corr.) **QE-1**, 183–184.
Fox, A. G. and Li, T. (1966). *I.E.E.E. J.Q.E.* **QE-1**, 774.
Fox, A. G. and Smith, P. W. (1967). *Phys. Rev. Lett.* **18**, 826–828.
Francis, G. (1956). *Handbuch der Physik* **22**, 53–208.
Francis, G. (1960). "Ionisation Phenomena in Gases." Butterworth, London.
Freiberg, R. J. and Halsted, A. S. (1969). *Appl. Opt.* **8**, 355–362.
Fridrikhov, S. A. and Fotiadi, A. E. (1971). Proc. 10th Int. Conf. on Ions. Phen. in Gases, p. 153. Oxford.
Frova, A., Duguay, M. A., Garrett, C. G. B. and McCall, S. L. (1969). *J. Appl. Phys.* **40**, 3969–3972.
Fry, E. S. and Williams, W. L. (1969). *Rev. Sci. Inst.* **40**, 1141–1143.
Fuhr, J. R. and Wiese, W. L. (1971). "Bibliography on Atomic Transition Probabilities." N.B.S. Special Pub. 320, Suppl. 1.
Gaddy, O. L. and Schaefer, E. M. (1966). *Appl. Phys. Lett.* **9**, 281–282.
Gadetskii, N. P., Tkach, Yu. V., Slezov, V. V., Bessarab, Ya. Ya. and Magad, I. I. (1971). *J.E.T.P. Lett.* **14**, 101–103.

Gagliano, E. P., Lumley, R. M. and Watkins, L. S. (1969). *Proc. I.E.E.E.* **57**, 114–146.
Galehouse, D. C., Ingard, U., Ryan, T. J. and Ezekiel, S. (1971). *Appl. Phys. Lett.* **18**, 13–15.
Gallardo, M., Garavaglia, M., Tagliaferri, A. A. and Gallego Lluesma, E. (1970). *I.E.E.E. J.Q.E.* **QE-6**, 745–747.
Gallego Lluesma, E., Tagliaferri, A. A., Massone, C. A., Garavaglia, M. and Gallardo, M. (1973). *J. Opt. Soc. Am.* **63**, 362–364.
Garscadden, A., Bletzinger, P. and Friar, E. M. (1964). *J. Appl. Phys.* **35**, 3432–3433.
Geneux, E. and Wanders-Vincenz, B. (1959). *Phys. Rev. Lett.* **3**, 422–423.
Gerritson, H. J. and Goedertier, P. V. (1964). *J. Appl. Phys.* **35**, 3060–3061.
Giallorenzi, T. G. and Ahmed, S. A. (1971). *I.E.E.E. J.Q.E.* **QE-7**, 11–17.
Gibbs, H. M. and Slusher, R. E. (1970). *Phys. Rev. Lett.* **24**, 638–641.
Gilles, M. (1931). *Ann. Phys.* [10], **15**, 267–410.
Gilson, T. R. and Hendra, P. J. (1970), "Laser Raman Spectroscopy." Wiley-Interscience, New York.
Gilson, V. A. and Stoll, H. M. (1969). *Appl. Opt.* **8**, 717–718.
Glazunov, U. K., Kitaeva, V. F., Ostrovskaya, L. Ya and Sobolev, N. N. (1967). *J.E.T.P. Lett.* **5**, 215–218.
Godard, B. (1973). Conf. Laser Eng. and Appl., Washington, paper 4.5.
Goldman, L. (1967). "Biomedical Aspects of the Laser." Springer-Verlag.
Goldman, L. and Rockwell, R. J., Jr. (1972). "Lasers in Medicine." Gordon and Breach.
Goldsborough, J. P. (1964). *Appl. Opt.* **3**, 267–275.
Goldsborough, J. P. (1966). *Appl. Phys. Lett.* **8**, 218–219.
Goldsborough, J. P. (1969). *Appl. Phys. Lett.* **15**, 159–161.
Goldsborough, J. P. and Bloom, A. L. (1968). *I.E.E.E. J.Q.E.* **QE-5**, 459–460.
Goldsborough, J. P., Hodges, E. B. and Bell, W. E. (1966). *Appl. Phys. Lett.* **8**, 137–139.
Gordeyev, D. V., Grimblatov, V. M., Ostapchenko, E. P. and Teselkin, V. V. (1969). *Rad. Eng. Elec. Phys.* **14**, 1420–1422.
Gordon, E. I. and Labuda, E. F. (1964). *Bell Sys. Tech. J.* **43**, 1827–1829.
Gordon, E. I., Labuda, E. F. and Bridges, W. B. (1964). *Phys. Lett.* **4**, 178–180.
Gordon, E. I., Labuda, E. F., Miller, R. C., and Webb C. E. (1965). Proc. Phys. Quan. Elec. Conf. (Eds. Kelley, P. L., Lax, B. and Tannenwald, P. E.), pp. 664–673. McGraw-Hill, 1966.
Gordon, E. I. and Rigden, J. D. (1963). *Bell Sys. Tech. J.* **42**, 155–179.
Gordon, E. I. and White, A. D. (1964). *Proc. I.E.E.E.* **52**, 206–207.
Gordon, E. I., White, A. D. and Rigden, J. D. (1963). Proc. Symp. Optical Masers, Polytechnic Press, New York, 309–319.
Gorog, I. (1971). *R.C.A. Rev.* **32**, 88–114.
Gorog, I. and Singer, J. R. (1966). *J. Appl. Phys.* **37**, 4141–4147.
Gorog, I. and Spong, F. W. (1966). *Appl. Phys. Lett.* **9**, 61–63.
Gorog, I. and Spong, F. W. (1967). *RCA. Rev.* **28**, 38–57.
Gorog, I. and Spong, F. W. (1969). *R.C.A. Rev.* **30**, 277–284.
Goto, T., Kawahara, A., Collins, G. J. and Hattori, S. (1971b). *J. Appl. Phys.* **42**, 3816–3818.
Goto, T., Kano, H. and Hattori, S. (1972). *J. Appl. Phys.* **43**, 5064–5068.
Goto, T., Kawahara, A. and Hattori, S. (1971a). *I.E.E.E. J.Q.E.* **QE-7**, 555–560.
Goto, T., Nakamura, K. and Hattori, S. (1970). *I.E.E.E. J.Q.E.* **QE-6**, 159–160.
Green, J. M., Collins, G. J. and Webb, C. E. (1973). *J. Phys. B.* **6**, 1545–1550.

Gregg, D. W. and Thomas, S. J. (1966). *Appl. Phys. Lett.* **8**, 316–318.
Griem, H. R. (1964). "Plasma Spectroscopy." McGraw-Hill, New York.
Gurskii, T. R. (1969). *Appl. Phys. Lett.* **15**, 5–6.
Habegger, M. A. (1967). *Appl. Phys. Lett.* **10**, 103–105.
Habegger, M. A., Harris, T. J. and Max, E. (1966). *I.B.M. J. Res. Dev.*, **10**, 346–350.
Hadley, R., Lynch, D. W., Stanek, E. and Rosauer, E. A. (1971). *Appl. Phys. Lett.* **19**, 145–147.
Hall, J. L. (1968). *I.E.E.E. J.Q.E.* **QE-4**, 638–641.
Hammer, J. M. and Wen, C. P. (1965). *Appl. Phys. Lett.* **7**, 159–161.
Hammer, J. M. and Wen, C. P. (1967). *J. Chem. Phys.* **46**, 1225–1230.
Handrup, M. B. and Mack, J. E. (1964). *Physica* **30**, 1245–1275.
Harris, S. E. (1966). *Proc. I.E.E.E.* **54**, 1401–1413.
Harris, S. E. and McDuff, O. P. (1964). *Appl. Phys. Lett.* **5**, 205–206.
Harris, S. E. and McDuff, O. P. (1965). *I.E.E.E. J.Q.E.* **QE-1**, 245–262.
Harris, S. E., Oshman, M. K., McMurtry, B. J. and Ammann, E. O. (1965). *Appl. Phys. Lett.* **7**, 184–186.
Harris, S. E. and Targ, R. (1964). *Appl. Phys. Lett.* **5**, 202–204.
Harrison, G. R. (1969). M.I.T. Wavelength Tables. M.I.T. Press, Cambridge, Mass.
Harting, C. M., Wiener, E. and Porto, S. P. S. (1970). *Bull. Ann. Phys. Soc.* **15**, 327.
Hashino, Y., Katsuyama, Y. and Fukuda, K. (1972). *Jap. J. Appl. Phys.* **11**, 907.
Hashino, Y., Katsuyama, Y. and Fukuda, K. (1973). *Jap. J. Appl. Phys.* **12**, 470.
Hattori, S. and Goto, T. (1968a). *Appl. Phys. Lett.* **12**, 131–133.
Hattori, S. and Goto, T. (1968b) *J. Appl. Phys.* **39**, 5998–6003.
Hattori, S. and Goto, T. (1969a). *I.E.E.E. J.Q.E.* **QE-5**, 531–538.
Hattori, S. and Goto, T. (1969b). *Jap. J. Appl. Phys.* **8**, 1159–1161.
Haug, A., Kohler, B. E., Priestley, E. B. and Robinson, G. W. (1969). *Rev. Sci. Inst.* **40**, 1429–1444.
Heard, H. G., Makhov, G. and Peterson, J. (1964). *Proc. I.E.E.E.* (Corr.) **52**, 414.
Heard, H. G. and Peterson, J. (1964). *Proc. I.E.E.E.* **52**, 1050.
Heising, S. J., Jarrett, S. M. and Kuizenga, D. J. (1971a). *Appl. Phys. Lett.* **18**, 516–518.
Heising, S. J., Jarrett, S. M. and Kuizenga. D. J. (1971b). *I.E.E.E. J.Q.E.* **QE-7**, 205–207.
Hercher, M. M. and Pike, H. A. (1971). *Opt. Comm.* **3**, 65.
Hernqvist, K. G. (1969). *R.C.A. Rev.* **30**, 429–434.
Hernqvist, K. G. (1970a). *Appl. Phys. Lett.* **16**, 464–467.
Hernqvist, K. G. (1970b). *Appl. Opt.* **9**, 2247–2249.
Hernqvist, K. G. (1972). *I.E.E.E. J.Q.E.* **QE-8**, 740–743.
Hernqvist, K. G. and Fendley, J. R. (1967). *I.E.E.E. J.Q.E.* **QE-3**, 66–72.
Hernqvist, K. G. and Pultorak, D. C. (1972). *Rev. Sci. Inst.* **43**, 290–292.
Herziger, G. and Seelig, W. (1968). *Z. Phys.* **215**, 437–465.
Herziger, G. and Seelig, W. (1969a). *I.E.E.E. J.Q.E.* **QE-5**, 364.
Herziger, G. and Seelig, W. (1969b). *Z. Phys.* **219**, 5–31.
Herziger, G. and Theiss, F. J. (1970). *Z. Angew. Physik.* **29**, 157–159.
Hesser, J. E. (1968). *Phys. Rev.* **174**, 68–74.
Hibler, G., Lippert, J. and Peticolas, W. L. (1971). *Spex Speaker*, 16, No. 1.
Hodara, H. and George, N. (1966). *I.E.E.E. J.Q.E.* **QE-2**, 337–340.
Hodges, D. T. (1970). *Appl. Phys. Lett.* **17**, 11–13.
Hodges, D. T. (1971). *Appl. Phys. Lett.* **18**, 454–456.
Hodges, D. T., Morantz, H. and Tang, C. L. (1970). *J. Opt. Soc. Am.* **60**, 192–199.
Hodges, D. T. and Tang, C. L. (1970). *I.E.E.E. J.Q.E.* (Corr.) **QE-6**, 757–758.
Hodgson, R. T. and Dreyfus, R. W. (1972a). *Phys. Lett.* **38A**, 213–214.

Hodgson, R. T. and Dreyfus, R. W. (1972b). *Phys. Rev. Lett.* **28**, 536–539.
Hoffman, V. and Toschek, P. (1970). *I.E.E.E. J.Q.E.* (Notes and Lines) **QE-6**, 757.
Hofmeister, V. P., Desai, S. K. and Kagan, Yu. M. (1969). Proc. 9th Int. Conf. on Ion. Phen, in Gases, p. 167.
Hohimer, J. P., Kelly, R. C. and Tittel, F. K. (1972). *Appl. Opt.* **11**, 626–629.
Holstein, T. (1947). *Phys. Rev.* **72**, 1212–1233.
Holstein, T. (1951). *Phys. Rev.* **83**, 1159–1168.
Holzer, W., Murphy, W. F. and Bernstein, H. J. (1970). *J. Chem. Phys.* **52**, 469–470.
Hopkins, B. D. and Fowles, G. R. (1968). *I.E.E.E. J.Q.E.* **QE-4**, 1013–1015.
Horrigan, F. A., Koozekanani, S. H. and Paananen, R. A. (1965). *Appl. Phys. Lett.* **6**, 41–43.
Hotop, H., Niehaus, A. and Schmeltekopf, A. L. (1969). *Z. Phys.* **229**, 1–13.
Hoyaux, M. F. (1968). *Am. J. Phys.* **36**, 726–734.
Humphreys, C. J. (1936). *J. Res. N.B.S.* **16**, 639–648.
Humphreys, C. J. (1939). *J. Res. N.B.S.* **22**, 19–53.
Humphreys, C. J (1965). Unpublished data supplied to W. B. Bridges and A. N. Chester.
Husain, S. A. (1968). *Elec. Lett.* **4**, 248–249.
Hutchital, D. A. and Rigden, J. D. (1967). *I.E.E.E. J.Q.E.* **QE-3**, 378–379.
Hyman, H. A. (1971). *Chem. Phys. Lett.* **10**, 242–244.
Illingworth, R. (1970). *J. Phys. D.* **3**, 924–930.
Illingworth, R. (1972). *J. Phys. D.* **5**, 686–692.
Imre, A. I., Dashchenko, A. I., Zapesochnyi, I. P. and Kelman, V. A. (1972). *Sov. Phys. J.E.T.P. Lett.* **15**, 503–507.
Ivanov, I. G. and Sem, M. F. (1973). *Sov. Phys. Tech. Phys.* **17**, 1234–1235.
Jackson, D. A. and Paul, D. M. (1969). *J. Phys. E.* **2**, 1077–1080.
Jarrett, S. M. and Barker, G. C. (1968). *J. Appl. Phys.* **39**, 4845–4846.
Jarrett, S. M. and Barker, G. C. (1969). *I.E.E.E. J.Q.E.* **QE-5**, (Notes and Lines), 166.
Javan, A., Bennett, W. R., Jr. and Herriott, D. R. (1961). *Phys. Rev. Lett.* **6**, 106–110.
Jennings, W. C., Noon, J. H., Holt, E. A. and Buser, R. G. (1970). *Rev. Sci. Inst.* **41**, 322–326.
Jensen, R. C., Collins, G. J. and Bennett, W. R. Jr., (1969). *Phys. Rev. Lett.* **23**, 363–367.
Jensen, R. C., Collins, G. J. and Bennett, W. R., Jr. (1971). *Appl. Phys. Lett.* **18**, 50–51.
Jensen, R. C. and Fowles, G. R. (1964). *Proc. I.E.E.E.* **52**, 1350.
Johnson, A. M. and Gerardo, J. B. (1973). Opt. Soc. Am. Meeting, Denver, Mar 14.
Johnson, A. M. and Webb, C. E. (1967). *I.E.E.E. J.Q.E.* (Notes and Lines), **QE-3**, 369.
Kadomtsev, B. B. and Nedospasov, A. V. (1960). *J. Nucl. Energy* **1**, 230–235.
Kagan, Yu. M. and Perel, V. I. (1957),* *Optika i Spektroskopsia* **2**, 298–303.
Kagan, Yu. M. and Perel, V. I. (1958a),* *Optika i Spektroskopsia* **4**, 3–8.
Kagan, Yu. M. and Perel, V. I. (1958b),* *Optika i Spektroskopsia* **4**, 285–288.
Kagan, Yu. M. (1958).* *Izv. Akad. Nauk SSSR*, Ser. Fiz., **22**, 702–707.
Karabut, E. K., Mikhalevskii, V. S., Papakin, V. F. and Sem, M. F. (1970). *Sov. Phys. Tech. Phys.* **14**, 1447–1448.
Katiyar, R. S., Ryan, J. F. and Scott, J. F. (1971). *Phys. Rev.* **B4**, 2635–2638.
Keefe, W. M. and Graham, W. J. (1965). *Appl. Phys. Lett.* **7**, 263–264.
Keefe, W. M. and Graham, W. J. (1966). *Phys. Lett.* **20**, 643.
Kernahan, J. A., Lin, C. C. and Pinnington, E. H. (1970). *J. Opt. Soc. Am.* **60**, 898–901.
Kitaeva, V. F., Odintsov, A. N. and Sobolev, N. N. (1970). *Sov. Phys. Usp.* **12**, 699–730.
Kitaeva, V. F. and Osipov, Yu. I. (1968). *Sov. Phys. Tech. Phys.* **13**, 282–284.

* Translations of these papers are available from the Librarian, U.K.A.E.A., Harwell, Oxford.

Kitaeva, V. F., Osipov, Yu. I., Pavlova, L. S., Polyakov, V. M., Sobolev, N. N. and Fedorov, L. S. (1972). *Sov. Phys. Tech. Phys.* **16**, 1509–1511.
Kitaeva, V. F., Osipov, Yu. I., Rubin, P. L. and Sobolev, N. N. (1969). *I.E.E.E. J.Q.E.* **QE-5**, 72–77.
Kitaeva, V. F., Osipov, Yu. I. and Sobolev, N. N. (1966a). *I.E.E.E. J.Q.E.* **QE-2**, 635–637.
Kitaeva, V. F., Osipov, Yu. I. and Sobolev, N. N. (1966b). *J.E.T.P. Lett.* **4**, 146–148.
Kitaeva, V. F., Osipov, Yu. I. and Sobolev, N. N. (1967a). *Sov. Phys. Dokl.* **12**, 55–56.
Kitaeva, V. F., Osipov, Yu. I., Sobolev, N. N. and Rubin, P. L. (1967b). *Sov. Phys. Tech. Phys.* **12**, 850–852.
Kitaeva, V. F., Osipov, Yu. I. and Sobolev, N. N. (1971). *I.E.E.E. J.Q.E.*, **QE-7**, 391–396.
Klein, M. B. (1970). *Appl. Phys. Lett.* **17**, 29–32.
Klein, M. B. and Maydan, D. (1970). *Appl. Phys. Lett.* **16**, 509–511.
Klein, M. B. and Silfvast, W. T. (1971). *Appl. Phys. Lett.* **18**, 482–485.
Kobayashi, S., Izawa, T., Kawamura, K. and Kamiyama, M. (1966). *I.E.E.E. J.Q.E.* **QE-2**, 699–700.
Koenig, J. L. (1971). *Appl. Spec. Rev.* **4**, 233–305.
Koenig, J. L. (1972). *J. Pol. Sci.* Pt. D. *Macromolecular Reviews*, **6**, 59–177.
Kogelnik, H. and Patel, C. K. N. (1962). *Proc. Inst. Rad. Eng. N.Y.* **50**, 2366.
Kohl, W. H. (1960). "Materials and Techniques for Electron Tubes." Reinhold Pub. Co. New York & Chapman and Hall, London.
Kohn, R. L., Shank, C. V., Ippen, E. P. and Dienes, A. (1971). *Opt. Comm.* **3**, 177–178.
Kolesnikov, V. N. and Obukhov-Denisov, V. V. (1962). *Sov. Phys. J.E.T.P.* **15**, 692–698.
Konkov, I. D., Rovinskii, R. E. and Cheburkin, N. V. (1969). *Rad. Eng. Elec. Phys.* **14**, 1793–1796.
Konkov, I. D., Rovinskii, R. E., Rozanov, A. G. and Cheburkin, N. V. (1968). *Rad. Eng. Elec. Phys.* **13**, 2008–2009.
Koozekanani, S. H. (1966). *I.E.E.E. J.Q.E.* **QE-2**, 770–773.
Koozekanani, S. H. (1967a). *I.E.E.E. J.Q.E.* **QE-3**, 206.
Koozekanani, S. H. (1967b). *Appl. Phys. Lett.* **11**, 107–108.
Koozekanani, S. H. (1968a). *I.E.E.E. J.Q.E.* **QE-4**, 59–60.
Koozekanani, S. H. (1968b). *Phys. Rev.* **176**, 160–163.
Korolev, F. A., Lebedeva, V. V., Odintsov, A. I. and Salimov, V. M. (1969). *Rad. Eng. Elec. Phys.* **14**, 1318–1320.
Koster, G. F., Statz, H., Horrigan, F. A. and Tang, C. L. (1968). *J. Appl. Phys.* **39**, 4045–4046.
Koval'chuk, V. M. and Petrash, G. G. (1966). *J.E.T.P. Lett.* **4**, 144–146.
Kretschmer, C. B., Boeschoten, F. and Demeter, L. J. (1968). *Phys. Fluids* **11**, 1050–1056.
Kreye, W. C. (1971). *J. Opt. Soc. Am.* **61**, 833–835.
Krishnamurty, S. G. and Rao, K. R. (1935). *Proc. Roy. Soc.* **A149**, 56–70.
Kucherenko, E. T. and Yavorskiy, I. A. (1965). *Rad. Eng. Elec. Phys.* **10**, 629–634.
Kuizenga, D. J. (1971). *Appl. Phys. Lett.* **19**, 260–263.
Kuizenga, D. J. and Siegman, A. E. (1970a). *I.E.E.E. J.Q.E.* **QE-6**, 694–708.
Kuizenga, D. J. and Siegman, A. E. (1970b). *I.E.E.E. J.Q.E.* **QE-6**, 709–715.
Kulagin, S. G., Likhachev, V. M., Markuzon, E. V., Rabinovich, M. S. and Sutovskii, V. M. (1966a). *J.E.T.P. Lett.* **3**, 6–8.
Kulagin, S. G., Likhachev, V. M., Rabinovich, M. S. and Sutovskii, V. M. (1966b). *J. Appl. Spect.* **5**, 398–399.

Kurzel, R. B. and Steinfeld, J. I. (1970). *J. Chem. Phys.* **53**, 3293–3303.
Labuda, E. F., Gordon, E. I. and Miller, R. C. (1965). *I.E.E.E. J.Q.E.* **QE-1**, 273–279.
Labuda, E. F. and Johnson, A. M. (1967). *I.E.E.E. J.Q.E.* **QE-3**, 164–167.
Labuda, E. F., Johnson, A. M. and Miller, R. C. (1966a). Conf. on Electron Device Research, California Institute of Technology, Pasadena. June 1966.
Labuda, E. F., Webb, C. E., Miller, R. C. and Gordon, E. I. (1966b). *Bull. Am. Phys. Soc.* **11**, 497.
Lamb, W. E., Jr. (1964). *Phys. Rev.* **134**, 1429–1450.
La Macchia, J. T., Lin, L. M. and Burckhardt, C. B. (1969). *J. Opt. Soc. Am.* **59**, 490.
Landon, D. O. and Reed, P. R. (1972). *Spex Speaker* 17, No. 4.
Langmuir, I. (1923). *J. Franklin Inst.* **196**, 751–762.
Latimer, I. D. (1968). *Appl. Phys. Lett.* **13** 333–335.
Latimer, I. D. and St. John, R. M. (1970). *Phys. Rev.* **A1**, 1612–1615.
Laures, P. (1968). *Phys. Lett.* **10**, 61–62.
Laures, P., Dana, L. and Frapard, C. (1964a). *C.R. Acad. Sci. Paris* **258**, 6363–6365.
Laures, P., Dana, L. and Frapard, C. (1964b). *C.R. Acad. Sci. Paris* **259**, 745–747.
Lazay, P. D. and Fleury, P. A. (1971). *In* "Light Scattering in Solids" (Ed. M. Balkanski), pp. 406–410. Flammarion Sciences, Paris.
Lebedeva, V. V., Odintsov, A. I. and Salimov, V. M. (1968). *Rad. Eng. Elec. Phys.* **13**, 655–658.
Lebedeva, V. V., Mashtakov, D. M. and Odintsov, A. I. (1970). *Opt. Spec.* **28**, 187–189.
Leiby, C. C., Jr., and Oskam, H. J. (1967). *Phys. Fluids* **10**, 1992–1996.
Leonard, E. T., Yaffee, M. A. and Billman, K. W. (1970). *Appl. Opt.* **9**, 1209.
Leonov, R. K., Protsenko, E. D. and Sapunov, Yu. M. (1966). *Opt. Spec.* **21**, 141–142.
L'Esperance, F. A., Jr. (1968). *Trans. Am. Ophthal. Soc.* **66**, 827.
L'Esperance, F. A., Jr. and Kelley, G. R. (1969). *Arch. Ophthal.* **81**, 583.
Levinson, G. R., Papulovskiy, and Tychinskiy, V. P. (1968). *Rad. Eng. Elec. Phys.* **13**, 578–582.
Li, H. and Andrew, K. L. (1971). *J. Opt. Soc. Am.* **61**, 96–109.
Lidsky, L. M., Rothleder, S. O., Rose, D. J., Yoshikawa, S., Michelson, C. and Makin, R. J. (1962). *J. Appl. Phys.* **33**, 2490–2497.
Likhachev, V. M., Rabinovich, M. S. and Sutovskii, V. M. (1967a). *J.E.T.P. Lett.* **5**, 43–45.
Likhachev, V. M., Rabinovich, M. S. and Sutovskii, V. M. (1967b). Proc. 8th Int. Conf. Ionis. Phen. In Gases. Vienna, p. 259 (Vienna: International Atomic Energy Agency).
Lin, L. H., Pennington, K. S., Stroke, G. S. and Lebeyrie, A. E. (1966). *Bell Sys. Tech. Jour.* **45**, 659–660.
Ling, H., Colombo, J. and Fisher, C. L. (1970). *Rev. Sci. Inst.* **41**, 1436–1437.
Lippert, J. L., Hibler, G. W., Small, E. W. and Peticolas, W. L. (1971). *In* "Light Scattering in Solids" (Ed. M. Balkanski), pp. 342–345. Flammarion Sciences, Paris.
Lippert, J. L. and Peticolas, W. L. (1972). Am. Chem. Soc. Meeting, Boston, April, 1972.
Little, H. L., Zweng, H. C. and Peabody, R. R. (1970). *Trans. Am. Acad. Ophth.* **74**, 85.
Lomnes, R. K. and Taylor, J. C. W. (1971). *Rev. Sci. Inst.* **42**, 766–769.
Luyken, B. F. (1972). *Physica* **60**, 432–458.
Macdonald, R. E. and Beck, J. W. (1959). *J. Appl. Phys.* **40**, 1429–1435.
Mack, M. E. (1968). *J. Appl. Phys.* **39**, 2483–2485.
MacNair, D. (1969). *I.E.E.E. J.Q.E.* **QE-5**, 460–470.
Maecker, H. (1956). *Z. Naturforsch.* **11A**, 457–479.

Maischberger, K. (1971). *I.E.E.E. J.Q.E.* **QE-7**, 250–252.
Maitland, A. (1969). *J. Phys. D*, **2**, 535–539.
Maitland, A. (1971). *J. Phys. D*, **4**, 907–915.
Maitland, A. and Cornish, J. C. L. (1972). *J. Phys. D*, **5**, 1807–1814.
Manger, H. and Rothe, H. (1963). *Phys. Lett.* **7**, 330–331.
Marantz, H. (1968). Ph.D. Thesis. Cornell University.
Marantz, H., Rudko, R. I. and Tang, C. L. (1966). *Appl. Phys. Lett.* **9**, 409–411.
Marantz, H., Rudko, R. I. and Tang, C. L. (1969). *I.E.E.E. J.Q.E.* **QE-5**, 38.
Martin, D. C. (1935). *Phys. Rev.* **48**, 938–944.
Martin, W. C. and Kaufman, V. (1970). *J. Res. N.B.S.* (*U.S.*) **74A**, 11–22.
Martin, W. C. and Tech, J. L. (1961). *J, Opt. Soc. Am.* **51**, 591–594.
Massey, G. M. (1966). *Appl. Opt.* **5**, 999–1001.
Massey, G. M., Oshman, M. K. and Targ, R. (1965). *Appl. Phys. Lett.* **6**, 10–11.
Mathias, L. E. S. and Parker, J. T. (1963). *Phys. Lett.* **7**, 194–196.
Matick, R. E. (1972). *Proc. I.E.E.E.* **60**, 266–289.
Maydan, D. (1970). *J. Appl. Phys.* **41**, 1552–1559.
McClintock, M. and Balling, L. C. (1969). *J. Quant. Spec. Radiative Trans.* **9**, 1209–1214.
McClintock, M., Demtröder, W. and Zare, R. N. (1969). *J. Chem. Phys.* **51**, 5509–5521.
McClure, R. E. (1965). *Appl. Phys. Lett.* **7**, 148–150.
McDaniel, E. W. (1964) "Collision Phenomena."
McDuff, O. P. and Harris, S. E. (1967). *I.E.E.E. J.Q.E.* **QE-3**, 101–111.
McFarlane, R. A. (1964a). *Appl. Phys. Lett.* **5**, 91–93.
McFarlane, R. A. (1964b). *Appl. Opt.* **3**, 1196.
McGuff, P. E. (1966). "Surgical Applications of the Laser." C. C. Thomas.
McMahon, W. H. (1968). *Appl. Phys. Lett.* **12**, 383–385.
Medicus, G. and Friar, E. M. (1966). *I.E.E.E. J.Q.E.* **QE-2**.
Meneely, C. T. (1967). *Appl. Opt.* **6**, 1434–1436.
Merkelo, H., Wright, R. H., Kaplafka, J. P. and Bialecke, E. P. (1968). *Appl. Phys. Lett.* **13**, 401–403.
Mercer, G. N. (1967). Ph.D. Thesis. Yale University.
Mercer, G. N., Chebotayev, V. P. and Bennett, W. R., Jr. (1967). *Appl Phys. Lett.* **10**, 177–179.
Meyerhofer, D. (1970). Laser Focus, 6, Feb., pp. 40–43.
Mielenz, K., Stephens, R. B., Gillilland, K. E. and Nefflen, K. F. (1966). *J. Opt. Soc. Am.* **56**, 156–162.
Mikaeliane, A. L., Bobrinev, V. I., Naumov, S. M. and Sokolova, L. Z. (1970). *I.E.E.E. J.Q.E.* **QE-6**, 193–198.
Miles, P. A. and Lotus, J. W. (1968). *I.E.E.E. J.Q.E.* **QE-4**, 811–819.
Miles, B. M. and Wiese, W. L. (1970). "Bibliography on Atomic Transition Probabilities." N.B.S. Special Publ. 320.
Miller, R. C., Labuda, E. F. and Webb, C. E. (1964–7). Unpublished.
Miller, R. C., Labuda, E. and Webb, C. E. (1967). *Bell. Sys. Tech. J.* **46**, 281–284.
Miller, R. C. and Webb, C. E. (1971). Unpublished data supplied to W. B. Bridges.
Minnhagen, L. (1963). *Ark. Fys.* **25**, 203–284.
Minnhagen, L., Strihed, H. and Petersson, B. (1969). *Ark. Fys.* **39**, 471–493.
Mooradian, A. (1970a). Proc. 6th Quan. Elec. Conf. Kyoto, Sept. 1970.
Mooradian, A. (1970b). *Science, N.Y.* **169**, 20–25.
Moore, C. B. (1971). *Ann. Rev. Phys. Chem.* **22**, 387–428.
Moore, C. E. (1949). Atomic Energy, Levels. N.B.S. Circular 467.

Moos, H. W. and Woodworth, J. R. (1973). *Phys. Rev. Lett.* **30**, 775–778.
Moran, J. M. (1971a). *Appl. Opt.* **10**, 412–415.
Moran, J. M. (1971b). *Appl. Opt.* **10**, 1909–1913.
Moskalenko, V. F., Ostapchenko, E. P., Pechurina, S. V., Stepanov, V. A. and Tsukunov, Yu. M. (1971). *Opt. Spec.* **30**, 201–202.
Moss, T. S., Killick, D. E. and De la Perreile, E. T. (1964). *Infra-red Physics*, **4**, 209–211.
Mountain, R. D. (1970). *Crit. Rev. Solid State Sciences* **1**, 5–46.
Nasini, M., Pratesi, R. and Redaelli, G. (1969). *Rev. Sci. Inst.* **40**, 1473–1474.
Neusel, R. H. (1966). *I.E.E.E. J.Q.E.* (Corr.) **QE-2**, 331–333.
Odintsov, A. I., Lebedeva, V. V. and Abrosimov, G. V. (1968). *Rad. Eng. Elec. Phys.* **13**, 650–653.
Osterink, L. M. and Targ, R. (1967). *Appl. Phys. Lett.* **10**, 115–117.
Paananen, R. A. (1966). *Appl. Phys. Lett.* **9**, 34–35.
Paananen, R. A. (1966). *I.E.E.E. Spectrum*, June, 88–99.
Paik, S. F. and Creedon, J. E. (1968). *Proc. I.E.E.E.* **56**, 2086–2087.
Palenius, H. P. (1966). *Appl. Phys. Lett.* **8**, 82–83.
Papayoanou, A. and Gumeiner, I. (1970). *Appl. Phys. Lett.* **16**, 5–8.
Parker, J. V. (1973). *Phys. Fluids* **6**, 1657–1658.
Pekarek, L. (1967). 8th Int. Conf. Ion. Phen. in Gases, Vienna.
Pekarek, L. (1968). *U.S.P. Fiz Nauk.* **49**, 463.
Pennington, K. S. and Lin, L. H. (1965). *Appl. Phys. Lett.* **7**, 56.
Perry, D. L. (1965). *Proc. I.E.E.E.* **53**, 76–77.
Perry, D. L. (1971). *I.E.E.E. J.Q.E.* **QE-7**, 102.
Peterson, O. G., Tuccio, S. A. and Snavely, B. B. (1970a). Proc. 6th Quan. Elec. Conf., Kyoto, Sept. 1970.
Peterson, O. G., Tuccio, S. A. and Snavely, B. B. (1970b). *Appl. Phys. Lett.* **17**, 245–247.
Peterson, D. G. and Yariv, (1966). *Appl. Opt.* **5**, 985–991.
Peticolas, W. L. (1972). *Adv. Pol. Sci., Fortschr. Hochpolym. Forsh*, **9**, 285–334.
Peticolas, W. L. (1972). *Ann. Rev. Phys. Chem.* **23**, 93–116.
Peticolas, W. L., Hibler, G. W., Lippert, J. L., Peterlin, A. and Oef, H. (1971). *Appl. Phys. Lett.* **18**, 87–89.
Petrash, G. G. (1972). *Sov. Phys. Usp.* **14**, 747–765.
Petrash, G. G., Isaev, A. A. and Kazanyan, M. A. (1973). Conf. Laser Eng. and Appl. Washington, paper 4.3.
Phelps, A. V. and Molnar, J. P. (1953). *Phys. Rev.* **89**, 1202–1208.
Pike, E. R. (1970). *In* "Quantum Optics" (Ed. S. M. Kay and A. Maitland), pp. 127–176. Academic Press, London and New York.
Piltch, M. and Gould, G. (1966). *Rev. Sci. Inst.* **37**, 925–927.
Piltch, M., Walter, W. D., Solimene, N., Gould, G. and Bennett, W. R., Jr. (1965). *Appl. Phys. Lett.* **7**, 309–310.
Piper, J. A. (1973). Private communication.
Piper, J. A., Collins, G. J. and Webb, C. E. (1972). *Appl. Phys. Lett.* **21**, 203–205.
Piper, J. A. and Webb, C. E. (1973a). *J. Phys. B.* **6**, L116–L120.
Piper, J. A. and Webb, C. E. (1973b). *J. Phys. D.* **6**, 400–407.
Pleasance, L. D. and George, E. V. (1971). *Appl. Phys. Lett.* **18**, 557–561.
Pole, R. V. and Myers, R. A. (1968). *I.E.E.E. J.Q.E.* **QE-2**, 182–184.
Powers, E. J. (1966). *Proc. I.E.E.E.* **54**, 804–805.
Prescott, L. J. and van der Ziel, A. (1964). *Phys. Lett.* **12**, 317–319.
Prescott, L. J. and van der Ziel, A. (1964). *Appl. Phys. Lett.* **5**, 48.

Preston, K., Jr. (1972). "Coherent Optical Computers." McGraw-Hill.
Preobrazhenski, N. G. and Shapanev, N. Ya. (1968). *Opt. Spec.* **25**, 172–173.
Prosnitz, D. and George, E. V. (1971). Quarterly Progress Report No. 103, Research Laboratory of Electronics. M.I.T.
Raff, G. J. and George, E. V. (1971). Quarterly Progress Report, No. 101, Research Laboratory of Electronics, M.I.T., pp. 109–114.
Rajchman, J. H. (1973). *Laser Focus*, **9**, April, p. 19.
Rao, A. B. (1938). *Ind. J. Phys.* **35**, 399–405.
Redaelli, G. (1970). *Appl. Opt.*, **9**, 2593–2594.
Redaelli, G. and Sona, A. (1967). *Alta Frequenza*, **36**, 150–152.
Renton, C. A. and Zelby, L. W. (1965). *Appl. Phys. Lett.* **6**, 167–169.
Rhodes, C. and Hoff, P. (1973). Opt. Soc. Am. Meeting, Denver, Mar. 14.
Rigden, J. D. (1965). *I.E.E.E. J.Q.E.* **QE-1**, 221.
Rigrod, W. W. (1963b). *J. Appl. Phys.* **34**, 2602–2609.
Rigrod, W. W. (1965). *J. Appl. Phys.* **36**, 2487–2490.
Rigrod, W. W. (1968). *Appl. Opt.* **7**, 2325.
Rigrod, W. W. and Bridges, T. J. (1965). *I.E.E.E. J.Q.E.* **QE-1**, 298–303.
Rigrod, W. W. and Johnson, A. M. (1967). *I.E.E.E. J.Q.E.* **QE-3**, 644–646.
Riseberg, L. A. and Schearer, C. D. (1971). *I.E.E.E. J.Q.E.* (Corr.) **QE-7**, 40–41.
Roberts, D. E. (1968). *J. Phys. B.*, **1**, 53–61.
Roberts, H. N., Watkins, J. W. and Johnson, R. H. (1973). *Laser Focus*, **9**, p. 22.
Rosenberger, D. (1969). *Z. Naturforsch.* **24a**, 867–869.
Rosenberg, R., Rubinstein, C. B. and Herriott, D. R. (1964). *Appl. Opt.* **3**, 1079–1083.
Rozgachev, K. I. and Yaroslavtseva, L. T. (1970) *Opt. Spec.* **28**, 1072.
Rudko, R. I. (1967). Ph.D. Thesis. Cornell University.
Rudko, R. I. and Tang, C. L. (1966). *Appl. Phys. Lett.* **9**, 41–44.
Rudko, R. I. and Tang, C. L. (1967). *J. Appl. Phys.* **38**, 4731–4739.
Ruttenauer, A. (1922). *Z. Physik.* **10**, 269–274.
Ryan, T. J., Youmans, D. G., Hackel, L. A. and Ezekiel, S. (1972). *Appl. Phys. Lett.* **21**, 320–322.
Sakurai, K. and Broida, H. P. (1969a). *J. Chem. Phys.* **50**, 557–558.
Sakurai, K. and Broida, H. P. (1969b). *J. Chem. Phys.* **50**, 2404–2410.
Sakurai, K. and Broida, H. P. (1970). *J. Chem. Phys.* **53**, 1615–1616.
Sakurai, K., Johnson, S. E. and Broida, H. P. (1970). *J. Chem. Phys.* **52**, 1625–1632.
Sandercock, J. R. (1970). *Opt. Comm.* **2**, 73–76.
Sandercock, J. R. (1972a). *Phys. Rev. Lett.* **28**, 237–240.
Sandercock, J. R. (1972b). *Solid State Comm.* **11**, 729–731.
Sasaki, W., Kumata, K., Saito, S. and Takemoto, S. (1973). *Jap. J. Appl. Phys.* **12**, 468–469.
Sayers, M. D. (1969). *Phys. Lett.* **29A**, 591–592.
Scavennec, A. (1971). *C.R. Acad. Sci. Paris* **272**, 1303–1306.
Schacter, H. and Skurnick, E. (1971). *Israel J. Tech.* **9**, 209–212.
Schaeffer, A. R. (1971). *J. Quant. Spect. Rad. Trans.* **11**, 197–201.
Schafer, G. and Seelig, W. (1970). *Z. Angew Physik*, **29**, 246–248.
Schaufele, R. F. (1970). *Macromolecular Reviews* **4**, 67–90.
Schawlow, A. L. and Townes, C. H. (1958). *Phys. Rev.* **112**, 1940–1949.
Schearer, L. D. and Holton, W. C. (1970). *Phys. Rev. Lett.* **24**, 1214–1217.
Schearer, L. D. and Padovani, F. A. (1970). *J. Chem. Phys.* **52**, 1618–1619.
Schoen, P. E. and Jackson, D. A. (1972). *J. Phys. E.* **5**, 519–521.
Schottky, W. (1924). *Phys. Z.* **25**, 635–645.
Schuebel, W. K. (1970a). *Appl Phys. Lett.* **16**, 470–472.

Schuebel, W. K. (1970b). *I.E.E.E. J.Q.E.* (Notes and Lines) **QE-6**, 654–655.
Schuebel, W. K. (1970c). *I.E.E.E. J.Q.E.* (Corr.) **QE-6**, 574–575.
Schuebel, W. K. (1971). *I.E.E.E. J.Q.E.* (Corr.) **QE-7**, 39–40.
Schwarz, H. and Hora, H. (1969). *Appl. Phys. Lett.* **15**, 349–351.
Scott, J. F. (1972). *Spex Speaker*, 17, June, No. 2.
Sedel'nikov, V. A., Sinichkin, Yu. P. and Tuchin, V. V. (1971). *Opt. Spec.* **31**, 408–409.
See, B. A., Garwoli, W. and Hughes, J. C. (1967). *I.E.E.E. J.Q.E.* **QE-3**, 169–170.
Seelig, W. H. and Banse, K. V. (1970). *Laser Focus*, Aug., pp. 33–37.
Self, S. A. (1963). *Phys. Fluids* **6**, 1762–1768.
Shair, F. H. and Remer, D. S. (1968). *J. Appl. Phys.* **39**, 5762–5767.
Shank, C. V., Ippen, E. P. and Dienes, A. (1972). Proc. 7th Int. Quan. Elec. Conf., Montreal, May 1972.
Shapiro, V. D. (1965). *Sov. Phys. J.E.T.P. Lett.* **2**, 291–294.
Sheidlin, A. E. and Asinovskii, E. I. (1963). Proc. 6th Symp. on Ion. Phen. in Gases. Paris, Vol. 2, pp. 379–380.
Shipman, J. D., Jr. (1967). *Appl. Phys. Lett.* **10**, 3–4.
Shirk, J. S. and Bass, A. M. (1970). *J. Chem. Phys.* **52**, 1894–1901.
Shulman, A. R. (1970). "Optical Data Processing." Wiley-Interscience.
Shumaker, J. B. (1961). *Rev. Sci. Inst.* **32**, 65–67.
Siegman, A. E. (1971). "An Introduction to Lasers and Masers," McGraw-Hill.
Silfvast, W. T. (1968). *Appl. Phys. Lett.* **13**, 169–171.
Silfvast, W. T. (1969). *Appl. Phys. Lett.* **15**, 23–25.
Silfvast, W. T. (1971). *Phys. Rev. Lett.* **27**, 1489–1492.
Silfvast, W. T. and Klein, M. B. (1970). *Appl. Phys. Lett.* **17**, 400–403.
Silfvast, W. T. and Klein, M. B. (1972). *Appl. Phys. Lett.* **20**, 501–503.
Silfvast, W. T. and Smith, P. W. (1970). *Appl. Phys. Lett.* **17**, 70–73.
Silfvast, W. T. and Szeto, L. H. (1970). *Appl. Opt.* **9**, 1484–1485.
Silfvast, W. T. and Szeto, L. H. (1971). *Appl. Phys. Lett.* **19**, 445–447.
Silfvast, W. T., Fowles, G. R. and Hopkins, B. D. (1966). *Appl. Phys. Lett.* **8**, 318–319.
Simmons, W. W. and Witte, R. S. (1970a). *I.E.E.E. J.Q.E.* **QE-6**, 466–469.
Simmons, W. W. and Witte, R. S. (1970b). *I.E.E.E. J.Q.E.* **QE-6**, 648–649.
Sinclair, D. C. (1966). *J. Opt. Soc. Am.* **56**, 1727–1731.
Sinclair, D. C. (1968). *Appl. Phys. Lett.* **13**, 98–100.
Sinclair, D. C. and Bell, W. E. (1969). "Gas Laser Technology." Holt, Rinehart and Winston, Inc.
Singer, J. R. (1968). *J. Appl. Phys.* **39**, 4124–4127.
Skurnick, E. and Schacter, H. (1972). *J. Appl. Phys.* **43**, 3393–3396.
Smith, A. L. S. and Dunn, M. H. (1968). *I.E.E.E. J.Q.E.* **QE-4**, 838–842.
Smith, D. C. (1969). *I.E.E.E. J.Q.E.* **QE-5**, 600–607.
Smith, H. A. and Mahr, H. (1970). Proc. 6th Quan, Elec. Conf., Kyoto, Sept. 1970.
Smith, P. W. (1965). *I.E.E.E. J.Q.E.* **QE-1**, 343–348.
Smith, P. W. (1966a). *I.E.E.E. J.Q.E.* **QE-2**, 62–68.
Smith, P. W. (1966b). *I.E.E.E. J.Q.E.* **QE-2**, 666–668.
Smith, P. W. (1967). *I.E.E.E. J.Q.E.* **QE-3**, 627–635.
Smith, P. W. (1968). *I.E.E.E. J.Q.E.* **QE-4**, 485–490.
Smith, P. W. (1970). *Proc I.E.E.E.* **58**, 1342–1357.
Smith, P. W. (1972). *I.E.E.E. J.Q.E.* **QE-8**, 704–710.
Smith, P. W., Schneider, M. V. and Danielmeyer, H. G. (1969). *Bell. Sys. Tech. J.* **48**, 1405–1419.
Sosnowski, T. P. (1969). *J. Appl. Phys.* **40**, 5138–5144.

Spitzer, L., Jr. (1961). "Physics of Fully Ionised Gases." Princeton Univ. Press, Princeton, New Jersey.
Statz, H. (1967). *J. Appl. Phys.* **38**, 4648–4655.
Statz, H., de Mars, G. A. and Tang, C. L. (1969a). *Appl. Phys. Lett.* **15**, 125–127.
Statz, H., de Mars, G. A. and Tang, C. L. (1969b). *Appl. Phys. Lett.* **15**, 428–430.
Statz, H., Horrigan, F. A., Koozekanani, S. H., Tang, C L. and Koster, G. F. (1965). *J. Appl. Phys.* **36**, 2278–2286.
Stefanov, V. J. and Petrova, M. D. (1971). *Phys. Lett.* **35A**, 424–425.
Stetser, D. A. and De Maria, A. J. (1966) *Appl. Phys. Lett.* **9**, 118–120.
Striganov, A. R. and Sventitskii, N. S. (1968). "Tables of Spectral Lines of Neutral and Ionized Atoms", translated from Russian. IFI/Plenum Press, New York.
Stuart, R. S. and Wehner, G. K. (1962). *J. Appl. Phys.* **33**, 2345–2352.
Sugawara, Y. and Tokiwa, Y. (1970). *Jap. J. Appl. Phys.* **9**, 588–589.
Sugawara, Y., Tokiwa, Y. and Iijima, T. (1970a). VIth Int. Quan. Elec. Conf., Kyoto, Digest of Tech. Papers, 320–321.
Sugawara, Y., Tokiwa, Y. and Iijima, T. (1970b). *Jap. J. Appl. Phys.* **9**, 1537.
Suzuki, N. (1965). *Jap. J. Appl. Phys.* **4**, 452–457.
Suzuki, T. (1971). *Jap. J. Phys.* **10**, 1419–1424.
Sze, R. C., Antropov, Ye. T. and Bennett, W. R., Jr. (1972). *Appl. Opt.* **11**, 197–198.
Sze, R. C. and Bennett, W. R., Jr. (1972). *Phys. Rev.* **A5**, 837–853.
Szmansky, H. A. (1970). (Ed.) "Raman Spectroscopy Theory and Practice," Vol. 20. Plenum Press, New York.
Tandler, W. S. W. (1969). Proc. Int. Coll. on Lasers, Paris. Unpublished.
Targ, R. (1972). *I.E.E.E. J.Q.E.* (Corr.) **QE-8**, 726–728.
Targ, R., French, J. M. and Yarborough, J. M. (1968). *I.E.E.E. J.Q.E.* **QE-4**, 644–648.
Targ, R., Osterink, L. M. and French, J. M. (1967). *Proc. I.E.E.E.* **55**, 1185–1192.
Targ, R. and Yarborough, J. M. (1968). *Appl. Phys. Lett.* **12**, 3–4.
Taylor, J. E., Kiang, Y. C. and Unterleitner, F. C. (1966). Proc. IVth Quan. Elec. Conf., Phoenix.
Tien, P. K., MacNair, D. and Hodges, H. L. (1964). *Phys. Rev. Lett.* **12**, 30–33.
Tkach, Yu. V., Fainberg, Ya. B., Bolotin, L. I., Bessarab, Ya. Ya., Gadetskii, N. P., Chernen'kii, Yu. N. and Berezin, A. K. (1967). *J.E.T.P. Lett.* **6**, 371–373.
Tkach, Yu. V., Fainberg, Ya. B., Bolotin, L. I., Bessarab, Ya. Ya., Gadetskii, N. P., Magda, I. I., Bogdanovich, A. V. and Chernen'kii, Yu. N. (1969). *Ukr. Fiz. Zh.* **14**, 1470–1475.
Tkach, Yu. V., Fainberg, Ya. B., Bolotin, L. I., Bessarab, Ya. Ya., Gadetskii, N. P. (1972). *Sov. Phys. J.E.T.P.* **35**, 886–892.
Tobias, I. and Wallace, R. A. (1964). *Phys. Rev.* **134**, A549-–552.
Tombers, R. B. and Chanin, L. M. (1970). *J. Appl. Phys.* **41**, 2433–2438.
Tomlinson, W. J., Kaminow, I. P., Chandross, E. A., Fork, R. L. and Silfvast, W. T. (1970). *Appl. Phys. Lett.* **16**, 486–489.
Tonks, L. (1939). *Phys. Rev.* **56**, 360–373.
Tonks, L. and Langmuir, I. (1929). *Phys. Rev.* **34**, 876–922.
Toyoda, K. and Yamanaka, C. (1967). Tech. Rep. Osaka Univ. (Japan), Vol .17, 407–414.
Troitskii, Yu. V. and Goldina, N. D. (1968). *J.E.T.P. Lett.* **7**, 36–38.
Tsytovich, V. N. and Shapiro, V. D. (1964). *Sov. Phys. Tech. Phys.* **9**, 583–585.
Tuccio, S. A. and Strome, F. C., Jr. (1972). *Appl. Opt.* **11**, 64–73.
Turner-Smith, A. R., Green, J. M. and Webb, C. E. (1973). *J. Phys. B*, **6**, 114–130.
Uchida, T. and Ueki, A. (1967). *I.E.E.E. J.Q.E.* **QE-3**, 17–30.

Vainshtein, L. A. and Vinogradov, A. (1967). *Opt. Spect.* **23**, 101–102.
Van der Sijde, B. (1972a). *Phys. Lett.* **38A**, 89–90.
Van der Sijde, B. (1972b). *J. Quant. Spec. Rad. Trans.* **12**, 703–729.
Van der Sijde, B. (1972c). *J. Quant. Spec. Rad. Trans.* **12**, 1497–1516.
Van der Sijde, B. (1972d). *J. Quant. Spec. Rad. Trans.* **12**, 1517–1538.
Vandilla, M. A., Trujillo, T. T., Mullaney, P. F. and Coulter, J. R. (1969). *Science, N.Y.* **163**, 1213–1214.
Van Dyck, R. S., Jr., Johnson, C. E. and Shugart, H. A. (1971). *Phys. Rev.* **A4**, 1327–1336.
Van Dyck, R. S., Jr., Johnson, C. E. and Shugart, H. A. (1972). *Phys. Rev.* **A5**, 991–993.
Vasil'eva, A. N., Likhachev, V. M. and Sutovskii, V. M. (1969). *Sov. Phys. Tech. Phys.* **14**, 245–249.
Vladimirova, N. M., Konkov, I. D., Rovinskii, R. E. and Cheburkin, N. V. (1970a). *Sov. Phys. J.E.T.P.* **30**, 813–816.
Vladimirova, N. M., Konkov, I. D., Rovinskii, R. E. and Cheburkin, N. V. (1970b). *Rad. Eng. Elec. Phys.* **15**, 2339–2341.
Vladimirova, N. M., Konkov, I. D., Rovinskii, R. E. and Cheburkin, N. V. (1971). *Opt. Spec.* **31**, 91–93.
Walker, K. G. and St. John, R. M. (1972). *Phys. Rev.* **A6**, 240–250.
Walker, W. J. and Weber, A. (1971). *J. Mol. Spec.* **39**, 57–64.
Walter, W. T., Solimene, N., Piltch, M. and Gould, G. (1966). *I.E.E.E. J.Q.E.* **QE-2**, 474–479.
Wang, C. P. and Lin, S. C. (1972). *J. Appl. Phys.* **43**, 5068–5073.
Watanabe, S., Chihara, M. and Ogura, I. (1972). *Jap. J. Appl. Phys.* **11**, 600.
Waynant, R. W. (1972). *Phys. Rev. Lett.* **28**, 533–535.
Waynant, R. W. (1973). *Appl. Phys. Lett.* **22**, 419–420.
Webb, C. E. (1968a). *I.E.E.E. J.Q.E.* (Corr.), **QE-4**, 426–427.
Webb, C. E. (1968b). *J. Appl. Phys.* **39**, 5441–5470.
Webb, C. E., Turner-Smith, A. K. and Green, J. M. (1970). *J. Phys. B.* **3**, L134–L138.
Weber, A. and Porto, S. P. S. (1965). *J. Opt. Soc. Am.* **55**, 1033–1034.
Weber, A., Porto, S. P. S., Cheesmen, L. E. and Barrett, J. J. (1967). *J. Opt. Soc. Am.* **57**, 19–28.
Weber, A. and Schlupf, J. (1972). *J. Opt. Soc. Am.* **62**, 428–432.
Webster, J. M. and Graham, R. W. (1968). *J. Photo. Sci.* **16**, 49–56.
Wehner, G. K. (1954). *Phys. Rev.* **93**, 633–634.
Wehner, G. K. (1955). *Adv. Elec. Elec. Phys.* **7**, 239–298.
Wehner, G. K. (1959). *J. Appl. Phys.* **30**, 1762–1765.
Wen, C. P., Hammer, J. M., Gorog, I., Spong, F. W. and Van Raalte, J. A. (1966). *I.E.E.E. J.Q.E.* **QE-2**, 711–713.
Wexler, B. L. (1972). *Rev. Sci. Inst.* **43**, 1853–1854.
White, A. D., Gordon, E. I. and Rigden, J. D. (1963). *Appl. Phys. Lett.* **2**, 91–93.
White, A. D. (1965). *I.E.E.E. J.Q.E.* **QE-1**, 349–357.
White, A. D. and Rigden, J. D. (1962). *Proc. I.R.E.* **50**, 1697.
Wiese, W. L., Smith, M. W. and Glennon, B. M. (1966). "Atomic Transition Probabilities," Vol. 1. NSRDS-NBS 4.
Wiese, W. L., Smith, M. W. and Miles, B. M. (1969). "Atomic Transition Probabilities", Vol. II. NSRDS-NBS 22.
Wiggins, R. L., Vaughan, K. D. and Friedmann, **G. B.** (1972). *Appl. Opt.* **11**, 179–181.
Willenberg, G. R. and Carruthers, J. A. (1970). *J. Appl. Phys.* **41**, 5040–5041.
Willett, C. S. (1970). *I.E.E.E. J.Q.E.* (Corr.) **QE-6**, 469–471.

Willett, C. S. (1971). Laser Lines in Atomic Species, Vol. 1, Pt. 5, Prog. in Quan. Elec. Pergamon Press.
Willett, C. S. and Heavens, O. S. (1966). *Opt. Act.* **13**, 271–273.
Wolbarsht, M. L. (Ed.) (1971). "Laser Applications in Medicine and Biology", Vol. 1. Plenum Press.
Wright, G. B. (Ed.) (1969). Proc. Int. Conf. on "Light Scattering Spectra of Solids". New York 1968, Springer-Verlag, New York.
Yamada, Y., Yamamoto, M. and Nomura, S. (1970). Proc. 6th. Quan. Elec. Conf., Kyoto, Sept. 1970.
Yarborough, J. M. and Hobart, J. L. (1968). *Appl. Phys. Lett.* **13**, 305–307.
Yariv, A. (1963). *Proc. I.E.E.E.* **51**, 1723–1731.
Yariv, A. (1965). *J. Appl. Phys.* **36**, 388–391.
Yariv, A. (1967). "Quantum Electronics," Wiley, New York.
Zakharov, P. N. and Pekar, Yu. A. (1971). *Sov. Phys. Tech. Phys.* **15**, 1294–1297.
Zarowin, C. B. (1966). *Appl. Phys. Lett.* **9**, 124–242.
Zarowin, C. B. (1969). *Appl. Phys. Lett.* **15**, 36–38.
Zarowin, C. B. and Williams, C. K. (1967). *Appl. Phys. Lett.* **11**, 47–48.
Zory, P. (1966). *J. Appl. Phys.* **37**, 3643–3644.
Zory, P. (1967). *I.E.E.E. J.Q.E.* **QE-3**, 390–398.
Zory, P. S., Jr. and Lynch, G. W. (1971). *Proc. I.E.E.E.* **59**, 684–689.

AUTHOR INDEX

Numbers in italics show the page on which the complete reference is listed.

C

D

E

F

G

L

Q

R

T

SUBJECT INDEX

L

M

N

O

P

Q

R